THE ROUTLEDGE HANDBOOK OF WASTE, RESOURCES AND THE CIRCULAR ECONOMY

The *Handbook* introduces, contextualises, critiques, and discusses a range of perspectives associated with the concept of the circular economy. These perspectives span an array of subjects including economics, environmental policymaking, sociology, environmental science, environmental and industrial engineering, management, international development, and human geography.

A fundamental underpinning of the *Handbook* is that it takes account of a wide range of sectors, as well as geographical perspectives that incorporate both a Global North and Global South world context. This approach is crucial because it is only within such a holistic perspective that the circular economy concept can truly be examined. In addition, these issues are examined both from a theoretical as well as a practical perspective, using real-world case studies for illustration.

Given its wide subject, sectoral, and geographical areas of focus, the *Handbook* should be of value not only for those undertaking research in the field of circular economy but also stakeholders involved in policymaking, as well as decision-making on the front line.

Terry Tudor has worked in the fields of sustainable waste management and circular economy since 2003. This work has been undertaken both in the United Kingdom as well as abroad (e.g., in Italy, Brazil, Barbados, and India). He works as a consultant. Previously, he was an associate professor at the University of Northampton, UK.

Cleber JC. Dutra is an associate professor and research fellow in sustainability management at the Federal University of Rio Grande do Sul (Brazil) and the Technische Universität Berlin (Germany). He has been examining approaches to sustainability through multiparadigmatic perspectives as a mechanism to overcome its understanding shortcomings. His main areas of interest include innovation, sustainability, circular economy, multiorganisational programmes, resource efficiency, pro-environmental/sustainable behaviour, and CSR.

THE ROUTLEDGE HANDBOOK OF WASTE, RESOURCES AND THE CIRCULAR ECONOMY

Edited by Terry Tudor and Cleber JC. Dutra

Routledge
Taylor & Francis Group
LONDON AND NEW YORK

First published 2021
by Routledge
2 Park Square, Milton Park, Abingdon, Oxon OX14 4RN

and by Routledge
52 Vanderbilt Avenue, New York, NY 10017

Routledge is an imprint of the Taylor & Francis Group, an informa business

British Library Cataloguing-in-Publication Data
A catalogue record for this book is available from the British Library

Library of Congress Cataloging-in-Publication Data
Names: Tudor, Terry, editor. | Dutra, Cleber JC, editor. |
Routledge (Firm)
Title: The Routledge handbook of waste, resources and the circular
economy / edited by Terry Tudor and Cleber JC. Dutra.
Description: First Edition. | New York : Routledge, 2020. |
Includes bibliographical references and index.
Identifiers: LCCN 2020028023 (print) | LCCN 2020028024 (ebook)
Subjects: LCSH: Sustainable development. | Conservation of natural
resources. | Recycling industry. | Environmental policy.
Classification: LCC HC59.72.E5 R68 2020 (print) | LCC HC59.72.E5
(ebook) | DDC 333.7—dc23
LC record available at https://lccn.loc.gov/2020028023
LC ebook record available at https://lccn.loc.gov/2020028024

ISBN: 978-0-367-36464-9(hbk)
ISBN:978-0-367-63759-0(pbk)
ISBN: 978-0-429-34634-7 (ebk)

Typeset in Bembo
by Apex Covantage, LLC

CONTENTS

FIGURES

TABLES

CONTRIBUTORS

Stewart Barr is Professor of Geography at the University of Exeter (UK) and has worked for over 20 years in the field of pro-environmental behavioural change and sustainability. His research has explored a range of everyday sustainability behaviours, including household waste management, energy saving, water conservation, everyday consumption, and travel behaviour. More recently he has focused on the policy and politics of behavioural change through understanding the ways in which publics can engage as collaborative partners for promoting meaningful change to reduce our impact on the climate. His current research is seeking to understand the ways in which everyday sustainability behaviours link to the built environment and how longer-term planning of cities, infrastructure, and services can promote positive change.

Martha Bicket is a Senior Research Fellow at the University of Surrey, UK. Martha works at the interface of research and policy, working closely with senior policymakers and analysts across local, national, and international levels of government towards the goal of better-informed decision-making and policy. She is part of the core research team at the UK Centre for Evaluating Complexity Across the Nexus (CECAN) – a collaboration between academics, UK government, and practitioners to generate novel, cutting-edge policy evaluation methods fit for a complex world. Her interests include mapping interactions within the food-water-energy-environment nexus and managing transitions to sustainable resource use.

Annabelle Boulay graduated with a BSc (Hons) in environmental science and a PhD in human geography with a specialisation in rural geography from Kingston University London. She is currently a senior lecturer in human and development geography at the University of Derby. Her research and lecturing expertise lies in human/development geography, with particular focus on rural development, agricultural geography (urban agriculture, dairy farming, organic and sustainable agriculture, consumption), waste management (recycling), and social and cultural geography (population, migration, rural places, homelessness, livelihoods), as well as research methods (quantitative and qualitative). She was a lead researcher on a research project entitled "Unpacking the Household: Understanding the Domestic Dynamics of Recycling," which observed home recycling behaviour in France and the United Kingdom. This project used an innovative methodology (diaries, interviews, and observation).

Gianmarco Bressanelli is a Post Doctoral Researcher at the Department of Mechanical and Industrial Engineering of the University of Brescia, Italy. He obtained his PhD in 2019, discussing the dissertation "Assessing the Sustainability Impacts of Circular Economy Supply Chains: A New Framework and Simulation Tool for the Washing Machine industry." He does research in circular economy, focusing on its implications on operations and supply chain management, combining case study research with simulation and assessment methodologies. At the same time, he collaborates with IQ Consulting to carry out advisory projects.

Marco Caniato is a humanitarian worker specialising in solid waste management, WASH, and climate change. Marco worked in the Balkans, Africa, and the Middle East and authored scientific papers about SWM and access to energy in humanitarian context. He was based in Palestine between 2016–20, first as Cesvi Head of Mission and then as UN OCHA West Bank Field Coordinator.

Paul Cox is a Senior Lecturer in the Faculty of Arts, Science and Technology at the University of Northampton, having joined the university in 1999. He holds a University Teaching Fellowship, gained in 2011, for work on a "Technology Enhanced Transition Programme" for distance learners and is a fellow of the Higher Education Academy. Areas of interest include resource recovery processes, waste minimisation, and sustainability. He has collaborated with colleagues in the development of a range of waste and resource management distance learning courses at the university.

Gia Daprano is a PhD candidate at the University of Exeter, UK. Her research centres on an exploration of urban food foraging practices within Plymouth city centre, asking participants to engage in conversation with her on a foraging walk, exploring why we forage and what it is we learn from the practice. In this way, the research attempts to account for the political ecology of a post-humanist re-signification of urban green spaces into learning spaces for eco-literacy through a food-commons. She currently works as a project co-ordinator for a food sharing project, is involved in several community gardening initiatives, and in the past has worked as a research assistant on "Food as a Lifestyle Motivator," examining how food can influence lifestyle changes in hard-to-reach populations within the city.

Rory Doherty is a Senior Lecturer in the School of Natural and Built Environment at Queen's University Belfast. He applies chemistry, geophysics, and microbiology to environmental problems to produce viable engineering solutions. His interests include the circular economy, reuse and recovery of hazardous and valuable materials, restoration of peatlands, sustainable remediation of pollution, and the identification and apportionment of potentially toxic elements.

Johannes Fresner is a chemical engineer specialising in process analysis and optimisation with special emphasis on resource-efficient and cleaner production and energy efficiency. He is Managing Director of STENUM GmbH, an Austrian consultancy working in resource efficiency and environmental management. He has worked with more than 200 companies in 40 countries and done a variety of projects for Austrian government, UNIDO, and UNEP.

Charlie Farrell is a PhD student on the Renewable Engine Program at Queen's University Belfast and South West College. He does research in photovoltaics, materials separation, chemical kinetics, pyrolysis, chemical thermodynamics, and the circular economy.

Elizabeth Gilligan is a PhD student in architecture at Queen's University Belfast and the Director of Concrete Jungle, a company that manufactures bioreceptive facades for the outside of buildings using waste materials.

Marvin Henry is an academic based at Utrecht University, the Netherlands, pursuing a PhD at the Innovation Studies Group of Copernicus Institute for Sustainable Development. In his research, he examines sustainability transitions. Most of his current work investigates the circular economy concept as a vehicle for such transitions, focusing on the role of circular start-ups in the circular economy in Europe. He supervises several student and research projects on the circular economy and sustainable business models. He supports the Ashoka Globalizer program as a pro-bono advisor and is actively engaged for integration in Germany through an affiliation as mentor for Deutschlandstiftung Integration.

Sabine Hielscher works as a research fellow at the Technische Universitaet Berlin, Germany, and the Science Policy Research Unit (SPRU), University of Sussex, Brighton, UK. She is interested in the politics, processes and materialisations of grassroots innovations and the dynamics of everyday (sustainable) consumption patterns. Her research work has been trans- and interdisciplinary and draws upon theories and methodologies from design, sociology, innovation studies, and science and technology studies.

Eleni Iacovidou is a Lecturer in Environmental Management at Brunel University London. Her research focuses on the sustainability assessment of resources and waste management systems based on a systems approach. She combines environmental science and engineering with an understanding of the political, organisational, structural, and cultural aspects that act synergistically in a resource recovery system to highlight areas of intervention and develop metrics for the evaluation of multi-dimensional positive and negative impacts. She works with industry and policy-makers, trying to integrate research findings into the policy-making process and improve communication between different stakeholders. She is the Head of Strategy of the newly established Society of Circular, Regenerative and Sustainable Systems (CRES) that seeks to enable the re-invention of systems to deliver a sustainable future.

Melanie Jaeger-Erben is Professor of Transdisciplinary Sustainability Research in Electronics at Technische Universitaet Berlin and a member of the board of the Institute for Social Innovation. She studied psychology and sociology in Goettingen, Germany, and Uppsala, Sweden. Melanie does research in systems and practices of sustainable consumption and production, sociotechnical evolution, social innovation, and sustainability transformation. Her current projects cover issues like obsolescence and product lifetimes, social practices of repair and making, and the role of consumers in a circular economy.

Thomas Jalili Tanha is an ESRC-funded doctoral researcher at Lancaster University Management School, UK. He is interested in researching "concerned" markets, their making of climate (in)action and its associated market politics. His PhD focuses on the making of a circular economy for electric vehicle batteries. In his work with car manufacturers, energy storage providers, battery and environmental scientists, underwriters, and others, he focuses on the role that the production of managerial and scientific technologies plays in the strategic efforts that make, shape, and transform those markets. His work has included two projects supported by the European Regional Development Fund.

Sally-Anne Käsner is a Director of Circular-Vision, a circular economy strategy and design consultancy. Her focus is on driving economic, social, and environmental prosperity through connecting across the value chain; collaborating on projects; and creating new business models that respond to the circular economy principles, leading to a thriving and inclusive economy.

She was previously an Executive Associate at JG Afrika (an engineering and environmental consultancy), a founding member of EcoStandard (083-558-NPO – HYPERLINK "http://www.ecostandard.co.za" www.ecostandard.co.za), a founding member and Director of the African Circular Economy Network (Non profit Company Registration number: K2020115399 – www.acen.africa), and a Non-Executive Director of Polyco (HYPERLINK "http://www.polyco.co.za" www.polyco.co.za).

Julian Kirchherr is an Associate Partner at McKinsey & Company and a Researcher at the Copernicus Institute of Sustainable Development, Utrecht University, the Netherlands. His research examines sustainability transitions. Most of his current work investigates the circular economy concept as a vehicle for such transitions. For instance, he is the Principle Investigator (PI) of the NWO-funded research project Transitioning Towards a Circular Economy: The Role of Circular Start-up Hubs. He also supervises several PhD projects on the circular economy. He has authored several of the most highly cited scholarly works on the circular economy, most notably "Conceptualizing the Circular Economy: An Analysis of 114 Definitions."

Anton Kleshkov graduated from the Polytechnical University in Kiev. He works for the National Resource Efficient and Cleaner Production Center within the network of UNIDO centres with a focus on energy optimisation and energy management.

Christina Krenn is an Energy and Environmental Manager and Auditor at STENUM GmbH in Austria. She has more than 15 years of experience in energy optimisation and management systems, systems engineering, project management of research and development projects, energy optimisation of processes and buildings, and waste management concepts.

Elena Lovat is a Research Assistant at Brunel University London. She received an MSc in environmental engineering from the University of Padua (Italy) and an MSc in water sanitation and health engineering from the University of Leeds (UK). Elena is an expert on solid waste management, having worked for the past three years in the Middle East, leading projects on strategic solid waste management planning and systems design. From January 2020, she will be leading technical cooperation projects in Africa in the field of water, sanitation, and solid waste management.

Federico Magalini is a mechanical engineer. He holds a PhD in management, economics, and industrial engineering. He has more than 15 years practical experience on WEEE management, capacity building, and policy and 4 years' experience as an operations and quality manager in a take-back scheme for WEEE, batteries, photovoltaics, and other waste streams. Federico has a solid track record in implementation of take-back policies, development and enforcement of standards, business model development and financing systems, policy development, and technology transfer. He is currently Managing Director of the UK branch of Sofies, a consulting group headquartered in Geneva with offices in Zurich, London, and India.

Patrick J. Mahon is Policy Lead at the Waste and Resources Action Programme (WRAP), a UK-based circular economy charity. He has 20 years' experience in UK and European policy on resources, waste and the circular economy. He holds an MSc in Environmental Management, is a Chartered Resource and Waste Manager, and is the Chair of the Chartered Institution of Wastes Management's (CIWM) Resources and Waste Strategy Special Interest Group. During 2018, he was seconded to the UK Government part-time, to help draft the Resources and Waste Strategy for England. Prior to joining WRAP, he worked for the UK's Environment Ministry and for a UK trade association.

Katy Mason is Professor of Markets and Management at Lancaster University Management School, UK. Following a career in marketing, Katy's research adopts an engaged approach to understanding how different market actors organise and manage to make markets. Drawing on practice and performativity theories, she has engaged in collaborative research projects with blue chip companies and catalysts, as well as open innovation initiatives in policy-oriented setting, that develop and perform market-making strategies. Her work has been published in leading journals: *Journal of Management Studies, Organisation Studies, Long Range Planning, Marketing Theory, Journal of Marketing Management,* and *Industrial Marketing Management.* Katy is an AcSS Fellow, AIM Management Practices Fellow and Cabinet Office Policy Fellow, and Chair of The British Academy of Management.

Louise Maxwell is a Senior Lecturer in Wastes Management at the University of Northampton, UK. She has lectured in the field of waste management for over 18 years. Louise is a member of the Chartered Institution of Wastes Management (CIWM).

Marco Meloni holds an MSc in industrial ecology and biology. He is currently a consultant at the UK branch of Sofies. Prior to joining the company, he worked in the non-profit sector, where he helped companies and governments implement sustainable and innovative processes around the circular economy concept in the fields of electronics and food systems. Marco started his career in aquatic ecology research and, looking to apply his skills in industrial ecosystems, moved to the Netherlands for an industrial ecology MSc at TU Delft and Leiden University, where he refined his environmental and circular economy skills.

Alan Metcalfe holds a doctorate in sociology from the University of Lancaster, UK. He has worked on several major research projects, including a Coca Cola-funded project as part of the company's "Recycle for the Future" campaign; "Lifestyles and Life-Courses: The Social Context of Household Waste Management" funded by the Leverhulme Trust; "The Social Technologies of Waste" as part of the ESRC-funded "Waste of the World" programme; "Men, Children, Food" within the Leverhulme Trust-funded "Changing Families, Changing Food" programme; and an ESRC-funded project, "Disposal, Devaluation, Consumerism." He has held positions at the Universities of Nottingham, Sheffield, Portsmouth, and Exeter.

Karen Miller is a Researcher in the Centre for Industrial Sustainability at the University of Cambridge, UK. Her work focuses on helping organisations develop more sustainable approaches to business through sustainable business model innovation (SBMI). SBMI involves firms innovating novel ways of reconfiguring their capabilities in order to alter how they generate and capture value for all stakeholders. She has a PhD from Cambridge for her thesis: "Understanding the Characteristics of Design Leaders in Large Fashion Retailers." Before joining academia, Karen had an industrial career with over 25 years of experience as a design and innovation director in leading international retail organisations.

Mario Augusto P. Monteiro is Professor in Economics and Finance at Estácio University (Fortaleza, Brazil). He heads an Economy Section at the Regulatory Agency of Public Services of the State of Ceará, with a BSc degree in economics (Federal University of Ceará), an MBA in finance (Brazilian Institute of Capital Markets), and a PhD in business management (University of

Fortaleza). Mario is also an active consultant on regulatory issues concerning network industries and providing advice to governments, as well as to private and public utilities. Besides regulatory issues, his research interests include sustainability and the performance of firms and industries.

Fabio Morea works at AREA Science Park, a national research institution in Trieste, Italy, with a focus on innovation. He is currently engaged in research on sustainable energy and mobility, specifically the harmonisation of SEAPs (SECAPs) and SUMP.

Ruth Morrow is Professor of Biological Architecture, Planning and Landscape at the University of Newcastle Upon Tyne, UK.

Ananya Mukherjee is an IOD Parc/CECAN Fellow based at CRESS/Sociology, at the University of Surrey, UK. She is an environmental sociologist with interdisciplinary training in social sciences and with experience working on complex environmental and biodiversity conservation issues. She believes in the interconnectedness of nature, environment and the several layers of impact it has on humans and social-ecological systems and is interested in using social innovation in addressing issues of sustainability and development. Ananya currently works at the Welsh Centre for Public Policy as a Research Associate on the sustainability and productivity challenges facing economic growth in Wales, based at the University of Cardiff.

Sreejith Nanukuttan is a Senior Lecturer in Civil Engineering at the Queen's University Belfast. His research focuses on the performance of the built infrastructure and concrete.

Tania Nunes da Silva is a Professor at the Federal University of Rio Grande do Sul (UFRGS) in Brazil and Deputy Coordinator of the Manager (Study Group on Organisations), registered with CNPq. Her experience is in business administration, focusing on agribusiness, cooperatives, sustainability, collective entrepreneurship, cooperation, social strategy and innovation, social technology, social entrepreneurship, and sustainable energy. She was previously Coordinator of PPG-Agribusiness/UFRGS from 07/2003 to 04/2008 and PPG-Administração/UFRGS from 01/2011 to 12/2012.

Sarah O'Carroll serves on the Executive of the African Circular Economy Network. She is passionate about job creation and the development of sustainable cities in Africa – cities that are low carbon, resource efficient, and socially inclusive. She is also the Government and Cities Network Manager at the Ellen MacArthur Foundation. Sarah facilitates networking and knowledge sharing between the Foundation's government and cities members, working with them to build collaborations with each other and the Foundation's broader network. Prior to working at the Foundation, Sarah worked at a green economy NPO in Cape Town, South Africa.

Eugenio Avila Pedrozo is an Associate Professor and researcher at the Federal University of Rio Grande do Sul (UFRGS), in Brazil. His main interests include sustainability, complexity (Morinian's perspective), innovation, technology, BOP, and societal discussions linking individuals, organisations, and interorganisational and societal levels. He has special interests in inter/transdisciplinarity analysis using multiple points of view.

Marco Perona is Professor of Industrial Logistics at the University of Brescia, Italy. He is the Scientific Director of the Research and Innovation for Smart Enterprises (RISE) Laboratory. He holds a degree in Mechanical Engineering. He does research, teaching, and transfer activities on operations, supply chains, circular economy and service management. He is author of several scientific publications in such fields. He has also taken part in several company transfer projects on such topics.

Erik Poppe is the research assistant to the Professor of Transdisciplinary Sustainability Research in Electronics at the Technische Universitaet Berlin and a board member of the research and consulting institute SUSTAINUM in Berlin. He studied political science and economics at the MLU Halle (Saale) (BA) and at the FU Berlin (MA) with a focus on modernisation theories and transformation processes. His research focuses on the analysis and mitigation of premature product obsolescence, rebound effects, and product profusion. He has expertise in life-cycle assessments (LCAs) and research for products. In addition to his work, he is actively involved at the Roundtable Repair (RTR) for the improvement of reparability in Germany.

Elizabeth Ragonga is a PhD student in the School of Management and Law at the University of Bradford, UK. She has an MSc in international business and management (University of Bradford, UK), an MBA in strategic management (Daystar University, Kenya), and a second-class upper division in Bachelor of International Business (Makerere University, Uganda). Her areas of research interest are circular economy, digital technologies, and operations management. She is also interested in international business in emerging economies, foreign direct investment, and mergers and acquisitions.

Guy M. Robinson is a human geographer with over 45 years' experience of research in environmental and waste management and rural development. He is currently an adjunct Professor in the School of Geography, Environment and Population in the School of Social Sciences at the University of Adelaide, South Australia, and is also a departmental associate in the Department of Land Economy, University of Cambridge, UK. He has previously held positions in the Universities of Oxford, Edinburgh, Kingston London, and South Australia. He is the Editor-in-Chief of the journal *Research in Globalization* and Editor Emeritus of *Land Use Policy*. He has published over 200 academic papers and 12 books, including *Methods and Techniques in Human Geography*, *Sustainable Rural Systems*, *Conflict and Change in the Countryside*, and *Handbook on the Globalisation of Agriculture.*

Nicola Saccani is Associate Professor at the Department of Industrial and Mechanical Engineering at the University of Brescia (Italy), where he is part of the RISE laboratory (Research and Innovation for Smart Enterprise, www.rise.it). He is member and past coordinator of the ASAP Service Management Forum, a community involving academics and practitioners to develop the culture and excellence in service management and servitisation (www.asapsmf.org). His research activities concern service and supply chain management, with particular reference to the impact of phenomena such as the digital transformation, servitisation, and circular economy on business models, supply chain configuration and operations management. He is author of several scientific publications in such fields. He has also taken part to several company transfer projects on such topics.

Joe Segal holds an MEng aeronautical engineering and has eight years of experience leading companies in the off-grid energy and emerging technology sectors. In five years at a fast-growing start-up, he led the manufacturing operations, digitalised the logistics network, and expanded the business into eight new markets. Since then, he has worked for companies operating in the healthcare, agriculture, and energy sectors, supporting ESG programmes and financial raises and establishing business models for nascent companies. He is the Chief Operations Officer of Trend

Solar a key off-grid energy company focusing on providing digital services through their proprietary smartphone and has over 7 years experience of expanding clean tech companies in the sector.

Gareth Shaw is Professor of Retail and Tourism Management at the University of Exeter Business School, UK. He has published numerous papers and books on both retailing and tourism. He has worked on several research projects relating to innovations in retailing and their impacts on consumer behaviour. In addition, he has led projects on innovation in the hotel sector and aspects of knowledge transfer. More recently, he completed a large ESRC-funded project on British management in the interwar period.

Paul Sheeran is a Senior Lecturer in Macroeconomics in the Department of Accounting, Finance, Mathematics and Economics at the University of Winchester, UK. Paul has recently completed an international trade project, SSHRC-ESRC Understanding the future of Canada-UK Trade, from a circular economy perspective and is currently writing a book and related papers on ecological economics and the circular economy.

Uthayasankar Sivarajah is the Head of Business Analytics, Circular Economy and Supply Chain (BACES) Research Centre and a Chair in Technology Management and Circular Economy at the School of Management, University of Bradford, UK. His broad area of research interest and expertise is interdisciplinary, focusing on the use of emerging digital technology for the betterment of society, be it in a business or government context. He is the Deputy Editor of Journal Enterprise Information Management (JEIM) and actively publishes in leading high impact factor journals such as *Journal of Business Research, Computers in Human Behaviour and Government Information Quarterly*. His research has also been featured in reputable media/trade publications such as Computer Weekly, Public Sector Focus, London School of Economics academic blog, etc. He is a Fellow of the UK Higher Education Academy (FHEA), a member of the British Academy of Management (BAM) and Association for Information Systems (AIS).

Isadora do Carmo Stangherlin is a PhD student in innovation and sustainability at the Business School of the Federal University of Rio Grande do Sul, Brazil. Between September 2019 and December 2020, she visited the Department of Management, Aarhus University, Denmark. Her areas of research interest are sustainable consumption, food waste, circular economy and consumer practices.

V. Madha Suresh is a Professor at the University of Madras in Chennai, India, and is the Head of the Centre for Natural Hazards and Disaster Studies. He is also associated with the Departments of Geography and Education. He has worked with a number of international universities, including the University of Bonn, Germany; York University, Canada; Staffordshire University; and the University of Northampton, UK. He has contributed to both text and reference books and reputed journals. Besides academic and research activities, he was Director of IQAC. Prior to the current organisation, he served as faculty in the National Council of Educational Research and Training, where he was state coordinator for both Tamil Nadu and Pondicherry, carrying out various education programmes, and as Assistant Director, Inland Waterways Authority of India, where he led the automation of cartographic cells created to prepare navigational charts.

John Thøgersen is Professor of Economic Psychology, Department of Management, Aarhus University, Denmark, where he heads the Marketing and Sustainability Research Group. He has more than 100 journal articles and more than 35 book chapters, mostly in the area of sustainable consumption.

Sean Thomas is a Reader at The York Law School, University of York, UK. His research interests concern, broadly, the transfer of ownership of personal property. He is particularly interested in the implications of property regimes for circular economics.

Valérie Thöni holds two MScs in international law and environmental studies. She has 15 years' experience in international cooperation (UN agencies and other international organisations). For the UIAA (the International Federation for Mountaineering and Climbing) she led projects promoting environment protection and green solutions in mountain areas. Over the last two years she has been working on solid waste issues and conducted research about solid waste management in the Occupied Palestinian Territories (published in 2019). Her field of interests also include nature-based solutions in urban environments in developing countries.

Fabio Tomasi is the Head of Internal Projects at AREA Science Park in Trieste, Italy. He has additional experience in working in the training department coordinating courses on innovation, TRIZ, sustainable development, and energy efficiency. Fabio has coordinated EU projects like TETRIS (www.tetris-project.org), IUSES (www.iuses.eu) and PATRES (www.patres.net).

Mentore Vaccari is an Associate Professor of Sanitary and Environmental Engineering at the University of Brescia, Italy. He received his MSc in civil engineering at the University of Brescia, followed by a PhD in sanitary engineering at the Polytechnic of Milan (Italy). His research includes municipal, industrial, medical, and electronic solid waste management; wastewater treatment; contaminated site remediation; health risk assessment; circular economy; and appropriate technologies for developing countries. He has more than 230 scientific publications.

Eva Verdejo holds a bachelor's degree in chemistry from the University of Valencia (Spain). She joined AIMPLAS in 1993 and is currently Leader of the Chemical Recycling Area and the former Head of Department of Sustainability and Industrial Valorization of AIMPLAS. Her expertise is focused on sustainable development, materials from natural resources, biodegradable plastics, and recycling. She has significant experience in proposal writing and participation/coordination of relevant national and international R+D projects related to her expertise. She has spoken at numerous seminars and conferences and is the author of technical articles for scientific journals and magazines and a professor in different sustainability and recovery-related courses.

Vicente B. Vert holds a PhD in chemical engineering from Universtitat Politècnica de València (Spain) and a master's degree in polymeric materials and composites from the same university. He has 15 years experience in the execution and technical management of national and international research projects, based on the development of materials, electrochemistry, the circular economy, recycling, biodegradability, and the renewable origin of plastics. He has worked in the areas of sustainability and industrial recovery with AIMPLAS since 2017, where he also leads the circular economy strategy. He provides technical advice to companies interested in reducing the environmental impact of plastic-based products, proposing product alternatives, suitability of materials, and environmental assessment studies, such as life-cycle analysis or carbon footprinting, among other eco-labels and certifications.

Giovanni Vinti received his MSc in civil and environmental engineering at the University of Palermo in Italy. Since 2017, he has been a PhD student in the programme in Civil,

Environmental, International Cooperation and Mathematical Engineering at the University of Brescia, in Italy. His research topic concerns solid waste and health issues, with a focus on developing countries. Over the years, he has followed field activities in developing settings dealing with health and environmental aspects. Since October 2019, he has also been a research fellow at the University of Brescia and is working on a case study concerning biowaste management among rural communities in Ghana.

Doroteya Vladimirova is Director of the Regenerative Business Models Research Programme and a senior research associate at the Centre for Industrial Sustainability at the University of Cambridge. She is a fellow of the Cambridge Institute for Sustainability Leadership. Her research focuses on how business is taking on the most pressing global challenges relating to finite natural resources, enhancing livelihoods, reducing poverty, protecting ecosystems, and tackling climate change to turn them into market-driving opportunities.

Mike Webster has worked for over 17 years in the waste and secondary resource sector, with frontline experience of planning, development, and delivery of effective waste and recycling systems. He has worked both in the United Kingdom and internationally, including the South Pacific, Latin America, and West Africa, working with numerous community-based organisations to deliver composting and recycling systems, including Solomon Island, The Gambia, Senegal, Ghana, Nigeria, Kenya, Somaliland, and DR Congo. He has experience in both consultancy and the charitable sector, having founded the NGO WasteAid UK and set up a training centre for waste livelihoods in The Gambia. He is also an experienced trainer, having developed numerous training courses around waste and recycling related topics around the world. Mike was co-author on the 2017 Making Waste Work, a toolkit for community waste management in lower- and middle-income countries. He is currently Chief of Waste Operations for Project STOP in Indonesia.

David C. Wilson has been an international waste and resource management consultant for 40+ years and a visiting professor at Imperial College London since 2000. He is well known for his extensive work on waste and resources policy, strategy, and evidence in both the EU and developing and emerging countries. He was the lead author of UNEP and ISWA's inaugural Global Waste Management Outlook (2015), prepared in parallel to the sustainable development goals.

Miying Yang is a Lecturer in Engineering Management at the University of Exeter, UK. Prior to this, she obtained her PhD degree at the Centre for Industrial Sustainability at University of Cambridge. Miying's research focuses on how firms move towards sustainability/circular economy through business model innovation and specialises in developing product-service system business models for this purpose. Her recent research interest includes sustainable business model innovation through digitalisation and how digital technologies affect supply chain performance.

Christian Zurbrügg is a Senior Researcher in the Department of Sanitation, Water and Solid Waste for Development (Sandec) at the Swiss Federal Institute of Aquatic Science and Technology (Eawag), appointed member of the Eawag directorate, and adjunct professor at the University of Agricultural Sciences (SLU) in Uppsala, Sweden. His research interests lie in the interdisciplinary approach of upgrading services and infrastructure in urban areas of low- and middle-income settings – with a special focus on solid waste management.

ACKNOWLEDGEMENTS

The editors humbly acknowledge the significant contribution of all of the authors, without whom this volume would not have been possible. We especially thank those authors who came on board somewhat later in the day. We also sincerely thank Egle Zigaite and Andrew Mould at Routledge for their guidance, support, and patience throughout the process. We also would like to thank our families, loved ones, friends, and colleagues for their support.

1
INTRODUCTION
Examining the concept of the circular economy

Terry Tudor and Cleber JC. Dutra

Background

The rise in the global population to 10 billion people, coupled with the need for resources to satisfy this growth, is resulting in an unsustainable increase in the extraction and processing of raw materials. It is predicted that the largest rise in extraction and processing will be in minerals, especially in developing countries (OECD, 2018). The impacts of this escalation are expected to be higher pollution rates, a 'significant' contribution to climate change, and two to three planets' worth of natural resources by 2030 and 2050. With nations in the Global South expected to account for more than 90% of the world's people (Goyal et al., 2018), these impacts will jeopardise the gains made in global health and well-being. Reserves of available natural resources are expected to decrease, while the costs of sourcing and extracting virgin materials will significantly rise. Globally, material use will double from 79 Gt in 2011 to 167 Gt in 2060. While there is evidence of some decoupling of economic growth and material use, with a material intensity of 1.3%, reducing the extraction of raw materials and increasing the recovery of value from existing resources is therefore, becoming exponentially important.

Evidently, our current rate of resource consumption means that we face significant challenges with regard to sustainability, as well as with our environment, health, and economies. These concerns are particularly acute for developing economies, which often lack the resources and governance mechanisms to enable them to cope with and withstand any systemic shocks. The concept of a circular economy (CE) offers the potential to mitigate these risks by facilitating the cyclic use of resources and increasing value creation.

It is within this context of the need for change, particularly to benefit countries in the Global South, that this *Handbook* has been written. It provides both theoretical perspectives as well as practical examples and case studies of how the CE principles could be, and indeed, have already been employed to effect real change in policies and practice. These perspectives and case studies cover a range of disciplines, sectors, and geographies from the Global South and the Global North.

The *Handbook* utilises a critical lens in examining the issues and suggested ways forward. It employs a multi/interdisciplinary approach to establish a rich, holistic context that will be of value to academics but also to policy- and decision-makers and industry.

The circular economy

It is generally held that the general concept of a CE resulted from the work of Boulding (1966: 5), who proposed that the Earth was a closed system with "limited assimilative capacity and as such the economy and environment must coexist in equilibrium" (i.e., a Space Ship Earth) (Ghisellini et al., 2016; Andersen, 2007; Lieder and Rashid, 2016). However, some contend that actually, due to the limitations in resources, countries in the Global South have practiced the concept long before those in the Global North (e.g., Schröder et al., 2019). These differences in the rationales for and approaches to implementation of the concept are explored throughout the *Handbook*.

According to Millar et al. (2019), Boulding's work influenced that of others from a range of schools of thought (e.g., environmental economics [Pearce and Turner, 1990], industrial ecosystems [Jelinski et al., 1992], cleaner production [Stevenson and Evans, 2004], product-service systems [Tukker, 2015], biomimicry [Benyus, 1997], the performance economy [Stahel, 2010], eco-efficiency [Haas et al., 2015], cradle-to-cradle design [Braungart et al., 2007], regenerative design [Lyle, 1996], and industrial ecology [Graedel and Allenby, 2003]) around the concept of a closed-loop economy. For example, Walter Stahel in the late 1970s came to the view that a linear economic model was not sustainable, as consumption would lead to resource limitations in the future. Stahel proposed the need to close material cycles and reform the economy. In 1972, the Club of Rome published their ground-breaking report, *Limits to Growth* (Meadows et al., 1972), which came to a similar conclusion that the existing economic model was leading to increased demands for resources and by extension generating high levels of waste.

A CE approach is an umbrella concept. It aims to promote the responsible and cyclical use of resources, to reduce material input and waste generation, and to decouple economic growth from natural resource use (Blomsma and Brennan, 2017; Homrich et al., 2018; EASAC, 2016; EEA, 2016; Pauliuk, 2018). It requires an optimised closing of the 'loop' and a reduction in the need for the extraction of virgin raw materials by minimising waste, extending product life, maintaining materials at their highest level, optimising reuse, and utilising energy sources such as biomass (Webster, 2015; Stahel, 2016). For example, the EU's Circular Economy Action Plan (EC, 2015) notes that a CE occurs when "the value of products, materials and resources is maintained in the economy for as long as possible, and the generation of waste minimized" in such a way so as to "develop a sustainable, low carbon, resource efficient and competitive economy." It is associated with other concepts, such as industrial symbiosis (Chertow and Ehrenfeld, 2012) and eco-cities, a concept that has grown rapidly in countries such as Japan and Singapore (Dong et al., 2016).

Products and materials keep circulating in a high-value state of use, through supply chains, for as long as possible (ZWS, 2019). These processes require all stakeholders (e.g., designers, businesses, communities, governments, and individuals) to be involved (Bocken et al., 2016; Jiang and Zheng, 2014; Kirchherr and Piscicelli, 2019). For example, some 80% of environmental impacts are made at the design stage (EMF, 2019). However, if we can 'get it right,' then the benefits are significant. For example, it is thought that a CE will be worth US$4.5 trillion by 2030, create 500,000 jobs in France alone, and reduce 37% of the energy consumption in the EU (WBCSD, 2016).

In 2008, China became one of the first countries to enact specific legislation for a CE (CIRAIG, 2015). Germany and Japan have also been recognised as pioneering countries in the development of policies (Geng et al., 2013).

The concept has been introduced for differing reasons in different countries. For example, in the early 1990s, the concept was embedded into German environmental policy to address issues associated with raw material and natural resource use for sustained economic growth (Geng and Doberstein, 2008). In China, in the late 1990s, an eco-industrial park model was promoted, and in the mid-2000s, the application of the concept was introduced in line with Hu Jintao's concept of a 'harmonious society,' with a focus on post-consumption waste recycling (Sakai et al., 2011). In China, the CE has been used for some time as a mechanism for profitable product development, new technology development, upgrading equipment, and improving industry management (Yuan and Moriguichi, 2006; Jiang and Zheng, 2014). In the United States and Europe, application of the CE concept is primarily to enhance waste reduction, reuse, and recycling and to conduct product-level life-cycle studies (CEC, 1975; EC, 2011; Unilever, 2015; EC, 2018; Lee, 2018). In many countries in the Global South, it serves as a fundamental way of life, especially among the informal sector, supporting livelihoods, job creation, and community development (Schröder et al., 2019; Wellesley et al., 2019).

Situating circular economy principles in the field of waste and resource management

The CE concept involves four technical cycles, namely (EMF, 2013):

- Maintaining product
- Re-using/redistributing (used) product
- Upgrading/remanufacturing product
- Recycling product

In the smallest cycle, the product/material retains the highest value, as it can be applied longer in accordance with its original purpose. In the longest cycle, the value or residual value of the product is lowest, and a different application must be found.

The concept creates value in four different ways (Accenture, 2014):

- Renewable sources continuously regenerated in the course of time
- Quick markets allowing for optimal use and access of products and possessions
- Increased product lifespan by designing products that are made to last
- Linked value chains that do not produce waste

The theory behind the CE is applied in various related movements, such as in the cradle-to-cradle philosophy, biomimicry, green economy, and blue economy. These movements also focus on smarter and more effective use of commodities.

The waste hierarchy forms a critical component in the CE concept (Pires and Martinho, 2019). It originated in Holland in the late 1970s but was made popular in the United States (Overcash, 2002; Parto et al., 2007; van Ewijk and Stegemann, 2016). The principles of the hierarchy were adopted in the EU Waste Framework Directive 2008/98/EC (European Council, 2008). In 2015, the EU Circular Economy Strategy (COM/2015/0614) (European Council, 2015) emphasised the importance of the waste hierarchy in protecting the environment and recovery of value. In 2016, it was included in Sustainable Development Goal (SDG) 12, with the aim by 2030 to "substantially reduce waste generation through prevention, reduction, recycling and reuse."

Challenges associated with the concept of a circular economy

At a fundamental level, the complexity of the CE concept, coupled with the fact that it is meant to 'manage' resources, makes it difficult to implement (Geng et al., 2009; Winans et al.,

2017; Kirchherr et al., 2018). Unlike that for sustainable development, no universal definition of the concept of a CE exists in the literature (Millar et al., 2019). One of the challenges faced, therefore, is how can progress towards a CE best be measured if there is no common definition? Indeed, what factors should be measured, and how (Moraga et al., 2019)?

Millar et al. (2019) also contend that there are several contradictions and knowledge gaps with the concept. For example, continual economic development is invariably linked to environmental degradation. In addition, closed material loops are practically and theoretically impossible. Korhonen et al. (2018) note that much of the research on CE is focused around metrics, tools, and computations, with limited attention paid to factors such as values, societal structures, cultures, worldviews, and the paradigmatic potential of the concept. Thus, the authors contend that any evaluation of the concept should take account of all levels (i.e., from 'global' to individual).

The structure of the *Handbook*

This *Handbook* presents a variety of perspectives, views, case studies, and initiatives. We do not seek to take sides or to state that any one perspective is 'better.' Indeed, it is for you, the reader, to come to your own (hopefully more informed) conclusions based on what is presented.

It covers a range of views, sectors, and geographies. An introduction sets out the rationale, aims, and structure of the *Handbook*. It is organised into seven interrelated parts. Part I introduces the context, including the growth in materialism, resource consumption, and waste in Chapters 2 and 3. Socioeconomic and health risk factors and the Sustainable Development Goals are discussed in Chapters 4–7. And Chapters 8–10 call for a shift towards more sustainable and circular approaches. Part II then discusses some of the interpretations of the concept of circularity, including design tools and modularity in Chapters 11–13, and ends with critiques of the use of 'systems approaches' in Chapters 14 and 15. The challenges of and dichotomies in the interpretation and implementation of the concept are highlighted in these initial chapters. Indeed, these issues form an underlying theme in several of the case study chapters in the final four sections of the handbook. Part III provides a brief evaluation of the policy and legislative context for a more circular approach. For example, challenges around the development and implementation of circularity within policies and trade agreements are discussed in Chapters 16–18, while Chapter 19 provides some international perspectives. Part IV is concerned with analysing how the circularity concept has been used to develop a 'sharing economy' in both the Global North (e.g., Chapters 20–22) and the Global South (Chapter 23), for example, within the informal sector (e.g., Chapters 24 and 25). The discussion covers resources such as food and energy. Part V provides examples of recycling that are being employed in different sectors, primarily from Europe (Chapters 26–28) and Africa (Chapters 29 and 30). Part VI outlines case studies of the utilisation of reuse in Europe (Chapters 31–33) and India (Chapter 34). Finally, Part VII explores the use of technologies for circularity, for example, the use of energy and fourth industrial technologies (Chapters 35–37), and ends with the downstream implications of the use of technologies (Chapters 38 and 39). Chapter 40 provides an overall conclusion for the *Handbook*.

Conclusions

With global consumption trends rising to unsustainable levels, and the impacts this is having on the environment, health, and economies, the utilisation of the concept of a CE offers a mechanism to address these issues. Using a number of views and case studies from both the Global

South and Global North, this *Handbook* explores the contextual, policy, legislative, logistical, and behavioural issues faced in the effective utilisation of the concept. Without action, the impacts will be catastrophic. However, in the next ten years, key international milestones in global climate change, the Sustainable Development Goals, and agreements on environmental protection offer a unique opportunity. It is an opportunity that must be seized to integrate the CE concept into existing global political policies and agendas and individual practices to catalyse real change.

References

Accenture (2014). *Circular advantage.* Available at: www.accenture.com/t20150523t053139__w__/us-en/_acnmedia/accenture/conversion-assets/dotcom/documents/global/pdf/strategy_6/accenture-circular-advantage-innovative-business-models-technologies-value-growth.pdf (Accessed 13/10/2019).

Andersen M. (2007). An introductory note on the environmental economics of the circular economy. *Sustainability Science.* 2 (1): 133–140.

Benyus JM. (1997). *Biomimicry – innovation inspired by nature.* New York: William Morrow and Company.

Blomsma F, Brennan G. (2017). The emergence of circular economy: a new framing around prolonging resource productivity. *Journal of Industrial Ecology.* 21: 603–614.

Bocken NMP, De Pauw I, Bakker C, van der Grinten B. (2016). Product design and business model strategies for a circular economy. *Journal of Industrial Production Engineering.* 33: 308–320.

Boulding KE. (1966). The economics of the coming spaceship Earth. In: Jarrett, H. (Ed.), *Environmental quality in a growing economy: essays from the sixth RFF forum.* New York, NY: RFF Press. 3–14.

Braungart M, McDonough W, Bollinger A. (2007). Cradle-to-cradle design: creating healthy emissions – a strategy for eco-effective product and system design. *Journal of Cleaner Production.* 15 (13–14): 1337–1348.

CEC (1975). *Council directive 75/442/EEC on waste.* Brussels: The Council of the European Communities.

Chertow M, Ehrenfeld J. (2012). Organizing self-organizing systems. *Journal of Industrial Ecology.* 16 (1): 13–27.

CIRAIG (2015). *Circular economy: a critical literature review of concepts.* Montreal, Canada: Centre for the Life Cycle of Products Processes and Services.

Dong H, Fujita T, Geng Y, Dong L, Ohnishi S, Sun L, Fujii M. (2016). A review on eco-city evaluation methods and highlights for integration. *Ecological Indicators.* 60: 1184–1191.

EASAC (2016). *Indicators for a circular economy.* Halle: European Academies' Science Advisory Council.

EC (2011). *Road map to a resource efficient Europe.* COM (2011) 571 final. Brussels: European Commission.

EC (2015). *Commission recommendation of 2 December 2015 on closing the loop – an EU action plan for the circular economy.* COM (2015) 614 final. Brussels: European Commission.

EC (2018). *Measuring progress towards circular economy in the European Union – key indicators for a monitoring framework – SWD (2018) 17 final.* Strasbourg: European Commission.

EEA (2016). *Circular economy in Europe – developing the knowledge base: report 2.* Copenhagen: European Environment Agency.

EMF (Ellen MacArthur Foundation) (2013). *Towards the circular economy.* Cowes: Ellen MacArthur Foundation.

EMF (Ellen MacArthur Foundation) (2019). *Designing for reuse and circulation of products and materials.* Available at: https://www.ellenmacarthurfoundation.org/assets/downloads/2_Products_Designing_Mar19.pdf

European Council (2008). *Directive 2008/98/EC of the European parliament and of the council of 19 November 2008 on waste and repealing certain directives.* Brussels, Belgium.

Geng Y, Doberstein B. (2008). Developing the circular economy in China: challenges and opportunities for achieving' leapfrog development'. *International Journal of Sustainable Development & World Ecology.* 15 (3): 231–239.

Geng Y, Sarkis J, Ulgiati S, Zhang P. (2013). Measuring China's circular economy. *Science.* 339 (80): 1526–1527.

Geng Y, Zhu Q, Doberstein B, Fujita T. (2009). Implementing China's circular economy concept at the regional level: a review of progress in Dalian, China. *Waste Management.* 29 (2): 996–1002.

Ghisellini P, Cialani C, Ulgiati S. (2016). A review on circular economy: the expected transition to a balanced interplay of environmental and economic systems. *Journal of Cleaner Production.* 114: 11–32.

Goyal S, Esposito M, Kapoor A. (2018). Circular economy business models in developing economies: lessons from India on reduce, recycle, and reuse paradigms. *Thunderbird International Business Review.* 60: 729–740.

Graedel N, Allenby B. (2003). *Industrial ecology.* London: Pearson College Division. 2nd Ed.

Haas W, Krausmann F, Wiedenhofer D, Heinz M. (2015). How circular is the global economy? An assessment of material flows, waste production and recycling in the European Union and the world in 2005. *Journal of Industrial Ecology*. 19 (5): 765–777.
Homrich AS, Galvão G, Abadia LG, Carvalho MM. (2018). The circular economy umbrella: trends and gaps on integrating pathways. *Journal of Cleaner Production*. 175: 525–543.
Jelinski IW, Graedel TE, Laudise RA, McCall DW, Patel CKN. (1992). Industrial ecology: concepts and approaches. *Proceedings of the National Academy of Science*. 89: 793–797.
Jiang Y, Zheng X. (2014). Private sector participation and performance of urban water utilities in China. *Economic Development and Cultural Change*. 63: 155–189.
Kirchherr J, Piscicelli L. (2019). Towards an education for the circular economy (ECE): five teaching principles and a case study. *Resources, Conservation and Recycling*. 150: 104406.
Kirchherr J, Piscicell, L, Bour R, Kostense-Smit E, Muller J, Huibrechtse-Truijens A, Hekkert M. (2018). Barriers to the circular economy: evidence from the European Union (EU). *Ecological Economics*. 150 (C): 264–272.
Korhonen J, Nuur C, Feldmann A, Birkie SE. (2018). Circular economy as an essentially contested concept. *Journal of Cleaner Production*: 544–552.
Lee G. (2018). *The state of the circular economy in America: trends, opportunities, and challenges*. San Francisco, CA: Circular Colab.
Lieder M, Rashid A. (2016). Towards circular economy implementation: a comprehensive review in context of manufacturing industry. *Journal of Cleaner Production*. 115: 36–51.
Lyle J. (1996). *Regenerative design for sustainable development*. New York, NY: John Wiley & Sons.
Meadows DH, Meadows DL, Randers J, Behrens WW. (1972). *Limits to growth*. New York, NY: Universe Books.
Millar N, McLaughlin E, Börger T. (2019). The circular economy: swings and roundabouts? *Ecological Economics*. 158: 11–19.
Moraga G, Huysveld S, Mathieux F, Blenginic GA, Alaerts L, Acker KV, de Meester S, Dewulf J. (2019). Circular economy indicators: what do they measure? *Resources, Conservation and Recycling*. 146: 452–461.
OECD (2018). *Global material resources outlook to 2060: economic drivers and environmental consequences*. Available at: www.oecd.org/environment/waste/highlights-global-material-resources-outlook-to-2060.pdf (Accessed 20/5/2020).
Overcash M. (2002). The evolution of US pollution prevention, 1976–2001: a unique chemical engineering contribution to the environment: a review. *Journal of Chemical Technology and Biotechnology*. 77: 1197–1205.
Parto S, Loorbach D, Lansink A, Kemp R. (2007). Transitions and institutional change: the case of the Dutch waste subsystem. In: Parto S, Herbert-Copley B. (Eds.), *Industrial innovation and environmental regulation: developing workable solutions*. New York: United nations University Press. 233–257.
Pauliuk S. (2018). Critical appraisal of the circular economy standard BS8001:2017 and a dashboard of quantitative system indicators. *Resources, Conservation and Recycling*. 129: 81–92.
Pearce D, Turner R. (1990). *Economics of natural resources and the environment*. Baltimore, MD: Johns Hopkins University Press.
Pires A, Martinho G. (2019). Waste hierarchy index for circular economy in waste management. *Waste Management*. 95: 298–305.
Sakai SI, Yoshida H, Hirai Y, Asari M, Takigami H, Takahashi S, Douvan AR. (2011). International comparative study of 3R and waste management policy developments. *Journal of Material Cycles and Waste Management*. 13 (2): 86–102.
Schröder P, Anantharaman M, Anggraeni K, Foxon T. (2019). *The circular economy and the Global South*. London: Routledge.
Stahel WR. (2010). *The performance economy*. London: Palgrave Macmillan.
Stahel WR. (2016). The circular economy. *Nature*. 531: 7595.
Stevenson R, Evans J. (2004). Editorial to: cutting across interests: cleaner production, the unified force of sustainable development. *Journal of Cleaner Production*. 12 (3): 185–187.
Tukker A. (2015). Product services for a resource-efficient and circular economy – a review. *Journal of Cleaner Production*. 97: 76–91.
Unilever. (2015). *Reducing our environmental impact: how we harness the latest science to minimise our environmental footprint*. Available at: https://www.unilever.com/about/innovation/safety-and-environment/reducing-our-environmental-impact/

van Ewijk S, Stegemann JA. (2016). Limitations of the waste hierarchy for achieving absolute reductions in material throughput. *Journal of Cleaner Production*. 132: 122–128.

WBCSD (2016). *CEO guide to the circular economy*. Available at: https://docs.wbcsd.org/2017/06/CEO_Guide_to_CE.pdf (Accessed 5/10/2019).

Webster K. (2015). *The circular economy: a wealth of flows*. Cowes: Ellen MacArthur Foundation Publishing.

Wellesley L, Preston F, Lehne J. (2019). *An inclusive circular economy: priorities for developing countries*. Chatham House. Available at: www.chathamhouse.org/publication/inclusive-circular-economy-priorities-developing-countries (Accessed 4/6/2020).

Winans K, Kendall A, Deng H. (2017). The history and current applications of the circular economy concept. *Renewable and Sustainable Energy Reviews*. 68: 825–833.

Yuan Z, Bi J, Moriguichi Y. (2006). The circular economy: a new development strategy in China. *Journal of Industrial Ecology*. 10 (1–2): 4–8.

ZWS (2019). *Why we need a circular economy*. Available at: www.zerowastescotland.org.uk/circular-economy/why-it-is-required (Accessed 25/11/2019).

PART I

The need for and challenges surrounding circularity

2

NATURAL RESOURCES. CONSUMPTION, POLLUTION, AND HEALTH RISKS

Developed versus developing economies

Giovanni Vinti and Mentore Vaccari

Introduction

Throughout human history, natural resources have been essential to make human life easier. However, the need for and exploitation of natural resources has often been accompanied by wars and injustices, as well as environmental and health problems (Sciortino, 2003, 2008).

A number of environmental issues related to the consumption of natural resources exist. In addition, these resources and their benefits are often not equally distributed worldwide. Indeed, as will be shown in this chapter, on the one hand, several natural resources are mainly used in the so-called 'developed economies'; however, they are extracted in low- and middle-income countries. On the other hand, while environmental technologies to face pollution are more accessible in developed countries, in low- and middle-income countries, this is not the same, often resulting in poor environmental quality.

A further element to consider is the growing pressure on the Earth due to the rise in the human population. Indeed, according to the United Nations (UN, 2019), the human population in 2050 will stand at between 9.4 and 10.1 billion people. It is significant to note that in 1950, there were around 2.5 billion people, three times less than today.

As a consequence, if our global society does not reduce its consumerist approach and increase the efficient use of resources, in future, the need for natural resources will be much greater, sharpening the already significant global resources competition. Gulley et al. (2018) even asserted that, unlike the past and present period in which resource conflicts have often been related to fuel minerals, future conflicts could be more related to competition for the nonfuel minerals needed for emerging technologies. Rivalry between the United States and China might, therefore, become more contentious. In this scenario, the waste hierarchy principle, based on prioritising reduction, recycling, and reuse of waste, could be very helpful (Pires and Martinho, 2019).

Many classifications related to natural resources exist, reflecting the different physical characteristics of the resources as well as the geographic diversity of the producing areas. Among these, the United Nations' framework classification (UN, 2010) includes fossil fuels and mineral reserves and resources located on or below the Earth's surface. This chapter focuses on the exploitation of these natural resources, analysing and comparing the related environmental pollution both in high-income and low-income countries. As will be shown, it is significant to note in comparing European countries with low-income countries, that while European countries are very low regarding the production of natural resources, they are at the top of the list in terms of gross domestic product (GDP), as well as, not surprisingly, environmental indexes, while the situation is often the opposite in low-income countries. A detailed series of case studies is presented, discussing both the environmental and health issues related to natural resource exploitation.

The use of natural resources

Natural resources: distribution and consumption around the world

Natural resources are neither distributed nor consumed evenly. Indeed, starting with oil (or petroleum), the British Petroleum Statistical Review of World Energy (BP, 2018) shows the main producers worldwide are the United States, Saudi Arabia, and Russia, all with a similar rate (between 11 and 13 million barrels [Mb] daily in 2017), more than double that compared to Iran, Canada, and Iraq. At the same time, considering the proven oil reserves, the situation changes considerably. Venezuela is the first (ca. 300 billion barrels [Gb] of oil), followed by Saudi Arabia (266), Canada (168), Iran (157), and Iraq (149). Proven reserves may be defined as those quantities of petroleum which, by analysis of geological and engineering data, can be estimated with a high degree of confidence. It is notable that the main oil consumers worldwide are the United States, with a rate of almost 20 Mb daily in 2017, China (less than 13 Mb), India (less than 5 Mb), Japan (4), and Saudi Arabia (3.9).

Considering natural gas, according to BP (2018), the main producers worldwide in 2017 were the United States, with about 730 billion m^3 (Gm^3), followed by Russia (630), Iran (220), Canada (176), and Qatar (176). Considering natural gas reserves, the situation changes a bit. Russia is first with 35 trillion m^3 (Tm^3), followed by Iran (33), Qatar (25), Turkmenistan (19.5), and the United States (8.7). In terms of consumption, the United States also in this case is the first (ca. 740 Gm^3 in 2017), followed some distance behind by Russia (425), China (240), Iran (214), and Japan (117).

Regarding coal, the British Petroleum data (BP, 2018) highlight the main producer in 2017 was China, with about 1750 million tonnes oil equivalent (Mtoe), several times greater than the United States (371), Australia (297), India (294), and Indonesia (271). The scenario changes considerably focusing on coal reserves. The United States is the first, with about 250 billion tonnes (Gt), followed by Russia (160); Australia (145); and only fourth China, with about 140 Gt, while fifth is India (100). At the same time, in terms of consumption, China is the main consumer worldwide (ca. 1890 Mtoe), followed by India, the United States, Japan, and Russia. The sum of the consumption of these last four is equal to about half of the total Chinese consumption.

Regarding mineral resources, the World Mining Data 2019 (Reichl and Schatz, 2019), skipping the already discussed mineral fuels, gives information about the following groups as well:

- Iron and ferro-alloy metals
- Non-ferrous metals
- Precious metals
- Industrial minerals (e.g., salt, gypsum, gemstone diamonds)

Among iron and ferro-alloy metals, the most-produced metals in 2017 were iron (more than 1.5×10^9 tonnes), followed at a great distance by manganese (ca. 1.8×10^7 tonnes) and chromium (ca. 1.4×10^7 tonnes). In this group, it is worth remembering tantalum as well (known also as coltan), which is one of the main components of capacitors (U.S. Geological Survey, 2019), although production levels are low compared with the previous (less than 2000 tonnes in 2017). Regarding non-ferrous metals, the most produced in 2017 was bauxite (ca. 3.1×10^8 tonnes), followed by aluminium (ca. 6.0×10^7 tonnes) and copper (ca. 2.0×10^7 tonnes). Rare-earth metals are also included in this group, but, as expected, their production is lower (ca. 1.2×10^5 tonnes). Considering precious metals, the most produced in 2017 was silver (ca. 2.6×10^7 tonnes), followed by gold (3.3×10^6 tonnes) and palladium (ca. 2.0×10^5 tonnes). Finally, considering industrial minerals, the most produced in 2017 were salt (ca. 2.8×10^8 tonnes), gypsum (ca. 1.5×10^8), and gemstone diamonds (9.2×10^7, expressed in carats).

Also the aforementioned minerals are not equally distributed worldwide, as shown by their production. The main producers of iron are Australia, China, and Brazil, producing around two-thirds of the total, followed by India and Russia. While the main producer of manganese is South Africa, with almost one-third of the total, followed by Australia, China, Gabon, and Brazil. Considering chromium, about half of that produced worldwide comes from South Africa, followed by Kazakhstan, Turkey, India, and some distance behind by Finland. Regarding tantalum, the main producers are Democratic Republic of Congo and Rwanda, with almost two-thirds of the global production, followed some way behind by China, Australia, and Brazil.

Regarding non-ferrous metals, the main producer of bauxite is Australia, with almost one-third of the total production, followed by China, Guinea, Brazil, and India. The main producer of aluminium is China (with about half of the global production), followed by Russia, India, United Arab Emirates, and Australia. The main producers of copper are Chile, Peru, China, Democratic Republic of Congo, and Australia, with more than half of global production. For rare-earth metals, the undisputed producer is China, with more than 80%. Regarding precious metals, the main producers of silver are Mexico, Peru, China, Chile, and Russia, which produce more than 50% of the total. The main producer of gold is China, with more than 10% of the global production, followed by Australia, Russia, the United States, and Peru. Palladium is produced mainly by Russia and South Africa (around 80%), followed by Canada, the United States, and Zimbabwe. Regarding industrial minerals, the main producers of salt are China, the United States, and India, with about 50% of the total production worldwide, followed by Germany and Australia. Gypsum is mainly produced by China, the United States, Iran, Thailand, and Turkey, which produce about 50% of the total. Finally, the main producers of gemstone diamonds are Russia, Canada, Botswana, Angola, and Australia, with about 90% of the global production (in carats).

All these data demonstrate how heterogeneous the distribution of natural resources is throughout the world. At the same time, though, it is not possible to note a particular region of the world with the highest reserves or production. It is also noteworthy that Europe has limited availability of natural resources.

Environmental issues due to the exploitation of natural resources: the status in developed and developing economies

Among the 20 countries with the highest GDP (IMF, 2019), most of the aforementioned main producers of the natural resources are missing. For example, there is no African country and only Saudi Arabia from the Middle East. At the top of the list are the United States and China.

At the same time, it is possible to find many European countries that are also on the list. Considering the GDP per capita, the situation is similar but even more unbalanced. This is a very relevant aspect which highlights how the presence of natural resources to exploit does not necessarily represent economic benefits for local communities. Furthermore, it has to be considered that advanced intensive environmental technologies to reduce the effects of pollution related to resource exploitation usually need a lot of money (Popp et al., 2010), and as a consequence, many low-income countries have to involve international actors (such as multinational corporations) to overcome their economic and technical weaknesses, which also concern the skills needed to exploit their resources. This last point is controversial, as it does not always really help local actors to improve their conditions (Rudra and Jensen, 2011).

It is interesting to analyse the Environmental Democracy Index (EDI), which is based on the concept that public participation is fundamental to ensuring that land and natural resource decisions adequately address the interests of citizens. The EDI considers the following rights: the right to freely access information on environmental quality, to participate in decision-making, and to seek enforcement of environmental laws (TAI and WRI, 2015). Observing the EDI, which actually considers only 70 countries worldwide, it is possible to note how the highest-ranked countries come from all the continents, excluding Australia and Africa (with the exception of 'rich' South Africa), but the last 20 come only from Africa, Asia, and Latin America, with none from Europe or North America.

A further detailed way to quantify the environmental performances of a country is through the Environmental Performance Index (EPI) (Wendling et al., 2018), which is based on two policy objectives: Environmental Health, which measures threats to human health, and Ecosystem Vitality, which measures natural resources and ecosystem services. In the EPI, 24 indicators have been grouped within the following ten issue categories: Air Quality, Water and Sanitation, Heavy Metals, Biodiversity and Habitat, Forests, Fisheries, Climate and Energy, Air Pollution, Water Resources, and Agriculture. EPI values are available for 180 countries. In the lowest positions are the aforementioned natural resource–rich countries such as Democratic Republic of Congo (178th position), India (177th), and Afghanistan (168th). In the top positions are mainly European countries, and just after the 20th position, there are natural resource–rich countries with high GDP, namely Australia (21st), Canada (25th), and the United States (27th).

In the last decades, rapid technological evolution has led not only to increasing production of new equipment, but also to a growing generation of waste electrical and electronic equipment (WEEE), which are exported from industrialised to developing countries, serving as a secondary source of valuable materials for the local population (Vaccari et al., 2019). Indeed, depending on the specific devices, a wide range of both precious metals and rare-earth elements can be found in addition to base metals (Cesaro et al., 2019). In this scenario, Breivik et al. (2014) asserted that 23% of WEEE generated in the countries from the Organisation for Economic Co-Operation and Development (OECD) is exported to seven preferred non-OECD countries: China, India, Nigeria, Ghana, Côte d'Ivoire, Benin, and Liberia. Furthermore, roughly 80% of the entire amount of the WEEE globally produced is transferred to Asia and Africa, whereas the main source countries are in Europe (Vaccari et al., 2019).

The recycling of WEEE in these developing regions of the world is often carried out without any environmental and health protection, causing a lot of problems for people working or living in those areas. Indeed, high concentration of organic and inorganic contaminants (toxic or carcinogenic) produced during these informal activities have been found in the environment (soil, dust, sediments, air, and water), with traces of the pollutants present in the bodies of people in these areas (Vaccari et al., 2019). Observing this scenario, it can be stated that the rich countries often start a cycle using the natural resources of the poorest ones and close the same cycle by sending back the waste deriving from those resources, contributing to an increase in their environmental and health problems.

Environmental and health issues due to the exploitation of natural resources: case studies

It is possible to analyse the pollution which affects many areas of the world due to the exploitation of natural resources thanks to the scientific literature, which documents it with several case studies. In some of them, the health risk is estimated as well.

In China, the Hunan province is known as the "hometown of non-ferrous metals" because of the over 140 kinds of minerals in the country, 111 have been found there. Consequently, there are thousands of active and abandoned mines. A study conducted by Lei et al. (2015) assessed the pollution situation around seven mining districts (including mainly mines of lead, zinc, antimony, gold, and manganese). The concentrations of Pb, Cd, Cu, and Zn in paddy soils were measured. The ranges of values found were significantly higher than Hunan background values. Furthermore, human health risk related to these heavy metals was assessed to measure the toxicity for adults, considering the exposure to contaminated white rice by ingestion. Due to the effective concentration of heavy metals in white rice, the results showed the hazard index (HI), which is related to exposure to all the found metals, was always higher than one, which is the limit value, suggesting consumption of such contaminated white rice caused a health risk.

A further study from China (Xu et al., 2013) assessed the environmental pollution and the related human health risks in the Anhui province, an area in which there are many mines exploiting polymetallic iron–copper–gold-sulphide deposits. Soil samples were collected to evaluate the concentration of heavy metals such as Cd, Cu, Pb, and Zn. From the sites, 12 vegetable species, representing the most widespread in the study area, were collected. In this case, the concentration of heavy metals in the edible tissues of the plants was calculated through the transfer factor (TF), used for trace metal transfer from soil to vegetables. Also in this case, with a similar hazard parameter (i.e., the target hazard quotient; THQ), the hazard value exceeded one, meaning there were toxic risks for human health.

Cai et al. (2015) investigated the levels of Cd, Cu, Pb, and As in soil, crop, well, and fish samples collected close to the Tonglushan mine, in the northeast of Daye city (China), where there is one of the largest copper basins in China, with a raw copper yield beyond 10,000 t/year. The results indicated that soils near the mine were heavily contaminated by all the metals and metalloids analysed. Indeed, their concentrations in crop samples grown in mine-affected soils were significantly higher than those of the reference soils. In particular, Cd and Cu concentrations exceeded the soil quality standard values regulated by the State Environmental Protection Administration of China. Furthermore, in the edible plant portions of plants as well as in many species of fish grown in mine-affected areas, concentrations of Cd and Cu were significantly higher than in reference areas. In the case of water wells, all parameters except Pb were significantly greater in the mine-affected area than those in the reference areas.

Studies on the pollution related to mineral exploitation have also been conducted in Africa. For example, Bortey-Sam et al. (2015a) analysed borehole drinking water from 18 communities in Tarkwa, a town in the southwest of Ghana, which is a noted centre for gold and manganese mining. The geometric mean concentration of Co, Fe, and Mn resulted in higher than recommended limit values of WHO (2011) for drinking water in 5%, 11%, and 61% of the sampled communities, respectively. Not surprisingly, the average cancer risk values of As both for adults and children was higher than 10^{-5}, which represents a high level of risk. Considering the HI, the value was higher than one just in one community.

A further study, conducted in the same gold mining area (Bortey-Sam et al., 2015b), assessed the concentrations of Cr, Mn, Fe, Co, Ni, Cu, Zn, As, Cd, and Pb in foodstuffs from agricultural soils to estimate the human health risk associated with their consumption. The results showed that for both children and adults, the THQ of Pb in the cassava plant was almost always above one. Furthermore, 30% of cassava samples collected contained higher concentrations of Pb when compared to Codex Alimentarius Commission standard values of FAO/WHO (2001).

In Indonesia, a study by Castilhos et al. (2006) investigated the effects on fish from Hg released in two gold mining areas (Tatelu and Galangan) and assessed the human health hazards as well. The mean concentration of total Hg in the muscles of the freshwater fish from one of these areas (Tatelu) showed that more than 45% of them had Hg levels above the WHO guidelines for human consumption. Furthermore, the hazard quotient (HQ), a further hazard parameter, was applied to both areas to determine the threat of Methylmercury (MeHg) exposure for communities in both areas. Samples from Tatelu (excluding marine species) and Galangan had an HQ above one, which is the acceptable value of risk, with a value of 9.9 even in a sub-area of Galangan, suggesting potentially harmful fish consumption by the local population.

Shifting towards oil and gas exploitation, it is important to consider there are tens of thousands oil fields globally, and they have been estimated to potentially impact the health and environment of over 600 million people (O'Callaghan-Gordo et al., 2016). A recent review (Johnston et al., 2019) assessed the association between cancer and oil extraction, taking as reference several previous studies. An alarming situation was found in the United States, in which a Colorado registry-based case-control study showed that children and young people (ages 5–25) living near active oil and gas extraction wells diagnosed with acute lymphocytic leukaemia were 4.3 times more than the control population. A further study conducted in a community in New Mexico (United States) near an oil-drilling site and an oil waste pit identified elevated prevalence of neurological and respiratory symptoms and cardiovascular problems compared to a more distant community. Moving to Latin America, comparing cancer incidence in an Ecuadorian community in the Amazon basin impacted by oil extraction to a reference population, high incidences and mortality for all types of cancer were found. In the Northern Peruvian Amazon, oil has been extracted for over four decades, causing several issues. In 2013 and 2014, the Peruvian government declared an environmental and health state of emergency in some of these areas (Rosell-Melé et al., 2018). One of the causes of concern was the high level of Pb in the blood among the indigenous population of the Corrientes River. A subsequent meta-analysis of chemical data suggested a correlation between oil and gas extraction and increased concentration of Pb, as well as other heavy metals in the rivers of the region. Furthermore, there has been the occurrence of hydrocarbon pollution in soils and sediments from sites accessed regularly by the local indigenous populations for subsistence hunting or fishing in the vicinity of oil infrastructure, which poses a significant threat for the local population. A further study (Boothroyd et al., 2016) analysed the fugitive emissions of CH_4 from former oil and gas exploration and production wells onshore in the United Kingdom. The results showed that 30% of the 102 wells considered had soil gas CH_4 at the soil surface significantly greater than their respective ambient control values.

Studies on health effects due to coal mining pollution have been conducted as well. For example, to investigate the contamination level and health risks related to heavy metals in

Huainan, a coal-mining city in China, Tang et al. (2017) analysed the concentrations of eight heavy metals (As, Cd, Co, Cr, Cu, Hg, Pb, Sb) in street dust. The concentrations of Cd and Hg were associated with considerable health risks in more than half of the sites. At the same time, a source assessment indicated that metals in dust from Huainan were derived from many factors and not only coal mining, namely vehicular-related activities, coal combustion, industrial emissions, weathering of coal dust, and natural soils. From the other side of the world, a study (Hendryx and Ahern, 2008) merged data from a survey conducted in West Virginia with county-level coal production to investigate the relations between health indicators and residential proximity to coal mining. The authors noted that as coal production increased, rates of a series of diseases – namely cardiopulmonary, lung, cardiovascular, and kidney diseases and diabetes – increased as well. At the same time, the authors highlighted the need for further, more detailed studies.

Conclusions

In this chapter, the exploitation of natural resources has been discussed, along with its implications. The environmental, economic, and health aspects both in rich and poor countries worldwide have been compared. It has been highlighted how the existing need for natural resources is conditioned by several factors, including emerging technologies as well as the increase of the human population. Furthermore, unlike past and present periods characterised by conflicts often ascribable to the need for fossil fuels, future conflicts could be more related to competition for nonfuel minerals. In this scenario, different parallel solutions should be looked for, following a holistic approach. The waste hierarchy principle, based on prioritising reduction, recycling, and reuse of waste, could be very helpful, as well as an improvement in international relations, aiming to reduce also global inequalities. It has been shown that minerals are unevenly distributed. The main producers of the main resource reserves are not European countries, which are the countries at the top of the list in terms of GDP and environmental indices (i.e., EDI and EPI). This is an important aspect. In addition, the United States and China, which are the highest countries in terms of GDP, are not in a very favourable situation in terms of environmental aspects. Indeed, considering the EPI, they are only at the 27th (United States) and 120th (China) positions out of 180. Furthermore, the rapid technological evolution of the last decades has led to growing generation of WEEE, which has been exported from industrialised to developing countries. These exported wastes have caused significant environmental and health issues to local communities of developing areas of the world. The detailed list of case studies presented highlighted the pollution and health issues associated with exploitation of natural resources. It is evident that humankind requires the use of many of these resources, but further environmental efforts are needed to reduce the related pollution, in particular considering nonfuel minerals. Regarding fossil fuels, the situation requires further consideration. Indeed, also taking into account the global warming related to them and the already existing alternatives, it is clearly desirable that efforts of individuals should go towards a change of course, aimed at clean renewable energies.

References

The Access Initiative (TAI) World Resources Institute (WRI) (2015). *Environmental democracy index (EDI) – rank countries.* Available at: https://environmentaldemocracyindex.org/rank-countries#all (Accessed 16/6/2019).

Boothroyd IM, Almond S, Qassim SM, Worrall F, Davies RJ. (2016). Fugitive emissions of methane from abandoned, decommissioned oil and gas wells. *Science of the Total Environment.* 547: 461–469.

Bortey-Sam N, Nakayama SMM, Akoto O, Ikenaka Y, Fobil JN, Baidoo E, Mizukawa H, Ishizuka M. (2015b). Accumulation of heavy metals and metalloid in foodstuffs from agricultural soils around Tarkwa area in Ghana and associated human health risks. *International Journal of Environmental Research and Public Health*. 12: 8811–8827.

Bortey-Sam N, Nakayama SMM, Ikenaka Y, Akoto O, Baidoo E, Mizukawa H, Ishizuka M. (2015a). Health risk assessment of heavy metals and metalloid in drinking water from communities near gold mines in Tarkwa, Ghana. *Environmental Monitoring and Assessment*. 187: 1–12.

Breivik K, Armitage JM, Wania F, Jones KC. (2014). Tracking the global generation and exports of e-waste. Do existing estimates add up? *Environmental Science and Technology*. 48: 8735–8743.

British Petroleum (BP) (2018). *BP statistical review of world energy*. London: British Petroleum.

Cai LM, Xu ZC, Qi JY, Feng ZZ, Xiang TS. (2015). Assessment of exposure to heavy metals and health risks among residents near Tonglushan mine in Hubei, China. *Chemosphere*. 127: 127–135.

Castilhos ZC, Rodrigues-Filho S, Rodrigues AP, Villas-Bôas RC, Siegel S, Veiga MM, Beinhoff, C. (2006). Mercury contamination in fish from gold mining areas in Indonesia and human health risk assessment. *Science of the Total Environment*. 368: 320–325.

Cesaro A, Belgiorno V, Gorrasi G, Viscusi G, Vaccari M, Vinti G, Jandric A, Dias MI, Hursthouse A, Salhofer S. (2019). A relative risk assessment of the open burning of WEEE. *Environmental Science and Pollution Research*. 26: 11042–11052.

FAO/WHO (2001). *Joint FAO/WHO food standards programme – codex Alimentarius commission*. Report of the 33rd Session of the Codex Committee on Food Additives and Contaminants. The Hague, The Netherlands: FAO/WHO.

Gulley AL, Nassar NT, Xun S. (2018). China, the United States, and competition for resources that enable emerging technologies. *PNAS*. 115 (16): 4111–4115. https://doi.org/10.1073/pnas.1717152115.

Hendryx M, Ahern M. (2008). Relations between health indicators and residential proximity to coal mining in West Virginia. *American Journal of Public Health*. 98: 669–671.

International Monetary Fund (IMF) (2019). *World economic outlook database*. Washington, DC: International Monetary Fund, April 2019. Available at: https://www.imf.org/external/pubs/ft/weo/2019/01/weodata/index.aspx

Johnston JE, Lim E, Roh H. (2019). Impact of upstream oil extraction and environmental public health: a review of the evidence. *Science of the Total Environment*. 657: 187–199.

Lei M, Tie B, Song Z, Liao B, Lepo JE, Huang Y. (2015). Heavy metal pollution and potential health risk assessment of white rice around mine areas in Hunan province, China. *Food Security*. 7: 45–54.

O'Callaghan-Gordo C, Orta-Martínez M, Kogevinas M. (2016). Health effects of non-occupational exposure to oil extraction. *Environmental Health*. 15: 56.

Pires A, Martinho G. (2019). Waste hierarchy index for circular economy in waste management. *Waste Management*. 95: 298–305.

Popp D, Newell RG, Jaffe AB. (2010). Energy, the environment, and technological change. In: Hall B, Rosenberg N. (Eds.), *Handbook of the economics of innovation*. Oxford: Elsevier. Vol. 2. 873–937.

Reichl, C, Schatz, M. (2019). *World mining data 2019*. Vienna, Austria: Federal Ministry for Sustainability and Tourism.

Rosell-Melé A, Moraleda-Cibrián N, Cartró-Sabaté M, Colomer-Ventura F, Mayor P, Orta-Martínez M. (2018). Oil pollution in soils and sediments from the Northern Peruvian Amazon. *Science of the Total Environment*. 610: 1010–1019.

Rudra N, Jensen NM. (2011). Globalization and the politics of natural resources. *Comparative Political Studies*. 44 (6): 639–661.

Sciortino A. (2003). *Prima della globalizzazione. La genesi del mercato mondiale e le origini del sottosviluppo*. Rome, Italy: Edizioni Associate.

Sciortino A. (2008). *L'Africa in guerra. I conflitti africani e la globalizzazione*. Milan: Baldini Castoldi Dalai Editore.

Tang Z, Chai M, Cheng J, Jin J, Yang Y, Nie Z, Huang Q, Li Y. (2017). Contamination and health risks of heavy metals in street dust from a coal-mining city in Eastern China. *Ecotoxicology and Environmental Safety*. 138: 83–91.

United Nations (UN) (2010). *United Nations framework classification for fossil energy and mineral reserves and resources 2009*. New York, NY and Geneva: United Nations.

United Nations (UN) (2019). *World population prospects 2019: highlights (ST/ESA/SER.A/423)*. New York, NY: United Nations.

US Geological Survey (2019). *Mineral commodity summaries 2019: US geological survey*. Reston, VA.

Vaccari M, Vinti G, Cesaro A, Belgiorno V, Salhofer S, Dias MI, Jandric A. (2019). WEEE treatment in developing countries: environmental pollution and health consequences – an overview. *International Journal of Environmental Research and Public Health*. 16: 1595.

Wendling ZA et al. (2018). *Environmental Performance Index*. New Haven, CT: Yale Center for Environmental Law & Policy.

WHO (World Health Organization) (2011). *Guidelines for drinking-water quality*. Geneva: World Health Organization. 4th Ed.

Xu DC, Zhou P, Zhan J, Gao Y, Dou CM, Sun QY. (2013). Assessment of trace metal bioavailability in garden soils and health risks via consumption of vegetables in the vicinity of Tongling mining area, China. *Ecotoxicity and Environmental Safety*. 90: 103–111.

3

CONSUMPTION AND MATERIALISM

From acquisitive to responsible materialism

Isadora do Carmo Stangherlin and John Thøgersen

Introduction

Our current model of consumption and production is unsustainable (Steffen et al., 2015; Thøgersen, 2014b). Humans now dominate the earth's continental and oceanic ecosystems (Braje and Erlandson, 2013) to an extent that has led to the crossing of some planetary boundaries of safe operating space and the rapid approaching of others (Rockström et al., 2009). Changes in production and consumption in order to mitigate negative impacts on the planet have begun but are progressing much too slowly (Hoekstra and Wiedmann, 2014; IPCC, 2018). Suggestions for more sustainable forms of consumption often take their point of departure in a critique of current, "acquisitive" materialism, with material possessions being a central aspect of consumers' lives and a source of satisfaction or dissatisfaction (Belk, 1985). Suggested alternatives include access-based consumption (Bardhi and Eckhardt, 2012), collaborative consumption (Scaraboto, 2015), sharing (Belk, 2009), and sufficiency (Gorge et al., 2015). Companies are also under increasing pressure to make their production more sustainable and to offer more sustainable options to their customers (Kiron et al., 2012; Kotler, 2011).

An important part of the problem is the current linear model of consumption and production, which is built on a "take, make, and dispose" view (Ghisellini et al., 2016), resulting in waste and over-consumption of resources (Biggs et al., 2015; Prothero et al., 2011; Thøgersen, 2014a; White et al., 2019). Hence, a change to a circular economy is urgently needed (Geissdoerfer et al., 2017; MacArthur, 2013). A circular economy is an economic model based on cycles of closed production, increased efficiency in the use of resources, reduced use of virgin materials, and minimising waste generation (Ghisellini et al., 2016; Sauvé et al., 2016). The circular economy concept integrates the idea of cyclical closed-loop systems (Kirchherr et al., 2017; Murray et al., 2017) and the use of recovered materials and recycled products (Singh and Ordoñez, 2016). Within a closed-loop resource system, recycling is important to secure the circular flow of resources (Bocken et al., 2016). Stahel (1982) suggested four loops of circularity: reuse, repair, reconditioning, and recycling. All of these loops are integrated in a circular economic system. Company strategies to achieve a circular economy include remanufactured products (Abbey et al., 2015; Baxter et al., 2017; Camacho-Otero et al., 2018; Sun et al., 2018), access-based consumption (Bardhi and Eckhardt, 2012; Catulli et al., 2013), refurbished products (Holmström and Böhlin, 2017; Mugge et al., 2017; van Weelden et al., 2016), use of recycled components (Rozin et al., 2015), amongst others. A successful circular economy strategy requires the involvement of a range of stakeholders, including suppliers, producers, and consumers (Jabbour et al., 2019).

Radical changes in consumer behaviour are needed for the transition to a circular economy (Jurgilevich et al., 2016; Merli et al., 2018; Mont and Heiskanen, 2015; Singh and Ordoñez, 2016). By definition, a circular economy means reusing, repairing, reconditioning, and recycling products (Bocken et al., 2016; Stahel, 1982). Consumers make essential decisions in this connection, such as whether to purchase new, used, or remanufactured products; when, where, and how to dispose (rest) products after use; and whether and how they will adopt new ways of consuming (Hazen et al., 2017). Hence, for a circular economy to work, consumers are required to change their perceptions of possession, ownership, and product newness and to make substantial changes in everyday routines, engaging more in behaviours, such as repairing, returning, and recycling products (Camacho-Otero et al., 2018). Therefore, it is important to study the role and requirements of consumer behaviour in the circular economy transition (Farooque et al., 2019; Ghisellini et al., 2016; Kirchherr et al., 2017; Merli et al., 2018; Mugge et al., 2017).

Among other things, a circular economy requires a change from the "acquisitive" materialism we know, with its focus on the acquisition of material goods as a means to achieve distinction and status (Kasser et al., 2004), to a new materialism (Scott et al., 2014) that we could call "responsible materialism," built on the insights of circular economy thinking and focusing on the responsible use of limited material and other resources. Materialism as a consumer value (Richins and Dawson, 1992) is conceptually close to self-enhancement in Schwartz's value system (Schwartz, 2007) and negatively related to pro-environmental attitudes and behaviour (Hurst et al., 2013). In contrast, pro-environmental attitudes and behaviour, including valuing products with environmentally friendly attributes, are related to self-transcendence values in Schwartz's value system (de Groot and Steg, 2009; Thøgersen and Ölander, 2002). Hence, motivations to acquire, possess, and collect (Belk, 1982) may change by the inclusion of reused materials. The usual conceptions of possessions and ownership are also challenged by circular economy principles (Singh and Ordoñez, 2016). Hence, the transition to a circular economy requires a radical change from an "acquisitive" to a responsible materialism.

This chapter focuses on consumer reactions to a restorative system, where there is no "end-of-life" of material products, where waste is eliminated through superior design of materials, products, and systems (Bovea et al., 2018) and where production materials are reused (MacArthur, 2013). For the end consumer, the circular economy implies more recycling (Hamzaoui-Essoussi and Linton, 2014), more remanufactured products (Abbey et al., 2015; Hazen et al., 2017), extension of product life (Selvefors et al., 2019), and perhaps also new and extended ways of sharing (Belk, 2009). Selvefors et al. (2019) discuss four design strategies for a circular economy: (1) design for extended use (to increase products' utility), (2) design for pre- and post-use (to handling a product prior to and after use), (3) design for exchange (to exchange the products with others agents), and (4) design for multiple use-cycles (to put the product back to new product cycles).

In this chapter, we will especially focus on challenges related to multiple use-cycles, where waste products and rest materials are reused for the creation of different products. Hence, we discuss consumer responses to the use of recycled, recovered, or used materials, where products are remanufactured or made of recycled content from post-consumer products and post-consumer waste. The main goal is insight into what the use of reused materials in the product means to consumer motivation to buy it, identifying the most important barriers and opportunities regarding the acceptance of products with reused materials. We also discuss mechanisms that explain the ambiguity of positive and negative associations of products made with recycled and reused components. Finally, we discuss a few cases where recycled products have been successful.

The role of consumers in the circular economy

To be part of the circular economy, consumers can change behaviour in different ways, such as using products for longer time, repairing products that are broken, using products that have outlived their original purpose for other purposes, and recycling products. These behaviour changes challenge the "acquisitive" materialistic mindset (Richins and Dawson, 1992) and studying them can provide valuable insights for the design of interventions to promote and facilitate a circular economy (Singh and Ordoñez, 2016). Several consumer practices of relevance for the transition to a circular economy have been studied by past research, as illustrated in Table 3.1. In Table 3.1, the studied practices are organised in stages of the consumption cycle: acquisition, use, maintenance, and post-use.

As reflected in these practices, consumers have different roles in the circular economy: as maintainer, repairer, sharer or collaborator, purchaser, and engaging with waste sorting and reuse (Maitre-Ekern and Dalhammar, 2019); see Figure 3.1. Each of these roles implies different activities, which can be overlapping, for example, when a "repairer" buys spare parts or tools. Also, a "purchaser" may acquire access to products in different ways than buying, for example, renting, leasing, or sharing.

Table 3.1 Examples of researched consumer practices in the circular economy

Stage	*Practices*
Acquisition	Reducing overall consumption levels (Tunn et al., 2019) Participating in access-based consumption (Bardhi and Eckhardt, 2012; Catulli et al., 2013) and product service systems (Camacho-Otero et al., 2018) Participating in sharing economy and collaborative consumption solutions (Camacho-Otero et al., 2018) Buying circular products (Borrello et al., 2017), such as: upcycled food (Bhatt et al., 2018), refurbished smartphones (Mugge et al., 2017; van Weelden et al., 2016), remanufactured products (Camacho-Otero et al., 2018; Hazen et al., 2017; Singhal et al., 2019; Wang and Hazen, 2016; Wang et al., 2013), products made of recycled plastic (Magnier et al., 2019) Buying products with recycled packaging (Geueke et al., 2018)
Use	Reducing food waste (Borrello et al., 2017; Jurgilevich et al., 2016; Mylan et al., 2016) Dealing with unwanted food (Mylan et al., 2016) Using and reusing leftovers (Mylan et al., 2016) Reusing products and using for different purposes (Camacho-Otero et al., 2018) Using the product for a longer time (Bovea et al., 2018; Tunn et al., 2019) Reusing packages (Marsh and Bugusu, 2007)
Maintenance	Repairing products (Bovea et al., 2018; Camacho-Otero et al., 2018; Gwozdz et al., 2017; McCollough, 2009)
Post-use	Returning packages (Borrello et al., 2017; Marsh and Bugusu, 2007) Returning goods to recycle (Camacho-Otero et al., 2018) Source-separating household waste (Camacho-Otero et al., 2018; Maitre-Ekern and Dalhammar, 2019)

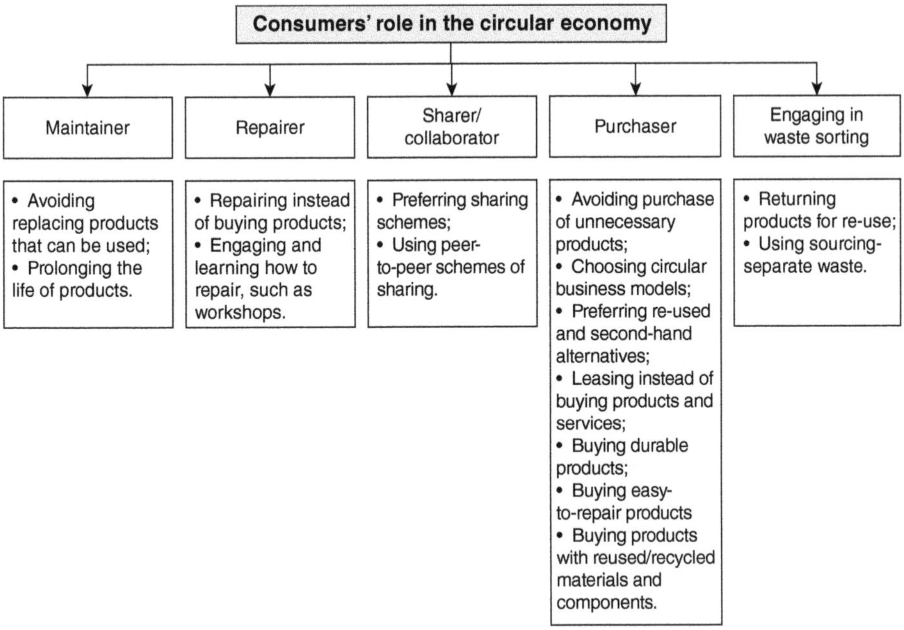

Figure 3.1 The role of consumers in the circular economy

For companies to be able to produce products made with recycled components, consumers need to return end-of-use products for reuse, refurbishing, or recycling (Ylä-Mella et al., 2015). Hence, companies, governments, and/or NGOs need to provide opportunities for consumers to return their used products. However, perhaps even more important for the evolution of a circular economy is consumer acceptance of products with reused materials. Therefore, we focus mostly on the consumer as purchaser in this chapter. Consumer acceptance requires that consumers perceive products made with recycled components favourably. Products with reused materials are not evaluated as new products (Mobley et al., 1995), and consumers often hold negative beliefs about the quality of used products (Matsumoto et al., 2018). However, not all individuals have equally negative or only negative beliefs about circular products.

Camacho-Otero et al. (2018) reviewed a large number of studies on factors influencing consumer acceptance of circular products. They found that acceptance is reduced by beliefs about lower quality and risk of contamination, among other things. Also, personal characteristics that are negatively related to acceptance include materialism, status concern, and lack of knowledge and understanding. The most important factors increasing acceptance are perceived environmental benefits and lower prices (Hazen et al., 2017).

Overall, research suggests an ambiguity in consumers acceptance of products with reused materials. Several studies report positive attitudes and intentions to buy these products (Holmström and Böhlin, 2017; van Weelden et al., 2016), which increase with environmental knowledge (Sun et al., 2018), pro-environmental attitudes (Harms and Linton, 2016), environmental consciousness (Gaur et al., 2015), and price consciousness (Matsumoto et al., 2018). However, other studies report negative consumer responses due to the belief that products made with recycled components are inferior. Research has identified concern about product performance (Abbey et al., 2015; Camacho-Otero et al., 2018; Holmström and Böhlin, 2017; Mugge et al.,

2017), contagion (Bardhi and Eckhardt, 2012; Baxter et al., 2017; Catulli et al., 2013; Rozin et al., 2015), and disgust regarding the acceptance of products made with reused materials (Abbey et al., 2015; Bardhi and Eckhardt, 2012; Camacho-Otero et al., 2018; Rozin et al., 2015).

In the next section, we discuss the most important positive and negative consumer associations with products made with reused materials identified by extant research and their implications for consumer motivation to buy these products.

Consumer acceptance of circular products: barriers and opportunities

Barriers

Various factors influence the acceptance of reused or remanufactured products and products made with recycled materials or components. Most studies find that consumers evaluate these products less favourably than a "new" version of the same product (Abbey et al., 2015). Moreover, when comparing remanufactured products with products that have different environmental advantages, such as an energy-efficient product, consumers generally prefer the latter (Khor and Hazen, 2017). Consumers generally perceive higher risks and lower benefit/value of remanufactured products (Matsumoto et al., 2018; Mugge et al., 2017; van Weelden et al., 2016; Wang and Hazen, 2016) and products with recycled components (Sun et al., 2018). These negative inferences lead to lower willingness to pay for products that are made with used materials (Guide and Li, 2010; Harms and Linton, 2016). Hence, consumers often perceive a trade-off between the possible environmental benefits of products with reused components and other quality attributes.

The most important reservation consumers express regarding remanufactured products is lower quality (Abbey et al., 2015; Achabou and Dekhili, 2013; Gan and Chen, 2019; Kuah and Wang, 2019; Vafadarnikjoo et al., 2018; Wang and Hazen, 2016). Abbey et al. (2017) found that the low perceived quality is due to a high perceived probability of functionality and cosmetic defects, which translates into a higher perceived risk (Magnier et al., 2019; van Weelden et al., 2016). Perceived risks include performance, financial, time, obsolescence, physical risks, and social and resource risks, with performance risks (Holmström and Böhlin, 2017; Matsumoto et al., 2018; Mugge et al., 2017) and financial risks (Holmström and Böhlin, 2017; Wang and Hazen, 2016) having the biggest negative effects on consumers motivations to buy products with recycled components.

Performance risks are associated with functionality expectations (Magnier et al., 2019). They represent consumers' doubts about the quality performance of remanufactured products (Wang and Hazen, 2016) and other products containing recycled/reused materials (Sun et al., 2018). Consumers may fear that these products will break down more quickly due to the fact that they or materials in them already had a previous use (van Weelden et al., 2016). Therefore, perceived performance risks are linked to lower perceived product quality. However, lower performance risk is generally perceived for refurbished products than for secondhand products (van Weelden et al., 2016).

Product quality and functionality are often difficult to assess before purchasing and using the product (van Weelden et al., 2016). It appears that uncertainty about the use history of a reused or refurbished product can strengthen the fear of functional failure (Ovchinnikov, 2011; van Weelden et al., 2016). The increased uncertainty has a negative impact on consumers' quality perceptions and willingness to pay for products with reused components (Hazen et al., 2012). In addition, consumers sometimes fear that remanufactured products and products containing reused components can quickly become technically obsolete (van Weelden et al., 2016).

Research suggests that consumer motivation to buy a remanufactured product is higher if the product has a superior brand image (Achabou and Dekhili, 2013; Gaur et al., 2015; Guide and Li, 2010; Vafadarnikjoo et al., 2018). In general, consumers express more confidence in refurbished products offered by a well-known company (van Weelden et al., 2016). However, some studies found no relationship between brand image and consumers perceptions of these products (Abbey et al., 2015).

A reduced price is a common strategy to compensate for consumers' negative beliefs regarding remanufactured products. However, the results of discounts for remanufactured products are equivocal. Some scholars found that a lower price is an important reason to buy these products (Guide and Li, 2010; van Weelden et al., 2016; Wang and Hazen, 2016). However, a discount can also amplify negative associations with the product (Abbey et al., 2015), leading to lower quality perceptions and increased perceived risks (Ovchinnikov, 2011; van Weelden et al., 2016). Therefore, it is important to carefully consider the right price strategy for these products.

Widely different willingness to pay for circular products has been found in different product categories. For example, Hamzaoui-Essoussi and Linton (2014) found a higher willingness to pay for recycled paper than for branded new paper. However, a lower willingness to pay was found for other products, including food products produced using recycled irrigation water (Savchenko et al., 2018) and remanufactured cameras (Michaud and Llerena, 2011). Hence, it appears that reused or recycled content is more acceptable in some product categories than in others (Hamzaoui-Essoussi and Linton, 2014; Magnier et al., 2019). For example, consumers seem to have lower quality perceptions and purchase intentions for textile products with reused or recycled content than for durables and fast-moving consumer goods (Magnier et al., 2019). Magnier et al. (2019) found differences between running shoes, sweaters, dishwashing soap, and hand cream. They also found that consumers respond less favourably to packaging made of recycled ocean plastic when the product it is used for is absorbed through the skin, such as creams. Hence, it appears that consumers are less willing to accept such a product when they perceive it as unsafe.

A key consumer reservation with products made of recycled ocean plastic is risk of contamination (Magnier et al., 2019). Similarly, Abbey et al. (2015) found that consumers describe remanufactured products with words such as used, old, broken, dirty. Hence, it seems that remanufactured products and products made with recycled materials are sometimes perceived as dirty and disgusting, mainly because they were previously used by a different person (Abbey et al., 2015). This is at least partly due to consumer uncertainty and confusion (Abbey et al., 2015; Holmström and Böhlin, 2017; Ovchinnikov, 2011). Hence, consumers' lack of knowledge and misunderstanding of what refurbishment and recycling entail are important barriers to the acceptance of these products (van Weelden et al., 2016).

When the product or parts of it were used before, it may lack the thrill of newness and also desirable innovative features (van Weelden et al., 2016), which are additional barriers to the adoption of these products (Mugge et al., 2017). Singh and Ordoñez (2016) explain consumers' rejection of circular products with reference to psychological essentialism. Recycled items lack authenticity (Singh and Giacosa, 2019), which reduces their uniqueness and the perceived value of the product.

Another barrier is lack of availability. Because consumers are not willing to make an extra effort to find a store with products made with recycled materials, they are less likely to buy these products if they don't find them in the purchasing environment (Tunn et al., 2019; van Weelden et al., 2016).

Opportunities

Some research found positive consumer responses to products with reused materials (Holmström and Böhlin, 2017; Magnier et al., 2019). For example, Mobley et al. (1995) found that consumers rated recycled paper products more favourably than products made of virgin fibres. Also, Sun et al. (2018) found a positive attitude, intention to purchase, and self-reported purchase of products with recycled content. Some consumer segments are more willing than others to accept products with reused materials (Mugge et al., 2017). They perceive lower risk and trust these products' quality and performance more (van Weelden et al., 2016). They are also characterised by higher environmental consciousness (Gaur et al., 2015) and expertise and demographically by higher age, education, and income (Magnier et al., 2019). Hence, it seems that products with reused materials represent a market opportunity for companies targeting consumers with these characteristics (Mobley et al., 1995).

Knowledge and information are essential to consumer acceptance of circular products (Mugge et al., 2017; Vafadarnikjoo et al., 2018), and consumer quality perceptions can be improved by educational means (Abbey et al., 2015). The better consumers understand these products, the lower risk they perceive (Wang et al., 2013). An important impediment is that consumers often cannot assess product quality before using it (Michaud and Llerena, 2011; Wang and Hazen, 2016). However, when they can experience the product before buying it, even in a retail store, this has a positive impact on quality perceptions (van Weelden et al., 2016). In addition, social norms, including how the country values recycled and remanufactured products, affect consumer acceptance (Gaur et al., 2015). Consumers evaluation of these products is also influenced by their beliefs about important others' perceptions (van Weelden et al., 2016). In this way, well-crafted communication is needed to increase attention to and positive perceptions of products with reused materials as well as their perceived social acceptance.

Research suggests that both generic and product-related information is needed (van Weelden et al., 2016). Generic information concerns the process of production and the benefits associated with circular production. Product-related information, on the other hand, should aim to reduce perceived risks and promote the product. For example, for a refurbished smartphone, this information might include details on the battery life, guaranteed software updates, and upgraded performance scores (Mugge et al., 2017). Moreover, it is important to include clear and precise information about the state of the product (such as age, possible damage, battery condition, and results of performance tests) and often also its history (such as how the product was used, the reasons for discarding it, and use of protective tools) (van Weelden et al., 2016). Sometimes it is possible to offer features that are better than in a new product, such as better batteries, which can increase the attractiveness of a refurbished product (van Weelden et al., 2016). Obviously, such benefits should always be communicated at the point of purchase.

Consumers are more willing to accept circular products when doubts and knowledge deficits are reduced (van Weelden et al., 2016). Research find that consumer acceptance of circular products increases with their knowledge (Matsumoto et al., 2018) and experience with the products (Guide and Li, 2010; Hamzaoui-Essoussi and Linton, 2014). Communication should be designed to inform consumers in a balanced way about possible risks and imperfections (van Weelden et al., 2016). Balanced information has been found to increase consumers' willingness to pay, for example, for food products produced with recycled water (Savchenko et al., 2018). Information about the nature of the production process is important to avoid or reduce exaggerated risk perceptions (Abbey et al., 2017). Also, making the past history of the product transparent appears to increase the acceptance of products with reused components (Kamleitner et al., 2019). Winterich et al. (2019) found a positive side effect of making consumers think about the transformation of recyclables into new products is that it also makes them recycle more.

Together with interventions to increase knowledge about a circular product (cost, quality, and greenness), it is useful to implement risk-mitigating strategies (Wang and Hazen, 2016), such as quality assurance and warranties (Magnier et al., 2019; Michaud and Llerena, 2011; Ovchinnikov, 2011; Vafadarnikjoo et al., 2018), quality labels (van Weelden et al., 2016), and performance classification systems (Mugge et al., 2017). Consumer acceptance of circular products can be improved by means of quality certification (Abbey et al., 2017) and a warranty (van Weelden et al., 2016). Consumers in general evaluate remanufactured products more favourably when they are remanufactured by the original manufacturer or its authorised factories than by a third party (Subramanian and Subramanyam, 2012). In contrast, information about the history of the product is more effective at reducing uncertainty about products when it comes from independent third parties (Ovchinnikov, 2011), including third-party eco-labels (Harms and Linton, 2016). The same is true for a system providing verification of the product quality (Ylä-Mella et al., 2015).

Consumers' intention to buy a refurbished product is positively related to perceived environmental benefits (Kuah and Wang, 2019; Mugge et al., 2017). However, environmental benefits are usually not the most important determinant of consumer choices (Gan and Chen, 2019; Holmström and Böhlin, 2017; van Weelden et al., 2016), among other things because consumers often lack necessary knowledge to judge which product is more environmentally friendly (Khor and Hazen, 2017). For this and other reasons, consumers do not necessarily view remanufactured products as more environmentally friendly (Abbey et al., 2015). Still, new buyers express that the environmental benefits are an important motivation to choose these products (van Weelden et al., 2016). Hence, it is important to clearly communicate the environmental benefits associated with these products.

Magnier et al. (2019) found that recognisability increases the willingness to pay for products made from recycled ocean plastic. This suggests that consumer acceptance of circular products can be increased by making recycled components clearly visible (cf. Griskevicius et al., 2010). As a caveat, Achabou and Dekhili (2013) found that consumers responded positively to recycled packages of luxury products but not to recycled materials in the product itself, such as luxury shirts. Recycled packaging generally benefits from low perceived risks in terms of quality, functionality, and contamination, which increases consumer acceptance (Magnier et al., 2019).

Some practical cases

Companies increasingly reuse materials for their products. For example, Nike collects waste products to use in new products (https://purpose.nike.com/reuse-a-shoe), such as recycled PET for lace tips and recycled factory waste rubber outsole (Mestre and Cooper, 2017), something which they highlight in market communication (https://purpose.nike.com/flyleather-earth-day-pack). In a similar way, Patagonia and the North Face sell outdoor products that are "made from other clothes" (https://wornwear.patagonia.com and www.thenorthfacerenewed.com), and Elvis and Kresse sells luxury bags with reclaimed materials (www.elvisandkresse.com).

One of Fairphone's key selling propositions is that its mobile phones are produced with recycled components (https://shop.fairphone.com), and the fabric in the new edition of Renault's popular electric car, ZOE, introduced in 2019, is made from 100% recycled materials (https://group.renault.com/en/news-on-air/news/new-zoe-recycled-fabrics-nothing-is-lost/). These are just some of the many examples of companies that reuse materials in their products and use this fact actively in their marketing while trying to overcome consumers quality concerns and perceived risks. All of these companies explicitly communicate that their products contain reused or recycled materials, leading to reduce waste and other environmental benefits. Hence, they all seem to agree that a well-crafted communication strategy is a key to consumers accepting, even valuing, products with reused materials. Supporting this, Kamleitner et al. (2019) found that a communication strategy making the past "identity" of an upcycled product salient (e.g., a backpack made from an airbag) can spur the demand for an upcycled product by allowing the buyer to feel special.

It is obviously also important to engage consumers in returning the end-of-use products for reuse, reprocessing, or recycling (Kamleitner et al., 2019; Ylä-Mella et al., 2015). The companies mentioned previously and many others have programs to collect consumers' post-use products as part of an integrated strategy to increase production with reused materials. Winterich et al. (2019) showed that information about the transformation of materials into final products can motivate consumers to participate in recycling programs. In sum, well-crafted communication is needed to both increase the acceptance of products with reused materials and motivate consumers to participate in recycling programs.

Conclusions

Driven by mounting waste problems and even more by dwindling virgin resources, a transition from a linear to a circular economy is emerging, including the increasing use of reused or recycled materials in the production of new products. This chapter has reviewed research on factors that influence consumer acceptance of circular products. It appears that products with reused materials are sometimes viewed favourably by consumers, but they are more often perceived negatively. Hence, it is important to develop effective strategies to promote and facilitate the broad acceptance of circular products. The research reviewed in this chapter suggests that a combination of means is needed for this purpose.

A transition to a circular economy requires several changes by consumers, including engaging in different models of ownership, changing their perception of product newness, and reducing their obsession with possessions and ownership (Scott et al., 2014). Current "acquisitive" materialism should be replaced by a new responsible materialism, built on an understanding of circular economy principles, respect for material resources, and environmental values. There is a movement in the direction of a higher acceptance of products with reused parts (Mugge et al., 2017), especially among people with higher environmental consciousness (Gaur et al., 2015) and expertise (Magnier et al., 2019). Social norms aligned with the importance of valuing reused materials also seem to be emerging (Gaur et al., 2015; Griskevicius et al., 2010; van Weelden et al., 2016). However, it is difficult to change materialistic consumers' view of ownership and possessions (Scott et al., 2014). Hence, it is also important that the perceived quality of the products improve and that the perceived risks be reduced through improved production processes and, when needed, assurances and warranties. In addition, well-crafted communication is needed to inform and educate consumers about the products, the production processes, and sometimes the past history of reused materials, to reduce consumer uncertainty and misperceptions.

Governments play an important role for increasing consumer acceptance of these products as well, for example, by means of ecolabels certifying and communicating products' environmental impact from a life-cycle perspective (Darnall Ji and Vázquez-Brust, 2018). Third-party ecolabels are effective means to inform consumers about products' environmental characteristics and help them make better choices (Thøgersen, 2002). They can also be instrumental in educating consumers about their choice options and providing assurance of product quality.

Several factors need to be considered to craft an effective communication strategy for products with reused materials (Kamleitner et al., 2019). The environmental benefits are important for consumers' choice of products with reused materials. However, only stressing the environmental benefits may backfire, since they are rarely the main reason consumers buy a product (Abbey et al., 2015; van Weelden et al., 2016). Especially, it is essential to deal effectively with consumer concerns about performance and financial risks. The reuse of components and materials can become a competitive advantage for companies who know how to position their products and have a good understanding of their target customers.

At a more basic level, the transition to a circular economy with effective reuse of materials for new products in a never-ending cycle requires that we challenge the current "acquisitive" materialism with its obsession with possessions and ownership. Instead, we need a new, responsible materialism, built on an understanding and appreciation of circular economy principles and intelligent and respectful use of limited resources. Responsible companies and public policy need to work together to accomplish this value change. Alone, consumers cannot build a sustainable future, but this goal can also not be achieved without their contribution in many different roles. Equally indispensable are companies and private and public institutions organizing and regulating circular economic systems and motivating and facilitating the participation of all stakeholders in the transition to a circular economy.

References

Abbey JD, Kleber R, Souza GC, Voigt G. (2017). The role of perceived quality risk in pricing remanufactured products. *Production and Operations Management*. 26 (1): 100–115.

Abbey JD, Meloy MG, Guide Jr, VDR, Atalay S. (2015). Remanufactured products in closed-loop supply chains for consumer goods. *Production and Operations Management*. 24 (3): 488–503.

Achabou, MA, Dekhili, S. (2013). Luxury and sustainable development: is there a match? *Journal of Business Research*. 66 (10): 1896–1903.

Bardhi F, Eckhardt GM. (2012). Access-based consumption: the case of car sharing. *Journal of Consumer Research*. 39 (4): 881–898.

Baxter W, Aurisicchio M, Childs P. (2017). Contaminated interaction: another barrier to circular material flows. *Journal of Industrial Ecology*. 21 (3): 507–516.

Belk RW. (1982). Acquiring, possessing, and collecting: fundamental processes in consumer behavior. *Marketing theory: Philosophy of Science Perspectives*. 185–190.

Belk RW. (1985). Materialism: trait aspects of living in the material world. *Journal of Consumer Research*. 12 (3): 265–280.

Belk RW. (2009). Sharing. *Journal of Consumer Research*. 36 (5): 715–734.

Bhatt S, Lee J, Deutsch J, Ayaz H, Fulton B, Suri R. (2018). From food waste to value-added surplus products (VASP): consumer acceptance of a novel food product category. *Journal of Consumer Behaviour*. 17 (1): 57–63.

Biggs EM, Bruce E, Boruff B, Duncan JMA, Horsley J, Pauli N, . . . Imanari Y. (2015). Sustainable development and the water–energy–food nexus: a perspective on livelihoods. *Environmental Science & Policy*. 54: 389–397.

Bocken NMP, de Pauw I, Bakker C, van der Grinten B. (2016). Product design and business model strategies for a circular economy. *Journal of Industrial and Production Engineering*. 33 (5): 308–320.

Borrello M, Caracciolo F, Lombardi A, Pascucci S, Cembalo L. (2017). Consumers' perspective on circular economy strategy for reducing food waste. *Sustainability*. 9 (1): 141.
Bovea MD, Ibáñez-Forés V, Pérez-Belis V, Juan P, Braulio-Gonzalo M, Díaz-Ávalos C. (2018). Incorporation of circular aspects into product design and labelling: consumer preferences. *Sustainability*. 10 (7): 2311.
Braje TJ, Erlandson JM. (2013). Human acceleration of animal and plant extinctions: a late pleistocene, holocene, and anthropocene continuum. *Anthropocene*. 4: 14–23.
Camacho-Otero J, Boks, C, Pettersen I. (2018). Consumption in the circular economy: a literature review. *Sustainability*. 10 (8): 2758.
Catulli M, Lindley JK, Reed NB, Green A, Hyseni H, Kiri S. (2013). What is mine is not yours: further insight on what access-based consumption says about consumers. *Consumer Culture Theory*. 185–208.
Darnall N, Ji H, Vázquez-Brust DA. (2018). Third-party certification, sponsorship, and consumers' ecolabel use. *Journal of Business Ethics*. 150: 953–969.
de Groot JIM, Steg L. (2009). Mean or green? Which values can promote stable pro-environmental behavior? *Conservation Letters*. 2 (2): 61–66.
Farooque M, Zhang A, Thürer M, Qu T, Huisingh D. (2019). Circular supply chain management: a definition and structured literature review. *Journal of Cleaner Production*. 228: 882–900.
Gan Q, Chen S. (2019). *Assessing consumers' motivations for purchasing remanufactured products*. Kybernetes. Ahead-of-print. doi: https://doi.org/10.1108/K-03-2019-0206.
Gaur J, Amini M, Banerjee P, Gupta R. (2015). Drivers of consumer purchase intentions for remanufactured products: a study of Indian consumers relocated to the USA. *Qualitative Market Research: An International Journal*. 18 (1): 30–47.
Geissdoerfer M, Savaget P, Bocken NMP, Hultink EJ. (2017). The circular economy – a new sustainability paradigm? *Journal of Cleaner Production*. 143: 757–768.
Geueke B, Groh K, Muncke J. (2018). Food packaging in the circular economy: overview of chemical safety aspects for commonly used materials. *Journal of Cleaner Production*. 193: 491–505.
Ghisellini P, Cialani C, Ulgiati S. (2016). A review on circular economy: the expected transition to a balanced interplay of environmental and economic systems. *Journal of Cleaner Production*. 114: 11–32.
Gorge H, Herbert M, Özçağlar-Toulouse N, Robert I. (2015). What do we really need? Questioning consumption through sufficiency. *Journal of Macromarketing*. 35 (1): 11–22.
Griskevicius V, Tybur JM, Van den Bergh B. (2010). Going green to be seen: status, reputation, and conspicuous conservation. *Journal of Personality and Social Psychology*. 98: 392–392.
Guide JV. Daniel R, Li J. (2010). The potential for cannibalization of new products sales by remanufactured roducts. *Decision Sciences*. 41 (3): 547–572.
Gwozdz W, Steensen Nielsen K, Müller T. (2017). An environmental perspective on clothing consumption: consumer segments and their behavioral patterns. *Sustainability*. 9: 762.
Hamzaoui-Essoussi L, Linton JD. (2014). Offering branded remanufactured/recycled products: at what price? *Journal of Remanufacturing*. 4 (1): 9.
Harms R, Linton, JD. (2016). Willingness to pay for eco-certified refurbished products: the effects of environmental attitudes and knowledge. *Journal of Industrial Ecology*. 20 (4): 893–904.
Hazen BT, Mollenkopf DA, Wang Y. (2017). Remanufacturing for the circular economy: an examination of consumer switching behavior. *Business Strategy and the Environment*. 26 (4): 451–464.
Hazen BT, Overstreet RE, Jones-Farmer LA, Field HS. (2012). The role of ambiguity tolerance in consumer perception of remanufactured products. *International Journal of Production Economics*. 135 (2): 781–790.
Hoekstra AY, Wiedmann TO. (2014). Humanity's unsustainable environmental footprint. *Science*. 344 (6188): 1114–1117.
Holmström S, Böhlin H. (2017). *Towards a circular economy: a qualitative study on how to communicate refurbished smartphones in the Swedish market*. Available at: https://pdfs.semanticscholar.org/b7ec/66821293095593163466e4f490978081a437.pdf?_ga=2.98008024.637448970.1584138589-1061879393.1584138589 (Accessed 13/12/2019).
Hurst M, Dittmar H, Bond R, Kasser T. (2013). The relationship between materialistic values and environmental attitudes and behaviors: a meta-analysis. *Journal of Environmental Psychology*. 36: 257–269.
IPCC. (2018). *Global warming of 1.5 C*. Cambridge and New York: IPCC & Cambridge University Press.
Jabbour CJC, de Sousa Jabbour ABL, Sarkis J, Filho MG. (2019). Unlocking the circular economy through new business models based on large-scale data: an integrative framework and research agenda. *Technological Forecasting and Social Change*. 144: 546–552.
Jurgilevich A, Birge T, Kentala-Lehtonen J, Korhonen-Kurki K, Pietikäinen J, Saikku, L, Schösler H. (2016). Transition towards circular economy in the food system. *Sustainability*. 8 (1): 69.

Kamleitner B, Thürridl C, Martin BAS. (2019). A Cinderella story: how past identity salience boosts demand for repurposed products. *Journal of Marketing*. 83 (6): 76–92.

Kasser T, Ryan RM, Couchman CE, Sheldon KM. (2004). Materialistic values: their causes and consequences. *Psychology and Consumer Culture: The Struggle for a Good Life in a Materialistic World*. 1 (2): 11–28.

Khor KS, Hazen BT. (2017). Remanufactured products purchase intentions and behaviour: evidence from Malaysia. *International Journal of Production Research*. 55 (8): 2149–2162.

Kirchherr J, Reike D, Hekkert M. (2017). Conceptualizing the circular economy: an analysis of 114 definitions. *Resources, Conservation and Recycling*. 127: 221–232.

Kiron D, Kruschwitz N, Haanaes K, von Streng Velken I. (2012). Sustainability nears a tipping point. *MIT Sloan Management Review*. 53 (2): 69–74.

Kotler P. (2011). Reinventing marketing to manage the environmental imperative. *Journal of Marketing*. 75 (4): 132–135.

Kuah ATH, Wang P. (2019). Circular economy and consumer acceptance: an exploratory study in East and Southeast Asia. *Journal of Cleaner Production*. 119097.

MacArthur E. (2013). *Towards the circular economy, economic and business rationale for an accelerated transition*. Cowes: Ellen MacArthur Foundation.

Magnier L, Mugge R, Schoormans J. (2019). Turning ocean garbage into products – consumers' evaluations of products made of recycled ocean plastic. *Journal of Cleaner Production*. 215: 84–98.

Maitre-Ekern E, Dalhammar C. (2019). Towards a hierarchy of consumption behaviour in the circular economy. *Maastricht Journal of European and Comparative Law*. 26 (3): 394–420.

Marsh K, Bugusu, B. (2007). Food packaging – roles, materials, and environmental issues. *Journal of Food Science*. 72 (3): R39–R55.

Matsumoto M, Chinen K, Endo H. (2018). Remanufactured auto parts market in Japan: historical review and factors affecting green purchasing behavior. *Journal of Cleaner Production*. 172: 4494–4505.

McCollough J. (2009). Factors impacting the demand for repair services of household products: the disappearing repair trades and the throwaway society. *International Journal of Consumer Studies*. 33 (6): 619–626.

Merli R, Preziosi M, Acampora A. (2018). How do scholars approach the circular economy? A systematic literature review. *Journal of Cleaner Production*. 178: 703–722.

Mestre A, Cooper T. (2017). Circular product design: a multiple loops life cycle design approach for the circular economy. *The Design Journal*. 20(sup1): S1620–S1635.

Michaud C, Llerena, D. (2011). Green consumer behaviour: an experimental analysis of willingness to pay for remanufactured products. *Business Strategy and the Environment*. 20 (6): 408–420.

Mobley AS, Painter TS, Untch EM, Rao Unnava H. (1995). Consumer evaluation of recycled products. *Psychology & Marketing*. 12 (3): 165–176.

Mont O, Heiskanen E. (2015). Breaking the stalemate of sustainable consumption with industrial ecology and a circular economy. In: Reisch LA, Thøgersen J. (Eds.), *Handbook of research on sustainable consumption*. Cheltenham: Edward Elgar. 33–47.

Mugge R, Jockin B, Bocken N. (2017). How to sell refurbished smartphones? An investigation of different customer groups and appropriate incentives. *Journal of Cleaner Production*. 147: 284–296.

Murray A, Skene K, Haynes K. (2017). The circular economy: an interdisciplinary exploration of the concept and application in a global context. *Journal of Business Ethics*. 140 (3): 369–380.

Mylan J, Holmes H, Paddock J. (2016). Re-introducing consumption to the 'circular economy': a sociotechnical analysis of domestic food provisioning. *Sustainability*. 8 (8): 794.

Ovchinnikov A. (2011). Revenue and cost management for remanufactured products. *Production and Operations Management*. 20 (6): 824–840.

Prothero A, Dobscha S, Freund J, Kilbourne WE, Luchs MG, Ozanne LK, Thøgersen J. (2011). Sustainable consumption: opportunities for consumer research and public policy. *Journal of Public Policy & Marketing*. 30 (1): 31–38.

Richins ML, Dawson S. (1992). A consumer values orientation for materialism and its measurement: scale development and validation. *Journal of Consumer Research*. 19 (3): 303–316.

Rockström J, Steffen W, Noone K, Persson Å, Chapin FS, Lambin EF, . . . Foley JA. (2009). A safe operating space for humanity. *Nature*. 461 (7263): 472–475.

Rozin P, Haddad B, Nemeroff C, Slovic P. (2015). Psychological aspects of the rejection of recycled water: contamination, purification and disgust. *Judgment and Decision Making*. 10 (1): 50–63.

Sauvé S, Bernard S, Sloan P. (2016). Environmental sciences, sustainable development and circular economy: alternative concepts for trans-disciplinary research. *Environmental Development*. 17: 48–56.

Savchenko OM, Kecinski M, Li T, Messer KD, Xu H. (2018). Fresh foods irrigated with recycled water: a framed field experiment on consumer responses. *Food Policy*. 80: 103–112.

Scaraboto D. (2015). Selling, sharing, and everything in between: the hybrid economies of collaborative networks. *Journal of Consumer Research*. 42 (1): 152–176.

Schwartz SH. (2007). Cultural and individual value correlates of capitalism: a comparative analysis. *Psychological Inquiry*. 18 (1): 52–57.

Scott K, Martin DM, Schouten JW. (2014). Marketing and the new materialism. *Journal of Macromarketing*. 34 (3): 282–290.

Selvefors A, Rexfelt O, Renström S, Strömberg H. (2019). Use to use – a user perspective on product circularity. *Journal of Cleaner Production*. 223: 1014–1028.

Singh P, Giacosa E. (2019). Cognitive biases of consumers as barriers in transition towards circular economy. *Management Decision*. 57 (4): 921–936.

Singh J, Ordoñez I. (2016). Resource recovery from post-consumer waste: important lessons for the upcoming circular economy. *Journal of Cleaner Production*. 134: 342–353.

Singhal D, Tripathy S, Jena SK. (2019). Acceptance of remanufactured products in the circular economy: an empirical study in India. *Management Decision*. 57 (4): 953–970.

Stahel WR. (1982). The product life factor. In Orr SG. (Ed.), *An inquiry into the nature of sustainable societies: the role of the private sector*. Houston: Houston Area Research Center. 72–105.

Steffen W, Richardson K, Rockström J, Cornell SE, Fetzer I, Bennett EM, . . . de Wit CA. (2015). Planetary boundaries: guiding human development on a changing planet. *Science*. 347 (6223): 1259855.

Subramanian R, Subramanyam R. (2012). Key factors in the market for remanufactured products. *Manufacturing & Service Operations Management*. 14 (2): 315–326.

Sun H, Teh, PL, Linton JD. (2018). Impact of environmental knowledge and product quality on student attitude toward products with recycled/remanufactured content: implications for environmental education and green manufacturing. *Business Strategy and the Environment*. 27 (7): 935–945.

Thøgersen J. (2002). Promoting green consumer behavior with eco-labels. In: Dietz T, Stern P. (Eds.), *New tools for environmental protection: education, information, and voluntary measures*. Washington, DC: National Academy Press. 83–104.

Thøgersen J. (2014a). Unsustainable consumption. *European Psychologist*. 19 (2): 84–95.

Thøgersen J. (2014b). Unsustainable consumption: basic causes and implications for policy. *European Psychologist*. 19: 84–95.

Thøgersen J, Ölander F. (2002). Human values and the emergence of a sustainable consumption pattern: a panel study. *Journal of Economic Psychology*. 23: 605–630.

Tunn VSC, Bocken NMP, van den Hende EA, Schoormans JPL. (2019). Business models for sustainable consumption in the circular economy: an expert study. *Journal of Cleaner Production*. 212: 324–333.

Vafadarnikjoo A, Mishra N, Govindan K, Chalvatzis K. (2018). Assessment of consumers' motivations to purchase a remanufactured product by applying fuzzy Delphi method and single valued neutrosophic sets. *Journal of Cleaner Production*. 196: 230–244.

van Weelden E, Mugge R, Bakker C. (2016). Paving the way towards circular consumption: exploring consumer acceptance of refurbished mobile phones in the Dutch market. *Journal of Cleaner Production*. 113: 743–754.

Wang Y, Hazen BT. (2016). Consumer product knowledge and intention to purchase remanufactured products. *International Journal of Production Economics*. 181: 460–469.

Wang Y, Wiegerinck V, Krikke H, Zhang H. (2013). Understanding the purchase intention towards remanufactured product in closed-loop supply chains: an empirical study in China. *International Journal of Physical Distribution & Logistics Management*. 43 (10): 866–888.

White K, Habib R, Hardisty DJ. (2019). How to SHIFT consumer behaviors to be more sustainable: a literature review and guiding framework. *Journal of Marketing*. 83 (3): 22–49.

Winterich KP, Nenkov GY, Gonzales GE. (2019). Knowing what it makes: how product transformation salience increases recycling. *Journal of Marketing*. 83 (4): 21–37.

Ylä-Mella J, Keiski RL, Pongrácz E. (2015). Electronic waste recovery in finland: consumers' perceptions towards recycling and re-use of mobile phones. *Waste Management*. 45: 374–384.

Further reading

Abbey JD, Meloy MG, Blackburn J, Guide Jr, VDR. (2015). Consumer markets for remanufactured and refurbished products. *California Management Review*. 57 (4): 26–42. (Discussion on issues about remanufactured consumer products.)

Belk RW. (1985). Materialism: trait aspects of living in the material world. *Journal of Consumer Research*. 12 (December): 265–280. (Definition of consumer materialism and how to measure this trait.)

Camacho-Otero J, Boks C, Pettersen I. (2018). Consumption in the circular economy: a literature review. *Sustainability*. 10 (8): 2758. (A review of research on the consumption phase in the circular economy.)

Kamleitner B, Thürridl C, Martin BAS. (2019). A Cinderella story: how past identity salience boosts demand for repurposed products. *Journal of Marketing*. 83 (6): 76–92. (An analysis of making repurposed products' past identity salient.)

MacArthur, E. (2013). *Towards the circular economy, economic and business rationale for an accelerated transition*. Cowes: Ellen MacArthur Foundation. (An influential report about the circular economy.)

Magnier L, Mugge R, Schoormans J. (2019). Turning ocean garbage into products – Consumers' evaluations of products made of recycled ocean plastic. *Journal of Cleaner Production*. 215: 84–98. (An analysis of consumers' responses to products made of recycled ocean plastic.)

Mugge R, Jockin B, Bocken N. (2017). How to sell refurbished smartphones? An investigation of different customer groups and appropriate incentives. *Journal of Cleaner Production*. 147: 284–296. (An examination of perceived risks and benefits of refurbished phones.)

Scott K, Martin DM, Schouten JW. (2014). Marketing and the new materialism. *Journal of Macromarketing*. 34 (3): 282–290. (A suggestion for and discussion about a new materialism.)

van Weelden E, Mugge R, Bakker C. (2016). Paving the way towards circular consumption: exploring consumer acceptance of refurbished mobile phones in the Dutch market. *Journal of Cleaner Production*. 113: 743–754. (An analysis of the factors that influence consumer acceptance of refurbished products.)

Winterich KP, Nenkov GY, Gonzales GE. (2019). Knowing what it makes: how product transformation salience increases recycling. *Journal of Marketing*. 83 (4): 21–37. (An analysis of the effect of product transformation salience on recycling behaviour.)

4

EMBEDDING MORE CIRCULAR APPROACHES TO THE MANAGEMENT OF RESOURCES

Louise Maxwell

Introduction

There is a general acceptance that current development patterns have resulted in unsustainable consumption of resources, leading to economic, environmental, and social problems. Global consumption and the subsequent generation of waste are thought to exist in a general cycle. Raw materials are extracted in resource-rich countries. Using these materials, products are made in manufacturing hubs like China. These finished products are exported for consumption, mainly in the Global North. Finally, the post-consumer wastes are collected and sent for recycling, reuse, and disposal, mostly in developing countries (Geng et al., 2013; Stahel, 2016; Qu et al., 2019).

Two central principles generally tend to guide decision-making over the allocation and consumption of resources, namely equity (providing equal pay for equal work) and efficiency (not wasting resources). When these two principles conflict with one another, people will often waste resources to avoid inequity. When efficiency and equity are in conflict, individuals often preserve equity rather than ensuring efficiency (Shaw, 2013; Choshen-Hillel et al., 2015).

A critical challenge therefore is to achieve both prosperity in Global North and South countries and sustainable development (Hanumante et al., 2019). Agenda 2030 (United Nations General Assembly, 2015) aims to moderate anthropocentric resource consumption. The Sustainable Development Goals (SDGs) note the need to "Ensure sustainable consumption and production patterns," with SDG 12.5 specifically outlining the importance of reducing waste generation. In the past decade, over 50 countries have adopted bioeconomy policies (German Bioeconomy Council, 2018). Initially, driven by increased energy prices which triggered the expansion of the biofuels sector, more recently the shift has been due to innovations and technological opportunities, changing consumer preferences, and sustainability concerns (El-Chichakli et al., 2016). For example, through the Climate Change Act, the UK government introduced a demanding and legally binding target of an 80% reduction in greenhouse emissions by 2050, based on 1990 levels (TSO, 2008).

Challenges faced

Significant growth in consumption, population, and economic growth have driven the rapid rise in resource extraction and demand (UNEP, 2016). Indeed, over the last 40 years, annual extraction of materials has more than tripled. As the population grows and the middle class increases, demand for resources is expected to reach 130 billion tonnes by 2050, up from 50 billion in 2014, leading to a 400% use of the Earth's total capacity (Accenture, 2016). In addition, other factors such as shorter product replacement cycles as a result of technological innovation are also contributing to rising waste generation rates.

Approximately 1 billion people (16% of the global population) suffer from chronic hunger, despite there being more than enough food to feed everyone on the planet (McCarthy et al., 2018). It is predicted that there will be an increase of 1.7 billion in the world's population by 2050. This will place increasing pressure on existing resources. Food security (i.e., all people at all times having access to sufficient, safe, nutritious food to maintain a healthy and active life) is dependent on availability (i.e., having adequate quantities), access (physical and economic), utilisation (i.e., supply), and stability (i.e., the factors that impact supply, including weather and wastage) (FAO, 2015). Food security issues are linked to various factors (Filippini et al., 2019). For example, in the Global South, food security issues are frequently related to chronically insufficient access to food (Battersby, 2013). The continuous increase in the global population has led to the rapid expansion of urban settlements. These urban areas, particularly in the Global South, are being impacted by increasingly adverse climate conditions and natural disasters which are exacerbating the problems of food production and availability. This problem is particularly acute in the most urbanised areas (e.g., megacities), which also suffer from a lack of land availability and are widespread in Asian countries. Developing countries are also severely affected by volatility in global food prices and instability in governance (Szabo, 2016). At the same time, wasted food could potentially feed those in need and contribute to food security (Irani et al., 2018).

Global generation rates of municipal solid waste (MSW) are expected to increase to 2.2 billion tonnes per year by 2025. While in the Global North, utilisation of technologies has enabled value recovery from MSW (e.g., for heat and fuels), in the Global South, the collection, transport, and effective management of waste is still challenging (Moya et al., 2017). The South Asian region, which is home to nearly one-quarter of the world's population, is experiencing rapid urban growth, economic development, and consumerism which are leading to significant growth in waste generation (Véron et al., 2018).

Even with minimal technological improvements and resource efficiency, there will still be overuse of the resources by around 40 billion tonnes of natural resources every year by 2050. Resource price volatility has become the norm. The scale and speed of demand growth, coupled with short-term but frequent price fluctuations, could lead to higher long-term prices and greater supply insecurity (WBCSD, 2016). Addressing these issues therefore requires a slower and more circular management of resources (EMF, 2014; Laurenti et al., 2016; Singh et al., 2019).

Key resources

It is, therefore, accepted that there is a need to do something and which resources should we be especially focusing upon. According to WBCSD (2016), eight materials (steel, aluminium, plastics, cement, glass, wood, primary crops, and cattle) are thought to be responsible for 20% of the global greenhouse gas emissions, 95% of water use, and 88% of land use. Of these materials, 34% go into shelter and 33% into food. Implementing circular economy approaches targeted to these materials would therefore address challenges associated with climate change, water, and land use.

The mineral phosphorus (P) is non-substitutable in food production. Maximising the efficiency of use and recovery of P is therefore crucial for food security and economic growth globally (Rahman et al., 2019). Global reserves of phosphate rock from which mineral/chemical P fertilizers are sourced for food production are found only in a handful of countries, indicating geopolitical disputes in the future (Chowdhury et al., 2017). The largest inflow of P in waste streams is seen in China and the United States. Countries throughout East Asia and Western Europe tend to show moderate P in flows, while Northern European nations (e.g., Denmark, Sweden, and Norway) have low P in flows (Rahman et al., 2019). Recovery of P is mostly seen in the food production subsystem from both the organic fraction of municipal solid waste and wastewater.

Waste electrical and electronic equipment (WEEE) is traded between the Global North and South either as waste or for metal recovery. Quantities of WEEE are increasing globally (StEP, 2018; Balde et al., 2017; Ilankoon et al., 2018), partly due to the fact that across existing and legacy electronics, the innovation cycle (i.e., the time between a product's market entry and peak sales) is steadily decreasing over time. However, in 2016, only 20% of global WEEE was recycled, leading to a significant loss of value, as WEEE contains a number of metals and critical earth materials (e.g., Au, Zn, Pt, Ni, Mn, Al, and Rh) (Ilankoon et al., 2018). Many of these critical materials are also widely used in electric vehicles and clean energy technologies, leading to concerns about their long-term supply security (Gaustad et al., 2018). In addition to the loss of value, while practices in developed countries are generally well regulated, those in developing countries tend to be largely unregulated and labour intensive and pose risks to the environment, health, and well-being. However, using circular economy approaches to managing WEEE faces various challenges, including an increasing number, type, and diversity of devices (Ryen et al., 2014); increasing complexity and lack of standardisation, making effective repair, upgrade, disassembly, and material identification and segregation difficult and labour intensive (Cucchiella et al., 2015); increasing difficulty in finding high-value items due to product lightweighting (Kasulaitis et al., 2019); the presence of hazardous materials (e.g., Pb, Cd, and Hg) in older devices (Kiddee et al., 2013); and shrinking product lifespans, reducing opportunities for innovation to be deployed in new devices (Babbitt et al., 2017).

In recent times, there has been an increasing focus on plastics. Global use of plastics has increased 20-fold in the past half-century and is expected to double again in the next 20 years, and their production is expected to double again over the next two decades. However, according to EMF (2017), some 95% of the value of plastic packaging materials, equivalent to US$80–120B annually, is lost to the economy. Only 14% of plastic packaging is recycled. Approximately 32% of the material escapes collection systems, generating significant economic costs by polluting the natural environment and clogging urban infrastructure. The cost of such after-use externalities, plus the cost associated with greenhouse gas emissions from its production, is conservatively estimated at US$40B annually. Over 90% of plastics produced are derived from virgin fossil feedstocks, representing about 6% of global oil consumption, which is equivalent to the oil consumption of the global aviation sector. At the current rate, the plastics sector will account for 20% of total oil consumption and 15% of the global annual carbon budget by 2050. In terms of waste, there could be more plastic than fish (by weight) in the ocean by 2050. Plastics often contain various chemical substances, some of which can potentially adversely affect human health and the environment.

The plastics value chain is highly fragmented (Sri, 2016). The rate at which new materials, labels, and types of plastics are being produced makes it difficult for collection schemes and sorting and reprocessing systems to keep up. There is also a disconnect between the materials and formats that are designed and utilised and the collection/management systems and infrastructure employed to manage them. Waste along the value chains is high. The development of effective markets and standards, as well as user engagement, is also made challenging. The resulting confusion limits investment in the required systems and infrastructure.

Another key resource is organic materials and their associated energy potential. For example, Mittal et al. (2019) contend that biogas generation from waste (e.g., crop residues, animal manure, MSW, and municipal and industrial wastewater) in India could be in the range of 310–655 billion m^3/yr, by 2040. This target would represent approximately 35% of India's total primary energy supply, using 2015 as the base year. In Africa, especially in the sub-Sahara, use of waste as an energy source is generally limited (Surroop et al., 2019). However, if the energy potential were exploited, not only would there be a recovery of energy but also creation of jobs, reduction in poverty, improved public health, and increased economic outputs. Through the Africa Biogas Partnership Program that was established in 2009, more than 60,000 small-scale biodigesters have been installed in five African countries (Ethiopia, Burkina Faso, Kenya, Tanzania, and Uganda) (Clemens et al., 2018; ABPP, 2018). Similar facilities have also been established in various other countries, including Rwanda, Zimbabwe, Senegal, Cameroon, and Egypt (Surroop et al., 2019). There is significant potential for bioenergy production in Africa. For example, Mugodo et al. (2017) estimated an electricity generation potential of 148 GWh/yr from anaerobic digestion of agricultural residues and wastes.

Adopting a more circular approach to resource management

Hanumante et al. (2019) suggest that the circular economy (CE) concept can deliver benefits to the manner in which we consume and manage our resources. However, the approach adopted should be balanced and within a timeframe of circa 70–90 years. Implementation after this time, or an 'aggressive' implementation before, would lead to system collapse, due to the depletion of commonly accessed ecosystem services caused by continued high consumption.

CE strategies therefore should be agile and responsive to the changing markets and be based on more accurate resource forecasting models (Althaf et al., 2019). These strategies might include green product design, creation of reuse markets, development of material recovery technologies, and policies to effectively engage multiple stakeholders in resource conservation and recovery activities (Bocken et al., 2016; Gaustad et al., 2018; O'Connor et al., 2016). Velenturf and Jopson (2019) note that the business case for resource recovery should support multidimensional growth that would enable the redistribution of economic, social, and environmental benefits. There is a need for innovations for resource (e.g., food) security and resource protection policies that enhance a sustainable utilisation of natural resources (von Braun, 2018). Governments should set up the policy, legal, and institutional frameworks to incentivise uptake of biogas technologies, while at the community level, the development of demonstration schemes to promote awareness and overcome sociocultural concerns is needed.

Various writers call for the use of sustainability-driven business model innovations to reduce the overall material and energy demands and facilitate the uptake of more circular approaches (e.g., Cooper, 2010; Tukker, 2015; Brocken and Short, 2016; Evans et al., 2017; Martin, 2016; Yang et al., 2017; Singh et al., 2019). For example, increasing the intrinsic durability of a product and expanding its lifespan beyond one life cycle would serve to reduce resource consumption and thus reduce wastage (Cooper, 2010). Similarly, the provision of repair and maintenance for products (Gnanapragasam et al., 2017), as well as the efficient recovery of components (or parts) and recycling (Cole et al., 2019; Singh and Ordoñez, 2016), are also potential approaches to ensure greater resource security.

Specifically for plastics, a number of countries have implemented policies and legislation to reduce plastic waste and encourage recovery. For example, China and Malaysia have banned the import of low-value mixed plastics for recycling, and the Indian government has pledged to ban all single-use plastics by 2022 (PWC, 2019). There is an EU Plastics Strategy (EC, 2018). The

Resource and Waste Strategy for England (Defra, 2018) calls for the development of standards for bio-based and biodegradable plastics and the banning of single use plastics from the central government estate by 2020. The UK Plastics Pact (WRAP, 2018), which is a collaborative initiative to create a circular system for plastics, was launched in 2018. It is led by the charity Waste and Resources Action Programme (WRAP), in partnership with the Ellen MacArthur Foundation, and is composed of a coalition of members covering the entire plastics value chain. Its ambitious targets for 2025 for plastic packaging are:

- 100% to be reusable, recyclable, or compostable
- 70% to be effectively recycled
- 30% average recycled content across all plastic packaging
- Action taken to eliminate problematic or unnecessary single-use plastic packaging items

Plastics Europe (2017a) estimated that banning plastics and other recyclables from landfill by 2025 would lead to savings of 7 million tonnes of CO_2 and the creation of 300,000 permanent jobs related to sorting, recycling, and energy recovery. In 2016, the Ellen MacArthur Foundation launched the New Plastic Economy Initiative. Amongst other things, the New Plastic Economy Initiative aimed to establish a cross-value chain dialogue and scale up the adoption of reusable packaging as well as industrially compostable plastic packaging, establish a Global Packaging Protocol, utilise innovation and an evidence base, and decouple plastics from a fossil feedstock. Ledsham (2018) suggests the need to:

- Boost demand for recyclable materials by encouraging companies to commit to using post-consumer resin (PCR) wherever possible, thus creating end markets that make the collection of the materials viable.
- Design for recyclability. Most plastic items/packaging tend to be composed of a composite of plastics, making disassembly and recycling difficult. Designing for recyclability in mind would overcome these challenges.
- Encourage a holistic, collaboration between resin suppliers, packaging manufacturers, brands, waste contractors, and recyclers. For example, CEFLEX, a European consortium of companies representing the entire value chain of flexible packaging, is working to develop robust design guidelines for both flexible packaging and the infrastructure used to collect, sort, and recycle it. Facilities using chemical recycling, which enables not only plastic bottles but other waste (e.g., from oceans and textiles), are also being developed (e.g., Loop Recycling).

Plastics Europe (2017b) calls for the use of:

- Full life-cycle thinking (i.e., adopting a holistic approach to the consumption of resources [including energy]) to preserve resources
- Environmental protection and well-being. Ensure that products – produced from virgin or recycled raw materials such as plastics – are safe for their intended uses and do not damage human health or the environment
- Awareness building to create behaviour change (e.g., for effective segregation of plastics waste)

At a wider level, community/citywide initiatives are also important. For example, through its Smart Cities Mission, India is seeking to incorporate circular economy principles into the design of the infrastructure needed to provide water, sanitation, and waste services at scale (GoI, 2019). The initiative, which was launched in 2015, is an urban renewal and retrofitting program by the government of India with the mission to develop 100 smart cities across the country, making them citizen friendly and sustainable. More systemic planning of city spaces, integrated with circular mobility solutions, can contribute to higher air quality, lower congestion, and reduced urban sprawl. Flexible use of buildings and urban spaces, enabled by digital applications, are also being employed to increase utilisation rates, thus getting more value out of the same assets.

References

ABPP (Africa Biogas Partnership Programme) (2018). *Blue flames golden fertiliser: about ABPP: where we work*. Available at: www.africabiogas.org (Accessed 15/11/2019).

Accenture (2016). *Circular advantage*. Available at: www.accenture.com/us-en/insightcircular-advantage-innovative-businessmodels-value-growth%20

Althaf S, Babbitt CW, Chen R. (2019). Forecasting electronic waste flows for effective circular economy planning. *Resources, Conservation and Recycling*. 151: 104362.

Babbitt CW, Althaf S, Chen R. (2017). *Sustainable materials management for the evolving consumer technology ecosystem*. Summary report of phase 1 research. Available at: https://www.rit.edu/gis/ssil/docs/Final%20Report%20SMM%20Phase%201%202017.pdf

Balde CP, Forti V, Gray V, Kuehr R, Stegmann P. (2017). *The global e-waste monitor 2017: quantities. Flows and resources*. Bonn, Geneva and Vienna: United Nations University, International Telecommunication Union and International Solid Waste Association.

Battersby J. (2013). *Urban food security and climate change: a system of flow. Climate change assets and food security in southern African cities*. Abingdon: Earthscan. 35–56.

Bocken, NMP, De Pauw, I, Bakker, C, Van Der Grinten, B. (2016). Product design and business model strategies for a circular economy. *Journal of Industrial Production Engineering*. 1015: 1–12.

Brocken NMP, Short SW. (2016). Towards a sufficiency-driven business model: experiences and opportunities. *Environmental Innovation and Social Transitions*. 18: 41–61.

Choshen-Hillel C, Shaw A, Caruso EM. (2015). Waste management: how reducing partiality can promote efficient resource allocation. *Journal of Personality and Social Psychology*. 109 (2): 210–231.

Chowdhury RB, Moore GA, Weatherley AJ, Arora M. (2017). Key sustainability challenges for the global phosphorus resource, their implications for global food security and options for mitigation. *Journal of Cleaner Production*. 140: 945–963.

Clemens H, Bailis R, Nyambane A, Ndung'u V. (2018). Africa biogas partnership program: a review of clean cooking implementation through market development in East Africa. *Energy and Sustainable Development*. 46: 23–31.

Cole C, Gnanapragasam A, Cooper T, Singh J. (2019). An assessment of achievements of the WEEE Directive in promoting movement up the waste hierarchy: experience in the UK. *Waste Management*. 87: 417–427.

Cooper T. (2010). The significance of product longevity. In: Cooper T. (Ed.), *Longer lasting products: alternatives to the throwaway society*. Oxfordshire: Routledge. 3–36.

Cucchiella F, Adamo ID, Koh SCL, Rosa P. (2015). Recycling of WEEEs: an economic assessment of present and future e-waste streams. *Renewable and Sustainable Energy Review*. 51: 263–272.

Defra (Department for Environment, Food and Rural Affairs) (2018). *The resources and waste strategy for England*. Available at: https://www.gov.uk/government/publications/resources-and-waste-strategy-for-england

EC (2018). *An European strategy for plastics in a circular economy*. Available at: http://ec.europa.eu/environment/circular-economy/pdf/plastics-strategy-brochure.pdf

El-Chichakli B, von Braun J, Lang C, Barben D, Philp J. (2016). Policy: five cornerstones of a global bioeconomy. *Nature*. 14, 535 (7611): 221–223.

EMF (Ellen MacArthur Foundation) (2014). *Towards the circular economy. 3: accelerating the scale-up across global supply chains*. Cowes: Ellen MacArthur Foundation.

EMF (Ellen MacArthur Foundation) (2017). *The new plastics economy: rethinking the future of plastics and catalytic action*. Available at: www.ellenmacarthurfoundation.org/assets/downloads/publications/NPEC-Hybrid_English_22-11-17_Digital.pdf.

Evans S, Vladimirova D, Holgado M, van Fossen K, Yang M, Silva E, Barlow CY. (2017). Business model innovation for sustainability: towards a unified perspective for creation of sustainable business models. *Business Strategy and the Environment*. 26 (5): 597–608.
FAO (2015). *FAO statistical pocketbook*. United Nations. Available at: www.fao.org/3/a-i4691e.pdf (Accessed 16/12/2019).
Filippini R, Mazzocchi C, Corsi S. (2019). The contribution of urban food policies toward food security in developing and developed countries: a network analysis approach. *Sustainable Cities and Society*. 47: 101506.
Gaustad G, Krystofik M, Bustamante M, Badami K. (2018). Circular economy strategies for mitigating critical material supply issues. *Resources, Conservation and Recycling*. 135: 24–33.
Geng Y, Joseph S, Sergio U, Zhang P. (2013). Measuring China's circular economy. *Science*. 339: 1526–1527.
German Bioeconomy Council (2018). *Bioeconomy policy (Part III) update report of national strategies around the world*. Available at: https://biooekonomierat.de/fileadmin/Publikationen/berichte/GBS_2018_Bioeconomy-Strategies-around-the_World_Part-III.pdf (Accessed 13/12/2019).
Gnanapragasam A, Oguchi M, Cole C, Cooper T. (2017). Consumer expectations of product lifetimes around the world: a review of global research findings and methods. In: Bakker C, Mugge R. (Eds.), *Product lifetimes and environment*. Amsterdam, The Netherlands: IOS Press.
GoI (Government of India) (2019). *Smart cities*. Available at: http://smartcities.gov.in/content/ (Accessed 16/12/2019).
Hanumante NC, Shastri Y, Hoadley A. (2019). Assessment of circular economy for global sustainability using an integrated model. *Resources, Conservation and Recycling*. 151: 104460.
Ilankoon IMSK, Ghorbani Y, Chong MN, Herath G, Moyo T, Petersen J. (2018). E-waste in the international context – a review of trade flows, regulations, hazards, waste management strategies and technologies for value recovery. *Waste Management*. 82: 258–275.
Irani Z, Sharif AM, Lee H, Aktas E, Topaloglu Z, van Woult T, Huda S. (2018). Managing food security through food waste and loss: small data to big data. *Computers and Operations Research*. 98: 367–383.
Kasulaitis BV, Babbitt CW, Krock AK. (2019). Dematerialization and the circular economy dematerialization and the circular economy comparing strategies to reduce material impacts of the consumer electronic product ecosystem. *Journal of Industrial Ecology*: 1–14.
Kiddee P, Naidu R, Wong MH. (2013). Electronic waste management approaches: an overview. *Waste Management*. 33: 1237–1250.
Laurenti R, Sinha R, Singh J, Frostell B. (2016). Towards addressing unintended environmental consequences: a planning framework. *Sustainable Development*. 24: 1–17.
Ledsham N. (2018). *Creating a circular economy for plastics*. Available at: https://sustainability.com/our-work/insights/creating-a-circular-economy-for-plastics/
Martin CJ. (2016). The sharing economy: a pathway to sustainability or a nightmarish form of neoliberal capitalism? *Ecological Economics*. 121: 149–159.
McCarthy U, Uysal I, Badia-Melis R, Mercier S, O'Donnell C, Ktenioudaki A. (2018). Global food security – issues, challenges and technological solutions. *Trends in Food Science & Technology*. 77: 11–20.
Mittal S, Ahlgren EO, Shukla PR. (2019). Future biogas resource potential in India: a bottom-up analysis. *Renewable Energy*. 141: 379–389.
Moya D, Aldás C, López G, Kaparaju P. (2017). Municipal solid waste as a valuable renewable energy resource: a worldwide opportunity of energy recovery by using waste-to-energy technologies. *Energy Procedia*. 134: 286–295.
Mugodo K, Magama PP, Dhavu K. (2017). Biogas production potential from agricultural and agro-processing waste in South Africa. *Waste Biomass Valorization*. 8: 2383–2392.
O'Connor MP, Zimmerman JB, Anastas PT, Plata DL. (2016). A strategy for material supply chain sustainability: enabling a circular economy in the electronics industry through green engineering. ACS Sustain. *Chemical Engineering*. 4: 5879–5888.
Plastics Europe (2017a). *Plastics contribution to a sustainable future*. Available at: www.plasticseurope.org/en/focus-areas/circular-economy (Accessed 15/11/2019).
Plastics Europe (2017b). *Plastics Europe's views on a strategy on plastics – increasing circularity and resource efficiency*. Available at: www.plasticseurope.org/application/files/4515/1700/9733/20170907_final_views_on_a_strategy_on_plastics_updated_sept.pdf (Accessed 15/11/2019).
PWC (2019). *Plastics and the circular economy: how we can reduce our impact*. Available at: www.pwc.co.uk/services/sustainability-climate-change/insights/plastics-and-circular-economy.html (Accessed 15/11/2019).
Qu S, Guo Y, Ma Z, Chen WQ, Liu J, Liu G, Wang Y, Xu M. (2019). Implications of China's foreign waste ban on the global circular economy. *Resources, Conservation and Recycling*. 144: 252–255.

Rahman S, Chowdhury RB, D'Costa NG, Milne N, Bhuiyanc M, Sujauddin M. (2019). Determining the potential role of the waste sector in decoupling of phosphorus: a comprehensive review of national scale substance flow analyses. *Resources, Conservation and Recycling*. 144: 144–157.
Ryen EG, Babbitt CW, Tyler AC, Babbitt GA. (2014). Community ecology perspectives on the structural and functional evolution of consumer electronics. *Journal of Industrial Ecology*. 18: 708–721.
Shaw (2013). Beyond "To share or not to share": the impartiality account of fairness. *Current Directions in Psychological Science*. 22: 413–417.
Singh J, Cooper T, Cole C, Gnanapragasam A, Shapley M. (2019). Evaluating approaches to resource management in consumer product sectors – an overview of global practices. *Journal of Cleaner Production*. 224: 218–237.
Singh J, Ordonez I. (2016). Resource recovery from post-consumer waste: important lessons for the upcoming circular economy. *Journal of Cleaner Production*. 134: 342–353.
Sri M. (2016). *Informal approaches towards a circular economy, learning from India*. Available at: www.sustainable-recycling.org/informal-approaches-towards-a-circular-economy-learning-from-india/ (Accessed 21/11/2019).
Stahel WR. (2016). Circular economy. *Nature*. 531: 435–438.
StEP (Solving the e-waste problem) (2018). Available at: www.step-initiative.org (Accessed 8/11/2019).
Surroop D, Bundhoo ZMA, Raghoo P. (2019). Waste to energy through biogas to improve energy security and to transform Africa's energy landscape. *Current Opinion in Green and Sustainable Chemistry*. 18: 79–83.
Szabo, S. (2016). Urbanisation and food insecurity risks: assessing the role of human development. *Oxford Development Studies*. 44 (1): 28–48.
TSO (2008). *The climate change act 2008*. London: The Stationery Office.
Tukker A. (2015). Product services for a resource-efficient and circular economy – a review. *Journal of Cleaner Production*. 97: 76–91.
UNEP International Resource Panel (2016). *Global material flows and resource productivity assessment report*. Available at: https://mahb.stanford.edu/library-item/globalmaterial-flows-resource-productivity/.
United Nations General Assembly (2015). *Transforming our world: the 2030 agenda for sustainable development*. New York, NY: UN.
Velenturf APM, Jopson JS. (2019). Making the business case for resource recovery. *Science of the Total Environment*. 648: 1031–1041.
Véron R, Fernando N, Narayanan NC, Upreti B, Ambat B, Pallawala R, Rajbhandari S, Dhananka SR, Zurbrügg C. (2018). Social processes in post-crisis municipal solid waste management innovations: a proposal for research and knowledge exchange in South Asia. *Research Ideas and Outcomes*. 4: e31430.
von Braun J. (2018). Bioeconomy – the global trend and its implications for sustainability and food security. *Global Food Security*. 19: 81–83.
WBCSD (World Business Council on Sustainable Development) (2016). *CEO guide to the circular economy*. Geneva, Switzerland: WBSCD.
WRAP (2018). *The UK plastics pact*. Available at: http://www.wrap.org.uk/content/the-uk-plastics-pact
Yang M, Evans S, Vladimirova D, Rana P. (2017). Value uncaptured perspective for sustainable business model innovation. *Journal of Cleaner Production*. 140: 1794–1804.

5

ENVIRONMENTAL JUSTICE, WASTE MANAGEMENT, AND THE CIRCULAR ECONOMY

Global perspectives

Paul Cox

Introduction

Environmental justice relates to the fair distribution of health and environmental risks weighed against socioeconomic benefits (Cotton, 2018). The urban environmental justice discourse is considered to have begun in the late 1970s in the United States, where the focus was around the presence of 'negative urban environmental burdens,' primarily due to the siting of facilities (e.g., landfills, hazardous waste treatments, intensive agricultural and resource extraction sites) in poor areas where non-whites generally tended to live (Agyeman, 2002; Pollock and Vittes, 1996; Massey, 2004; Myers, 2008; Taylor, 2002). Indeed, the concept had its roots in the civil rights movement, with notable cases including Love Canal (Levine, 1982) and Warren County (Bullard, 1989). In recent times, particularly in Europe, the debate has shifted also to include unsustainable models of development and the unequal power dynamics within and across communities and nation states (Dominelli, 2013). This has led to the coining of the concept of 'just sustainabilities,' which links environmental quality, sustainable development outcomes, and social inclusion (Agyeman, 2002).

Empirical research into the subject has generally tended to focus on communities facing social deprivation and/or economic and political peripheralisation due to pollution, planning, and resource extraction (e.g., Briggs et al., 2008; Bickerstaff and Agyeman, 2009; Blowers, 2010; Cotton, 2014; Groves, 2015; Walker et al., 2005). However, an important point is that with the increasing export of waste from the Global North to the Global South results also in the export of environmental injustices. Indeed, environmental injustices are particularly prevalent in urban areas in developing countries, where the majority of the residents have limited access to basic socioeconomic services such as adequate housing, infrastructure, clean water, and sanitation facilities (Binns et al., 2012; Kubanza and Simatele, 2016; Véron et al., 2018). While the focus has traditionally been on sub-Saharan countries (e.g., Din and Cohen, 2013; Otang-Ababio et al., 2013; Kubanza and Simatele, 2016), growing globalisation, urbanisation, and linear approaches to consumption mean that this is now very much a global concern. Indeed, these challenges are policy and planning issues that are being faced by rapidly developing countries and regions globally, for example, in Southeast Asian countries and in various regions of China where the economies have been rising in recent decades (Baek, 2016; Oktavilia et al., 2017; Wu et al., 2017).

Exporting environmental injustices

Globally, hazardous waste is one of the main ecological hazards and a primary cause of various environmental disasters (Xu et al., 2019). Improper management, storage, or transportation of hazardous waste can result in human health hazards and environmental air and soil pollution (Higgins, 2017). The growth in both resource consumption levels and the global population has led to a commensurate increase in hazardous waste. In the past few decades, around 90% of the waste has been generated from only 30 highly developed or industrialised Organization for Economic Co-operation and Development (OECD) countries (USEPA, 2018). However, due to differences in the availability of resources (i.e., funding, equipment, and skills), significant differences exist in the manner in which the waste is managed in developed and developing countries (Tang et al., 2015). In addition, there has been a persistent trend by Global North nations of dumping their waste in Global South countries (Xu et al., 2019). However, the developing countries often lack the necessary facilities or technology to properly treat or manage the waste, resulting in an increase in the associated environmental disasters. At the same time, however, a heavy reliance on export by some developed countries has meant that some (e.g., the United Kingdom) lack enough domestic treatment capacity to adequately recycle, reuse, and dispose of some of their solid waste. These issues therefore raise challenges both for developed and developing countries alike.

In 2016, over 15 million metric tonnes (MMTs) of waste plastics, 16 MMTs of scrap papers, and 2 MMTs of discarded textile materials were exported globally, with about 40% to China (UN Contrade, 2018). This export was generally due to five key reasons (Qu et al., 2019): (1) environmental regulations and enforcement are generally less stringent in developing countries; (2) separation of recycled materials is labour intensive and therefore cheaper in developing countries; (3) developing countries are also likely to be manufacturing hubs (e.g., China), thus making reusing recycled materials more convenient and economic; (4) shipping companies often offer competitive rates for return trips after shipping finished goods to developed countries in order to avoid empty return cargos; and (5) materials from certain types of solid waste are used to substitute raw materials for low-value, low-quality goods, particularly in developing countries.

In July 2017, the Chinese government announced a ban on 24 types of solid waste in four categories: waste plastics, unsorted scrap papers, discarded textile materials, and vanadium slag (Ministry of Environmental Protection, 2017). Bans on additional types of solid waste were also scheduled for 2018 and 2019, including scrap metals, scrap ships, compressed pieces of scrap automobiles, waste plastics from industrial sources, and wood waste and scrap (Ministry of Ecology and Environment, 2018). This decision was based on the country's adoption of a more sustainable approach; with an improved environment and lower demand for raw materials in certain industries due to a slowing of the economy, quantities of domestic wastes have grown, therefore becoming sufficient to replace the void left by externally sourced wastes. While the ban has resulted in some disruption, as well as price fluctuations and debate about the practice of exporting waste, it has also led to the flow of the waste to other countries in the region with growing economies (e.g., Thailand, Malaysia, Indonesia, and India).

In South Asia, ineffective waste management has become a key concern, as it is linked to drinking water quality, sanitation, and human health, affecting the urban poor the most (Véron et al., 2018). Waste management policies also directly impact employment opportunities and livelihoods, particularly those in the informal sector (Wilson et al., 2013; Véron et al., 2018).

Waste, pollution, and socioeconomic development

There is a school of thought that as economic growth in countries takes place, environmental degradation results. As growth continues, the realisation of the need for action results in the implementation of environmental regulations and the decoupling of this growth from the degradation (Ward et al., 2016; Destek and Sarkodie, 2019; Alvarado and Toledo, 2017). This phenomenon is known as the Environmental Kuznets Curve (EKC) (Grossman and Krueger, 1991, 1995). Using chromium (Cr) as a proxy for environmental pollution, Chen and Taylor (2020) demonstrated that the link between economic development and pollution did follow the EKC. Levels of Cr increased rapidly during the 1960s–1980s, at the height of economic growth, but then subsequently stabilised during a period of de-industrialisation and economic modernisation. A trade-off of the decoupling is that as growth in these countries continues, there is 'exportation' of environmentally damaging industries and practices to those with less stringent legislation and regulations, the so-called pollution haven effect (Gill et al., 2018). However, there are those who contend that more stringent environmental regulations actually result in innovation that offsets the costs of the regulations (Wang et al., 2019a; Porter and Van der Linde, 1995). Some note that stricter environmental standards may actually attract polluting firms (Ben Kheder and Zugravu, 2012; Dou and Han, 2019). Indeed, Shao et al. (2019) found that the pollution haven hypothesis does not stand in Brazil, Russia, India, and China (BRICS) and Mexico, Indonesia, Nigeria, and Turkey (MINT). Yet others suggest that environmental regulations do not play a role in the locational decisions of pollution-intensive firms (Manderson and Kneller, 2012). For example, Mulatu and Wossink (2014) found that environmental regulations did not have a measurable effect on the pig industry location in 43 European regions. Rather than location, the key factors are the type of environmental policy being implemented and industrial characteristics (Zheng and Shi, 2017; Wang et al., 2015a), as well as international production fragmentation and factor endowment (Cai et al., 2018).

While mining and hazardous waste recovery have economic benefits, the practices can also negatively impact the environment (e.g., release of dust, deposition of heavy metals, and pollution of the water and soil) (Johnson et al., 2019; Ao et al., 2019; Wang et al., 2019b) and also result in social issues such as marginalisation, child labour, and displacement of communities (Akabzaa, 2000; Filer, 1998). These challenges are especially prevalent in developing countries (Saedpanah and Amanollahi, 2019).

When pollution takes place in poorer regions or countries, it reduces life expectancy and longevity (Song et al., 2016). By extension, individuals living in more developed areas and countries tend to have higher life expectancy (Wang et al., 2015b). A number of studies have also found links between increasing urbanisation, socioeconomic factors related to low deprivation, and high levels of the particulate matter $PM_{2.5}$ (Han et al., 2018; Jin et al., 2019; Ouyang et al., 2019; Li et al., 2018). Emissions were found to be due to both natural and anthropogenic factors (e.g., population rises and economic development, leading to increasing energy consumption and use of vehicles). A key underlying factor is that in some countries, some policies and legislation have prioritised economic development at the expense of environmental protection (e.g., infrastructure that encourages urban green spaces).

Individuals working with waste, particularly in developing countries, are often exposed to toxic materials due to not having access to appropriate protective measures. Hazardous materials are sometimes also dumped on land and thus seep into water courses (Ewijk et al., 2018; Gu et al., 2017). Generally, various previous studies suggest that there is a link between respiratory illnesses and even lung cancer in individuals who live near landfills/dumpsites, whether the exposure is short, medium, or long term (Mataloni et al., 2016; Macklin et al., 2011). Children and other vulnerable segments of the population are especially at risk (Ma et al., 2007). However, there are other studies that have found no association (e.g., Kret et al., 2018).

Sellafield, in the North of England, has been the site of a nuclear waste facility since the 1980s. As a result, the communities (particularly Whitehaven) surrounding the site and the individuals living there are now perceived as being 'dirty' (Castán Broto et al., 2010). This stigmatisation has also resulted in other industries (e.g., coastal tourism) relocating from the area (Cotton, 2018). However, Qu et al. (2019) note that while China's ban in 2017 on the importation of certain waste had short-term negatives for players, it also offered an opportunity for their domestic recycling industries in the long term. The stricter environmental policies could also serve to enhance resource productivity and competitiveness of industries in the country. Indeed, China may even become like Japan and become an exporter of waste in future.

Facilitating a more circular approach and reducing environmental injustices

The development and more effective enforcement of policies and global standards for national waste management capacity are required to reduce the transfer of environmental burdens through trade (Brooks et al., 2018; Qu et al., 2019). While international policies such as the Basel Convention (2018) have been developed to prevent the transboundary movements of hazardous wastes and disposal, the enforcement of these policies has always been a challenge for both developed and developing countries.

Local pressure can and does play a role. For example, using case studies from Mexico (Moore, 2009), Senegal (Fredericks, 2018), and West Bengal (Cornea et al., 2016), Véron et al. (2018) note the interconnectedness of physical-material and sociopolitical processes and the fact that garbage can also represent a political 'resource.' For example, cities in West Bengal in India were cleaned, particularly before municipal elections, as a strategy to build political capital (Cornea et al., 2016), while in July 2012, the Mavallipura landfill, which was located around 15 km from Bangalore, was closed due to the efforts of the community, organised in part by the Bangalore-based NGO the Environment Support Group (ESG).

Environmental justice is achieved only when the polluter successfully justifies the imposition of additional environmental risks and burdens upon a community and then rectifies the imbalances produced (Shrader-Frechette, 2002; Cotton, 2018). In addition, the community should have access to the decision-making processes through open dialogue and available information through informed and active capacity for consent and influence in the decision-making processes (Schlosberg, 2007).

Since the 1990s, across Europe, there has been a shift in the ways in which infrastructure is planned for and governed. Cotton (2018) notes that there has been a move towards the regulatory state (as characterised by processes of privatisation and deregulation) and away from the 'dirigiste state' of the past (whereby the state exerts a strong directive influence over investment). For example, in the United Kingdom, planning decisions for large-scale infrastructure facilities such as energy from waste facilities are based on national interest (i.e., as a crucial element of economic growth and industrial development beyond the immediate economic benefits to developers or host communities) (Cotton, 2018). Democratic legitimacy in decision-making is derived from an aggregate or representative democratic model (i.e., from elected representatives) rather than a deliberative democratic model. However, this does mean that project development frequently exacerbates environmental injustices to local communities, as these groups effectively have no power to question the need for the infrastructure (Cotton, 2014).

Using the Managing Radioactive Waste Safely (MRWS) process (Defra, 2001; Mackerron and Berkhout, 2009), Cotton (2018) illustrates that via the use of 'scalar parity,' success can be gained through an overt commitment to dialogue through partnership, exploration of bottom-up community perspectives through a well-funded engagement process, and providing (potentially) long-lasting community benefits. The key to the success was the in-built capacity to simultaneously scale up local community concerns and scale down national concerns within a voluntary-participatory decision-making process, thus allowing 'meaningful, low-conflict political dialogue' to occur. Similarly, Véron et al. (2018) contend that there is a need to develop socially acceptable and politically feasible municipal waste management systems that are adapted to local and national political and sociocultural conditions rather than externally promoted and imposed approaches. In addition, hierarchical structures should be decentralised to facilitate more horizontal accountabilities, with shorter, closed-loop waste chains that benefit the local communities.

References

Agyeman J. (2002). Constructing environmental (in)justice: transatlantic tales. *Environmental Politics*. 11 (3): 31–53.

Akabzaa TM. (2000). *Boom and dislocation. The environmental and social impacts of mining in the Wassa West district of Ghana*. Accra: Third World Network e-Africa.

Alvarado R, Toledo E. (2017). Environmental degradation and economic growth: evidence for a developing country. *Environmental Development and Sustainability*. 19 (4): 1205–1218.

Ao M, Qiu G, Zhang C, Xu X, Zhao L, Feng X, Qin S, Meng B. (2019). Atmospheric deposition of antimony in a typical mercury-antimony mining area, Shaanxi provience, Southwest China. *Environmetal Pollution*. 245: 173–182.

Baek J. (2016). A new look at the FDI-income-energy-environment nexus: dynamic panel data analysis of ASEAN. *Energy Policy*. 91: 22–27.

Basel Convention (2018). *Control of transboundary movements of hazardous wastes and their disposal*. Available at: www.basel.int.

Ben Kheder S, Zugravu N. (2012). Environmental regulation and French firms location abroad: an economic geography model in an international compartive study. *Ecological Economics*. 77: 48–61.

Bickerstaff K, Agyeman J. (2009). Assembling justice spaces: the scalar politics of environmental justice in North-East England. *Antipode*. 41 (4): 781–806.

Binns T, Dixon A, Nel E. (2012). *Africa: diversity and development* (1st ed). London: Routledge.

Blowers A. (2010). Why dump on us? Power, pragmatism and the periphery in the siting of new nuclear reactors in the UK. *Journal of Integrative Environmental Sciences*. 7 (3): 157–173.

Briggs D, Abellan JJ, Fecht D. (2008). Environmental inequity in England: small area associations between socio-economic status and environmental pollution. *Social Science and Medicine*. 67: 1612–1629.

Brooks AL, Wang S, Jambeck JR. (2018). The Chinese import ban and its impact on global plastic waste trade. *Science Advances*. 4 (6): EAAR0131.

Bullard RD. (1989). *Race and the siting of hazardous waste sites*. Berkeley: University of California Press.

Cai X, Che X, Zhu B, Zhao J, Xie R. (2018). Will developing countries become pollution havens for developed countries? An empirical investigation in the Belt and Road. *Journal of Cleaner Production*. 198: 624–632.

Castán Broto V, Burningham K, Carter C, Elghali L. (2010). Stigma and attachment: performance of identity in an environmentally degraded place. *Society and Natural Resources*. 23 (10): 952–968.

Chen Q, Taylor D. (2020). Economic development and pollution emissions in Singapore: evidence in support of the environmental Kuznets curve hypotesis and its implications for regional sustainability. *Journal of Cleaner Production*. 243: 118637.

Cornea N, Véron R, Zimmer A. (2016). Clean city politics: an urban political ecology of solid waste in West Bengal, India. *Environment and Planning A*. 49 (4): 728–744.

Cotton M. (2014). Environmental justice challenges in United Kingdom infrastructure planning: lessons from a Welsh incinerator project. *Environmental Justice*. 7 (2): 39–44.

Cotton M. (2018). Environmental justice as scalar parity: lessons from nuclear waste management. *Social Justice Research*. 31: 238–259.

Defra (Department for Environment Food and Rural Affairs) (2001). *Managing radioactive waste safely: proposals for developing a policy for managing solid radioactive waste in the UK*. London: Routledge.

Destek MA, Sarkodie SA. (2019). Investigation of environmental Kuznets curve for ecological footprint: the role of energy and financial development. *Science of the Total Environment*. 650: 2483–2489.

Din G, Cohen E. (2013). Modelling municipal solid waste management in Africa: case study of Matadi, the Democratic Republic of Congo. *Journal of Environmental Protection*. 4: 435–445.

Dominelli L. (2013). Environmental justice at the heart of social work practice: greening the profession. *International Journal of Social Welfare*. 22: 431–439.

Dou J, Han X. (2019). How does the industry mobility affect pollution industry transfer in China: empirical test on pollution haven hypothesis and Porter hypothesis. *Journal of Cleaner Production*. 217: 105–115.

Ewijk S, Park JY, Chertow MR. (2018). Quantifying the system-wide recovery potential of waste in the global paper life cycle. *Resources, Conservation and Recycling*. 134: 48–60.

Filer C. (1998). *Mining in the South Pacific*. Available at: www.antenna.nl/ecsiep/bulletin.htmlo.

Fredericks R. (2018). *Garbage citizenship: vital infrastructures of labor in Dakar*. Senegal: Duke University Press.

Gill FL, Viswanathan KK, Karim MZA. (2018). The critical review of the pollution haven hypothesis. *International Journal of Energy and Economic Policy*. 8 (1): 167–174.

Grossman GM, Krueger AB. (1991). *From environmental impacts of a North American free trade agreement*. Available at: www.nber.org/papers/w3914.

Grossman GM, Krueger AB. (1995). Economic growth and the environment. *Quarterly Journal of Economics*. 110 (2): 353–377.

Groves C. (2015). The bomb in my backyard, the serpent in my house: environmental justice, risk, and the colonisation of attachment. *Environmental Politics*. 24 (6): 853–873.

Gu F, Guo J, Zhang W, Summers PA, Hall P. (2017). From waste plastics to industrial raw materials: a life cycle assessment of mechanical plastic recycling practice based on a real-world case study. *Science of the Total Environment*. 601: 1192–1207.

Han L, Zhou W, Li W, Qian Y. (2018). Urbanization strategy and environmental changes: an insight with relationship between population change and fine particulate pollution. *Science of the Total Environment*. 642: 789–799.

Higgins TE. (2017). *Hazardous waste minimization handbook*. Boca Raton, FL: CRC Press.

Jin JQJ, Du Y, Xu LJ, Chen ZY, Chen JJ, Wu Y, Ou CQ. (2019). Using Bayesian spatio-temporal model to determine the socio-economic and meteorological factors influencing ambient PM2.5 levels in 108 Chinese cities. *Environmental Pollution*. 254: 113023.

Johnson AI, Dipnall JF, Dennekamp M, Williamson GJ, Gao CX, Carroll MTC, Dimitriadis C, Ikin JF, Johnston FH, McFarlane AC, Sim MR, Stub DA, Abramson MJ, Guo Y. (2019). Fine particulate matter exposure and medication dispensing during and after a coal mine fire: a time series analysis from the Hazelwood Health Study. *Environmental Pollution*. 246: 1027–1035.

Kret J, Dame LD, Tutlam N, DeClue RW, Schmidt S, Donaldson K, Lewis R, Rigdon SE, Davis S, Zelicoff A, King C, Wang Y, Patrick S, Khan F. (2018). A respiratory health survey of a subsurface smoldering landfill. *Environmental Research*. 166: 427, 436.

Kubanza NS, Simatele S. (2016). Social and environmental injustices in solid waste management in sub-Saharan Africa: a study of Kinshasa, the Democratic Republic of Congo. *Local Environment*. 21 (7): 866–882.

Levine AG. (1982). *Love Canal: science, politics, and people*. Lexington: Lexington Books.

Li VOK, Han Y, Lam JCK, Zhu Y, Bacon-Shone J. (2018). Air pollution and environmental injustice: are the socially deprived exposed to more $PM_{2.5}$ pollution in Hong Kong. *Environmental Science and Policy*. 80: 53–61.

Ma J, Kouznetsova M, Lessner L, Carpenter DO. (2007). Asthma and infectious respiratory disease in children – correlation to residence near hazardous waste sites. *Paediatric and Respiratory Review*. 8 (4): 292–298.

Mackerron G, Berkhout F. (2009). Learning to listen: institutional change and legitimation in UK radioactive waste policy. *Journal of Risk Research*. 12 (7–8): 989–1008.

Macklin Y, Kibble A, Pollitt F. (2011). *Impact on health of emissions from landfill sites*. Advice from the Health Protection Agency. Available at: https://assets.publishing.service.gov.uk/government/uploads/system/uploads/attachment_data/file/334356/RCE-18_for_website_with_security.pdf

Manderson E, Kneller R. (2012). Environmental regulations, outward FDI and heterogeneous firms: are countries used as pollution havens? *Environmental & Resource Economics, Springer. European Association of Environmental and Resource Economists*. 51 (3): 317–352.

Massey R. (2004). *Environmental justice: income, race, and health.* Available at: http://www.ase.tufts.edu/gdae/education_materials/modules/Environmental_Justice.pdf.

Mataloni F, Badaloni C, Golini MN, Bolignano A, Bucci S, Ancona C. (2016). Morbidity and mortality of people who live close to municipal waste landfills: a multisite cohort study. *International Journal of Epidemiology*. 45 (3): 806–815.

Ministry of Ecology and Environment (2018). Available at: www.mee.gov.cn/xxgk/hjyw/201707/t20170728_418674.shtml.

Ministry of Environmental Protection (2017). Available at: http://zfs.mee.gov.cn/gz/bmhb/gwygf/201707/t20170728_418692.shtml.

Moore S. (2009). The excess of modernity: garbage politics in Oaxaca, Mexico. *The Professional Geographer.* 61 (4): 426-437.

Mulatu A, Wossink A. (2014). Environmental regulation and location of industrialized agricultural production in Europe. *Land Ecology*. 90 (3): 509–537.

Myers GA. (2008). Sustainable development and environmental justice in African cities. *Geography Compass*. 2: 695–708.

Oktavilia S, Sugiyanto FX, Firmansyah F, Widodo W. (2017). Environmental degradation, trade openness, and economic growth in Southeast Asian countries. *Pertanika Journal of Social Science and Humanities.* 25: 175–185.

Otang-Ababio M, Melara Arguello JE, Cabbay O. (2013). Solid waste management in African cities: sorting the facts from the Fads in Accra, Ghana. *Habitat International.* 39: 96–104.

Ouyang X, Shao Q, Zhu X, He Q, Xiang C, Wei G. (2019). Environmental regulation, economic growth and air pollution: panel threshold analysis for OECD countries. *Science of the Total Environment.* 657: 234–241.

Pollock E, Vittes P. (1996). *Poverty, pollution, and solid and hazardous waste: the linkages for different sources.* Florida Centre for Solid and Hazardous Waste Management, University of Central Florida. Report No. 96–93. Florida, USA.

Porter ME, Van der Linde C. (1995). Toward a new conception of the environment-competitiveness relationship. *Journal of Economic Perspectives.* 9 (4): 97–118.

Qu S, Guo Y, Ma Z, Chen WQ, Liu J, Liu G, Wang Y, Xu M. (2019). Implications of China's foreign waste ban on the global circular economy. *Resources, Conservation and Recycling.* 144: 252–255.

Saedpanah S, Amanollahi J. (2019). Environmental pollution and geo-ecological risk assessment of the Qhorveh mining area in West Iran. *Environmental Pollution.* 253: 811–820.

Schlosberg D. (2007). *Defining environmental justice: theories, movements, and nature.* Oxford: Oxford University Press.

Shao Q, Wang X, Zhou Q, Balogh L. (2019). Pollution haven hypothesis revisited: a comparison of BRICS and MINT countries based on VECM approach. *Journal of Cleaner Production.* 227: 724–738.

Shrader-Frechette KS. (2002). *Environmental justice: creating equality, reclaiming democracy*. Oxford: Oxford University Press.

Song W, Li Y, Hao Z, Li H, Wang W. (2016). Public health in China: an environmental and socio-economic perspective. *Atmospheric Environment.* 129: 9–17.

Tang Z, Huang Q, Yang Y. (2015). China: over haul rules for hazardous chemicals. *Nature.* 525 (7570): 455.

Taylor DE. (2002). *Race, class, gender, and American environmentalism.* Portland, OR: U.S. Department of Agriculture, Forest Service, Pacific Northwest Research Station. General Technical Rep. PNWGTR-534.

UN Contrade (2018). *Database.* Available at: https://comtrade.un.org.

USEPA (United States Environmental Protection Agency) (2018). *Learn the basics of hazardous waste.* Available at: www.epa.gov/hw/learn-basics-hazardous-waste#hwid.

Véron R, Fernando N, Narayanan NC, Upreti B, Ambat B, Pallawala R, Rajbhandari S, Dhananka SR, Zurbrügg C. (2018). Social processes in post-crisis municipal solid waste management innovations: a proposal for research and knowledge exchange in South Asia. *Research Ideas and Outcomes.* 4: e31430.

Walker G, Mitchell G, Fairburn J, Smith G. (2005). Industrial pollution and social deprivation: evidence and complexity in evaluating and responding to environmental inequality. *Local Environment.* 10 (4): 361–377.

Wang Q, Xie X, Wang M. (2015a). Environmental regulation and firm location choice in China. *China Economic Journal.* 8 (3): 215–234.

Wang SB, Luo KL, Liu YL, Zhang SX, Lin XX, Ni RX, Tian XL, Gao X. (2015b). Economic level and human longevity: spatial and temporal variations and correlations analysis of per capita GDP and longevity indicators in China. *Archives Gerontology Geriatrics.* 61: 93–102.

Wang X, Zhang C, Zhang Z. (2019a). Pollution haven or porter? The impact of environmental regulation on location choices of pollution-intensive firms in China. *Journal of Environmental Management*. 248: 109248.

Wang Z, Liu X, Qin H. (2019b). Bioconcentration and translocation of heavy metals in the soil-plants system in Machangqing copper mine, Yunnan province China. *Journal of Geochemical Exploration*. 200: 159–166.

Ward JD, Sutton PC, Werner AD, Costanza R, Mohr SH, Simmons CT. (2016). Is decoupling GDP growth from environmental impact possible? *PLOS ONE*. 11 (10): 0164733.

Wilson D, Velis C, Rodic L. (2013). Integrated sustainable waste management in developing countries. *Proceedings of the Institution of Civil Engineers – Waste and Resource Management*. 166 (2): 52–68.

Wu H, Guo H, Zhang B, Bu M. (2017). Westward movement of new polluting firms in China: pollution reduction mandates and location choice. *Journal of Computational Economics*. 45 (1): 119–138.

Xu L, Deng Y, Mancl K. (2019). Environmental disaster risk reduction-oriented centralized treatment of hazardous wastes: a novel approach for production-distribution decision optimization in China. *International Journal of Disaster Risk Reduction*. 40: 101253.

Zheng D, Shi M. (2017). Multiple environmental policies and pollution haven hypothesis evidence from China's polluting industries. *Journal of Cleaner Production*. 141: 295–304.

6

RESOURCE CONSUMPTION AND THE ASSOCIATED HEALTH RISKS

A brief overview

Terry Tudor

Introduction

Discussions around the concept of the circular economy (CE) often tend to focus on the resources conserved, as well as their economic value. However, the approaches employed in the extraction and processing of raw materials, as well as the inherent biochemical and physical properties of selected waste materials, can lead to risks to public and environmental health. Thus, an examination of these risks and the role that a CE can play in reducing them is vital (WHO, 2016; OECD, 2018; World Bank, 2017).

Risks posed

Globally, the extraction, processing, and use of raw materials is predicted to double from 79 gigatonnes (Gt) in 2011 to 167 Gt in 2060 (OECD, 2018). With regard to waste, the annual generation rate is expected to increase by 70% from 2016 levels to 3.40 billion tonnes in 2050 (Kaza et al., 2018). Non-metallic minerals (e.g., sand, gravel, and limestone) are expected to represent more than half of the overall materials used. The use of fossil fuels, as well as the production of iron, steel, and construction materials, leads to large energy-related emissions of greenhouse gas (GHG) and air pollutants. The increase in the extraction and use of raw materials is expected to lead to GHG emissions rising to 50 Gt CO_{2eq} per year by 2060. The extraction and use of raw materials is much more polluting than using secondary (e.g., recycled) materials. For example, the extraction and use of metals can result in a wide range of pollution impacts (e.g., toxic effects on humans and ecosystems).

Within the community setting, around 2 billion people across the world have no waste collection and three billion no controlled waste disposal (CIWM and WasteAid UK, 2018). The resulting indiscriminate littering and use of dumpsites (which are particularly common in the Global South, where access to functional landfills is often limited) can lead to the contamination of the land, as well as surface and ground water (Vaccari et al., 2019). More than half (approximately 38 of 50) of the world's largest uncontrolled dumpsites are in coastal areas; thus, the waste and leachate go directly into the water courses and sea (ISWA, 2016). Open waste burning, especially in rural areas in the Global South, is a key contributor to respiratory health issues (Reyna-Bensusan et al., 2018).

Approximately nine million people die each year of diseases linked to either mismanagement of waste or pollutants. It is the vulnerable segments of the population (e.g., children and the elderly) who are often disproportionately affected by these risks. In addition, a higher percentage of women often tend to be involved in the most hazardous elements of waste management activities (e.g., the sorting of the waste). By extension, these risks have negative knock-on impacts on the livelihoods and development potential of families and communities (e.g., United Nations General Assembly, 2015). For example, illness amongst children reduces their potential to attend school. In turn, this limits their future career and livelihood opportunities. These factors are often compounded in many countries in the Global South by limitations in effective governance, enforcement of environmental policies, investment in, and public awareness of effective waste management (Ferronato et al., 2019).

Around 75–90% of the waste produced by healthcare facilities is comparable to municipal waste and is usually termed 'non-hazardous' or 'general healthcare waste' (WHO, 2014). This fraction of the waste arises primarily from areas such as offices, kitchens, and non-intensive care wards and can include items such as packaging and food waste. The remaining 10–25% of the waste is regarded as hazardous and may potentially pose significant risks to human health and the environment. Even though hazardous healthcare waste represents only a small percentage of the overall waste fraction, hazards such as infections (bacteria, viruses, parasites, or fungi), sharps (hypodermic needles, scalpels, and infusion sets), radioactive waste, and chemicals (e.g., cytotoxic and genotoxic waste) can pose significant risks to workers as well as the community (WHO, 2016).

Reducing the risks

Pro-environmental behaviour change is important to reduce the quantity of waste produced in the first place. There is a need for awareness building in order to effect change in values, attitudes, and habits and ultimately in behaviours. If individuals are more aware of the potential hazards and benefits, then they are more likely to engage. In addition, products should be designed for recycling/reuse/remanufacturing. In this way, fewer raw materials need to be extracted, and the inherent value in the existing products is kept within the system.

The use of more environmentally sound waste disposal approaches/technologies and the recovery of value from that produced, especially organic waste, are vital steps to reducing health risks (World Bank, 2019). For example, there is a need to reduce the use of dumpsites and especially the dumping of hazardous materials at these sites (ISWA, 2016). In addition, climate change impacts can be reduced through the adoption of treatment and disposal technologies that capture biogas and landfill gas.

For healthcare waste, there has been an increasing focus on identifying suitable approaches to reduce both the risks and the costs associated with managing these risks (Caniato et al., 2015; Viani et al., 2016; Vaccari et al., 2018). Effective segregation of hazardous and non-hazardous waste is one of the most cost-efficient and effective approaches to reducing risks. Green procurement has for some time been viewed as a crucial tool within the healthcare sector for moving to a more circular approach (Bergsma and Sevenster, 2013). For example, at the international level, Health Care without Harm and the United Nations Development Programme (UNDP) developed the Sustainable Health in Procurement Project (SHiPP), a four-year project in 10 lower and middle-income countries. The project aims to develop and pilot a set of sustainable health procurement practices and policies. It will promote these policies and practices in the health sector and in the United Nations (UN) agencies. The project focuses on the reduction of toxicity of chemicals and materials in health products, the reduction of GHGs in the supply chain, and the conservation of resources.

There are a number of organisations, particularly in Europe and North America, which facilitate the sharing and reuse of used medical equipment (e.g., beds), as well as the sharing of medical services and facilities. These entities usually employ an online platform to enable stakeholders to either offer excess capacity of resources or request these resources.

Polyvinyl chloride (PVC), which is extensively used in healthcare settings, can be recovered, as it is a consistent, high-quality single plastic rather than being a multilayered composite plastic (PVC Med Alliance, 2016). Indeed, a number of PVC-based medical products are safely recycled and reprocessed into various products (e.g., hoses for fire extinguishers).

Finally, in some countries (e.g., in Germany and some states in the United States), reprocessing and reuse of medical devices is permitted, while in others, it is not (e.g., in Italy, France, and the United Kingdom). The use of the concept of designing for reuse (with appropriate sterilisation) and/or remanufacture can serve to reduce the use of single-use instruments.

Conclusions

While the focus for the discussion on the concept of the CE is often on waste reduction and value recovery, the extraction of raw materials and the management of waste are both associated with public and environmental health risks. There is a need, therefore, for a more holistic discussion that takes account of the potential benefits of also addressing these risks.

References

Bergsma G, Sevenster M. (2013). *End-of-life best approach for allocating recycling benefits in LCAs of metal packaging.* CE Delft. Available at: https://www.cedelft.eu/publicatie/end-of-life_best_approach_for_allocating_recycling_in_lcas_of_metal_packaging/1371

Caniato M, Tudor T, Vaccari M. (2015). Understanding the perceptions, roles and interactions of stakeholder networks managing health-care waste: a case study of the Gaza Strip. *Waste Management.* 35: 255–264.

CIWM and WasteAid UK (2018). *From the land to the sea.* Northampton: CIWM.

Ferronato N, Rada EC, Gorritty Portillo MA, Cioca LI, Ragazzi M, Torretta V. (2019). Introduction of the circular economy within developing regions: a comparative analysis of advantages and opportunities for waste valorization. *Journal of Environmental Management.* 230: 366–378.

ISWA (International Solid Waste Association) (2016). *A roadmap for closing waste dumpsites: the world's most polluted places.* Available at: https://www.iswa.org/fileadmin/galleries/About%20ISWA/ISWA_Roadmap_Report.pdf

Kaza, S, Yao, LC, Bhada-Tata, P, Van Woerden, F. (2018). *What a waste 2.0: a global snapshot of solid waste management to 2050: urban development.* Washington, DC: World Bank.

OECD (2018). *Global material resources outlook to 2060: economic drivers and environmental consequences.* Available at: www.oecd.org/environment/waste/highlights-global-material-resources-outlook-to-2060.pdf (Accessed 20/5/2020).

Palmer-Jones D. (2020). *Raising standards: circular.* Northampton: The Chartered Institution of Wastes Management.

PVC Med Alliance (2016). *The role of hospitals in the circular economy: the contribution of PVC waste management plans.* Available at: http://pvcmed.org/hospitals-circular-economy-contribution-pvc-waste-management-plans/ (Accessed 25/5/2020).

Reyna-Bensusan N, Wilson DC, Smith SR. (2018). Uncontrolled burning of solid waste by households in Mexico is a significant contributor to climate change in the country. *Environmental Research.* 163: 280–288.

United Nations General Assembly (2015). *Transforming our world: the 2030 agenda for sustainable development.* Available at: https://sustainabledevelopment.un.org/post2015/transformingourworld/publication

Vaccari M, Tudor TL, Perteghella A. (2018). Costs associated with the management of waste from healthcare facilities: an analysis at national and site level. *Waste Management & Research.* 36 (1): 39–47.

Vaccari M, Tudor TL, Vinti G. (2019). Characteristics of leachate from landfills and dumpsites in Asia, Africa, and Latin America: an overview. *Waste Management*. 95: 416–431.

Viani C, Vaccari M, Tudor T. (2016). Recovering value from used medical instruments: a case study of laryngoscopes in England and Italy. *Resources, Conservation and Recycling*. 111: 1–9.

World Bank (2017). *Solid waste management*. Available at: www.worldbank.org/en/topic/urbandevelopment/brief/solid-waste-management.

World Bank (2019). *Solid waste management*. Available at: www.worldbank.org/en/topic/urbandevelopment/brief/solid-waste-management.

World Health Organization (WHO) (2014). *Safe management of wastes from health-care activities*. Geneva. Switzerland: The World Health Organization. 2nd Ed.

World Health Organization (WHO) (2016). *Waste and human health: evidence and needs*. WHO meeting report. Available at: https://www.euro.who.int/__data/assets/pdf_file/0003/317226/Waste-human-health-Evidence-needs-mtg-report.pdf?ua=1

7
THE SUSTAINABLE DEVELOPMENT GOALS AS DRIVERS FOR CHANGE

David C. Wilson

Evolution of drivers for waste and resource management

Solid waste management (SWM) is one of the essential utility services underpinning modern society. However, unlike water supply and sewerage, electricity and gas, transport and telephone, and broadband internet, it is often not recognised as a utility (Wikipedia, n.d.). Arguably, it is the victim of its own success: regular solid waste collection in Western Europe and North America dates back to a series of cholera epidemics from the 1830s and the resulting drive to protect public health and has been ubiquitous, at least in urban areas, for well over a century, so now it is taken for granted – that is, until something goes wrong, when it makes national or international news headlines.

Public health remained the major driver for more than a century (Wilson, 2007). It was not until the environmental movement started in the 1960s that new regulations to eliminate open dumping and burning of wastes were introduced from the 1970s, with standards gradually ramped up over the next few decades.

Resource value within waste pre-dates public health as a driver. In pre-industrial times, products were routinely repaired and reused and materials recycled. Food wastes were either fed to animals or composted as a soil improver. The Industrial Revolution concentrated people in cites, increasing waste quantities while disrupting natural cycles back to the soil. Many of the urban poor made their livelihoods as informal 'street buyers' of source-separated reusable products or recyclable materials, or as 'waste pickers,' retrieving materials from mixed waste.

Since the nineteenth century, rising living standards and cheap mass production led to a steady decline in municipal solid waste (MSW) recycling rates in developed countries. Both rising disposal costs since the 1980s and increased public opposition to new sites have led developed countries to 'rediscover' recycling for MSW, driven not so much by revenues but rather by recycling markets offering a competitive 'sink.' The resource driver has recently been supplemented by growing concerns over the depletion of virgin raw materials and resource scarcity, leading to a new focus on waste prevention and resource efficiency. Hence the focus of this *Handbook* and of this chapter on waste and resource management (WaRM), not just SWM.

Despite this progress, UNEP's inaugural *Global Waste Management Outlook* (GWMO1) in 2015 estimated that at least 3 billion people, more than 40% of the world's population, still lack basic SWM services (UNEP and ISWA, 2015). Extending waste collection to unserved neighbourhoods, and eliminating open dumping and burning, is essential to protect public health, prevent flooding, and clean up the local environment, with children being especially vulnerable. In terms of the resource driver, the informal recycling sector is thriving in many cities as a means of livelihood for the urban poor and often achieves moderately high recycling rates, albeit at a social and environmental cost.

One of the objectives of GWMO1 was to highlight SWM as a political priority. The three historic drivers of public health, local environment, and resource value, on their own, are often not enough to spark the policy-level interest and resulting budgetary allocations needed to sustain basic, controlled municipal SWM services. An early analysis of the development drivers for waste management sought to strengthen the case by linking those to the drivers for sustainable development (Wilson, 2007). GWMO1 took that several steps further (UNEP and ISWA, 2015), establishing a firm framework of linkages to the Sustainable Development Goals (SDGs), which were at that time being developed in parallel. Linkages between the wider circular economy and the SDGs have also been explored (Schroeder et al., 2018).

Improved waste and resource management as an entry point for achieving multiple Sustainable Development Goals

Linking waste and resource management to the Sustainable Development Goals

The 2030 Agenda for Sustainable Development was adopted by heads of state at the UN General Assembly in 2015 as a 'plan of action for people, planet and prosperity.' It sets out 17 high-level SDGs, each subdivided into targets, 169 in all. The SDGs 'are integrated and indivisible and balance the three dimensions of sustainable development: the economic, social and environmental.' The primary purpose is as drivers for action 'in areas of critical importance for humanity and the planet' (UN, 2015).

Negotiating the SDGs took many years and involved much compromise. The scope had to be comprehensive, but, equally, the absolute number of goals and targets had to be constrained if the whole was to be implementable. So, some strong candidates for the primary goals had to miss out. For this *Handbook*, the good news is that ensuring sustainable consumption and production patterns (i.e., the circular economy) was adopted as SDG12; the bad is that SWM is not a primary goal. A similar compromise is reflected in SDG13, taking urgent action to combat climate change, while controlling air pollution in general is not a primary goal.

However, there is no suggestion that either SWM or controlling air pollution is a low priority; rather, they are cross-cutting issues featuring in numbers of targets across multiple goals. SWM and air quality have each found a 'base' within the SDG11 indicator set on sustainable cities, while aspects of both are encountered in multiple SDGs.

GWMO1 argued that this could be viewed as a strength rather than as a weakness: 'The political case for action can be significantly strengthened when waste management is viewed as an 'entry point' to address a range of [SDGs], many of which are difficult to tackle' (UNEP and ISWA, 2015). However, the call for attention to improve SWM, and the transition to WaRM, is diffused in practice across many SDG goals and targets, leading to potential gaps and overlaps. So GWMO1 set out what could be termed a SWM- and WaRM-related 'virtual SDG,' that is, what such a goal might have looked like had all the relevant targets been collated under a single high-level SDG. In GWMO1, the five targets were labelled 'Global Waste Management Goals,' but to avoid confusion, they have been termed here 'global waste targets' and numbered 'target GW1' and so on.

The linkages between these five global waste targets and the 17 SDGs and their targets was first set out in GWMO1 and subsequently elaborated (Rodic and Wilson, 2017) (Lenkiewicz,

2016). The following discussion has also been informed by the second edition of the GWMO (GWMO2) (UNEP and ISWA' 2020).

Figure 7.1 and Table 7.1 together present a simplified overview of some of these linkages. In Figure 7.1, the rows are the 17 SDGs, while the columns are primarily the five global waste targets. The linkages are sub-divided into three groups: direct links, which are in principle measurable and/or explicit under an SDG target (six SDGs); direct links that are more difficult to measure (six SDGs); and indirect links (five SDGs). Table 7.1 expands on the 12 direct linkages, with these categorised according to the three overarching drivers discussed in the first section.

Direct entry points to the Sustainable Development Goals

Public health is the driver for the first global waste target, GW1: 'ensure access for all to adequate, safe and affordable solid waste collection services.' The wording reflects that in SDG targets 11.1 (under the goal sustainable cities) and 1.4 (end poverty), with solid waste collection included in their list of 'basic services.' There is also a clear and direct link to SDG 3 (good health and well-being), as uncollected waste causes both childhood and waterborne diseases, the latter via blocked drains and watercourses.

Local environmental protection is the driver for target GW2, 'eliminate uncontrolled disposal and open burning.' Both targets GW1 and GW2 are directly linked to two SDG targets: 11.6 on reducing the adverse environmental impact of cities is the most explicit in referring to both air quality and solid waste management, while these two global waste targets provide a necessary first step towards SDG target 12.4, 'environmentally sound management of . . . all wastes . . . in order to minimise their adverse impacts on human health and the environment,' the wording of which is reflected in target GW3. Target GW2 also links directly to SDG target 6.3 ('improve water quality by . . . eliminating dumping . . .') and both targets GW2 and GW3 to SDG15 (protect life on land).

Resource value is a major driver for global waste targets GW4 and GW5, but Table 7.1 highlights linkages to other drivers that traditionally command more political attention. Sustainable livelihoods (SDG8) is an explicit driver for target GW4: 'Substantially reduce waste generation through prevention and the 3Rs (reduce, reuse, recycle) and thereby create 'green' jobs,' which also reflects in part the wording of SDG target 12.5. End hunger (SDG2) is a driver for target GW5: 'Halve per capita global food waste at the retail and consumer levels and reduce food losses in the supply chain,' which echoes the wording of SDG target 12.3. Indeed, the UN Food and Agriculture Organisation has calculated that one-third of all the edible food the world grows goes to waste, 1.3 billion tonnes a year, which would be enough to feed all the hungry twice over (FAO, n.d.).

In developing countries, SWM services, both for collection and recovery, are often provided by individuals and small- and micro-enterprises, which is often referred to as the 'informal sector' (Wilson et al., 2006) (Scheinberg et al., 2010) (UNEP and ISWA, 2015). Similarly, much food is grown by small-scale producers who are particularly vulnerable to food losses prior to market. Table 7.1 highlights several SDG targets, under SDG2, SDG8, and SDG9, aimed at supporting such groups, which will improve livelihoods and contribute directly to SDG1 (end poverty).

Linkages to global environmental crises

Two of the linkages shown in Figure 7.1 and Table 7.1 are worthy of more detailed discussion, as improved WaRM can make a major contribution to mitigating both global heating (SDG13) and the quantity of plastics entering the oceans (SDG14, target 14.1).

SUSTAINABLE DEVELOPMENT GOALS

Key to the shaded boxes:
Direct link – in principle, measurable
Number: SDG target that explicitly requires decent SWM (1.4)
Direct link – but difficult to measure
Indirect link

	Global Waste Targets [1 2]					
	GW1. Ensure access for all to solid waste collection	GW2. Stop uncontrolled dumping and open burning	GW3. Achieve environmentally sound management	GW4. Reduce, reuse, recycle and create jobs	GW5. Halve food waste from markets, shops and homes	*Governance factors underpinning sustainable SWM*
1 END POVERTY[2]	1.4					
2 ZERO HUNGER						
3 GOOD HEALTH AND WELL-BEING						
4 QUALITY EDUCATION						
5 GENDER EQUALITY						
6 CLEAN WATER AND SANITATION		6.3				
7 AFFORDABLE AND CLEAN ENERGY						
8 DECENT WORK AND ECONOMIC GROWTH						
9 INDUSTRY, INNOVATION AND INFRASTRUCTURE						
10 REDUCED INEQUALITIES						
11 SUSTAINABLE CITIES AND COMMUNITIES	11.6 11.1	11.6				
12 RESPONSIBLE CONSUMPTION AND PRODUCTION			12.4	12.5	12.3	
13 CLIMATE ACTION						
14 LIFE BELOW WATER	14.1	14.1				
15 LIFE ON LAND						
16 PEACE, JUSTICE AND STRONG INSTITUTIONS						
17 PARTNERSHIPS FOR THE GOALS						

Figure 7.1 Linkages between solid waste management (SWM)/waste and resource management (WaRM), and the Sustainable Development Goals (SDGs)[2]

Source: Managing waste properly and improving SWM/ WaRM / developing the circular economy can help deliver all the SDGs.

[1] The five global waste targets were termed 'Global Waste Management Goals' in GWMO1 (UNEP and ISWA, 2015). They represent what a SWM- and WaRM-related 'virtual SDG' might have looked like had it been included as a high-level SDG. [2] Some wordings have been edited for clarity and conciseness.'

Table 7.1 Relationship between Sustainable Development Goals (SDGs) and solid waste management (SWM-)/waste and resource management (WaRM-)related global waste targets

Drivers	*Sustainable Development*		*SWM/WaRM*
	SDG[1]	*Specific SDG Target*[1,2]	*Global Waste Targets*[3]
Protection of public health	SDG11: Sustainable cities	**11.1** Ensure access for all to adequate, safe, and affordable housing and **basic services** and upgrade slums	→ **Target GW1. Ensure access for all to adequate, safe, and affordable solid waste collection services.** *Waste collection is a basic utility service. Uncollected waste is often dumped in waterways or burned in the open air, thus directly causing pollution and contamination. Waste also clogs the drains, which exacerbates floods, keeps water stagnant, and contributes to water-borne diseases and malaria. Children are among the most vulnerable, so they are affected the most.*
	SDG1: End poverty	**1.4** Ensure that all men and women, in particular the poor and vulnerable, have equal . . . access to **basic services** . . .	
	SDG3: Good health and well-being	3.2 End preventable deaths of children under 5 years 3.3 End malaria and combat water-borne diseases 3.9 Reduce illnesses from hazardous chemicals; air, water, and soil pollution; and contamination	
	SDG11: Sustainable cities	**11.6** Reduce the adverse per capita environmental impact of cities, including by paying special attention to air quality and **municipal and other waste management**	
	SDG12: Responsible consumption and production	**12.4 Environmentally sound management of** chemicals and **all wastes** . . . in order to minimise their adverse impacts on human health and the environment	→ **Target GW2. Eliminate uncontrolled dumping and open burning**, *as the first stepping-stone to achieving environmentally sound SWM practices.*
Protection of the environment LOCAL	SDG6: Clean water and sanitation	**6.3** Improve water quality by reducing pollution, **eliminating dumping**, and minimising release of hazardous materials	→ **Target GW3. Achieve the sustainable and environmentally sound management of all wastes, particularly hazardous wastes** *(both chemical and biological hazardous wastes).*
	SDG15: Protect life on land	15.1 Ensure the conservation of terrestrial and inland freshwater ecosystems and their services	

Drivers	*Sustainable Development*		*SWM/WaRM*
	SDG[1]	*Specific SDG Target*[1,2]	*Global Waste Targets*[3]
GLOBAL	SDG7: Affordable and clean energy	7.2 Increase the share of renewable energy in the global energy mix	→ **Target GW3.** *SWM technologies can derive renewable energy from (organic) waste.*
	SDG13: Climate action	SDG13: Take urgent action to combat climate change and its impacts	→ **Targets GW2 & GW3.** *Improved WaRM practices can mitigate large amounts of climate pollutants*[4].
	SDG14: Life below water	**14.1 Prevent marine pollution** of all kinds, **in particular from land-based activities, including marine debris**	→ **Targets GW1** & **GW2.** *Extending waste collection to all and eliminating uncontrolled dumping will prevent waste (particularly plastics) entering the oceans.*[4]
Resource value Sustainable livelihoods	SDG12: Responsible consumption and production	**12.5 Reduce waste through prevention, reduction, recycling, and reuse** 12.1 Implement the 10-Year Framework Programme on Sustainable Consumption and Production . . . 12.2 Achieve the sustainable management and efficient use of natural resources	→ **Target GW4. Substantially reduce waste generation through prevention and the 3Rs (reduce, reuse, recycle) and thereby create 'green' jobs** *Waste prevention is the highest-ranking option in the waste management hierarchy. It is followed by reuse of products or their parts, and then by recycling of component materials.*
	SDG8: Inclusive and sustainable economic growth and decent work for all	8.3 . . . support . . . decent job creation . . . encourage the formalisation and growth of micro . . . -sized enterprises, including through access to financial services 8.4 Improve . . . global resource efficiency in consumption and production and endeavour to decouple economic growth from environmental degradation . . .	Both this target and several targets under SDGs 8, 9 and 12 extend explicitly into the circular economy Reuse and recycling have significant potential for creating jobs.

(Continued)

Table 7.1 (Continued)

Drivers	*Sustainable Development*		*SWM/WaRM*
	SDG[1]	*Specific SDG Target*[1,2]	*Global Waste Targets*[3]
	SDG9: Build resilient infrastructure, promote inclusive and sustainable industrialisation and foster innovation	9.1 Develop quality . . . sustainable . . . infrastructure . . . with a focus on affordable and equitable access to all 9.3 Increase the access of small-scale . . . enterprises . . . to financial services, including affordable credit, and their integration into value chains and markets 9.4 . . . increase resource-use efficiency and . . . adopt . . . clean and environmentally sound technologies and industrial processes . . .	→ **Target GW1 Extend collection, Target GW4 3Rs and Target GW5 Halve food waste** *In developing countries, SWM services (both for collection and recycling) are often provided by individuals and small and micro-enterprises. Similarly, much food is grown by small-scale producers who are particularly vulnerable to food losses prior to market. Any measures applied to support them will improve livelihoods and directly contribute to SDG1 as well as to SDGs 8, 9 and 2.*
Resource value **End hunger**	SDG1: End poverty	1.4 Ensure that all men and women, in particular the poor & . . . vulnerable, have equal rights to economic resources . . . land . . . and financial services, including microfinance	
	SDG12: Responsible consumption and production	**12.3 Halve per capita global food waste at the retail and consumer levels** and reduce food losses along production and supply chains, including post-harvest losses	→ **Target GW5. Halve per capita global food waste at the retail and consumer levels and reduce food losses in** the supply chain. *See also comment above on small-scale producers, which relates to SDG target 2.3, which calls for the doubling of their agricultural productivity and incomes by 2030.*
	SDG2: Zero hunger	SDG2: End hunger, achieve food security and improved nutrition, and promote sustainable agriculture	

Source: Adapted extensively from (Rodic and Wilson, 2017)

1 Source of SDGs and SDG targets: (UN, 2015) (UNStats, n.d.). The wording has in some cases been shortened. Some of the Global Waste Targets in the last column apply to multiple SDGs and SDG Targets in the previous columns (i.e. some rows overlap). 2 Bold highlighting of SDG targets denotes more explicit referencing to SWM/WaRM. 3 The five global waste targets were termed 'Global Waste Management Goals' in GWMO1 (UNEP and ISWA, 2015). They represent what a SWM- and WaRM-related 'virtual SDG' might have looked like had it been included as a high-level SDG. The wording has in some cases been shortened. Explanations are given in *italics*. 4 Further discussion and references are provided in the Section on Linkages to global environmental crises.

Climate change

Global heating has been a major driver for improvements in municipal SWM in Europe since at least the 1990s, with legislative targets set for diversion of waste from landfills explicitly to reduce emissions of methane (CH_4), a powerful short-lived climate pollutant, from anaerobic decomposition of organic wastes. For example, changes in Germany's waste sector between 1990 and 2006 reduced the country's total greenhouse gas (GHG) emissions by 5% (Dehoust et al., 2010); this was in addition to the significant mitigation of methane emissions already achieved between the first introduction of regulatory controls on solid waste disposal in the 1970s and 1990.

But direct emissions of methane are just part of the story. Reuse and recycling can yield large indirect savings in industrial processes, reducing direct energy consumption as well as avoiding mining, processing, and transporting virgin raw materials – for example, recycling reduces GHG emissions from a tonne of aluminium by 97%. A life-cycle assessment (LCA) carried out for the German government examined four example countries – Germany, Turkey, Tunisia, and Mexico – and estimated that a 10–15% reduction in global GHG emissions could be achieved through improved WaRM, including landfill mitigation and diversion and energy from waste and recycling (Dehoust et al., 2010).

Extending the scope of WaRM to include waste prevention increases the impact further. Taking as just one example the FAO estimate of 1.3 billion tonnes a year of edible food waste, growing this food which is not eaten accounts for 8% of total GHG emissions (FAO, 2015).

Against this background, the official estimate of around 3% as the contribution of the waste sector to global GHG emissions, made by the UN's Intergovernmental Panel on Climate Change (IPCC) in its 2013 fifth assessment report, looks like a significant underestimate. The discrepancies are, however, easy to explain. The IPCC estimate refers to a 2010 baseline, so the huge progress made before then by the developed world (but not yet developing countries) in mitigating methane from land disposal is not accounted for. The IPCC is the custodian of climate science and is incredibly careful to avoid any double counting – so indirect savings in other agricultural or industrial sectors from prevention, reuse, and recycling are explicitly excluded.

There is one additional factor to be considered. Open burning of MSW is recognised as a potential contributor, but IPCC did not have sufficiently reliable data to include in its 2013 assessment. Open burning produces black carbon (soot), an extremely powerful short-term climate pollutant with a potency compared to carbon dioxide of around 3,000 (which compares to 30 for methane). Recent research at Imperial College London has aimed to plug the data gaps, resulting in a best estimate of the contribution of black carbon from the open burning of MSW to global heating in the range 2–10% of global CO_2e emissions (Reyna-Bensusan et al., 2018) (Reyna-Bensusan et al., 2019).

Finding 'low hanging fruit' which offers significant reductions in climate pollutants over a short time-frame is a major priority in the fight to mitigate global heating. Improving WaRM and moving towards a circular economy provides a compelling entry point for making significant reductions to climate pollutants across the world economy. Whether the mitigation potential is 15, 20, 25%, or even more will continue to be debated (UNEP and ISWA, 2015) (Wilson, 2018). The linkage is strong, direct, and in principle measurable, although the UN has allocated responsibility for such measurement to the IPCC rather than under SDG targets and indicators – hence no specific SDG target is shown here in Figure 7.1.

Marine plastics

Marine plastics have emerged relatively recently as a major cause for public concern and largely post-date the SDGs. Target 14.1 encompasses plastics but is not as explicit as it would likely be if being drafted today: 'Prevent and significantly reduce marine pollution of all kinds, in particular from land-based activities, including marine debris and nutrient pollution.'

Most plastics entering the ocean come from land-based activities. Modelling studies estimate that about 10 million tonnes of plastics enter the oceans from the land each year (Jambeck et al., 2015), with 20% of that from rivers (Schmidt et al., 2017) (Lebreton et al., 2017); this is at least an order of magnitude greater than estimates of the visible quantities in the oceans or on beaches (ISWA, 2017) – so the first priority is to 'turn off the tap.' Land-based sources include unmanaged or mismanaged municipal solid wastes in developing countries; litter from all countries; manufacturing wastes, including spillages of plastic pellets; micro- and nano-plastics in products; fibres from washing synthetic textiles; and micro-plastics from road vehicle tyre wear (ISWA, 2017). A recent review of evidence concluded that unmanaged or mismanaged MSW in developing countries likely accounts for 50–70% by weight of plastics entering the oceans (CIWM and Wasteaid UK, 2018). It follows that the quantity of plastics entering the oceans could at least be halved by extending MSW collection to all and eliminating uncontrolled disposal, which are the global waste targets GW1 and GW2.

Indirect linkages to all the remaining Sustainable Development Goals

Figure 7.1 and Table 7.1 show that improved WaRM provides a direct entry point into 12 of the 17 SDGs, sub-divided into two levels of impact. The less detailed Figure 7.1 extends the argument to include indirect effects, which extends the linkages to include all 17 SDGs.

SWM is an issue of human dignity and arguably of human rights and equality. Lack of access to such a basic utility service directly impacts inequality within a country (SDG10). Women and children do most of the work within the home, including bearing the main responsibility for SWM/WaRM. Self-management by dumping and burning is both more time consuming and more hazardous, so access to collection services both promotes gender equality (SDG5) and frees up time for education (SDG4). Reuse and recycling similarly supplement household livelihoods, again impacting both gender equality and opportunity to access education. Properly functioning SWM/WaRM in a city requires good governance and well-functioning institutions (SDG16), including control of corruption (target 16.5). Improving WaRM requires capacity building and improved access to international finance and technology transfer (SDG17).

Indicators for the Sustainable Development Goals

'If you don't measure it, you can't manage it.' Having set 17 high-level SDGs and 169 component targets, the next step for the UN was to assign indicators to monitor progress towards meeting them and then to agree methodologies to measure each indicator.

Data relevant to sustainable development has not traditionally been collected by national statistics agencies. So, in parallel to negotiation of the SDGs, the UN secretary general set up an Independent Expert Advisory Group on Data Revolution for Sustainable Development (IEAG, 2014). Data then needs to be turned into indicators in a 'form that is designed to synthesise, simplify and communicate information and to turn that into knowledge' (UNEP and ISWA, 2015). Negotiations again aimed to keep the number of SDG indicators manageable: having agreed on 165 targets, this meant that more than half of the targets were restricted to just one indicator. The list of indicators adopted by the UN General Assembly in July 2017 to act as a starting point of work on measurement protocols totalled 241 entries or, allowing for those listed against more than one target, 230 indicators. Thirty-six major changes to the framework in the form of replacements, revisions, additions, and deletions were agreed upon as part of the 2020 Comprehensive Review (UNStats, n.d.). To give a flavour, Table 7.2 shows indicators corresponding to some of the 'circular economy' SDG12 targets.

Each indicator on the agreed-upon list was assigned to a UN custodian agency, which generally convened a multiagency and expert working group to develop an agreed-upon measurement methodology/protocol and trialled the draft indicator where necessary before rolling it out. Both the protocols and the resulting statistics database are becoming available via the regularly updated SDG website (UNStats, n.d.).

Much of the work on SWM/WaRM indicators has focused on target 11.6: 'By 2030, reduce the adverse per capita environmental impact of cities, including by paying special attention to air quality and municipal and other waste management,' for which UN-Habitat is the custodian agency. Two indicators were defined, one relating to WaRM (11.6.1) and the other to air quality (11.6.2).

UN-Habitat's working group has developed the concept of indicator 11.6.1 as a *data hub*, aiming to provide primary source data on WaRM in cities around the world. The wording of the indicator was officially clarified in the 2020 review so that it now reads: 'Proportion of municipal solid waste collected and managed in controlled facilities, out of total municipal solid waste generated, by cities.' This has then been disaggregated into two main sub-indicators for each city: (a) proportion of MSW collected and (b) proportion of MSW managed in controlled recovery and disposal facilities, each measured against the total MSW generated in a city. A third sub-indicator has been added: (c) potential plastics emissions to the environment from uncollected and uncontrolled MSW in a city.

The detailed methodology for indicator 11.6.1 addresses the definition of MSW; household surveys to measure waste generation; collection of primary data on household waste amounts and composition, both at the point of collection and the point of disposal; estimation of commercial and institutional wastes and wastes from public spaces; interviews through the recovery value chain to estimate quantities managed by material type; and how to decide if a particular recovery or disposal facility is 'controlled.' The draft methodology was field tested in 2019 in Nairobi and Mombasa in Kenya and in Seychelles, and a 'step-by-step' guide (UN-Habitat, 2020) (UNStats, n.d.) was prepared in parallel.

Linkages have also been agreed upon between indicator 11.6.1 on SWM/WaRM in cities and indicators 12.5.1, national recycling rates, and 12.3.1 (b), a (global) food waste index. The interface between these SDG 12 indicators and indicator 11.6.1 is complex and will require further work. One issue is how to aggregate 11.6.1 data collected at the city level to the national and global levels used by the other indicators. Another is that 12.5.1 includes recycling rates from non-municipal sources, including various industry sectors, for which recycling rates have historically been higher than for MSW.

Having argued in the section "Improved waste and resource management as an entry point for achieving multiple Sustainable Development Goals" that *not* being assigned as a high-level SDG goal could be turned into an advantage for WaRM, in that it provides an entry point into many of the 17 SDGs, the process of assigning indicators to monitor progress could be seen to undermine that case. An average of less than 1.5 indicators per target makes it difficult to include indicators to monitor cross-cutting issues such as SWM/WaRM or air quality. Figure 7.1 highlighted six SDGs and eight SDG targets with a direct and potentially measurable link to SWM/WaRM. However, just three of those targets, 11.6. 12.3, and 12.5, have been discussed so far in relation to the official UN work on indicators.

Table 7.2 Targets and indicators for SDG 12: ensure sustainable consumption and production patterns

Presented as an example of the definition of indicators for one group of targets. This version corresponds to the 'final' version adopted in the 2020 Comprehensive Review.

To save space, only targets 12.1–12.7 have been included.

No.	*Target*	*No.*	*Indicators*
12.1	Implement the *10-Year Framework of Programmes on Sustainable Consumption and Production Patterns*, all countries taking action, with developed countries taking the lead, taking into account the development and capabilities of developing countries	**12.1.1**	Number of countries developing, adopting, or implementing policy instruments aimed at supporting the shift to sustainable consumption and production
12.2	By 2030, achieve the sustainable management and efficient use of natural resources	**12.2.1**	Material footprint, material footprint per capita, and material footprint per GDP
		12.2.2	Domestic material consumption, domestic material consumption per capita, and domestic material consumption per GDP
12.3	By 2030, halve per capita global food waste at the retail and consumer levels and reduce food losses along production and supply chains, including post-harvest losses	**12.3.1**	(a) Food loss index and (b) food waste index
12.4	By 2020, achieve the environmentally sound management of chemicals and all wastes throughout their life cycle, in accordance with agreed-upon international frameworks, and significantly reduce their release to air, water, and soil in order to minimize their adverse impacts on human health and the environment	**12.4.1**	Number of parties to international multilateral environmental agreements on hazardous waste and other chemicals that meet their commitments and obligations in transmitting information as required by each relevant agreement
		12.4.2	(a) Hazardous waste generated per capita; and (b) proportion of hazardous waste treated, by type of treatment
12.5	By 2030, substantially reduce waste generation through prevention, reduction, recycling, and reuse	**12.5.1**	National recycling rate, tonnes of material recycled
12.6	Encourage companies, especially large and transnational companies, to adopt sustainable practices and to integrate sustainability information into their reporting cycle	**12.6.1**	Number of companies publishing sustainability reports
12.7	Promote public procurement practices that are sustainable, in accordance with national policies and priorities	**12.7.1**	Number of countries implementing sustainable public procurement policies and action plans

Source: (UNStats, n.d.)

For 14.1, the official indicator focuses on floating plastic debris density; indicator 11.6.1 (c) complements that with a measure of the source term, that is, potential plastics emissions to the environment from uncollected and uncontrolled MSW in a city. Indicator 1.4.1 is the proportion of the population living in households with access to basic services; as SWM is recognised as one of those basic services, a link to 11.6.1 as the WaRM data hub would seem to make sense.

For three of the SDG targets highlighted in Figure 7.1, the official indicators unfortunately do not include any focus on SWM/WaRM. Target 6.3 is to 'improve water quality by reducing pollution, eliminating dumping . . .,' and the two selected indicators focus on wastewater treatment and water quality. Target 11.1 includes 'access for all to . . . basic services . . .,' but the corresponding indicator focuses only on transport. Target 12.4 is 'achieve the environmentally sound management of chemicals and all wastes . . .,' but the two indicators focus only on chemicals and hazardous wastes. One could argue that 11.6.1 already covers, at least partially, the SWM elements of 6.3 and 11.1, but that is not the case for 12.4. Referring to Figure 7.1, indicator 11.6.1 (a) corresponds to global waste target GW1 of access to waste collection for all, while (b) corresponds to target GW2 to stop uncontrolled dumping and open burning. But the third step, moving from controlled disposal to environmentally sound management (ESM), as set out in target GW3, corresponds to target 12.4, and there is currently no official indicator for the proportion of MSW handled in ESM facilities.

One requirement for an official SDG indicator is that it should use objective, measurable data which can be collected by a national statistics agency. This is particularly challenging for WaRM: measuring access to and quality of MSW collection services is inevitably more qualitative than quantitative. Similarly, assessing recovery and disposal facilities against a set of quality criteria to decide whether they are 'controlled' is not wholly objective, requiring the exercise of 'expert judgment.' Some would argue that it is just not possible to reduce the overall performance of a city's WaRM system to a set of objective quantitative indicators, much less combine those into a single 'index.'

Nevertheless, the work on developing SDG indicator 11.6.1 into a data hub for the collection of primary quantitative data is potentially a major first step towards compiling a reliable global baseline on SWM/WaRM systems at a city level. UN-Habitat has launched the Waste Wise Cities Campaign and plans to publish the 11.6.1 data on that website (UN-Habitat, n.d.) to showcase the diversity of both cities and their WaRM systems.

A useful second step would be to extend the basic 11.6.1 data by undertaking a mass flow analysis (MFA) to identify and quantify all of the material flows within a city's WaRM system, allowing a rapid assessment of collection, recycling, recovery, and disposal, as well as leakages to the environment which are otherwise not measured. GIZ's waste flow diagram (GIZ, University of Leeds, Eawag-Sandec, Wasteaware (2020), 2020) and the ISWA plastics pollution calculator ISWA and University of Leeds, 2020) are two such tools.

A third step would be to use the Wasteaware benchmark indicators (Wilson et al., 2015a), which combine quantitative measures of waste generation and of collection, recycling, recovery, and disposal with qualitative indicators to assess both the quality of the 'physical' aspects of the service and also the important 'governance' aspects which are key to a successful SWM/WaRM system. These were used for comparing cities in GWMO1 (UNEP and ISWA, 2015). A detailed user manual is available to ensure a level of consistency in how the user's expert judgment is exercised (Wilson et al., 2015b). Indeed, it has been suggested that the state of a city's SWM system, as measured by the city's cleanliness, could be used as a proxy indicator for good governance (Whiteman et al., 2001), which is extremely difficult to measure (SDG16, Figure 7.1).

Taken together, these three methodologies – the SDG indicator 11.6.1 data hub, the waste flow diagram or ISWA plastics calculator, and the Wasteaware benchmark indicators – are symbiotic; they capture what is needed in terms of data, diagnostics, and direction of travel of a city's WaRM system.

Conclusions

Making the case for SWM as a political priority has always been challenging. The three historical drivers of public health, local environmental protection, and resource value can sometimes be good enough. But the case can be strengthened significantly by extending the scope from SWM to WaRM and the circular economy and by using the SDGs as drivers to link into higher profile development issues (Figure 7.1).

The primary home for a city's municipal SWM within the SDGs is SDG11 (sustainable cities) and indicator 11.6.1. But SWM is also directly related to targets and indicators under SDG1 (end poverty, where target 1.4 requires access to basic services) and SDG6 (clean water and sanitation) and, in the context of WaRM and the circular economy, SDG12 (responsible consumption and production). The linkages to climate action (SDG13) and life below water (SDG14) are surprisingly strong: for example, extending waste collection to all and eliminating uncontrolled disposal would reduce the weight of plastics entering the oceans by more than half. Linkages to other high-priority SDGs are still direct but more difficult to measure (e.g., SDG8, decent work and economic growth through sustainable job creation, and SDG2, zero hunger through reducing food waste). Even the indirect linkages to equality and governance issues are also important, as these are difficult to tackle, and improving WaRM provides a useful entry point.

Ongoing work on SWM- and WaRM-related SDG indicators within the UN system has focused in particular on 11.6.1, which is seen as a data hub for compilation of a global baseline at a city level. Three sub-indicators measure the proportion of the total MSW generated in a city which is: (a) collected and (b) managed in controlled recovery and disposal facilities and (c) potential plastics discarded into the environment from uncollected and uncontrolled MSW. In addition, most of the SDG12 indicators are relevant to some aspect of the transition towards a circular economy (Table 7.2).

SWM and WaRM data are directly required for six SDGs, and they provide a significant entry point for tackling at least 12, and arguably all 17, high-level goals. The SDGs thus provide an important opportunity to build a strong political case and can be harnessed as powerful drivers for change in improving waste and resource management and moving towards a circular economy.

Acknowledgements

These ideas have evolved over many years. I would like to thank the many co-authors and collaborators with whom I have had the pleasure to work and who contributed enormously to the process, in particular, Andy Whiteman (Wasteaware/RWA), Dr Ljiljana Rodic (independent), Dr Costas Velis (University of Leeds), Zoe Lenkiewicz (WasteAid UK), and Chris Godlove (Think Cities).

References

CIWM and Wasteaid UK (2018). *From the land to the sea.* Available at: https://wasteaid.org/wp-content/uploads/2018/04/From-the-Land-to-the-Sea.pdf (Accessed 22/10/2019).

Dehoust G, Schüler D, Vogt R, Giegrich, J. (2010). *Climate protection potential in the waste management sector – examples: municipal waste and waste wood.* Available at: www.umweltbundesamt.de/publikationen/climate-protection-potential (Accessed 21/10/2019).

FAO (2015). *Food Wastage Footprint & Climate Change.* Author: Scialabba N. Available at: www.fao.org/save-food/resources/en/; http://www.fao.org/documents/card/en/c/7338e109-45e8-42da-92f3-ceb8d92002b0 (Accessed 21/09/2020).

IEAG (2014). *A world that counts: mobilising the data revolution for sustainable development.* Report of the UN Secretary General's Indendent Expert Advisory Group. Available at: www.undatarevolution.org/ (Accessed 22/10/2019).

ISWA (2017). *Prevent marine plastic litter – now! ISWA Marine Task Force report 2017.* Available at: https://marinelitter.iswa.org/reports (Accessed 22/10/2019).

ISWA and University of Leeds (2020). ISWA Plastic Pollution Calculator: A baseline and benchmarking quantification tool for waste-related city and regional plastic pollution. Version 1.0. September 2020. Developed for the ISWA Marine Litter Task Force. Authors: Velis C.A., Cottom J., Lerpiniere, D. Rotterdam: International Solid Waste Association. Available at: https://plasticpollution.leeds.ac.uk/toolkits/calculator/ (Accessed 21/09/2020)

Jambeck JR, Geyer R, Wilcox C, Siegler TR et al. (2015). Plastic waste inputs from land into the ocean. *Science.* 347 (6223): 768–770.

Lebreton LCM, van der Zwet J, Damsteeg J-W, Slat B et al. (2017). River plastic emissions to the world's oceans. *Nature Communications.* 8: 15611.

Lenkiewicz Z. (2016). *Waste and the sustainable development goals.* Available at: https://wasteaid.org/waste-sustainable-development-goals/ (Accessed 31/10/2019).

Reyna-Bensusan N, Wilson DC., Davy PM., Fuller G. et al. (2019). Experimental measurements of black carbon emission factors to estimate the global impact of uncontrolled burning of waste. *Atmospheric Environment.* 213: 629–639.

Reyna-Bensusan N, Wilson DC, Smith SR. (2018). Uncontrolled burning of solid waste by households in Mexico is a significant contributor to climate change in the country. *Environmental Research.* 163: 280–288.

Rodic L, Wilson DC. (2017). Resolving governance issues to achieve priority SDGs related to solid waste management. *Sustainability.* 9 (404).

Scheinberg A, Wilson DC, Rodic L. (2010). *Solid waste management in the world's cities.* Published by Earthscan for UN-Habitat. Available at: https://www.waste.nl/publications/ (Accessed 21/09/2020).

Schmidt C, Krauth T, Wagner S. (2017). Export of plastic debris by rivers into the sea. *Environmental Science and Technology.* 51: 12246–12253.

Schroeder P, Anggraeni K, Weber U. (2018). The relevance of circular economy practices to the sustainable development goals. *Journal of Industrial Ecology.* 23 (1): 77–95.

UN (2015). *Transforming our world: the 2030 agenda for sustainable development.* Sustainable Development Goals Knowledge Platform. Available at: https://sustainabledevelopment.un.org/post2015/transformingourworld (Accessed 21/10/2019).

UNEP and ISWA (2015). *Global waste management outlook* (referred to here as GWMO1). Editor-in-Chief: Wilson, DC. Osaka: UN-Environment (UNEP). Available at: www.unenvironment.org/resources/report/global-waste-management-outlook (Accessed 17/10/2019).

UNEP and ISWA, 2020. *Global waste management outlook 2nd Edition* (GWMO2). Editor-in-Chief: Mavropoulos, A. Osaka: UN-Environment (UNEP).

UN-Habitat (n.d.). *Waste wise cities campaign.* Available at: https://unhabitat.org/waste-wise-cities-campaign (Accessed 17/3/2020).

UNStats (n.d.). *Sustainable Development Goal indicators website.* Available at: https://unstats.un.org/sdgs# (Accessed 29/4/2020).

GIZ, University of Leeds, Eawag-Sandec, Wasteaware (2020). *Waste Flow Diagram (WFD): A rapid assessment tool for mapping waste flows and quantifying plastic leakage.* Version 1.0. February 2020. Authors: Velis CA, Cottom J, Zabaleta I, Zurbruegg C, Stretz J and Blume S. University of Leeds: Leeds, UK. Available at: https://plasticpollution.leeds.ac.uk/toolkits/wfd/ (Accessed 21/09/20).

Whiteman A, Smith P, Wilson DC. (2001). *Waste management: an indicator of urban governance.* Paper prepared for the UK Department for International Development (DFID) and submitted by DFID to the UN-Habitat Global Conference on Urban Development, New York, June 2001. Available at: http://davidcwilson.com/project/waste-management-an-indicator-of-urban-governance/ (Accessed 22/10/2019).

Wikipedia (n.d.). *Utilities*. Available at: https://en.wikipedia.org/wiki/Public_utility (Accessed 21/10/2019).
Wilson DC. (2007). Development drivers for waste management. *Waste Management & Research*. 25: 198–207.
Wilson DC. (2018). *Don't waste our climate*. Available at: ttp://davidcwilson.com/publication/dont-waste-our-climate/ (Accessed 21/10/2019).
Wilson DC, Rodic L, Cowing MJ, Velis CA et al. (2015a). 'Wasteaware' benchmark indicators for integrated sustainable waste management in cities. *Waste Management*. 35 (1): 329–343.
Wilson DC, Rodic L, Cowing MJ, Velis CA et al. (2015b). *'Wasteaware' benchmark indicators user manual*. Available at: https://rwm.global/utilities/documents/wabimanual.pdf (Accessed 21/09/2020).
Wilson DC, Velis C, Cheeseman C. (2006). Role of informal sector recycling in waste management in developing countries. *Habitat International*. 30: 797–808.
UN-Habitat (2020). Waste Wise Cities Tool: Step-by-Step Guide to Assess a City's MSWM Performance through SDG indicator 11.6.1 Monitoring. Prepared by UN Habitat, Wasteaware, EAWAG and University of Leeds. Authors: Takeuchi N, Zabaleta I, Whiteman A and Velis CA. Nairobi: UN-Habitat.

8

TRIGGERS FOR INDUSTRIAL SYMBIOSIS

Lessons learnt from twenty-five case studies

Karen Miller and Doroteya Vladimirova

Introduction

In circular economy literature, studies follow three main lines of action: macro level, looking into economic and social factors; micro level, enabling firms to implement circular principles and encouraging new consumer behaviours; and the third meso level discusses industrial symbiosis experiences and eco-industrial parks (Merli et al., 2018). Thus, we can consider that industrial symbiosis (IS) is firmly nested within the circular economy field.

Industrial symbiosis is defined as a system of interindustry exchanges that provide economic and environmental benefits (Bacudio et al., 2016). These exchanges may be in tangible forms, with waste or by-products such as energy, water, and materials from one industrial plant being utilised as an input material in other industry within close geographic proximity (Domenech et al., 2019). Symbiosis may, however, also be intangible in nature; for example, expert knowledge may transferred between industries (Chertow et al., 2008). In all cases, collaboration is fundamental to IS, according to Chertow (2007).

In order to understand IS more specifically, Yap and Devlin (2017) argued for and outlined a mixed-level (society, network, and enterprise) analytical framework for analysing IS emergence, development, and disruption. This argument emerged from two patterns observed in the post-1990s IS literature – that enablers and barriers exist at national, regional, network, enterprise, and personal levels and that the processes involved in IS emergence and development are fluid, with the boundaries of the different stages porous. The framework does not assume that the evolution of IS networks is one way, linear, or caused by a single combination of factors. The development of each IS network is path dependent and contingent because it is dependent on a broad range of conditions. Different networks may achieve IS as a result of quite different combinations of factors (Yap and Devlin, 2017).

A further element to understand is the triggers that provide the impetus to commence resource synergies, with the ultimate aim of enabling the systematic application of industrial symbiosis. Triggers may also be understood as the stimuli or drivers that stimulate IS (Boons et al., 2011).

It is the triggers of IS that we focus on in this chapter before turning to the enablers and barriers in Chapter 28.

Strategy

The overall aim of our research was to identify best practices and lessons learnt for scaling up industrial symbiosis. Analysis and synthesis of case studies that have already been conducted offer a valuable means of not only understanding the broader landscape of a phenomenon but also the underlying principles. It was therefore important to develop a robust approach to the selection of the case studies of IS. Twenty-five was identified as the number of case studies that allowed for in-depth knowledge to be gained regarding key criteria, developed in alignment with the objectives of the overall project. The factors considered in choosing this number of case studies were two-fold; first, 25 enabled a saturation point (beyond which no new substantive knowledge emerges and patterns in data are repeated) to be reached. Second, 25 was a feasible number to analyse while keeping within the time and resource constraints of the project. A number of criteria were used in the selection of which case studies to analyse.

The first of these criteria was the need for each case study to be published in a reputable, publicly accessible source. Therefore, these case studies were distinctly different from those that may be derived from anecdotal evidence. This criterion was important in order to give us strong evidential support for the analytical framework. The analytical framework shown in Figure 8.1 was used to conduct thematic analysis and identify the key concepts which appear to be relevant for catalysing initial and thereafter scaling up IS systems. The framework shows that there are triggers, enablers and barriers that act to influence the development, or, in some cases, demise of IS systems. In this chapter (part 1) we focus on the triggers, and in Chapter 28 (part 2) we concentrate on enablers and barriers.

The second criterion was for the case studies to cover a broad range of industries, for example, the leather tanning, sugar, and automotive industries, in addition to others.

A breadth of geographies was the third criterion, with examples drawn from Europe, including Italy, Spain, Sweden, Denmark, Germany, Portugal, and the United Kingdom, as well as Australia, Japan, South Korea, and China.

The fourth criterion was to ensure that there was a broad spectrum of IS maturity. At one end of the spectrum are emergent systems at an early experimental stage, potentially with just two stakeholders involved, and at the other, mature long-established systems. For example, the case study of

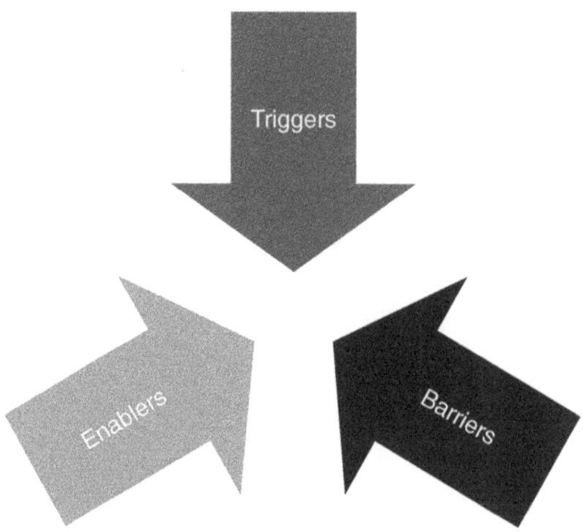

Figure 8.1 Analytical framework

the Kalundborg industrial cluster (Adamides and Mouzakitis, 2009) examines a very mature (in IS terms) system. This cluster started in 1961 and by the late 1980s was highly developed.

The fifth criterion was the need to include both` self-organised IS systems as well as those created by external stakeholders such as intermediaries, for example, the National Industrial Symbiosis Programme (NISP) in the United Kingdom.

In Table 8.1 is a summary of the case studies of industrial symbiosis that were reviewed. These case studies are for the remainder of this chapter and the subsequent one assigned a reference, for example, 'CS 2,' to assist the reader.

Table 8.1 Summary of 25 case studies of industrial symbiosis

Case study reference (CS)	*Author(s)*	*Focus of study*
CS 1	Notarnicola et al. (2016)	Taranto, Puglia, Italy – study centred on the company ILVA's steelworks and other facilities, including oil refineries, power plants, cement and construction companies, small agro industries, and a large beer factory. Also includes municipality of Taranto. Industrial district founded in 1960s
CS 2	Costa and Ferrão (2010)	Relvão Eco Industrial Park, Municipality of Chamusca, Portugal (founded 2006) – study of current/planned development, includes pulp and paper companies, agro industries, chemical companies (mainly fertiliser producers), and waste treatment facilities
CS 3	Adamides and Mouzakitis (2009)	Kalundborg industrial cluster, Denmark – established in 1960s – study of cluster that includes power stations, oil refineries, plasterboard producers, biotechnology companies, cement producers, soil remediation firms, and municipality of Kalundborg
CS 4	Velenturf (2016)	Study of Humber region, UK (initially started 1960s); main constituents include food processing, chemicals, furniture, and iron and steel and other metals, as well as oil and gas production facilities
CS 5	Vollmeier and Castorini (2015)	Spremberg, Germany – Hamburger Rieger GmbH paper mill study of exchanges between paper and pulp and energy producers – initiated in 2008
CS 6	Dall'Ara et al. (2015)	Italian National Agency for New Technologies, Energy and Environment (ENEA) joint Italian–Spanish study of poultry dejections as input material for leather tanning industry
CS 7	Organisation for Responsible Business (n.d.)	Fife, Scotland, UK – study of symbiosis between beverage and firewood producers
CS 8	Taddeo et al. (2012)	Long-established Italian cluster centred in Bussi sul Tirino, Abruzzo; study of cluster that includes basic chemicals, pesticides, silicates, power generation and distribution
CS 9	Mirata (2004)	Study of oil and gas synergies in Humberside, UK
CS 10	Mirata (2004)	West Midlands, UK, initiative involving automotive, metal production, plastic and rubber, and food processing companies

(*Continued*)

Table 8.1 (Continued)

Case study reference (CS)	*Author(s)*	*Focus of study*
CS 11	Mirata (2004)	Focus on chemical, oil, and gas sectors in addition to other process industries, that is, pulp and paper producers, located in Merseyside, UK
CS 12	Martin (2015)	Study of a well-established network composed of companies in forest and agro industry and electricity/gas production sector based in Norrköping, Sweden
CS 13	Organisation for Responsible Business (n.d.)	Focus on exchanges between food manufacturing and energy generation in Trowbridge, UK
CS 14	Zabaniotou et al. (2015)	Italian province of Foggia; symbiotic relationship study between agriculture and energy sectors
CS 15	Laybourn (2013)	Study examines several waste streams of major tyre producer in Ballymena (Northern Ireland) and synergies for materials such as carbon black and talcum powder
CS 16	Invest Northern Ireland (n.d.)	Focus on ThyssenKrupp Aerospace Ltd., aluminium alloy technology solutions provider and specialist manufacturer of tarpaulins in Newtown, Northern Ireland, UK
CS 17	Short et al. (2014)	Synergies study of AB Sugar (UK producer) and exchanges with landscaping, building construction, animal feed production, agro industries, cosmetics, biofuel, and beverage producers
CS 18	Celander et al. (2013)	Study of a long-established programme of energy production and waste treatment synergistic activities between Helsingborg City and Industry Park of Sweden
CS 19	Angren et al. (2012)	Focus on Lidköping, Sweden, with biogas plant and CHP plant development
CS 20	Yuan and Shi (2009)	Study of waste re-valourisation based upon Southwest China smeltery
CS 21	Yang and Feng (2008)	Study of Nanning Sugar Company Ltd. in Nanning, China
CS 22	Hashimoto et al. (2010)	Examination of Kawasaki Eco-town with 70 companies, including steel, chemical, cement, nonferrous metal processing, and paper making
CS 23	Dong et al. (2013)	Study of Liuzhou city, China, with closely related industries, including blast furnace slag, steel slag, and desulfurisation by-products
CS 24	Mathews and Tan (2011)	Ulsan; South Korean special industrial park (established in1962) – 700 companies, including energy generation, water supply and treatment, petrochemicals, chemicals, non-ferrous metals
CS 25	van Beers et al. (2007)	Study explores Kwinana and Gladstone Industrial Areas, Australia – synergies include alumina, nickel, iron, oil refinery, alumina and aluminium, pigment, cement, chemicals, fertiliser, power generation, water supply, and treatment

Triggers

To enable the triggers to be more clearly understood, they are presented at the national government level, regional/local government level, and then firm levels in line with Yap and Devlin (2017). However, it is important to acknowledge the interplay between these levels.

National level

At this macro level, there are strong global forces driving national governments' activities in relation to legislation, policymaking, and taxation. The most important are the Paris Accord, with the need to meet greenhouse gas (GHG) emissions targets, and the United Nations Sustainable Development Goals (SDGs). Both have activated trickle-down effects from national governments to regional/local governments and then on to firms, triggering the instigation or ramp-up of IS. These impacts are evident in CS 12, 16, 17, and 23. Other cases do not overtly refer to these factors, but they are implicit in most of the 25 analysed. In CS 14, which focuses on the Italian province of Foggia, for instance, more sustainable farming practices have important implications for GHG reductions.

Governmental activity is noticeable when they have sought to join global organisations where previously they have existed outside of these spheres of influence. For example, China's governmental policies to protect its sugar industry were weakened when it entered the World Trade Organization, due to stringent rules on subsidisation and the new competition with cheaper sugar from other countries. In effect, this caused the Chinese sugar industry to seek ways to organise itself more efficiently and thereby reduce costs and protect jobs (CS 21).

In order to meet the Paris targets, pressure to modify industrial activity is applied by national governments through the imposition of climate change levies (CS 10) and end-of-life regulations for products. In Europe, this is shifting the automotive industry and associated metals and plastics industries towards IS (CS 9). Many governments are also seeking to future-proof against shifts towards more stringent targets by tasking, for instance, the food production industry and regional and local authorities with organising waste collection schemes (CS 13). Policy levers have been applied successfully by Germany's government to drive firms with high outputs of waste, such as paper mills, to seek alternative routes for their waste products.

Tax incentivisation and subsidies, which work in the opposite way to punitive levies, have been successful in Sweden in encouraging the biofuel industry of Händelö to embrace IS in the Norrköping area (CS 12). Positive intervention is evident in Portugal (CS 2) and Japan (CS 22), as governments have actively intervened to slow economic erosion or to improve stagnating regions using subsidies.

When the Relvão (Portugal) project started (CS 2), a proposal for the setup of a large integrated centre for recovery, treatment, and elimination of hazardous waste was already in progress within the Chamusca region. This region was then selected by the national government for the implementation of an IS strategy to slow the decline in the area's economic and social fabric, which was compromising regional welfare. The Portuguese government, Chamusca municipality government, and the Technical University of Lisbon, together with industrial companies and entrepreneurs, collaborated to use highly targeted investment to provide social, economic, and environmental improvements through IS.

From these findings, it is evident that national governments, under pressure to meet global challenges and targets, and the need to stimulate growth or prevent economic decline, play significant roles in triggering and accelerating IS.

Regional/local, including city level

Many of the forces described previously have a trickle-down effect to drive regional/local governments and city authorities to instigate triggers that impact not only firms within their catchments but also their own activities. Most prominent are the pressures to meet the targets ratified in Paris and the SDGs. However, other factors also come into play at this level. Industrial pollution in some areas has caused significant negative impacts on the environment, in addition to serious implications for the local population. In Italy, there are examples (such as the Taranto district in CS 1) where local authorities have introduced controls on production within a particular region/location. In China, significant impacts on human health and the environment have led to direct interventions to stimulate IS systems, as seen in CS 21 with the Nanning Sugar Co. Ltd. Certain city authorities have particularly pressing needs to control pollution and reduce waste. In Lidköping in southern Sweden (CS 19), the city authority's ownership of the energy producer and waste treatment company provides evidence that direct control in an IS system is a valuable catalyst.

Following the 2008 global financial crisis, many areas and industries foundered, leading to a need for recovery strategies. CS 8 provides an example in which the local government supported chemical producers and power suppliers in the Abruzzo region of Italy. The Bussi chemical site, one of Italy's oldest industrial clusters, was identified as in dire need of economic and social revitalisation, which had increased unemployment. As a result, the local government stimulated interest in IS through its legislation entitled 'Ecologically Equipped Estates.' This legislation gave incentives to businesses that use coordinated systems in relation to the management of air, water, and soil pollution, in addition to service facilities, infrastructure, and technologies. State-owned companies in China have been similarly incentivised by local government to invest in and implement innovations to spur symbiotic development. In CS 20, a smeltery for zinc and tin production in Southwest China has developed innovations in technology, equipment, and production techniques that allow output materials to become feedstocks for a cement factory. These innovations enable a battery producer to join the system.

Regional bodies have offered a further stimulus with what may be described as spillover or knowledge transfer via one coordinating body or intermediary that operates in one region to another. An example is found in CS 10, with a Regional Development Authority's (RDA) support for the National Industrial Symbiosis Programme, which took learning from the Humberside IS programme (HISP) and began the setup of the IS programme within the West Midlands. The NISPs, which are located across multiple regions, were set up by the central UK government with a primary objective to catalyse IS – see CS 4, 7, 9, and 10. More recently, NISP has moved from being government financed to become a private enterprise (the impact of this shift is described later in the chapter). In a similar way to the NISP's inter- and intraregional linking role is that played by other UK-based intermediary organisations, for example, the Midlands Environmental Business Company (MEBC), which has an arm called Environmental Business Communications (EBC) that engages and builds firms' IS confidence and capacity through communications and workshops (CS 10).

Overall, the findings reveal the critical role played by this level not only in driving top-down governmental initiatives but also through lateral and downward action to encourage uptake of symbiosis. In this way, we can see how this level acts as a valuable IS conduit.

Firm level

At the firm level, there are numerous triggers, some as a direct consequence of those outlined previously at the national and regional/local levels. GHG reduction targets, for instance, featured as a fundamental driver in many of the case studies (CS 10, 12, 15, 16, 17, 20, and 21). In one case, CS 21, high environmental pollution from its plants compelled the Nanning Sugar Co. Ltd. in China to remodel its traditional, linear process and transition to a circular, symbiotic one (CS 21). Part of the remodelling process then allowed it to diversify successfully, which brought economic benefits.

Strongly economic triggers came to the fore in the form of cost-cutting and increasing competitiveness in several case studies. These include CS 1, where symbiotic relationships between players in Taranto, such as the ILVA steelworks, oil refiners, cement companies, and smaller agro firms, have become vital in order for the district to remain commercially viable. Similar financial drivers are evident in CS 6, as leather producers in Italy and Spain struggle in the face of global competition. With the help of the Italian National Agency for New Technologies, Energy and the Environment (ENEA), the producers collaborate with several agro industries, including poultry processing, that donate waste materials as inputs for use in the tanning process.

Closely related to the preceding trigger was a widespread incentive for firms to reduce the disposal costs of outputs. CS 7 offers an interesting example of a Scottish beverage manufacturer with textile packaging waste that was originally landfilled. Other firms have needed to prepare for/avoid specific landfill tax schemes – see CS 13, where a UK food producer has been spurred into action. Within CS 5, the symbiotic exchange between the Hamburger Rieger GmbH pulp and paper mill and the Spremberg Power Plant was triggered by the need not only to deal with the high disposal costs of pulp and paper mill wastes but also to meet Germany's regulation on waste. Germany, as in certain other EU countries, bans the direct disposal of waste into landfill.

Another prominent factor that emerged was the recognition by firms of the value in by-products previously considered waste. Furthermore, symbiotic exchanges of materials were identified as valuable to each player. In CS 10, companies located in the Mersey Banks industrial area within the chemical, oil, and gas sectors realised that symbiosis could be extended to broader synergies involving paper, automotive, aerospace, and metals companies.

Security from fluctuations in the supply and/or costs of energy and heat emerged as a stimulus in CS 12, which investigated the emergence of symbiotic relationships between companies on the island of Händelö. Constant energy is required to meet the needs of the Norrköping municipality; therefore, firms sought out a mutually conducive method of securing a continuous supply of electricity and heat at competitive prices, ensuring they were independent from market fluctuations and geopolitical instability. CS 4 in the Humber region of the United Kingdom provided similar insights on mutuality.

Resources also feature where firms need to find dependable suppliers of key raw materials or cheaper or additional quantities of input materials. For instance, in CS 7, one of the players had to establish a cheaper source of sustainable materials used as packaging for products, which led to a partnership with a beverage manufacturer. The exchange gave the beverage manufacturer an avenue for its waste hessian sacks that previously went to landfill. Industrial symbiosis in other case studies has been precipitated as firms seek to deliver a growth strategy where resources are scarce or limited. Water is a common constraint, and this has driven leather producers in specific locations to establish relationships with other industrial partners, as discussed in CS 6. Factors regarding water scarcity have acted as powerful motivational forces in other locations such as in the Kalundborg industrial cluster in Denmark (CS 3).

Diversification as a strategy to extract greater value from materials and raise barriers to entrants in core markets is a further motivational factor, as in CS 17. This case study explored the strategy of British Sugar (now part of AB Sugar), which is highly vertically integrated. The strategy focused on developing alternative business models using sugar refining by-products. The case study offered insights into the triggers for internal symbiosis, since only limited external synergies have been needed. The company had to navigate challenging times, particularly with the opening of the UK market to global competition and subsequent rivalry with low-cost sugar production in developing countries. Environmental protection only became a driver for symbiosis later.

Marketing benefits in the form of raised corporate social responsibility profiles have provided another incentive, evident in CS 12 and 16. By implementing a resource synergy, Thyssen-Krupp (CS 16) improved its environmental credentials in line with its company policy. This aluminium alloy technology solutions provider introduced a more sustainable (and cheaper) solution to dispose of tarpaulins, which were usually sent to landfill. Some firms have pursued an IS pathway with the strategic aim of sending zero waste to landfill; see, for example, CS 15 of Michelin in Northern Ireland.

Certain cases revealed long-standing IS practice for players in certain industries and particular locations. Of note was where there are clusters of firms operating in industries such as chemicals, oil and gas, and process industries, that is, pulp and paper. These clusters have evidently acted as best-practice examples that firms in other geographies have sought to emulate. Kalundborg has frequently been held up as a beacon, helping in the germination of the Relvão Eco Industrial Park project (CS 2). Spread of leading practice is cited in another case (CS 9), where a global player operating in oil and gas moved into another geographic location to operate alongside chemical processors. In so doing, IS knowledge was transferred from Tampico in Mexico to Humberside in the United Kingdom.

Conclusions

This chapter presented a set of triggers for industrial symbiosis based on a review of 25 case studies. The triggers are the first component (to be followed by the enablers and barriers in the next chapter) to support the move from anecdotal one-off resource synergies to a systematic application of resource synergies as a means to increase resource productivity and competitiveness.

This research was part of the wider EU H2020 SCALER research project (Scaling European Resources with Industrial Symbiosis), which sought to scale up industrial symbiosis throughout Europe.

References

Adamides ED, Mouzakitis Y. (2009). Industrial ecosystems as technological niches. *Journal of Cleaner Production*. 17 (2): 172–180. doi:10.1016/j.jclepro.2008.04.003.

Angren J et al. (2012). *Exploring industrial symbiosis in Lidköping, Sweden*. Available at: www.industriellekologi.se/documents/Lidkoping.pdf.

Bacudio LR, Benjamin MFD, Eusebio RCP, Holaysan SAK, Promentilla MAB, Yu KDS, Aviso KB. (2016). Analyzing barriers to implementing industrial symbiosis networks using DEMATEL. *Sustainable Production and Consumption*. 7: 57–65.

Boons F, Spekkink W, Mouzakitis Y. (2011). The dynamics of industrial symbiosis: a proposal for a conceptual framework based upon a comprehensive literature review. *Journal of Cleaner Production*. 19 (9–10): 905–911. doi.org/10.1016/j.jclepro.2011.01.003.

Celander F et al. (2013). *Industrial symbiosis in Helsingborg*. Available at: www.industriellekologi.se/documents/Helsingborg.pdf.

Chertow MR. (2007). Uncovering industrial symbiosis. *Journal of Industrial Symbiosis*. 11 (1): 11–30.

Chertow MR, Ashton WS, Espinosa JC. (2008). Industrial symbiosis in Puerto Rico: environmentally related agglomeration economies. *Regional Studies*. 42 (10): 1299–1312. doi:10.1080/00343400701874123.

Costa I, Ferrão P. (2010). A case study of industrial symbiosis development using a middle-out approach. *Journal of Cleaner Production*. 18: 984–992. doi:10.1016/j.jclepro.2010.03.007.

Dall'Ara A, Billi L, La Peruta MT, Cepparrone A, Roig M. (2015). *PODEBA: an industrial symbiosis case*. Available at: www.enea.it/it/seguici/pubblicazioni/pdfvolumi/ExperiencesofIndustrialSymbiosisinItaly_Proceedings.pdf.

Domenech T, Bleischwitz R, Doranova A, Panayotopoulos D, Roman L. (2019). Mapping industrial symbiosis development in Europe: typologies of networks, characteristics, performance and contribution to the circular economy. *Resources, Conservation and Recycling*. 141: 76–98. doi.org/10.1016/j.resconrec.2018.09.016.

Dong L, Fujita T, Zhang H, Dai M, Fujii M, Ohnishi S, Geng Y, Liu Z. (2013). Promoting low-carbon city through industrial symbiosis: a case in China by applying HPIMO model. *Energy Policy*. 61: 864–873. doi.org/10.1016/j.enpol.2013.06.084.

Hashimoto S, Fujita T, Geng Y, Nagasawa E. (2010). Realizing CO_2 emission reduction through industrial symbiosis: a cement production case study for Kawasaki. *Resources, Conservation and Recycling*. (54): 704–710. doi:10.1016/j.envpol.2010.10.022.

Invest Northern Ireland (n.d.). *Industrial symbiosis: improving productivity through efficient resource management*. Available at: https://secure.investni.com/static/library/invest-ni/documents/industrial-symbiosis-guide-for-businesses-in-northern-ireland.pdf.

Laybourn P. (2013). *Opportunities through industrial symbiosis: UK NISP and global experience*. Available at: www.endustriyelsimbiyoz.org/wp-content/uploads/2013/02/industrial-symbiosis_uk-nisp-and-global-experience_31.01.2013.pdf.

Martin M. (2015). Quantifying the environmental performance of an industrial symbiosis network of biofuel producers. *Journal of Cleaner Production*. 102: 202–212. doi:10.1016/j.jclepro.2015.04.063.

Mathews JA, Tan H. (2011). Progress toward a circular economy in China: the drivers (and inhibitors) of eco-industrial initiative. *Journal of Industrial Ecology*. 15 (3): 435–457. doi:10.1111/j.1530-9290.2011.00332.

Merli R, Preziosi M, Acampora A. (2018). How do scholars approach the circular economy? A systematic literature review. *Journal of Cleaner Production*. 178: 703–722. doi:10.1016/j.jclepro.2017.12.112.

Mirata M. (2004). Experiences from early stages of a national industrial symbiosis programme in the UK: determinants and coordination challenges. *Journal of Cleaner Production*. 12: 967–983. doi:10.1016/j.jclepro.2004.02.031.

Notarnicola B, Tassielli G, Renzulli PA. (2016). Industrial symbiosis in the Taranto industrial district: current level, constraints and potential new synergies. *Journal of Cleaner Production*. 122: 133–143. doi:10.1016/j.jclepro.2016.02.056.

Organisation for Responsible Business (n.d.). *NISP*. Available at: www.orbuk.org.uk/article/the-national-industrial-symbiosis-programme-nisp.

Short SW, Bocken NMP, Barlow CY, Chertow MR. (2014). From refining sugar to growing tomatoes: industrial ecology and business model evolution. *Journal of Industrial Ecology*. 18 (5): 603–618. doi:10.1111/jiec.12171.

Taddeo R, Simboli A, Morgante A. (2012). Implementing eco-industrial parks in existing clusters: findings from a historical Italian chemical site. *Journal of Cleaner Production*. 33: 22–29. doi:10.1016/j.jclepro.2012.05.011.

van Beers D, Corder G, Bossilkov A, van Berkel R. (2007). Industrial symbiosis in the Australian minerals industry: the cases of Kwinana and Gladstone. *Journal of Industrial Ecology*. 11 (1): 55–72.

Velenturf APM. (2016). Promoting industrial symbiosis: empirical observations of low-carbon innovations in the Humber region. *Journal of Cleaner Production*. 128: 116–130. doi:10.1016/j.jclepro.2015.06.027.

Vollmeier T, Castorini, A. (2015). *Case study Spremberg: RDF-fueled CHP plant for a paper mill*. Available at: www.enea.it/it/seguici/pubblicazioni/pdf-volumi/ExperiencesofIndustrialSymbiosisinItaly_Proceedings.pdf.

Yang S, Feng N. (2008). A case study of industrial symbiosis: Nanning Sugar Co. Ltd. in China. *Resources, Conservation and Recycling*. 52: 813–820. doi:10.1016/j.resconrec.2007.11.008.

Yap NT, Devlin JF. (2017). Explaining industrial symbiosis emergence, development, and disruption: a multilevel analytical framework. *Journal of Industrial Ecology*. 21 (1): 6–15. doi:10.1111/jiec.12398.

Yuan Z, Shi L. (2009). Improving enterprise competitive advantage with industrial symbiosis: case study of a smeltery in China. *Journal of Cleaner Production*. 17: 1295–1302. doi:10.1016/j.jclepro.2009.03.016.

Zabaniotou A, Rovas D, Libutti A, Monteleone M. (2015). Boosting circular economy and closing the loop in agriculture: case study of a small-scale pyrolysis – biochar based system integrated in an olive farm in symbiosis with an olive mill. *Environmental Development*. 14: 22–36. doi.org/10.1016/j.envdev.2014.12.002.

9

BOUGHT TODAY, GONE TOMORROW? FROM LINEAR TO CIRCULAR CONSUMPTION[1]

Melanie Jaeger-Erben

Introduction

In 2010, 3.5 million tonnes of waste were produced all over the world every day; by 2025, this amount is expected to almost double and to triple by 2100 (Hoornweg et al., 2013). In the EU alone, 13.5 kg of waste was generated per capita per day in 2015 (Eurostat, 2014). An increasing proportion of this is electrical devices that create social-ecological damage not only as waste (Wang et al., 2013) but also by the mining of resources, high emissions, and low occupational safety during production and transport (Hütz-Adams, 2012). The negative socioecological effects are exacerbated by the fact that many consumer items, in particular electrical appliances, are often used much less than the technical lifespan would allow and are often replaced, although they still work or can be easily repaired (Prakash et al., 2016; Jaeger-Erben and Hipp, 2018). At the same time, extending the useful life is an important lever for more sustainability in the consumption of everyday objects.

"We live in a throwaway society." This – often fatalistically connoted – statement is constantly expressed in interviews, surveys, or media discourses on the subject of modern consumer culture. Correspondingly, in a survey from 2017, 94% of the participants agreed with the statement (Jaeger-Erben and Hipp, 2018).

If one were to believe this perception, the question of the social drivers of product obsolescence, that is, the too-short useful life of consumer goods, would be answered very quickly: Disposable production and disposable consumption go hand in hand and are symptomatic of a modern, negligent use of finite resources. This perception has a counterpart in the material reality that presents itself in the form of "gigantic mountains of electronic trash" and an "incalculable trail of hazardous waste" (Oetzel, 2012, translation by author). Thus, is the diagnosis correct? The term "throw-away society" was framed in the second half of the 20th century (see, for example, Packard, 1960) and was constantly referred to in the context of growth- and consumption-critical debates that followed. There is talk of a cultural "throw-away mentality," and it is observed that the amount of consumer goods and particularly electronic devices in everyday life constantly increases (Oetzel, 2012).

However, terms such as "throw-away society" only scratch the surface of the modern consumer culture. In fact, the amount of waste per capita per year is increasing worldwide, but national trends vary greatly from one country to another (Trentmann, 2016). Nevertheless, a linear economy is based upon or inseparably entangled with different forms of linear consumption. Both are based on a more or less explicit but common definition or meaning of value: Value is created by taking natural resources and transforming them into products. As soon as a product enters the market, its value is usually the highest, be it monetary or symbolic, since new products are often favoured before used products. After a product is sold, its value is exploited until it is finally disposed of. Since refurbishment, reuse, or recycling schemes are far from being mainstream, the product is stored away, particularly if it is a small item, or it is landfilled or thermally processed.

In contrast, circular economy concepts not only consist of new forms of designing and processing or alternative business models, they also promote an alternative way of thinking and valorisation. Not only is the value of the product supposed to be preserved, but the value of ecosystemic reproduction and recovery is acknowledged. While the technological and organisational transformation of current systems of consumption and production into a circular economy is already a tremendous challenge, a change of mind-sets, social meaning, and societal narratives around value – or better: a profound cultural change – seems as unavoidable as it is demanding. It is argued here that a main prerequisite is a thorough understanding of how obsolescence is produced in our current material culture.

This chapter presents some conceptual reflections as well as empirical results from interdisciplinary research on the causes and drivers of obsolescence of consumer goods, particularly electronic devices. A basic conceptual differentiation is that obsolescence should be approached with a focus on:

- Material manifestations: The too-short life of products manifests on the one hand materially, for example, in the disintegration of the material and the loss of originally planned functionality, but also in the acceleration of material flows or the height of electronic waste mountains
- Manifestation in communication and discourses: On the other hand, the lifetime and useful life of products is the subject of social discourses and narratives dealing with social and technological change and the consequences of modern forms of production and consumption. Obsolescence is also a matter of negotiation and allocation of meanings, values, and norms.

Both manifestations of obsolescence are closely related, but sometimes the discourse is disconnected from material reality and tends to become independent, which can be challenging for overcoming obsolescence or promoting sustainable production and consumption, as we will see in the following. In this chapter, we will describe discursive and material practices in (re)producing obsolescence based on conceptual and empirical insights. Finally, the problem of one-sided or linear explanatory models and causal attributions is pointed out, and a systemic perspective on obsolescence is sketched.

Obsolescence as a social construct

The confrontation with the value of consumer objects and a certain discomfort in view of growing mountains of waste is not a modern phenomenon. While in the mid-20th century, it was rather a fear of decadence and waste, nowadays it is more of a concern over increasing pollution (Trentmann, 2016: 836) that causes discomfort.

The history of obsolescence is a history of the relationship between society and its consumer goods that is reproduced in social practices of consumption and production. Throughout history, the value of a product and the corresponding diagnosis "useful" or "obsolete" has changed almost independently from how the design or material properties of a product changed. For example, even though durable consumer goods such as an electronic mixer got more complex in their function and more resource intensive in their production, their value did not increase in general. On the contrary: Globalised mass production and the pervasive availability of electronic consumer goods have led to a massive decline in the monetary and practical value of the single device.

Becoming obsolete is not a natural process but a process of actively devaluating and invalidating an object until it is no longer useable or useful. What is considered useless and no longer worth preserving – for example, through repair, upcycling, or functional or aesthetic reuse – is therefore also the result of a social construct and not a predefined property of a consumer object that is determined by the product developer or manufacturer. Thus, in this social constructivist reading, there is no clear distinction between absolute or relative obsolescence – as Cooper (2004) suggests – because it is always relative in the sense of a comparison between obsolete/useless/dysfunctional and non-obsolete/used/desired/functional.

A more social constructivist perspective is also reflected in science and technology studies (STS) that define a consumer good as something that is actively produced or transformed to an object of consumption and commodity in the course of its useful life. "Consumer goods are neither finished nor invisible forms at the points of production and acquisition, but . . . continuously evolving, positioned within and affected by an ongoing flow of consumer practice" (Gregson et al., 2009: 250). Although certain usage modes and the range of object performance are to a certain extent inscribed into an object Akrich, 1992), it is the practice of usage that constitutes what an object becomes. Thus, the user or consumer might overwrite what was inscribed by design ("description," ibid.) as the object is used, and here it is decided whether the intended uses and services are actually realised. Product designers and users are thus connected. Their knowledge, their expectations, and their practices indirectly interact with each other via the object.

An analysis of the social drivers of obsolescence thus has to decipher the dynamic "triangular relationship" (Figure 9.1) of design, use, and object in a context-sensitive approach and needs to relate to both the discursive and the material manifestation of obsolescence. However, as described in the following public discourse, and sometimes even scholarly debate tends to hold more of a linear and one-dimensional understanding of obsolescence.

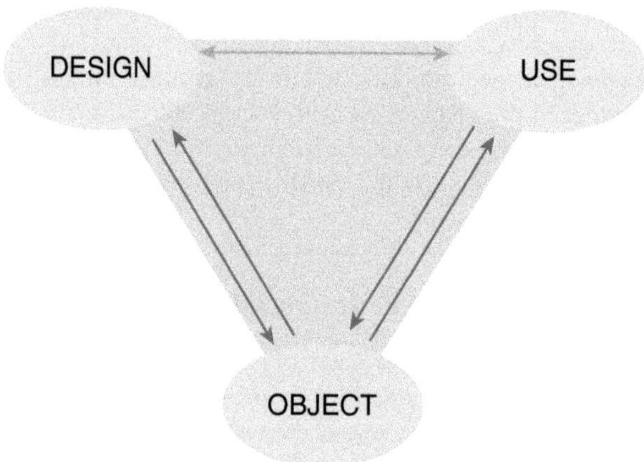

Figure 9.1 To understand social drivers of obsolescence, the "triangular relationship" of design, use, and consumer object must be taken into account

The communication of obsolescence – insights into media discourses

In public discourse, obsolescence is mostly perceived as "planned" or "enforced obsolescence," that is, a premature passing of a consumer good. The term itself was first mentioned in the early 20th century, particularly during the economic crises. In addition, it was frequently "reactivated" in public discourses as a societal problem in the decades that followed, particularly in times of crisis (Weber, 2014). However, a significant intensification of the debate and the emergence of civil society initiatives and campaigns[2] against obsolescence are particularly noticeable since 2011. This intensification coincides with the release of the documentary *The Lightbulb Conspiracy* by Cosima Dannoritzer[3] in 2011, which shows some historical and current cases of an active reduction of the lifetime of consumer goods, or at least the active acceleration of a new purchase. The documentary film points in particular to the agency of the manufacturers and the trade as the "planners" of obsolescence, a perspective that has prevailed in media discourses.

A content analysis of about 200 contributions from online archives of regional and national newspapers on themes such as a "useful life/lifetimes of consumer goods" or "obsolescence," since 2000,[4] shows that the obsolescence of consumer objects is mainly seen as a deliberate consequence of the "malicious intentions" of manufacturers. Many media contributions start with a presentation of planned obsolescence as a common everyday experience of consumers by introducing the issue with a sentence like "Like all of us have already experienced, the electric toothbrush/printer/mixer stops working as soon as the warranty has expired."

This perspective is underpinned in media presentations by repeatedly referring to the same examples. This includes cases like the so-called light bulb cartel, which was formed in the 1930s by the world's largest light bulb manufacturers and which ended in agreements to standardise – and at the same time reduce – the burning time of light bulbs.[5] Another example, which also is a main case in the previously mentioned documentary, is the case of inkjet printers with smart chips in their ink cartridges that prevent them from being used after a certain threshold (number of pages, time, etc.), even though the cartridge may still contain usable ink or could be refilled. Further examples are electric toothbrushes with glued-in rechargeable batteries or hand-held mixers with fast-wearing plastic wheels. Among these examples are cases like the light bulb cartel, which are vastly researched and relatively unequivocal, but in many cases, the burden of proof is more on anecdotes than on scientific sources and long-term comprehensive studies. The anecdotal evidence provides no proof but a presumed relevance to the lifestyle of readers. Furthermore, through the clear victim vs. perpetrator categories, the reader can feel like a potential victim of fraud. The "perpetrator–victim narrative" is also relatively resistant to the results of various investigations that have so far provided no clear or comprehensive evidence for a system of maliciously planned obsolescence[6] and even report rather increasing use and life cycles (Oguchi, 2017; Trentmann, 2017). In the course of the publication of results such as the studies of the German Environmental Agency in 2015 and 2016 (see Prakash et al., 2016), several contributions that call into question the generalised charges against manufacturers emerged. However, recent developments in the context of various lawsuits against manufacturers, in particular those based on laws on an "obsolescence programme" in France, have again led to a significant increase in contributions that view the guile of (many) manufacturers as certainty.

It is not surprising that media reports often use simplified representations and categorisations to create a gripping narrative. However, these narratives can be effective and become problematic when they lead to a distorted perception that tends to exacerbate the denounced problem, as discussed subsequently.

Perceptions and experiences of obsolescence – results of consumer research

The media representation of the causes of short-lived products can affect the perception of the consumers' own role in the productions of obsolescence and their perceived options for action. Different surveys of consumer perceptions show that manufacturers are mainly held responsible for the longevity of products (European Economic and Social Committee, 2016). Consumers even seem to expect that their products will have shorter lifespans due to deliberate misconstructions and cite this as justification for their short usage time (Wieser et al., 2015).

To explore consumer perspectives and practices in the context of product lifetimes and obsolescence, we conducted an online survey in 2017. Participants were recruited by a panel institution and received coupons for their participation. The sample was determined by screen-out conditions so that representativeness for the German population for age, education, income, and gender was given. The sample consisted of 2,000 participants between 14 and 66 years of age. The questionnaire contained 44 questions on consumption-related meanings and practices, with a focus on washing machines and smartphones, practical knowledge concerning electronic products, the perception of responsibility for products, and several attitudes around production and consumption.

Concerning the perception of responsibility for products lifetimes, our survey shows that 90% of respondents believe that some manufacturers purposefully build their devices in a way that they break shortly after the end of the warranty period of two years. Far more than half believe that it does not matter how much they spend on electrical appliances, since everything breaks down early anyway. However, when asked if they have ever experienced an owned device breaking prematurely, more than 60% of the survey respondents denied it. That is, the belief in planned obsolescence or the deliberate reduction in the lifetime by manufacturer negatively affects expectations towards products, although respondents' own experiences provide no proof for that (Figure 9.2).

The survey also asked about the respondents' experiences with broken devices and the presumed causes for their breakdown. When asked why the last device had reached the end of its life cycle from the point of view of respondents, the main cause was that they failed due to expected signs of wear and tear (except for smartphones and notebooks) (Figure 9.3).

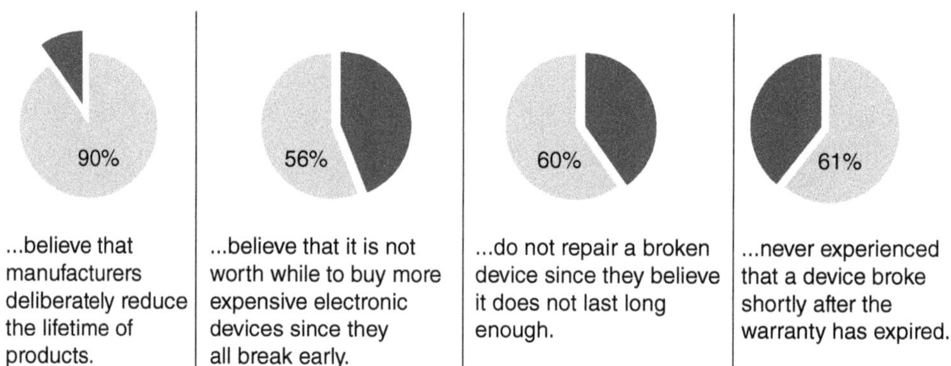

Figure 9.2 Communicative and material obsolescence in the perception and experience of consumers

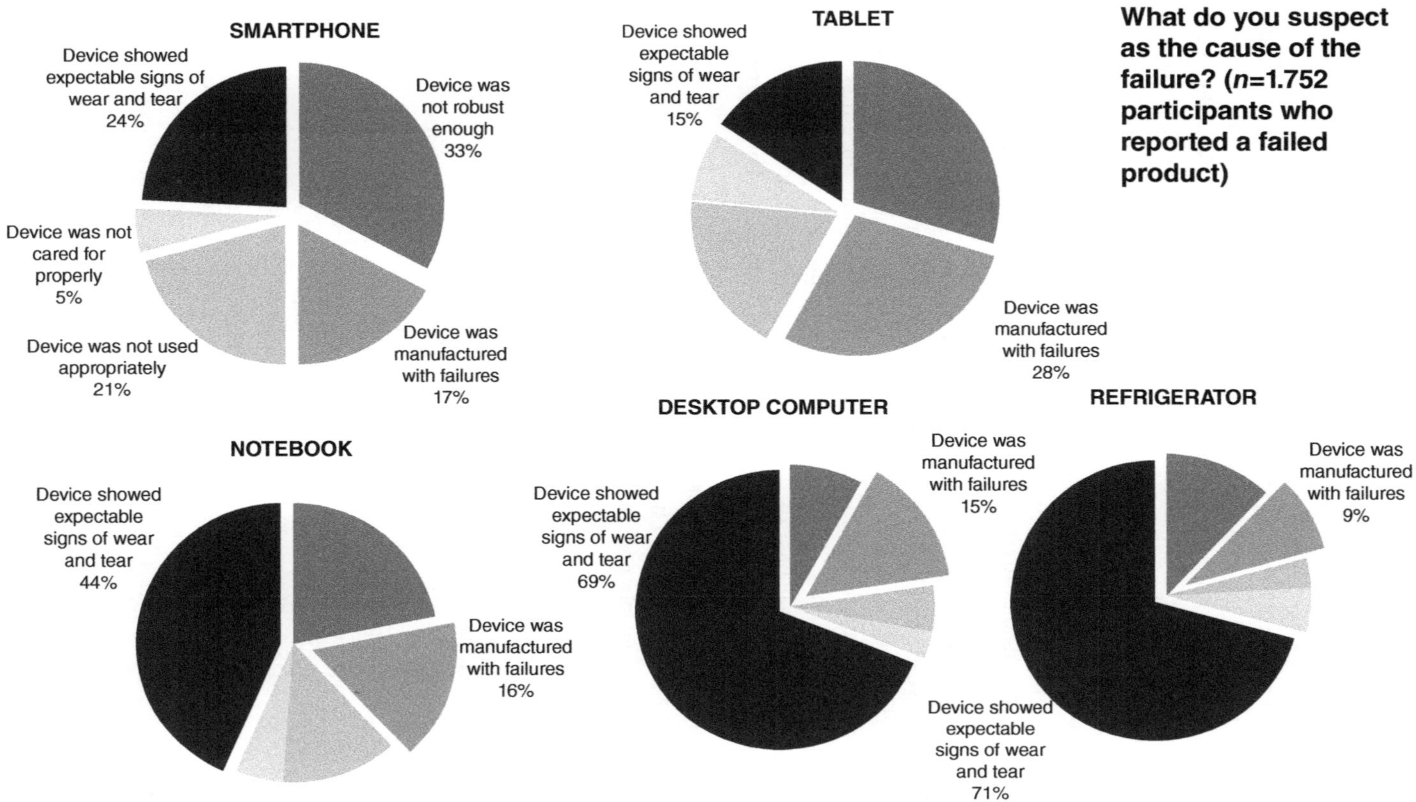

Figure 9.3 Suspected reasons for the last failure of a device

Thus, the representation in the media does not seem to be an all too common experience. – from the experiences of the survey participants, most devices reach the end of their life in a rather expected way because of anticipated wear and tear. Smartphones and notebooks are an exception, since in most cases, respondents suspect a lack of robustness. Both devices are among the most frequently used. This might be the reason the proportion of devices that have broken down due to improper use is comparatively high.

In line with the interactive relationship between design and usage described previously, it is also important to investigate how usage practices affect the functionality of products. The frequency of care practices or maintenance of washing machines, like decalcifying the machine or cleaning the drawer, can affect its durability. Moreover, the routines of mobile phone charging affect the long-term performance and functionality of the battery. These actions are embedded in everyday practices, which are automated and based on an implicit, practical knowledge. The decisive factor is whether the relevant knowledge for functionality and longevity exists. However, the know-how often seems to be limited. Among the survey participants, only about 40% knew how a lithium-ion battery should normally be recharged so that its performance is maintained for as long as possible (Jaeger-Erben and Hipp, 2018).

Another important question is whether enough time and space for care and maintenance are available in everyday life. Our survey points out that the proportion of people who do not care for or maintain their devices properly for lack of time is about one-third. Thirteen percent do not fix a broken device and immediately buy a new one, since they cannot do without the device. Successful conduct of everyday life seems to be based on the fact that devices are functional and available. At the same time, everyday life does not leave enough room to manoeuvre to ensure long-term functionality of the devices. This observation adds an alternative perspective to the common assumption that an early new purchase is primarily related to the "allure of the new" and a high sense of fashion or status (Ullrich, 2014; Klose, 2015; Harrell and McConocha, 1992). Nevertheless, various studies show that there are a considerable number of people preferring new products to caring for the old ones because they want to "go with fashion" (e.g., Cox et al., 2013). In our survey, we also found that about one-fifth of the participants see a broken device as a welcomed opportunity to buy something new. They relate quality of life to new devices and are happy to show that in their social environment.

To explore further the role of newness, we analysed the relation between the attractiveness of newness and the usage time of the last device The self-reported use time (material practice) of the previous electronic device was assessed by asking how many months the previous smartphone had been used and how many years the previous washing machine had been used. The attractiveness of newness was extracted from an exploratory factor analysis and based on seven items (α = .85), for example, "Having new devices means quality of life to me." Other factors were: responsibility for longevity attributed to the producer (10 items, α =.85, e.g., 'Producers of electronic devices should pay more attention to longevity when designing products') and to the user (6 items, α =.74, e.g., 'Consumers should be more careful with their products so they last longer'), satisfaction with the status quo of product longevity in society (2 items; e.g., 'I am satisfied with my rights as a consumer'), indifference about product longevity (2 items; e.g., 'I don't have time to maintain devices'), and the purchase criterion of longevity (2; e.g., 'I buy devices that are more expensive but last longer"). These attitudinal factors were entered in multiple regressions with the usage time of the last washing machine or smartphone. The analyses revealed that the length of the usage time of both devices was negatively predicted by the attractiveness of being new, whereas the other factors were not important, except for the user's perceived responsibility, which positively predicted the usage time of the smartphone.

That means that the attractiveness of a device being new is the best predictor of how long it is used (compare also Frick et al., 2019).

However, fashion, status, and novelty value are only drivers of obsolescence depending on the cultural context, and this context explains why there are differences between product categories. For example, it is often more important to own and demonstrate the latest fashion in the case of mobile phones but not in the case of large household appliances (Wieser et al., 2015; Evans and Cooper, 2010; Cox et al., 2013; Cooper and Mayers, 2000). In our survey, the proportion of respondents who bought their current device even though the old device was still working is also significantly higher in the case of smartphones (around 60%) than in washing machines (44%). Moreover, in this regard, it is worth looking at how the device is embedded in everyday practice. The smartphone as a "digital life companion" (Eisentraut, 2016) has an important role in the design of everyday life and social relationships and is constantly visible to the social environment. Devices that stand in the home and all have mainly one function, such as "washing" or "cooling," play an entirely different role in everyday life. The novelty thus serves only partly as an explanation; therefore, a more thorough analysis of social practice in dealing with different devices and the different forms of everyday integration is necessary. This consideration, in turn, needs to be embedded in the structural incentives and opportunities that favour a new purchase over function preservation and long-term use. Buying a new device seems much easier to many consumers than repair and is stimulated by appropriate offers. Not least the digitisation and the simplicity of online shopping contribute to the impression that everything is easy and simply available and that resources or material are worth nothing and therefore not worth preserving (Hilty and Bieser, 2017). In addition, care, maintenance, and repair are structurally not only neglected but sometimes even prevented. For example, the design – bonded parts, miniaturisation, and modularisation – can make repairs and care practices much more difficult. When selling, information about the technical function (capabilities) and maintenance requirements are not in the foreground, and consumers often feel a kind of learned helplessness in dealing with technically predictable devices (Echegaray, 2016).

The function and visibility of the device in everyday life and time and expertise for care, maintenance, and repair, as well as the product-related and structural incentives and barriers to the preservation of the old and the acquisition of the new therefore affect the freedom of action of consumers. The high dependence on functioning devices and the busy timing of everyday life complicate life-prolonging actions and accelerate functional obsolescence.

In addition, consumers' influence on product lifetimes was perceived as limited by the survey respondents, which corresponds to a low feeling of responsibility for functionality and durability. If the common media presentation is believed, the owners' possibilities – and thus the willingness to accept responsibility – are regarded as very limited, because "the market" controls the lifetime. The expected (short) lifetime can have a significant influence on the length of the useful life and the date of replacement (Cox et al., 2013; Wieser et al., 2015). In our survey, one-third of reported broken devices were not repaired because the assumption was that a repair would not be worthwhile because the device would not last long enough.

However, not all responsibility attributions point towards the manufacturer. As indicated previously, the vast majority (94%) agree with the statement "we live in a throw-away society." About 80% find it "awful that fellow citizens buy new electrical appliances, even though their old ones still work," and just as many believe that advertising significantly influences needs and encourages new purchases. Thus, the survey reflects in some ways the cultural pessimism that also characterises the media presentations of today's consumer society. Thus, a crucial question is: Who can bear responsibility for the product if the actors themselves do not have any room for decision and action?

Factors influencing life span from a holistic perspective

The preceding sections presented some empirical insight into the communicative and material reproduction of obsolescence. The insights into the discourse on obsolescence bear witness to a communication that works primarily with simplistic categories and linear attributions of cause and effect, often with the aim to tell an interesting story. This story draws some one-dimensional pictures of both the manufacturer or product designer and the consumer. However, it also reveals a very limited view of the products.

First, practical action by manufacturers and consumers is limited to decisions (for a particular design/business model for new purchase/discarding), and the decision itself is mainly presented as based on cost-benefit considerations. Manufacturers and consumers alike are said to make more or less calculated decisions for their own benefit maximisation – be it benefits through financial gain and competitive advantage or by status and novelty consumption. The practice itself and the relevant, often limited, know-how or the structural, often unquestioned, logic, which offer incentives for the one and hinder other actions, remain underexposed.

On the other hand, the products themselves attract little attention and appear instead as "passive puppets." They only become conspicuous when they can be presented as novelty, when they no longer function, or when they become threatening as "electronic waste mountains." Otherwise, being defined as passive, the objects can be used by companies to project their profit dreams onto them or by consumers to satisfy the desire for the new. To put it bluntly, producers and consumers are in some ways accomplices in the production of obsolescence, and the consumer products are the actual victims – and with them the natural resources used for that purpose, as well as the people and environments that are affected by manufacturing and disposal practices.

This escalation may initially be irritating. What should a consumer product be other than a human-made and manipulated object? Here we return to the triangular relationship of design, usage, and object outlined previously and add an extended understanding of consumer objects as "equal playing partners" (Miller, 2006) or "teammates in social practice" (Hörning, 2015). As co-players, consumer objects can be resistant and inscrutable. They can refuse their "service" and interrupt the flow of practice. This is even more true for modern electronic products and digital mobile technologies, which are becoming more and more important actors and role carriers in everyday life through an ever-expanding "life of their own" (Rammert, 2006; Rammert and Schulz-Schaeffer, 2002). Furthermore, modern production processes are complex chains and networks of globally widely distributed practices. It becomes an almost impossible challenge for product developers and manufacturers to keep an eye on all the components and functions – and thus possible malfunctions – of the devices. Samsung's effort to understand why Galaxy Note 7 batteries exploded shows that even market leaders have limited access and limited knowledge of the features of their device components. The emergence of a product and its quality involves a large number of actors who act according to their own logics. While designers primarily focus on the product itself, controlling and trading are above all about an increase in margins, which is the practice-relevant logic. In addition, not only does everyday life seem to accelerate, product developers also have less and less time to plan the quality and reliability of their products.

However, this independent role of consumer products usually only becomes the focus of attention when they do not work (van Hinte, 1997). Controllability and functionality are taken for granted, without thinking about how elementary it is for design and use in everyday life. The acceleration of everyday life by both product developers and users, the increasing complexity of manufacturing processes and devices, and the resulting decrease in know-how are the real challenges in the case of obsolescence.

As long as short lifetimes and usage times and the selling or purchase of new items is structurally more supported than prolonging the life of the old by repair, reuse, or upcycling, premature obsolescence will be reproduced. In our survey, robustness and long durability were nominally declared the most important criteria in the selection of a device. At the same time, in many cases, a new – as robust as possible – device is purchased, although the old device proves its robustness and durability and still works. These kinds of paradoxes cannot be understood by applying a linear cause and effect or attitude–behaviour scheme. Our final plea is therefore that the focus of consideration must be more on the generation of system knowledge. To develop options for action and transformation paths, it is important to decipher the "culture of obsolescence." In other words, when looking for the causes of short-lived electronic products and fast-moving consumption, we do not construct a detective story with clear perpetrator–victim categories but treat it more as an archaeological excavation. This would expose the fabric and layers of our material culture and allow us to ask ourselves the open question of why the short life of consumer products can be useful, practical, or simply the easiest way for different social actors. Only a systemic understanding of the causes, drivers, and stabilisers of obsolescence effectively answers the question of who is responsible for the lifetime of consumer products.

Notes

1 The research presented here is part of the research group "Obsolescence as a challenge for sustainability" (challengeobsolescence.info), funded 2016–2021 by the German Federal Ministry of Research and Education/Social Ecological Research Programme (www.bmbf.de/en/socialecologicalresearch).
2 An example is the French NGO "HOP – Halte à l'obsolescence programme" or the German organisation Murks NeinDanke e.V.
3 www.videoproject.com/Light-Bulb-Conspiracy-The.html
4 The research was based on terms such as obsolescence, lifespan, and product life. The body contained articles from 28 daily and weekly newspapers. The analysis of the articles was oriented towards a discourse analysis approach (Keller, 2005). A qualitative-reconstructive method is often used in the analysis of media to work out how social order and social meanings are constructed and constituted in discourses. The following are some initial results of the reported ongoing analyses.
5 While it is by arrangement that light-bulb manufacturers limit the burning time of incandescent lamps to about 1,000 hours, it is debatable whether this is a conscious, sales-increasing deception of consumers or an agreement to standardise the quality of light (Krajewski, 2014). The central parameters for light-bulb life, light output, and power consumption can each be optimised only to the detriment of the other parameters. For example, a more durable bulb will emit less light while consuming the same amount of power, so the desire for brighter bulbs will inevitably affect their burn time. For example, the light output of the centennial bulb, which has been in operation for over 100 years (see footnote 9), is only 4 watts.
6 See, for example, studies by the Öko-Institut on behalf of the Federal Environment Agency or the project LOiPE and Wieser et al., 2015

References

Akrich M. (1992). The description of technical objects. In: Bijker W, Law J. (Eds.), *Shaping technology, building society: studies in sociotechnical change*. Cambridge, MA: MIT Press. 205–224.

Cooper T. (2004). Inadequate life? Evidence of consumer attitudes to product obsolescence. *Journal of Consumer Policy*. 27: 421–449.

Cooper T, Mayers K. (2000). *Prospects for household appliances. E-SCOPE survey: an investigation of the implications and effectiveness of producer responsibility for the disposal of WEEE*. Halifax: Urban Mines Ltd.

Cox J, Sarah G, Sara G, King G. (2013). Consumer understanding of product lifetimes. *Resources, Conservation and Recycling*. 79: 21–29.

Echegaray F. (2016). Consumers' reactions to product obsolescence in emerging markets. The case of Brazil. *Journal of Cleaner Production*. 134: 191–203.

Eisentraut S. (2016). Das Mobiltelefon als digitaler Dauerbegleiter. In: Reuter J, Berli O. (Eds.), *Dinge befremden*. Wiesbaden: Springer Fachmedien Wiesbaden. 107–113.

European Economic and Social Committee (2016). *The influence of lifespan labelling on consumers*. Brussels. Available at: www.eesc.europa.eu/resources/docs/16_123_duree-dutilisation-des-produits_complet_en.pdf.

Eurostat (2014). *Waste generation by economic activities and household*. Available at: http://ec.europa.eu/eurostat/statistics-explained/index.php/File:Waste_generation_by_economic_activities_and_households,_2014-1.png.

Evans S, Cooper T. (2010). Consumer influences on product life-spans. In: Cooper T. (Ed.), *Longer lasting products: alternatives to the throwaway society*. Farnham and Burlington, VT: Ashgate. 319–350.

Frick V, Jaeger-Erben M, Hipp T. (2019). *The 'making' of product lifetime: the role of consumer practices and perceptions for longevity*. Plate Conference Proceedings. Berlin: Technische Universität Berlin.

Gregson N, Metcalfe A, Crewe L. (2009). Practices of object maintenance and repair. *Journal of Consumer Culture*. 9: 248–272.

Harrell GD, McConocha MD. (1992). Personal factors related to consumer product disposal tendencies. *Journal of Consumer Affairs*. 26: 397–417.

Hilty L, Bieser J. (2017). *Opportunities and risks of digitalization for climate protection in Switzerland*. Zurich: University of Zurich.

Hoornweg D, Bhada-Tata P, Kennedy C. (2013). Environment. Waste production must peak this century. *Nature*. 502 (7473): 615–617. doi:10.1038/502615a.

Hörning KH. (2015). Was fremde dinge tun. Sozialtheoretische herausforderungen. In: Hahn HP, Assmann J. (Eds.), *Vom eigensinn der dinge. Für eine neue perspektive auf die welt des materiellen*. Berlin: Neofelis Verlag. International Specialized Book Services. 163–176.

Hütz-Adams F. (2012). *Von der Mine bis zum Konsumenten: Die Wertschöpfungskette von Mobiltelefonen* [From mine to consumer: the value chain of mobile phones]. Siegburg: Südwind Verlag. Available at: https://suedwind-institut.de/files/Suedwind/Publikationen/2012/2012-41%20Von%20der%20Mine%20bis%20zum%20Konsumenten.%20Die%20Wertschoepfungskette%20von%20Mobiltelefonen.pdf (Accessed 24/08/2020).

Jaeger-Erben M, Hipp T. (2018). All the rage or take it easy – expectations and experiences in the context of longevity in electronic devices: descriptive analysis of a representative online survey in Germany. In: Obsolescence Research Group (Ed.), *OHA texts 1/2018*. Available at: https://challengeobsolescence.info/wp-content/uploads/2017/09/Research-Group-OHA_Description-Online-Survey_2018.pdf (Accessed 24/08/2020).

Keller R. (2005). Analysing discourse. An approach from the sociology of knowledge [33 paragraphs]. *Forum Qualitative Sozialforschung / Forum: Qualitative Social Research*. 6 (3), Art. 32. Available at: http://nbn-resolving.de/urn:nbn:de:0114-fqs0503327 (Accessed 24/08/2020).

Klose S. (2015). *Obsoleszenz – obsolet weil auskuratiert!? Obsoleszenz interdisziplinär. Vorzeitiger verschleiß aus sicht von wissenschaft und praxis*. Schriftenreihe des Instituts für Europäisches Wirtschafts- und Verbraucherrecht e.v, Bd. 37, 1. Auflage, Hrsg. Tobias Brönneke, und Andrea Wechsler. 169–184.

Krajewski M. (2014). Fehler-Planungen. Zur Geschichte und Theorie der industriellen Obsoleszenz. *Technikgeschichte*. 81 (1): 91–114.

Miller D. (2006). *Materiality* (2nd ed). Durham: Duke University Press.

Oetzel G. (2012). *Das globale müll-system. Vom verschwinden und wieder-auftauchen der dinge. Globale öffentliche güter in interdisziplinären perspektiven*. Schriftenreihe des Zentrums für Technik- und Wirtschaftsethik am Karlsruher Institut für Technologie. Karlsruhe: KIT Scientific Publishing. 79–98.

Oguchi M, Daigo I. (2017). Measuring the historical change in the actual lifetimes of consumer durables. In: Bakker CA, Mugge R. (Eds.), *Proceedings of PLATE 2017, November 2017*. Delft: Delft University of Technology. 319–332.

Packard V. (1960). *The waste makers*. New York: McKay Comp. 4th Ed.

Prakash S, Dehoust G, Gsell M, Schleicher T. (2016). *Einfluss der Nutzungsdauer von Produkten auf ihre Umweltwirkung: Schaffung einer Informationsgrundlage und Entwicklung von Strategien gegen "Obsoleszenz"*. Berlin: Umweltbundesamt.

Rammert W. (2006). *Technik, handeln und sozialstruktur. Eine einführung in die soziologie der technik*. Technical University Technology Studies. https://www.ts.tu-berlin.de/fileadmin/fg226/TUTS/TUTS_WP_3_2006.pdf

Rammert W, Schulz-Schaeffer I. (2002). *Technik und handeln – wenn soziales handeln sich auf menschliches verhalten und technische artefakte verteilt*. TUTS – Working Papers. 4–2002. Berlin: Technische Unversitaet Berlin.

Trentmann F. (2016). *Materielle kultur und energiekonsum. Verbraucher und ihre Rolle für eine nachhaltige Entwicklung.* Band 5. München: oekom.

Trentmann F. (2017). *Empire of things. How we became a world of consumers, from the fifteenth century to the twenty-first.* London: Penguin Books.

Ullrich W. (2014). *Habenwollen. Wie funktioniert die konsumkultur?* Bd. 16328. Ungekürzte Ausg, 4. Aufl. Frankfurt am Main: Fischer.

van Hinte E. (1997). *Eternally yours: visions on product endurance.* Rotterdam: 10 Publishers.

Wang F, Huisman J, Stevels, A, Baldé CP. (2013). Enhancing e-waste estimates: improving data quality by multivariate input-output analysis. *Waste Management.* 33 (11): 2397–2407. doi:10.1016/j.wasman.2013.07.005.

Weber H. (2014). *Einleitung. Entschaffen. Reste und das ausrangieren.* Zerlegen und Beseitigen des Gemachten. TG 81. H1. 3–32.

Wieser H, Tröger N, Renate H. (2015). *Die nutzungsdauer und obsoleszenz von Gebrauchtsgütern im Zeitalter der Beschleunigung.* Eine empirische untersuchung in österreichischen Haushalten, Arbeiterkammer Wien.

PART II

Measuring and implementing circularity

10

AFRICA – JUXTAPOSITION BETWEEN RAPID URBANISATION, INDUSTRIALISATION, AND THE NEED TO PRESERVE TRADITIONAL CIRCULAR SYSTEMS

Sally-Anne Käsner and Sarah O'Carroll

The circular economy in the African context

In the Global North, particularly Europe, the circular economy (CE) narrative is primarily focused on waste management through recycling, waste reduction, and cost savings; redesign of products and remanufacturing have only recently emerged (Desmond and Asamba, 2019). For example, CE principles are being seen in the development of business models in multinational organisations such as Philips ("Pay Per Lux"), Desso and Interface (renting carpet tiles using a cradle-to-cradle approach), and Renault (remanufacturing motor vehicle parts) (Ellen MacArthur Foundation, 2013).

In contrast, Africa's development path has focused on industrialisation in the conventional sense, and very little activity has been focused on high-value-added activities and methods of processing (United Nations Economic Commission for Africa, 2016). This translates into a continued reliance on imports of high-value products and weak linkages with products from the knowledge economy, research and development, and enhanced technology (ibid).

> The secret of the wealth of nations is clear: developed nations add value to everything they produce, while poor nations export raw materials. Africa must quit being at the bottom of the global value chains and move to rapidly industrialize, with value addition to everything that it produces. Africa must work for itself, its people, not exporting wealth to others.
>
> (African Development Bank Group, 2017)

Whereas the circular economy might have been used as a policy framework to deliver waste management and environmental outcomes, it is now increasingly being seen as a policy framework that can be used to maximise the value of resources for economic development and job creation. Pieterse (2015) notes that the focus should be on building adaptive cities, which is essentially the transition to a more circular approach.

Several African countries are shifting the focus to developing 'greener' economies through local product design and a renewed focus on reuse and repair, resulting in improved human well-being and social equity while reducing negative environmental impacts and ecological scarcities. How to reap benefits from circular approaches is still an emerging concept in Africa, although many would say that Africa has been naturally practicing CE for decades in repair and reuse, farming practices, and the beneficiation of materials.

Circular economy activities continue to increase across the continent, with many examples of circular economy models being practiced. Yet African case studies stay 'hidden,' as they have yet to be documented through academic research (Pieterse, 2015).

Africa's circular economy policy landscape

Unlike the European Union, Africa does not have any continent-wide circular economy policies as such, but there are opportunities within the African Union (AU) and country-level policy frameworks to promote CE. In addition to this, there also appears to be growing political will to support CE.

Agenda 2063 and the Durban Declaration: frameworks for a continental transition to circular economy

The AU's vision is an integrated, prosperous, and peaceful Africa, driven by its own citizens and representing a dynamic force in the global arena. Agenda 2063, officially adopted by the AU Assembly in 2015, provides a collective vision and roadmap to build a prosperous and united Africa based on shared values and a common destiny, a bold vision for the continent. Agenda 2063 is Africa's strategic framework that aims to deliver on the goals of inclusive and sustainable development and is a manifestation of the pan-African drive for unity, self-determination, freedom, progress, and collective prosperity pursued under Pan-Africanism and the African Renaissance. Agenda 2063 has 14 'flagship' projects that have been identified as key to accelerating Africa's economic growth and development as well as promoting a common identity by celebrating Africa's history and vibrant culture. Although Agenda 2063 does not explicitly mention CE, Actions D and F within the Agenda could be used as gateways for the circular economy:

- Action D: Transform, grow, and industrialise our economies through beneficiation and value addition of natural resources
- Action F: Act with a sense of urgency on climate change and the environment

(African Union, 2015)

There is also increasing political will for circular economy implementation on the continent. CE is being included in a number of pan-African policy and cooperation initiatives, and the African Alliance on the Circular Economy (AACE) was officially announced by South Africa's then-minister of environmental affairs in November 2017. This political alliance between South Africa, Rwanda, and Nigeria (UN Environment, 2019) aims to advance the circular economy in Africa through the implementation of policy frameworks and regulations. The alliance is supported by the World Economic Forum and the Global Environment Facility, as well as the African Development Bank (AfDB) (South African Department of Environmental Affairs, 2017).

Most recently, the Durban Declaration was agreed to by the African Ministerial Conference on Environment (AMCEN) in November 2019. The Durban Declaration is the first continent-wide policy that specifically includes the circular economy. Widely supported by the ACEA, the Durban Declaration:

- Recognises the value of the circular economy as a framework for economic development and job creation by improving the production and consumption of goods and services, reducing the generation of waste and pollution, reducing dependency on natural resources, and reducing pollution
- Aims to increase awareness of the circular economy through the development of policy, regulatory, and institutional arrangements
- Commits African countries to accelerate their transition to a circular economy as part of the African Union's Agenda 2063 through capacity-building programmes and initiatives
- Encourages the private sector and other actors to invest in CE

(African Ministerial Conference on the Environment, 2019)

Country-level circular economy policies and legislation

Government policy in Africa has a major role to play at both the national and local level to stimulate circular economy activities using policies to align stakeholders to a set of common circular economy goals to drive action. Although the legal and regulatory frameworks to foster circularity are still in their infancy, several African governments have made commitments and taken concrete steps to promote a variety of concepts or approaches related to the circular economy. These include 'green economy,' 'green growth,' 'sustainable production and consumption,' and 'resource efficiency,' amongst others.

There is little CE-specific legislation. Regulations and policies in operation are generally focused on climate change mitigation, the green economy (GE), and waste management. The UN 2030 Agenda for Sustainable Development and the United Nations Framework Convention on Climate Change (UNFCCC) COP 21 Paris Agreement appear to be the foundational global agreements for relevant policy and legislation developed to date. Proposals are often presented but are still awaiting promulgation into government policy and legislation.

Implementing GE policies is at varying degrees, with some countries only integrating aspects of it, whilst others, like Ethiopia, Kenya, and Rwanda, have implemented overarching GE strategies. The legal and regulatory framework to foster GE is still immature in many African countries, and mechanisms to realise the transition to GE are not yet in place (Circular Conversations, 2019). The most promising markets tend to be those related to agriculture (The World Bank, 2015), such as bio-trade, sustainable tourism, and renewable energy. This has been seen to be a result of the social and economic challenges in sub-Saharan Africa which make it unrealistic for a new concept to take root rapidly when it might come at the expense of development and growth (Klein and Reiher, 2016).

In Ghana, as in other African countries, the lack of an enabling policy environment, as well as financial and other incentives, is a major constraint for the creation of CE, particularly for entrepreneurs to set up informal repair businesses. The power exercised through the EU Extended Producer Responsibility (EPR) legislation by Global North manufacturers and their value chains will become a greater force for change, and localised EPR legislation will be less important.

There are few systematic studies of CE policies in Africa, and so identification of policies currently relies on informal research approaches. Table 10.1 summarises some of the CE-related

Table 10.1 CE-related policies, regulations, and initiatives in a selection of African countries

Country	*Name of policy and year*	*Implementing agencies*	*Short description*
Ethiopia	Climate Resilient Green Economy (2011)	Ministry of Environment	Green economy based on four pillars: agriculture; forestry; power (renewable energy for domestic and regional markets); transport, industrial sectors, and buildings.
Ghana	Ghana Goes for Green Growth (2010)	Ministry of Environment	Sustainable development and equitable low-carbon economic growth.
Kenya	Nationally Appropriate Mitigation Action (NAMA) on Circular Economy Solid Waste Management Approach for Urban Areas in Kenya	Ministry of Environment and Natural Resources of Kenya and UNDP Low Emission Capacity Building (LECB) Programme	Diversion of waste from disposal sites towards recycling
	Gazette Notice No. 2356 (Feb. 28, 2017), 119:31 *Kenya Gazette* (Special Issue) (Mar. 14, 2017), at 1077.)	Ministry of Environment and Natural Resources	Ban on "the use, manufacture and importation of all plastic bags used for commercial and household packaging" as of August 2017 – two categories of bags: the carrier bag, a "bag constructed with handles, and with or without gussets," and the flat bag, a "bag constructed without handles, and with or without gussets."
	Draft – National Sustainable Waste Management Policy (2018)	Ministry of Environment and Forestry	This policy will advance Kenya towards a more sustainable and circular economy. It will move the country towards realization of the zero waste principle, whereby waste generation is minimised or prevented.
Namibia	Green Economic Coalition Dialogue (2011)	Ministry of Labour and Social Welfare	Economic development, job creation and CO_2 reduction.
	National Solid Waste Management Strategy (2017)	Ministry of Environment and Tourism	To strengthen the institutional, organisational, and legal framework for solid waste management, including capacity development.
Nigeria	Extended Producer Responsibility Programme (2013)	National Environmental Standards and Regulations Enforcement Agency	Minimisation of industrial waste and promotion of recycling (specifically around electronic waste).

Rwanda	Ministerial instruction on the use and manufacturing of plastic bags in Rwanda – Plastic Bag Law 57 (2008); updated – Law relating to the prohibition of manufacturing, importation, use, and sale of plastic carry bags and single-use plastic items, Article 3, N° 17/2019 du 10/08/2019	Minister of Land, Environment, Forestry, Water and Natural Resources	Cabinet decision of 18/02/2004 prohibiting the importation and manufacturing of some plastic bags in Rwanda. Update: Law relating to the prohibition of manufacturing, importation, use, and sale of plastic carry bags and single-use plastic items.
	National Environment and Climate Change Policy (2019)	Ministry of Environment	Strategic policy direction designed to guide the environment and climate change (national and global linkages).
South Africa	National Environmental Management Act (1998) (includes various updates and amendments)	Department of Environment, Fisheries and Forestry (DEFF)	Minimisation of waste, pollution, and use of natural resources.
	Circular Economy Guideline and Briefing Note (in development)	United Nations Environment Programme and the South African National Department of Environment, Fisheries and Forestry (Department of Environmental Affairs, 2012)	As part of South African Greenfund focused on circular economy for sustainable consumption and production.

policies that are in existence for a selection of African countries. Further research is required to identify the extent and impact of sustainability legislation and policies such as waste management, recycling, extended producer responsibility, repair, and renewable energy.

The previous examples are only a small selection of government policies currently in existence. Much work still needs to be done to understand the circular economy policy landscape in Africa and the extent to which it is enabling for circular economy activities. After the development of circular economy policies, the major challenge in the future will be to ensure that legislation which has been implemented is enforced, a common challenge in Africa. Furthermore, regulations behind more complex elements of CE, such as designing products as services and remanufacturing, are yet to be created at scale in Africa.

Private-sector small and medium-sized organisations and non-profits drive the circular economy in Africa

Although large businesses, such as Unilever, Nestle, Veolia, and Caterpillar, are involved in circular economy activities in Africa, the circular economy transition in Africa is largely being championed by circular economy practitioners on the ground through various NGOs, consultancies, entrepreneurs and small and medium-sized enterprises (SMEs).

In fact, some of the continent's best circular economy innovations and technology developments have been created by entrepreneurs and SMEs. What sets these innovations and technologies apart is that they are developed with the local context specifically in mind: decentralised, labour intensive, and cost effective.

The private sector appears to be taking the lead in driving the circular economy transition, and, significantly, innovative business models appear to be led by SMEs rather than large global companies. Industrial symbiosis is possibly the fastest-growing circular economy activity on the continent, accelerated through the Switch Africa Green Programme and a British High Commission Prosperity Fund Project entitled 'Towards a National Industrial Symbiosis Programme for South Africa.' The Fund agreed to support eight projects in Southern Africa. The choices included projects to boost inclusive growth and sustainable economic development through interventions in power, infrastructure, cities, skills, and transparency in priority countries in the Southern African Development Community region.

The African Circular Economy Network (ACEN, 2019) is becoming an increasingly important organisation, bringing together CE practitioners across the continent to share knowledge with one another and to build a CE evidence base in Africa. ACEN currently has 31 country chapters located across all regions on the continent: southern, east, west, central, and north Africa.

As governments across the continent develop country and city CE roadmaps and strategies, it will be critical to ensure that the actors that have driven the CE transition to date remain central to the process in the future.

Other organisations involved in promoting the circular economy include the African Alliance on Circular Economy; the European Union; the governments of Nigeria, South Africa, Rwanda, Ghana (World Economic Forum, 2019), Morocco, and the District of Abidjan (Ivory Coast); the World Economic Forum and its Platform for Accelerating the Circular Economy (PACE),[1] Local Governments for Sustainability (ICLEI) Africa,[2] and United Nations Environment Programme (UNEP) Africa. By moving towards a CE strategy, Africa will be able to leapfrog to a more sustainable development approach by learning lessons and avoiding the pitfalls of resource-intensive practices of the linear economy which have been the primary driver of growth in the Global North (Desmond and Asamba, 2019).

Rapid urbanisation is eroding circular economy practices in Africa

Urbanisation is one of the defining forces of the planet's 21st-century development. In 1950, the urban share of the world's population was 30%, but by 2050, it may well be 66%. Nearly 90% of the increase will be in Africa and Asia, the fastest-urbanising global regions. Africa is the world's second-largest continent after Asia, comprising 54 sovereign countries (48 mainland and 6 island states), with an estimated population of 1.26 billion in 2017 (Ibid). The total population on the continent is projected to reach almost 2.5 billion people by 2050, with about 55% living in urban areas. This is a significant increase given that less than 10% of Africa's population resided in urban areas in 1950 (United Nations, 2018).

Africa's urbanisation reflects considerable variation, as countries at different income levels have urbanised at different rates, further adding complexity. Unlike global trends, urban–rural differences in welfare and living standards in Africa do not seem to narrow with increasing urbanisation. Many urban municipal authorities lack the capacity to plan for urban growth and accommodate migrants. Rapid unplanned and uncoordinated urbanisation provides many challenges (World Economic Forum, 2017).

Güneralp et al. (2017) provide an overview of the challenges faced by rapid urbanisation in many African cities. Basic urban services were already lacking in many cities even prior to rapid urbanisation due to poor governance; inadequate service infrastructure in urban areas can be traced back to colonial institutional arrangements and persistent political instability. Some African countries do not have any urban planning and development departments, and rigid zoning in central areas of many cities inadvertently contributes to proliferation of slums and sprawling development of residential areas.

The complexities of the previous provide direct challenges but also opportunities for the implementation of CE principles. Environmentally, urbanisation can offer benefits, reducing travel distances and preserving land. However, improper planning has the direct opposite effect. Africa's future cities need to be compact: dense, mixed use, and transit oriented, with planning for circular material flows (Ellen MacArthur Foundation, 2019). African cities are dominated by informal economies, often lacking many formal-sector and manufacturing jobs. Sixty-one percent of men and 74% of women working in non-agricultural sectors are informally employed, with a larger share (60%) of women in own-account self-employment (ILO, 2018).

Critical enablers for an African circular economy transition

There are a number of enablers that need to be developed to ensure that the CE develops in Africa in a manner that enhances economic productivity, supports job creation, provides substantive climate change benefits, and remains inclusive. These include:

- Developing country-level CE visions that are supported by political will;
- Developing regulatory frameworks that enable new ways of doings things;
- Cooperation at a global level that supports the development of system enablers such as capacity building, research, development and innovation, and investment.

Together these factors will ensure that circular activities remain at the heart of African economies rather than being eroded as the continent urbanises.

A circular economy vision supported by political will

Within some African countries, there are roadmaps that refer to CE activities, and the Durban Declaration has set the CE as a policy agenda for the continent; however, specific circular economy policies, plans, or strategies have yet to be developed in any African country. Roadmaps and strategies are possibly one of the most effective means of accelerating the CE transition in any country or city, as they align actors to a specific set of objectives and create opportunities for collaboration between actors. However, if these roadmaps and strategies are not supported politically, they are likely to be ineffective to catalyse any measurable action.

Going forward, it will be critical for circular economy awareness-raising and capacity-building activities to be delivered in Africa as stated in the Durban Declaration, particularly amongst policymakers, in order to gain buy-in and as a result drive the transition forward.

Enabling regulatory frameworks at the country, provincial, and city levels is needed

Many countries in Africa have developed or are in the process of developing robust waste management regulatory frameworks. However, ambiguity around definitions (including the definition of what "waste" actually constitutes to avoid the overregulation of managing both biological and technical nutrients) and misalignment between different acts (environment and waste management) are often cited as major barriers in the development of CE activities on the continent. Reform is needed to update regulations and legislation in order to create an enabling regulatory framework for CE activities.

Cooperation at the global level to support system enablers

Cooperation is needed at the global level to support African countries to maintain and improve circularity at the heart of their economies rather than letting it be eroded as the continent urbanises. Greater cooperation is needed at the international level to agree on common rules and standards for making fair and ethical trading part of global value chains (Ethical Corporation, 2019). This cooperation will be particularly important where there may be a risk of displacing informal-sector waste collectors and for materials streams like e-waste, where there continue to be significant environmental and health risks related to current processing and disposal methods. International institutions such as the AACE, OECD, UN Industrial Development Organisation (UNIDO), and UN Environment Programme can play critical roles:

- Ensuring that meaningful dialogue is created between developed countries and African governments;
- Supporting the piloting of CE solutions that have been developed in African countries by SMEs;
- Facilitating the proactive engagement of multinational companies with suppliers in African countries, including SMEs and those operating in the informal sector to scale up curricular activities in a manner that is inclusive and avoids the displacement of vulnerable workers.

Cognisance also needs to be taken that the strategies and materials required by a developed continent such as Europe to enter a renewable energy transition are reliant on an extraction-based economy, for example, for the necessary components and parts of battery storage. However, thought needs to be given as to how this relationship and business model will work within a CE context, as it cannot be undertaken at the expense of the African continent getting locked into a linear operating system as a supplier of the required extracted resources.

Capacity building to increase awareness of circular economies and create the skills needed for circular economies in Africa

Capacity building is needed to increase the level of understanding and awareness of the circular economy and to gain buy-in from critical stakeholders, policymakers in particular, to ensure that Africa is self-reliant, with the skills and knowledge to be able to develop circular practices. Importantly, capacity-building activities in Africa should take into account the local context such as the current state of waste management infrastructure, regulatory framework, and the role of the informal sector in the country.

Enhanced research, development, and innovation based on the African context

Research, development, and innovation (RDI) are needed to support the creation and utilisation of local technologies and business models that are fit and truly 'appropriate' for African contexts. Technologies are required that are decentralised, modular, cost effective, weather robust, and in many cases requiring basic operational skill levels that the African labour market can provide.

Financial support to ensure that circularity remains at the heart of African economies

Finally, financial support is needed to support African governments' investments in CE activities. There are numerous models to achieve this, including grants, equity investment, bonds, and crowdfunding. One popular approach is through micro-finance and small revolving loans, for example, Lend with Care (Guardian, 2013) and Grameen Bank (Yunus, 2013). Incubator and accelerator initiatives are also becoming more common.

An enhanced people-centric approach to circular economy development

For CE to be politically supported in Africa, it will need to develop in a manner that supports widespread job creation. An enhanced people-centric approach to the development of CE is suggested, one that considers the available workforce and the role of the informal sector to repair technical nutrients and regenerate biological nutrients. Repair-based, refurbishment, and remanufacturing activities would be preferred, as they are known to be high job-creating activities.

Conclusions

Many African countries have nodes of rapid urbanisation and population growth, placing increasing pressure on the natural, social, and economic resources. This pressure is leading to many challenges, including the provision of goods and services and a clean and healthy environment. This chapter has provided an overview for how, in many cases, Africa was practising CE; however, rapid urbanisation and industrialisation potentially threaten this model.

A number of policies and legislative changes have been made across the continent, and there appears to be political will at the highest level in the form of the leaders represented on the African Alliance on Circular Economy; the European Union; the governments of Nigeria, South Africa, Rwanda (Plastic Oceans, 2018), Ghana, Morocco, and the District of Abidjan (Ivory Coast); the World Economic African Circular Economy Network Forum and its PACE platform; ICLEI Africa; and UNEP Africa. By moving towards a CE strategy, Africa will be able to leapfrog to a more sustainable development approach by learning lessons and avoiding the pitfalls of resource-intensive practices of the linear economy, which has been the primary driver of growth in the Global North.

Notes

1 PACE is a public-private collaboration platform and project accelerator. http://www3.weforum.org/docs/WEF_PACE_Platform_for_Accelerating_the_Circular_Economy.pdf

2 ICLEI Africa is a global network of more than 1,750 local and regional governments committed to sustainable urban development. Active in 100+ countries, we influence sustainability policy and drive local action for low emission, nature-based, equitable, resilient, and circular development. Our members and team of experts work together through peer exchange, partnerships, and capacity building to create systemic change for urban sustainability. https://www.iclei.org/

References

African Circular Economy Network (2019). *About.* Accessed at: www.acen.africa/ (Accessed 2/10/2019).

African Development Bank Group (2017). *Industrialize Africa: strategies, policies, institutions, and financing.* Available at: www.afdb.org/fileadmin/uploads/afdb/Documents/Generic-Documents/industrialize_africa_report-strategies_policies_institutions_and_financing.pdf (Accessed 6/11/2019).

African Ministerial Conference on Environment (2019). *Draft Durban declaration on taking action for environmental sustainability and prosperity in Africa.* Available at: https://wedocs.unep.org/bitstream/handle/20.500.11822/30732/AMCEN_17Declaration.pdf?sequence=7 (Accessed 22/11/2019).

African Union (2015). *Agenda 2063: the Africa we want.* Available at: https://au.int/en/agenda2063/overview (Accessed 9/10/2019).

Circular Conversations (2019). *Circular Kenya: when the magic happens, a discovery journey into community-led circular innovations with Wekesa Zablon Wanyama.* Available at: www.circularconversations.com/blog/2019/5/20/circular-kenya-when-the-magic-happens (Accessed 8/10/2019).

Department of Environmental Affairs (2012). *National waste information baseline report.* Available at: http://sawic.environment.gov.za/documents/1880.pdf (Accessed 6/11/2019).

Desmond P, Asamba M. (2019). Accelerating the transition to a circular economy in South Africa – case studies from Kenya and South Africa. In: Schroeder P et al. (2019). *The circular economy and the Global South – sustainable lifestyles and green industrial development.* Abingdon: Routledge.

Ellen MacArthur Foundation (2013). *Towards the circular economy. Volume 1 – economic and business rationale for an accelerated transition.* Isle of Wight: Ellen MacArthur Foundation.

Ellen MacArthur Foundation (2019). *Urban buildings system summary.* Isle of Wight: Ellen MacArthur Foundation.

Ethical Corporation (2019). *How the circular economy could help developing countries grow sustainably.* Available at: www.ethicalcorp.com/how-circular-economy-could-help-developing-countries-grow-sustainably (Accessed 8/10/2019).

The Guardian (2013). *Lend with Care offers helping hand to world's poorer entrepreneurs.* Available at: www.theguardian.com/money/2013/aug/10/lend-with-care-help-poorer-entrepreneurs (Accessed 8/11/2019).

Güneralp B, Lwasa S, Masundire H, Parnell S, Seto KC. (2017). Urbanization in Africa: challenges and opportunities for conservation. *Environmental Research Letters.* 13 (1): 15002.

ILO (2018). *Women and men in the informal economy: a statistical picture.* Geneva: International Labour Office. 3rd Ed.

Klein J, Reiher S. (2016). Sub-Saharan Africa – potentials and barriers for green(er) economies. In: Carius A, Tänzler D, Semmling E. (Eds.), *The rise of the green economy a new paradigm for the developing world?*

Munich: Oekom verlag München. Available at: https://www.adelphi.de/en/system/files/mediathek/bilder/Towards%20a%20Green%20Economy%20in%20Emerging%20and%20Rapidly%20Growing%20Countries%20%E2%80%93%20The%20Way%20Forward%20after%20Rio%2B20.pdf
Pieterse E. (2015). Reaching for adaptive urbanism. In: Holm JH, Kallehauge MM. (Eds.), *Africa, architecture, culture, identity*. Humlebæk, Denmark: Louisiana Museum of Modern Art.
Plastic Oceans (2018). *Rwanda plastic bag ban*. Available at: https://plasticoceans.org/rwanda-plastic-bag-ban/ (Accessed 4/10/2019).
South African Department of Environmental Affairs (2017). *Minister Molewa's speech during launch of Africa Alliance on Circular Economy*. Available at: www.environment.gov.za/speech/molewa_cop23africaalliance_circular_economylaunch (Accessed 8/10//2019).
United Nations (2018). *World urbanisation prospects 2018*. Available at: www.un.org/development/desa/en/news/population/2018-revision-of-world-urbanization-prospects.html (Accessed 9/10/2019).
United Nations Economic Commission for Africa (2016). *Economic report on Africa*. Available at: www.un.org/en/africa/osaa/pdf/pubs/2016era-uneca.pdf (Accessed 6/11/2019).
UN Environment (2019). *Nigeria turns the tide on electronic waste*. Available at: www.unenvironment.org/news-and-stories/press-release/nigeria-turns-tide-electronic-waste (Accessed 8/10/2019).
The World Bank (2015). *Agriculture in Africa: telling facts from myths: overview*. Available at: www.worldbank.org/en/programs/africa-myths-and-facts (Accessed 4/10/2019).
World Economic Forum (2017). *Migration and its impact on cities*. Available at: http://www3.weforum.org/docs/Migration_Impact_Cities_report_2017_low.pdf (Accessed 9/10/2019).
World Economic Forum (2019). *Ghana becomes first African nation to join ambitious partnership to end plastic pollution*. Available at: www.weforum.org/press/2019/10/ghana-becomes-first-african-nation-to-join-ambitious-partnership-to-end-plastic-pollution/ (Accessed 8/10/2019).
Yunus M. (2013). *My journey to help the poor and build a more equitable world*. Available at: https://edition.cnn.com/2013/11/15/opinion/yunus-microfinance-grameen-bank/index.html (Accessed 8/10/2019).

11

CONCEPTUALISING CIRCULAR START-UPS

Marvin Henry and Julian Kirchherr

Introduction

Moving away from 'business as usual' appears to be necessary to break with the current trends of resource use. Extraction of fossil fuels, ores, minerals, and biomass rose 12-fold during the 20th century, amounting to 84.4 billion tonnes in 2015, with further doubling expected in 2050 (Circle Economy, 2018; Krausmann et al., 2009). This accelerating material use poses serious threats to the sustainability of society. A circular economy (CE) proposes a shift from this model to one in which materials are circulated in closed-loop production systems to reduce resource depletion and eliminate waste (Ellen MacArthur Foundation, 2012; Geng et al., 2009; Murray et al., 2017). There are many competing CE definitions (Kirchherr et al., 2017). We define a CE throughout this chapter as:

> An economic system that is based on business models which replace the "end-of-life" concept with reducing, alternatively reusing, recycling and recovering materials in production/distribution and consumption processes, thus operating at the micro-level (products, companies, consumers), meso level (eco-industrial parks) and macro level (city, region, nation and beyond), with the aim to accomplish sustainable development, which implies creating environmental quality, economic prosperity and social equity, to the benefit of current and future generations.
>
> (Kirchherr et al., 2017: 224–225)

The CE has become a widely discussed topic among policymakers, scholars, and industry professionals alike (Urbinati et al., 2017). Despite these numerous discussions, the implementation of CEs remains extremely limited. For instance, a study showed that in 2005, only 6% of waste materials were recycled at the global level, with the remainder incinerated, landfilled, or dispersed into the environment (Haas et al., 2005). While governments have been argued to hold pivotal roles in building a vision and steering market actors towards a CE, the roles that the private sector can play have been particularly highlighted in recent years (Lieder and Rashid, 2016; Lüdeke-Freund et al., 2018; Urbinati et al., 2017; Antikainen et al., 2017). Indeed, changes in business models are considered by many scholars a key enabler for the shift from a linear to a CE (van Renswoude et al., 2015; Ellen MacArthur

Foundation, 2012; McKinsey and Co, 2015; Ghisellini et al., 2016; Bocken et al., 2016; Lewandowski, 2016).

In principle, CE pioneers can be start-ups as well as incumbents – the drive towards CE is mainly determined by the respective organisations' circular spirits and entrepreneurial capabilities (De los Rios and Charnley, 2017). However, a common view in innovation research is that incumbents are seldom the source of innovations that break with existing dominant designs because they are locked into previous investments, existing supply chains, and business models that are hard to adapt once fully developed (Christensen, 1997; Henderson and Clark, 1990; Hill and Rothaermel, 2003; Johnson, 2010). Similarly, it has been questioned whether incumbents will or can fully embrace the more 'radical' business model approaches to CE, such as sharing platforms (e.g., peer-to-peer) or product-service systems (Lieder and Rashid, 2016). Empirical evidence indicates that large incumbents tend to focus on widespread strategies like recycling, which do not necessitate shifting their core business models (Bocken et al., 2017; Stewart and Niero, 2018). In contrast, start-ups, as new market entrants, have been argued to have a higher capacity to adopt more disruptive *circular business models* (CBMs; defined in the following section), due to their higher flexibility and responsivity to market changes (Bergset and Fichter, 2015; Hockerts and Wüstenhagen, 2010; Rizos et al., 2016). Entrepreneurship has been recognised as a major conduit for sustainable products and processes, and new ventures are being held up as an answer to many social and environmental concerns (Hall et al., 2010).

Definitions of sustainable and circular business models

This section outlines how sustainable and circular business models, innovations, and strategies have been defined in existing research and succinctly reviews the extant literature on the relevant concepts of environmental or sustainable entrepreneurship. In line with previous developments of the business model literature (Beattie and Smith, 2013; Osterwalder et al., 2010; Richardson, 2008), Bocken et al. (2014) highlight three main elements of sustainable business models, namely the value: proposition (i.e., the offer and the target customer segments); creation and delivery (i.e., the activities, resources, partners and distribution channels); and capture (i.e., the cost structure and the revenue model). Accordingly, a sustainable business model (SBM), when appropriately designed, should create, deliver, and capture economic value while simultaneously contributing to environmental and/or social sustainability (Boons and Lüdeke-Freund, 2013; Lüdeke-Freund et al., 2018; Schaltegger et al., 2016).

CBMs incorporate CE principles as guidelines for business model design (Heyes et al., 2018; Pieroni et al., 2019). They can be defined as circular operations at the micro level that aim at fully closing product or material loops and thereby making the 'end-of-life' concept obsolete or keeping resources in use for as long as possible through reducing or alternatively reusing, recycling, or recovering them (Kirchherr et al., 2017; Linder and Williander, 2017). Circular is a more concrete term than sustainable, since it describes a way to achieve (some form of) sustainability (Geissdoerfer et al., 2017). Indeed, while the concept of SBM is centred around the finality of creating economic, environmental, and social value without prescribing specific directions for achieving it (Lüdeke-Freund and Dembek, 2017; Stubbs and Cocklin, 2008), CBMs are more narrowly framed and establish specific ways to address the negative consequences of business operations leading to excessive waste generation and resource depletion.

The overlap of CBMs with SBMs may lead some authors to see the former as a subset of the latter (e.g., Bocken et al., 2014; Lüdeke-Freund et al., 2018). However, we disagree, because, as noted by Pieroni et al. (2019), SBMs can also generate unsustainable effects if design options for a CE are only partially in place (e.g., encouraging over-consumption or fast replacement when the respective reverse-logistics processes are not in place). Furthermore, while SBMs hold social relevance or work enrichment as a driver for value creation on top of environmental and economic impacts, these aspects have been described as secondary benefits instead of core drivers for value creation in the case of CBMs (Pieroni et al., 2019).

Closely related streams of literature have developed around sustainability-oriented innovations (Adams et al., 2016; Klewitz and Hansen, 2014; Schiederig et al., 2012) and, in particular, SBM and CBM innovations (Geissdoerfer et al., 2018). SBM innovations incorporate sustainability principles into the business model innovation process (Foss and Saebi, 2016; Massa and Tucci, 2014), seeking to conceptualise and implement new business models while also benefitting society and the environment (Boons and Lüdeke-Freund, 2013; Lüdeke-Freund et al., 2018; Pieroni et al., 2019). CBM innovations seek to create CBMs by integrating the more concrete circularity principles into the business model innovation process (Diaz Lopez et al., 2019; Geissdoerfer et al., 2018; Linder and Williander, 2017). An emblematic example of CBM innovations is the product-service system model, which entails a conversion from selling a product to providing leasing and sharing services, thereby decoupling the service provision from the ownership of physical goods (Tukker, 2004; Tukker and Tischner, 2006; Vezzoli et al., 2015). This model financially incentivises the company to invest in the durability and reuse of products and materials (Baines et al., 2007; Mont et al., 2006; Williams, 2007).

CBM innovations can occur at different positions along the value chain of goods and services. Urbinati et al.'s (2017) approach distinguishes between upstream, downstream, and full business model adoptions for circularity:

- *Downstream* circular companies adopt circular innovations related to their revenue model and customer interfaces, but they do not necessarily make relevant changes at the supplier level or in internal practices or product design
- *Upstream* circular companies are classified as adopting circular innovations internally and with a focus on interaction with their suppliers. Thus, they concern innovations during the pre-usage or pre-customer face of a product or service. This can happen through design practices (e.g., design for recycling/reuse/disassembly [Jawahir et al., 2007], source material innovation), through collaboration with suppliers to utilise external waste streams, or by selecting partners which can provide biodegradable materials
- *Full* circular companies are those which adopt both upstream and downstream CBM innovations.

In addition, CBM innovations should be distinguished from CBM strategies, which can be viewed as firms' general approach towards a CE. That is, CBM innovations are the processes by which firms practically implement their CBM strategies. Several conceptualisations of CBM strategies exist (e.g., Bocken et al., 2016; Ünal and Shao, 2019; Urbinati et al., 2017). Among those, the so-called R-framework distinguishes between different strategies to embrace circularity, known as R-strategies. Literature and practice offer different versions of the R-frameworks (Blomsma and Brennan, 2017; Sihvonen and Ritola, 2015; Yan and Wu, 2011), ranging from the three Rs (reduce, reuse, recycle) – in the Chinese government policy on CEs (Geng et al., 2009) – to ten Rs in a more recent literature review by Reike et al. (2018). While the number of Rs differs from list to list, most of the R lists establish a priority order for approaches to circularity, with the first R viewed as a higher priority than the second R, and so on. These R-strategies can be developed within the two types of material cycles characterising the CE (Ellen MacArthur Foundation, 2012; McDonough and Braungart, 2002): the biological cycle, which encompasses the flows of food and biologically based materials (e.g., cotton and wood) that are designed to return materials to the biosphere through processes like composting or anaerobic digestion, and the technical cycle, which relates to the flows of inorganic or synthetic materials. The 4R-framework, which is the most commonly used R framework in the CE literature according to Kirchherr et al. (2017), will be used throughout this chapter.

Adopting SBMs and CBMs and corresponding innovations and strategies is often seen first in new entrants, such as start-ups, due to their higher flexibility and their capacity to stimulate disruptive sustainability innovation (Cohen and Winn, 2007; Dean and McMullen, 2007; Hall et al., 2010; Hockerts and Wüstenhagen, 2010). Start-ups can be defined as 'new' (i.e., typically operating for four to six years) and 'independent' entrepreneurial ventures designed to effectively develop and validate a scalable, repeatable, and at least break-even business model (Brush and Vanderwerf, 1992; Luger and Koo, 2005; Zahra et al., 2000). The literature on innovation and entrepreneurship typically views incumbents in a stronger path dependency than start-ups (Christensen, 1997). Indeed, once the business model is established, it is typically highly difficult for a company to change it. As Chesbrough (2010) argues, companies may invest extensively in the development of new products and technologies that will be commercialised through their business model, but they often have little, if any, ability to innovate the business models through which these inputs will pass. Empirical evidence indicates that large incumbents tend to focus on lower CBM strategies, like recycling, and make changes at the margin instead of shifting their core business models (Bocken et al., 2016; Stewart and Niero, 2018).

By contrast, start-ups, as new market entrants, can adopt CBMs from the start, take a holistic perspective in their business model, and monetise design-to-last and maintenance efforts. In addition, they are not exposed to the risk of cannibalising the market share of their prior products or devaluing previous investment in manufacturing processes (Hockerts and Wüstenhagen, 2010). Furthermore, typical limitations in the implementation of CBMs – like company environmental culture, technical know-how, or administrative burden – only partly apply to start-ups. For the latter, more horizontal management styles, decreased bureaucratic structures, and more informal communication channels lead to higher flexibility and responsivity to changing market circumstances (Hockerts and Wüstenhagen, 2010; Rizos et al., 2016; Schaltegger et al., 2016).

Various streams of research have looked at the relationship between different aspects of sustainability and start-ups or entrepreneurship, including environmental entrepreneurship or ecopreneurship (Lenox and York, 2011; Schaltegger, 2002), social entrepreneurship (Borzaga and Defourny, 2001; Doherty et al., 2014), and sustainable entrepreneurship (Choi and Gray, 2008; Cohen and Winn, 2007). While environmental entrepreneurship exclusively focuses on the simultaneous creation of economic and environmental value (Jiang et al., 2018) by addressing environmentally relevant market failures, social entrepreneurship seeks to achieve social goals (e.g., work enrichment, ethical sourcing) and to secure funding for it. Sustainable entrepreneurship aims to embrace sustainability more comprehensively across a "triple bottom line" perspective, integrating its economic, environmental, and social dimensions (Schaltegger and Wagner, 2011; Thompson et al., 2011). The specificity of circular start-ups (CSUs) is that they have adopted a CBM.

The relationships between CSUs and closely related types of ventures are illustrated (Figure 11.1). CSUs have a large overlap with environmental start-ups[1] and environmental entrepreneurship, as both types of organisations simultaneously pursue the creation of economic and environmental value. However, CSUs are more specific in their operationalisation, focusing on closing product or material loops, while environmental start-ups include a broader range of business models, for example, renewable energy generation or biodiversity protection. Moreover, CSUs partly overlap with sustainable start-ups when, on top of economic and environmental

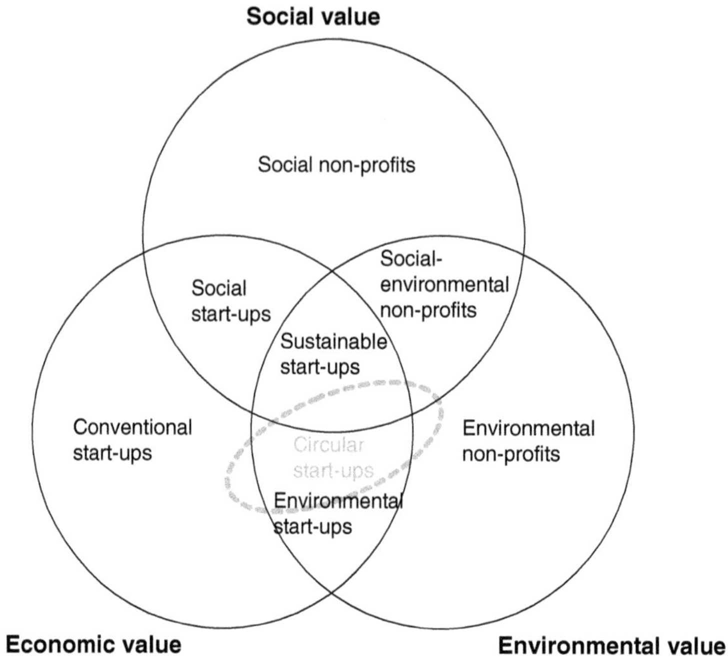

Figure 11.1 The relationships of circular start-ups with other types of enterprises

value, they also generate value to social relevance or work enrichment as secondary effects. This distinguishes them from social start-ups (e.g., fair-trade organisations, work integration social enterprises), for which this aspect is the essential driver for value creation. CSUs may overlap with conventional start-ups when the implementation of CBM strategies and innovations is only partial and leads to the creation of economic value but not environmental or social value as a primary focus. Finally, CSUs can be for profit (i.e., their main goal is to maximise economic value for those who control them), not for profit (i.e., there is limited creation of economic value for controlling stakeholders) or non-profit (i.e., the economic value is entirely retained by the organisation for the fulfilment of its social and/or environmental missions). For this reason, CSUs partly overlap with (social or environmental) non-profits.

Within the literature on environmental and sustainable entrepreneurship, which are the concepts closest to CSUs, there is a dearth of research on the actual business models developed by sustainable entrepreneurs, that is, what entrepreneurs practically do to achieve (environmental) sustainability. Instead, several studies have focused on defining the concepts (Schaltegger and Wagner, 2011; Shepherd and Patzelt, 2011) and on understanding the entrepreneurial opportunities and challenges arising from the existence of externalities and market inefficiencies (Cohen and Winn, 2007; Dean and McMullen, 2007; Jiang et al., 2018; Pacheco et al., 2010; York and Venkataraman, 2010). Other studies have explored the relationships between different actors (e.g., start-ups and incumbents; Hockerts and Wüstenhagen, 2010; Schaltegger et al., 2016) or strategic issues, such as the entrepreneurship process (Belz and Binder, 2017; Keskin et al., 2013) or the potential necessity of sustainable entrepreneurs to engage in institutional work to achieve their goals (Pinkse and Groot, 2015; Thompson et al., 2015). A research stream has looked at the skill sets and motivations of sustainable entrepreneurs and how they differ from those of conventional entrepreneurs (Kirkwood and Walton, 2010; Nhemachena and Murimbika, 2018; Parrish, 2010).

Similarly, while research on sustainable or circular innovations and strategies often cites entrepreneurship as a major conduit for these, it has rarely focused on the business models specifically developed by new entrants or start-ups (Boons and Lüdeke-Freund, 2013; Klewitz and Hansen, 2014). For instance, Geissdoerfer et al. (2018) mention sustainable start-ups as a way to develop SBM innovations without discussing it further. Several recent studies present taxonomies or typologies of circular strategies and innovations adopted by firms but do not distinguish between incumbents and new entrants (Lüdeke-Freund et al., 2018; Ünal and Shao, 2019). The typology of CSUs presented in this chapter seeks to address this research gap.

Theoretical framework

This section presents the theoretical framework that was developed to code the empirical data collected (see Chapter 27), based on the literature outlined. RQ1 is concerned with the CBM strategies adopted by CSUs. To code these, the 4R-framework as presented by Kirchherr et al. (2017) was chosen over the other versions of the R-framework, because the former has been used more frequently in previous literature than the latter (Kirchherr et al., 2017; Stewart and Niero, 2018). Therefore, this analysis is easier to compare with existing studies like Stewart and Niero's analysis of incumbents' approaches to circularity (2018).

The hierarchy among the R-strategies in the 4R-framework corresponds to different degrees of resource value retention, with higher strategies corresponding to higher value retention. The retention of resource value here means "conservation of resources closest to their original state, and in the case of finished goods retaining their state or reusing them with a minimum of entropy as to be able to give them consecutive lives" (Reike et al., 2018: 254). Hence, activities that explicitly deal with the resource streams after product usage (*Recover, Recycle*) are ranked lowest, because they tend to lose any of the original product structure and therefore correspond to low value retention. In addition, recycling materials often requires high energy inputs for collection and re-processing, which may supersede the retained value (Ghisellini et al., 2016; Reike et al., 2018). Recycling and energy recovery are also typically easier to implement than other strategies, since they require few changes to the fundamentals of the business model (Ranta et al., 2018) and are, therefore, still largely compatible with a linear economy (Potting et al., 2017). Following this logic, *Reuse*, which is aimed at extending the use phase of products, is ranked higher than Recycle and Recover, because products are generally used again for the same purpose for which they were conceived, thereby retaining most of the resource value (Castellani et al., 2015). However, it is not considered as desirable as *Reduce*, where potential residual resource streams are avoided before the product even gets into circulation.

A dimension that has not been well covered so far by the existing versions of the R-frameworks is the preservation and enhancement of natural/biological ecosystems to deliver products or services. This dimension is important to consider as far as CSUs are concerned, because some of them are engaged in the development of so-called nature-based solutions, and it is a widely discussed topic in theory and policy (Davies and Lafortezza, 2019; European Commission, 2015; Faivre et al., 2017; Kabisch et al., 2016). These companies use ecosystems with decreased input of non-renewable natural capital and increased investment in renewable natural processes (Maes and Jacob, 2017). Thus, beyond the existing 4R-framework, the CBM strategy *Regenerate* was inductively included to cover CSUs engaged in the development of solutions based on natural/biological ecosystems. This strategy was placed above Reduce because Regenerate seeks not only to use fewer resources in production and consumption processes but also to restore natural or modified ecosystems, thereby not only retaining resource value but also increasing it. The authors summarise the different R strategies relevant for this chapter in Table 11.1.

Table 11.1 Coding categories for RQ1

CBM strategy	*Definition*
Regenerate	Maintain and increase the delivery of biological ecosystem services (i.e., the benefits provided) to society, for instance, through urban agriculture, green roofs, or aquaponics
Reduce	Increase the efficiency of product design or manufacturing by preventing or minimising the use of specific hazardous materials or any virgin materials or allowing for more intensive product use
Reuse	Bring products back into the economy after initial use or extend the lifespan of products and their parts (through repair, second-hand markets etc.)
Recycle	Process materials through, for example, shredding or melting to obtain the same (upcycling) or lower (downcycling) quality
Recover	Incinerate residual flows with recovery of embodied energy

Sources: Definitions adapted from Brennan et al. (2015); European Commission (2008); Ghisellini et al. (2016); King et al. (2006); Kirchherr et al. (2017); Potting et al. (2017); Reike et al. (2018).

Looking at CBM strategies through the frameworks presented in Table 11.1 gives insight into the respective organisations' CBM strategies (i.e., their general approach towards CE) but is not informative about the types of CBM innovations they pursue.[2] In order to address this, and to gain insight into the modes of adoption of the CBM strategies, the authors added CSU innovation types to the analysis. The CBM innovations developed by CSUs were coded based on their positions along the value chain (supplier or customer) and on their organisational focus (external or internal; see Table 11.2). Adjusting the taxonomy of CBMs proposed by Urbinati et al. (2017), the authors introduced the additional 'source' category to differentiate within upstream activities – those conducted by the focal organisation internally (source) and those conducted in interaction with suppliers and partners (upstream). It can be argued, that – due to their collaborative nature – especially the upstream-oriented innovations require a strong business network and high social capital, which are fostered, and crosscutting relationships as a communal base for cooperation and collective action (Peng et al., 2018). Thus, early-on effort in (strategic and public) relationship-building activities to build up social capital is highly relevant for CSUs to enable upstream circular innovations (Paunescu et al., 2014; Saffer, 2019).

This differentiation in source and upstream is deemed necessary, especially against the backdrop of innovation ecosystem and systems building theory, where it is argued that new technologies or innovations require a favourable socioeconomic embedding to be successful (Hekkert et al., 2007; Planko et al., 2016). The downstream and upstream activities are, therefore, those directly related to sociological, institutional change because they entail modifications of the relationships with other actors in the value chain (e.g., in terms of changes in ownership, consumption habits, or collaborative practises along the supply chain) and thereby can create legitimacy for the focal organisation. Activities primarily determined by the direct goals of the focal organisation – which are subsequently achieved through market acceptance or favourable market structures for their developed goods and services – are mainly technological, that is, source (Carvalho et al., 2017). Since only one start-up's business model in the entire sample turned out to be service based (maintenance, etc.) and including ownership transfer to the customer (Amsterdam-based Fairphone), only archetypes where the producer retains ownership were included as product-service-systems (PSSs) in the CBM framework. In the following, integration-oriented PSSs, product-oriented PSSs, and service-oriented PSSs (Neely, 2008;

Table 11.2 Coding categories for RQ2: CBM innovation types

Key actor	*CBM Innovation Category*	*CBM Innovation Type*	*Description*
Downstream (consumers)	**PSS** (servitisation/ no customer ownership)	*Use phase–oriented*	Improved usage efficiency through renting, leasing, sharing, extended after-sales services
		Result/ performance-oriented	Improved usage efficiency through subscription-based business models or pay-per-use, that is, focusing on the functionality
	Active consumer involvement	*Return, repair, reuse*	Inclusion of consumers in after-use product/ resource life cycle, enabled by value recovery and take-back processes at the source
		Collaborative consumption	(Cultivation and) usage of shared assets within communities, potentially including shared ownership
		(Educational) consumer engagement	(Open-source) knowledge sharing to change customer preferences and/or diffuse distinct CE practices, materials, or processes
Source (focal organisation)	**Core technology**	*Source material*	Substitution of source materials with less resource-intensive, novel alternatives (bio-based, more durable, biodegradable, recyclable)
		Product design	Increase interchangeability, upgradability, modularity, energy efficiency, or maintainability of products and product components
		Key process	Novel production method or innovation of sub-processes enabling circularity
	Enabling technology	*Sharing platform*	Enables increased product utilisation rate and reduced material throughput through shared use/access/ownership
		Trading platform	(Web) platform to facilitate the exchange and resale of products and materials
		Asset tracking	Tracking of products/components to enable adequate end-of-life treatment or create transparency on resource availability and origin
Upstream (suppliers, partners)	**Industrial symbiosis (IS)**	*Input-oriented*	Structured interorganisational collaboration to create value from residual resource streams **of external organisations** or usage of shared assets
		Output-oriented	Structured interorganisational collaboration to create value from residual resource streams **of focal organisation** or usage of shared assets
	Circularity standards/ accreditation	*Sourcing, manufacturing, transport/ packaging*	Establishing of process/material standards with suppliers through knowledge sharing and backward integration of activities along the supply chain

Pereira Pesoa and Jauregui Becker, 2017; Tukker, 2004) are not included based on an inductive approach. (Customer) experience orientation is considered a value element (Tukker, 2004) or antecedent of value (Schallehn et al., 2019) of a PSS business model but not an archetypal model on its own.

The difference between core and enabling technology (Potting et al., 2017) within source activities is that the former is specific to a certain product or process, while the latter can be applied to many industries and across social, technological, economic, and cultural systems. Within downstream activities, the authors distinguished between consumers' active involvement and the adoption of PSS models (Ren et al., 2019; Tukker, 2004; Tukker and Tischner, 2006; Vezzoli et al., 2015). The major difference between these two innovation categories is that in PSS, the producers remain owners of the product (Tunn et al., 2019), while in active involvement, they could only facilitate the (shared) consumption and return of externally owned products. For this research, reverse logistics is understood as (technical) materials routed reversely to the main resource flow, which corresponds with a still-dominant view in academic literature (Bernon et al., 2017; Carter and Ellram, 1998; Lambert and Stock, 1982; Murphy, 1986; Murphy and Poist, 1988; Rogers and Tibben-Lembke, 1999). Therefore, reverse logistics are an enabler for a multiplicity of value creating CBM innovations (PSS, IS, Return/Repair/Reuse; Farooque et al., 2019; Spring and Araujo, 2017) and run transversally across the categories presented in Table 11.1 but are not a stand-alone innovation type. The following section describes the methodology followed to collect and analyse empirical data with the help of this theoretical framework to further illuminate the circular business model strategies and innovations developed by CSUs.

Discussion and Conclusion

The CSU typology based on a representative data sample shown in this chapter is a step towards better understanding the role of entrepreneurship in circularity and sustainability transitions. The envisaged findings based on extensive empirical analysis are of value to foster a link between scientific and practical approaches to CE and will contribute to deepening insight on business models of CSUs. The previously mentioned typology and CSU archetypes will help to guide a future scientific agenda through expounding and highlighting their business models and thereby inspiring and enabling more systematic research on CSUs (i.e., the actual business models developed by circular entrepreneurs). For firms interested in CE, applicable combinations of CBM strategies and respective CBM innovation paths are elucidated, which will serve as an important source of information and inspiration on business-model approaches towards CE. At the same time, by highlighting the characteristics, especially of those CSUs that have achieved high levels of circularity, lessons can be drawn for managers willing to increase the level of circularity of their businesses. Managers could, for instance, examine CSUs' activities to identify promising opportunities both within and between industries and adopt best practices in their own business models. The findings are also relevant to policymakers who intend to support circular (start-up) activity and circular innovation in the area under their administrations.

Notes

1 The terms "green start-up" or "clean-tech start-up" are also used in the literature as synonyms for environmental start-up (Bergset and Fichter, 2015).
2 The innovation types were not compared to incumbents' innovation types because the authors newly developed the innovation type framework for which there existed no comparative studies.

References

Adams R, Jeanrenaud S, Bessant J, Denyer D, Overy P. (2016). Sustainability-oriented innovation: a systematic review. *International Journal of Management Reviews*. 18: 180–205.

Antikainen M, Aminoff A, Paloheimo H, Kettunen O. (2017). *Designing CBM experimentation – case study*. ISPIM Innovation Forum. Available at: https://www.researchgate.net/publication/316046098_Designing_circular_business_model_experimentation_-_Case_study.

Baines TS et al. (2007). State-of-the-art in PSSs. Proceedings of the Institution of Mechanical Engineers. Part B: *Journal of Engineering Manufacture*. 221: 1543–1552.

Beattie V, Smith SJ. (2013). Value creation and business models: refocusing the intellectual capital debate. *The British Accounting Review*. 45: 243–254.

Belz FM, Binder JK. (2017). Sustainable entrepreneurship: a convergent process model. *Business Strategy and the Environment*. 26: 1–17.

Bergset L, Fichter K. (2015). Green start-ups – a new typology for sustainable entrepreneurship and innovation research. *Journal of Innovation Management*. 3: 118–144.

Bernon M, Tjahjono B, Ripanti EF. (2017). Aligning retail reverse logistics practice with circular economy values: an exploratory framework. *Production, Planning and Control*. 29 (6): 483–497.

Blomsma F, Brennan G. (2017). The emergence of circular economy: a new framing around prolonging resource productivity. *Journal of Industrial Ecology*. 21 (3): 603–614.

Bocken NMP et al. (2014). A literature and practice review to develop sustainable business model archetypes. *Journal of Cleaner Production*. 65: 42–56.

Bocken NMP et al. (2016). Product design and business model strategies for a circular economy. *Journal of Industrial and Production Engineering*. 33: 308–320.

Bocken NMP et al. (2017). The circular economy – exploring the introduction of the concept among S&P 500 firms. *Journal of Industrial Ecology*. 21 (3): 487–490.

Boons F, Lüdeke-Freund F. (2013). Business models for sustainable innovation: state-of-the-art and steps towards a research agenda. *Journal of Cleaner Production*. 45: 9–19.

Borzaga C, Defourny J. (2001). *The emergence of social enterprise*. London and New York: Routledge.

Brennan D, Tennant M, Blomsma F. (2015). Business and production solutions – Closing loops and the circular economy. In: Kopnina H, Blewitt J. (Eds.). *Sustainability: key issues*. London: Routledge.

Brush CG, Vanderwerf PA. (1992). A comparison of methods and sources for obtaining estimates of new venture performance. *Journal of Business Venturing*. 7: 157–170.

Carter CR, Ellram LM. (1998). Reverse logistics – a review of the literature and framework for future investigation. *Journal of Business Logistics*. 19 (1): 85–102.

Carvalho ADP et al. (2017). The role and contributions of sociological institutional theory to the socio-technical approach to innovation theory. *RAI Revista de Administração e Inovação*. 14 (3): 250–259.

Castellani V, Sala S, Mirabella N. (2015). Beyond the throwaway society: a life cycle-based assessment of the environmental benefit of reuse. *Integrated Environmental Assessment and Management*. 11: 373–382.

Chesbrough H. (2010). Business model innovation: opportunities and barriers. *Long Range Planning*. 43: 354–363.

Choi DY, Gray ER. (2008). The venture development processes of "sustainable" entrepreneurs. *Management Research News*. 31: 558–569.

Christensen C. (1997). *The innovator's dilemma: when new technologies cause great firms to fail*. Boston, MA: Harvard Business School Press.

Circle Economy (2018). *The circularity gap report: an analysis of the circular state of the global economy*. Available at: https://docs.wixstatic.com/ugd/ad6e59_ba1e4d16c64f44fa94fbd8708eae8e34.pdf.

Cohen B, Winn MI. (2007). Market imperfections, opportunity and sustainable entrepreneurship. *Journal of Business Venturing*. 22: 29–49.

Davies C, Lafortezza R. (2019). Transitional path to the adoption of nature-based solutions. *Land Use Policy*. 80: 406–409.

Dean TJ, McMullen JS. (2007). Toward a theory of sustainable entrepreneurship: reducing environmental degradation through entrepreneurial action. *Journal of Business Venturing*. 22: 50–76.

De los Rios IC, Charnley FJS. (2017). Skills and capabilities for a sustainable and circular economy: the changing role of design. *Journal of Cleaner Production*. 160: 109–122.

Diaz Lopez F, Bastein T, Tukker A. (2019). Business model innovation for resource-efficiency, circularity and cleaner production: what 143 cases tell us. *Ecological Economics*. 155: 20–35.

Doherty B, Haugh H, Lyon F. (2014). Social enterprises as hybrid organizations: a review and research agenda. *International Journal of Management Reviews*. 16: 417–436.

Ellen MacArthur Foundation (2012). *Towards the CE: economic and business rationale for an accelerated transition*. Available at: http://tinyurl.com/pv7q7l4.

European Commission (2008). *Directive 2008/98/EC of the European parliament and of the council of 19 November 2008 on waste and repealing certain directives*. Brussels, Belgium.

European Commission (2015). *Towards an EU research and innovation policy agenda for nature-based solutions and re-naturing cities*. Final Report of the Horizon 2020 expert group on "Nature-Based Solutions and Re-Naturing Cities." Brussels, Belgium.

Faivre N, Fritz M, Freitas T, de Boissezon B, Vandewoestijne S. (2017). Nature-based solutions in the EU: innovating with nature to address social, economic and environmental challenges. *Environmental Research*. 159: 509–518.

Farooque M, Zhang A, Huisingh D. (2019). Circular supply chain management: a definition and structured literature review. *Journal of Cleaner Production*. 228: 882–900.

Foss N, Saebi T. (2016). Fifteen years of research on business model innovation: how far have we come, and where should we go? *Journal of Management*. 43 (1): 200–227.

Geissdoerfer M, Savaget P, Bocken NMP, Hultink EJ. (2017). The circular economy – a new sustainability paradigm? *Journal of Cleaner Production*. 143: 757–768.

Geissdoerfer M, Vladimirova D, Evans S. (2018). Sustainable business model innovation: a review. *Journal of Cleaner Production*. 198: 401–416.

Geng Y, Zhu, Q, Doberstein B, Fujita, T. (2009). Implementing China's circular economy concept at the regional level: a review of progress in Dalian, China. *Waste Management*. 29 (2): 996–1002.

Ghisellini P, Cialani C, Ulgiati S. (2016). A review on CE: the expected transition to a balanced interplay of environmental and economic systems. *Journal of Cleaner Production*. 114: 11–32.

Haas W, Kausmann F, Wiedenhofer D, Heinz M. (2005). How circular is the global economy? An assessment of material flows, waste production, and recycling in the European Union and the world in 2005. *Journal of Industrial Ecology*. 19 (5): 765–777.

Hall JK, Daneke GA, Lenox MJ. (2010). Sustainable development and entrepreneurship: past contributions and future directions. *Journal of Business Venturing*. 25 (5): 439–448.

Hekkert MP, Suurs RAA, Negro SO, Kuhlmann S, Smits REHM. (2007). Functions of innovation systems – a new approach for analysing technological change. *Technological Forecasting and Social Change*.74 (4): 413–432.

Henderson R.; Clark K. (1990). Architectural innovation: the reconfiguration of existing product technologies and the failure of established firms. *Administrative Science*. 35 (1): 9–30.

Heyes G, Sharmina M, Mendoza JMF, Gallego Schmid A. (2018). Developing and implementing circular economy business models in service-oriented technology companies. *Journal of Cleaner Production*. 177: 621–632.

Hill CWL, Rothaermel FT. (2003). The performance of incumbent firms in the face of radical technological innovation. *Academy of Management Review*. 28 (2): 258–274.

Hockerts K, Wüstenhagen R. (2010). Greening Goliaths versus emerging Davids – theorizing about the role of incumbents and new entrants in sustainable entrepreneurship. *Journal of Business Venturing, Sustainable Development and Entrepreneurship*. 25: 481–492.

Jawahir IS, Rouch KE, Dillon OW, Holloway Jr, L, Hall A. (2007). Design for sustainability (DFS): new challenges in developing and implementing a curriculum for next generation design and manufacturing engineers. *International Journal of Engineering Education*. 23 (6): 1053–1064.

Jiang W, Chai H, Shao J, Feng T. (2018). Green entrepreneurial orientation for enhancing firm performance: a dynamic capability perspective. *Journal of Cleaner Production*. 198: 1311–1323.

Johnson MW. (2010). *Seizing the white space. Business model innovation for growth and renewal*. Boston, MA: Harvard Business School Press.

Kabisch N et al. (2016). Nature-based solutions to climate change mitigation and adaptation in urban areas: perspectives on indicators, knowledge gaps, barriers, and opportunities for action. *Ecology and Society*. 21 (2): 39.

Keskin D, Diehl JC, Molenaar N. (2013). Innovation process of new ventures driven by sustainability. *Journal of Cleaner Production*. 45: 50–60.

King AM, Burgess SC, Ijomah W, Mcmahon CA. (2006). Reducing waste: repair, recondition, remanufacture or recycle? *Sustainable Development*. 14: 257–267.

Kirchherr J et al. (2017). Conceptualizing the CE: an analysis of 114 definitions. *Resources, Conservation and Recycling*. 127: 221–232.
Kirkwood J, Walton S. (2010). What motivates ecopreneurs to start businesses? *International Journal of Entrepreneurial Behavior & Research*. 16 (3): 204–228.
Klewitz J, Hansen EG. (2014). Sustainability-oriented innovation of SMEs: a systematic review. *Journal of Cleaner Production*. 65: 57–75.
Krausmann F, Gingrich S, Eisenmenger N, Erb KH, Haberl H, Fischer-Kowalski M. (2009). Growth in global materials use, GDP and population during the 20th century. *Ecological Economics*. 68: 2696–2705.
Lambert DM, Stock JR. (1982). *Strategic physical distribution management*. Homewood, IL: Irwin.
Lenox M, York JG. (2011). *Environmental entrepreneurship*. Oxford: University Press.
Lewandowski M. (2016). Designing the business models for circular economy – towards the conceptual framework. *Sustainability*. 8 (1): 43.
Lieder M, Rashid A. (2016). Towards circular economy implementation: a comprehensive review in context of manufacturing industry. *Journal of Cleaner Production*. 115: 36–51.
Linder M, Williander M. (2017). CBM innovation: inherent uncertainties. *Business Strategy and Environment*. 26 (2): 182–196.
Lüdeke-Freund F, Carroux S, Joyce A, Massa L, Breuer H. (2018). The sustainable business model pattern taxonomy – 45 patterns to support sustainability-oriented business model innovation. *Sustainable Production and Consumption*. 15: 145–162.
Lüdeke-Freund F, Dembek K. (2017). Sustainable business model research and practice: emerging field or passing fancy? *Journal of Cleaner Production*. 168: 1668–1678.
Luger M, Koo J. (2005). Defining and tracking business start-ups. *Small Business Economics*. 24 (1): 17–28.
Maes J, Jacobs S. (2017). Nature-based solutions for Europe's sustainable development. *Conservation Letters*. 10: 121–124.
Massa L, Tucci C. (2014). *Business model innovation: the Oxford handbook of innovation management*. Oxford: Oxford University Press.
McDonough W, Braungart M. (2002). *Cradle to cradle: remaking the way we make things*. New York: North Point.
McKinsey & Co (2015). *Europe's circular-economy opportunity*. McKinsey & Company. Available at: https://www.mckinsey.com/~/media/McKinsey/Business%20Functions/Sustainability/Our%20Insights/Europes%20circular%20economy%20opportunity/Europes%20circulareconomy%20opportunity.pdf
Mont O, Tukker A. (2006). Product-service systems: reviewing achievements and refining the research agenda. *Journal of Cleaner Production*. 14 (17): 1452–1454.
Murphy PR. (1986). A preliminary study of transportation and warehousing aspects of reverse distribution. *Transportation Journal*. 25 (4): 12–21.
Murphy PR, Poist RF. (1988). Management of logistical retromovements: an empirical analysis of literature suggestions. *Journal of the Transportation Research Forum*. 29 (1): HS-040 801.
Murray A, Skene K, Haynes K. (2017). The circular economy: an interdisciplinary exploration of the concept and application in a global context. *Journal of Business Ethics*. 140: 369–380.
Neely A. (2008). Exploring the financial consequences of the servitization of manufacturing. *Operations Management Research*. 1 (2): 103–118.
Nhemachena C, Murimbika M. (2018). Motivations of sustainable entrepreneurship and their impact of enterprise performance in Gauteng province, South Africa. *Business Strategy & Development*. 1: 115–127.
Osterwalder A, Pigneur Y, Clark T. (2010). *Business model generation: a handbook for visionaries, game changers, and challengers*. Hoboken, NJ: Wiley.
Pacheco DF, Dean TJ, Payne DS. (2010). Escaping the green prison: entrepreneurship and the creation of opportunities for sustainable development. *Journal of Business Venturing, Sustainable Development and Entrepreneurship*. 25: 464–480.
Parrish BD. (2010). Sustainability-driven entrepreneurship: principles of organization design. *Journal of Business Venturing, Sustainable Development and Entrepreneurship*. 25: 510–523.
Paunescu C, Raluca Badea M. (2014). Examining the social capital content and structure in the pre-start-up planning. *Procedia Economics and Finance*. 15: 560–568.
Peng B, Tu Y, Wei G. (2018). Governance of electronic waste recycling based on social capital embeddedness theory. *Journal of Cleaner Production*. 187: 29–36.
Pereira Pesoa MV, Jauregui Becker JM. (2017). Overcoming the product-service model adoption obstacles. *Procedia CIRP*. 64: 163–168.

Pieroni MPP, McAloone TC, Pigosso DCA. (2019). Business model innovation for circular economy and sustainability: a review of approaches. *Journal of Cleaner Production*. 215: 198–216.

Planko J, Cramer J, Chappin M, Hekkert M. (2016). Strategic collective system building to commercialize sustainability innovations. *Journal of Cleaner Production*. 112: 2328–2341.

Pinkse J, Groot K. (2015). Sustainable entrepreneurship and corporate political activity: overcoming market barriers in the clean energy sector. *Entrepreneurship Theory and Practice*. 39: 633–654.

Potting J, Hekkert M, Worrell E, Hanemaaijer A. (2017). *Circular economy: measuring innovation in the product chain in policy report for PBL Netherlands environmental assessment agency*. The Hague, The Netherlands: PBL Netherlands Environmental Assessment Agency.

Ranta V, Aarikka-Stenroos L, Mäkinen SJ. (2018). Creating value in the circular economy: a structured multiple-case analysis of business models. *Journal of Cleaner Production*. 201: 988–1000.

Reike D, Vermeulen WJV, Witjes S. (2018). The circular economy: new or refurbished as CE 3.0? – exploring controversies in the conceptualization of the circular economy through a focus on history and resource value retention options. *Resources, Conservation and Recycling*. 135: 246–264.

Ren S, Zhang Y, Liu Y, Sakao T, Huisingh D, Almeida CMVB. (2019). A comprehensive review of big data analytics throughout product lifecycle to support sustainable smart manufacturing: a framework, challenges and future research directions. *Journal of Cleaner Production*. 210: 1343–1365.

Richardson J. (2008). The business model: an integrative framework for strategy execution. *Strategic Change*. 17: 133–144.

Rizos, V et al. (2016). Implementation of circular economy business models by small and medium-sized enterprises (SMEs): barriers and enablers. *Sustainability*. 8 (11): 1212.

Rogers DS, Tibben-Lembke RS. (1999). *Going backwards: reverse logistics trends and practices*. Pittsburgh, PA: Reverse Logistics Executive Council. Vol. 2.

Saffer AJ. (2019). Fostering social capital in an international multi-stakeholder issue network. *Public Relations Review*. 45 (2): 282–296.

Schallehn H, Seuring S, Strähle J, Freise M. (2019). Customer experience creation for after-use products: a product–service systems-based review. *Journal of Cleaner Production*. 210: 929–944.

Schaltegger S. (2002). A framework for ecopreneurship: leading bioneers and environmental managers to ecopreneurship. *Greener Management Journal*. 38: 45–58.

Schaltegger S, Lüdeke-Freund F, Hansen EG. (2016). Business models for sustainability: a co-evolutionary analysis of sustainable entrepreneurship, innovation, and transformation. *Organization & Environment*. 29: 264–289.

Schaltegger S, Wagner M. (2011). Sustainable entrepreneurship and sustainability innovation: categories and interactions. *Business Strategy and the Environment*. 20: 222–237.

Schiederig T, Tietze F, Herstatt C. (2012). Green innovation in technology and innovation management – an exploratory literature review. *R&D Management*. 42: 180–192.

Shepherd DA, Patzelt H. (2011). The new field of sustainable entrepreneurship: studying entrepreneurial action linking "what is to be sustained" with "what is to be developed." *Entrepreneurship Theory and Practice*. 35: 137–163.

Sihvonen S, Ritola T. (2015). Conceptualizing ReX for aggregating end-of-life strategies in product development. *Procedia CIRP*. 29: 639–644.

Spring M, Araujo L. (2017). Product biographies in servitization and the circular economy. *Industrial Marketing Management*. 60: 126–137.

Stewart R, Niero M. (2018). Circular economy in corporate sustainability strategies: a review of corporate sustainability reports in the fast-moving consumer goods sector. *Business Strategy and the Environment*. 27 (7): 1005–1022.

Stubbs W, Cocklin C. (2008). Conceptualizing a sustainability business model. *Organization & Environment*. 21: 103–127.

Thompson NA, Herrmann AM, Hekkert MP. (2015). How sustainable entrepreneurs engage in institutional change: insights from biomass torrefaction in the Netherlands. *Journal of Cleaner Production*. 106: 608–618.

Thompson NA, Kiefer K, York JG. (2011). *Distinctions not dichotomies: exploring social, sustainable, and environmental entrepreneurship*. Bingley: Social and Sustainable Entrepreneurship. Emerald Group Publishing Limited. 201–229.

Tukker A. (2004). Eight types of product–service system: eight ways to sustainability? Experiences from SusProNet. *Business Strategy and the Environment*. 13: 246–260.

Tukker A, Tischner U. (2006). Product-services as a research field: past, present and future. Reflections from a decade of research. *Journal of Cleaner Production*. 14: 1552–1556.

Tunn V, Fokker R, Luijkx K, de Jong S, Schoormans J. (2019). Making ours mine: increasing consumer acceptance of access-based PSS through temporary product customization. *Sustainability*. 11 (1): 274.

Ünal E, Shao J. (2019). A taxonomy of circular economy implementation strategies for manufacturing firms: analysis of 391 cradle-to-cradle products. *Journal of Cleaner Production*. 212: 754–765.

Urbinati A, Chiaroni D, Chiesa V. (2017). Towards a new taxonomy of circular economy business models. *Journal of Cleaner Production*. 168: 487–498.

van Renswoude K, ten Wolde A, Joustra DJ. (2015). *CBMs – Part 1: an introduction to IMSA's CBM scan*. Amsterdam: IMSA.

Vezzoli C, Ceschin F, Diehl JC, Kohtala C. (2015). New design challenges to widely implement 'sustainable product–service systems'. *Journal of Cleaner Production*. 97: 1–12.

Williams A. (2007). Product service systems in the automobile industry: contribution to system innovation? *Journal of Cleaner Production*. 15: 1093–1103.

Yan J, Wu N. (2011). *Technology supporting system of circular economy of mining cities*. 2011 Asia-Pacific Power and Energy Engineering Conference. IEEE. 1–5.

York JG, Venkataraman S. (2010). The entrepreneur–environment nexus: uncertainty, innovation, and allocation. *Journal of Business Venturing, Sustainable Development and Entrepreneurship*. 25: 449–463.

Zahra SA, Ireland RD, Hitt MA. (2000). International expansion by new venture firms: international diversity, mode of market entry, technological learning, and performance. *Academy of Management Journal*. 43: 925–950.

12

ECODESIGN AND CIRCULAR DESIGN OF PRODUCTS

Concepts, assessment, and strategies

Vicente B. Vert and Eva Verdejo

Introduction

The concept of ecodesign arose at the beginning of the 90s, following the movement of some sectors towards the use of products that are more environmentally friendly. It started from the need to minimise the potential environmental impacts of the product before they occurred, thus moving away from the traditional way of rectifying environmental damage after it had already taken place.

Ecodesign means the integration of environmental aspects into a product design with the aim of improving its environmental performance throughout its whole lifecycle, from the extraction of raw materials used for manufacturing to its end of life.

The lifecycle of a product is historically defined as the consecutive and interrelated stages of a product system. It is composed of the following stages (Figure 12.1):

- Extraction of raw materials. This includes extraction of raw materials and components, their associated energy use, and their processing
- Production. The transformation processes to which the materials and different components are subjected. These are considered until the finished product is obtained
- Distribution. All actions related to packaging, distribution, and marketing processes are included, both for finished products as well as for raw materials and components employed in the final product
- Use. This includes the useful life of the product and its interaction with the rest of the auxiliary resources (i.e., consumables, energy, maintenance, etc.) that make it possible for the product to perform its function. It also includes the processes for its maintenance
- End of life. These are the different disposal scenarios once the useful life of the product has been exhausted for the functions for which it was intended. The different scenarios include reuse, recovery, and disposal.

The need to change the traditional model from raw materials being used for producing products to be used then disposed, to a more circular model, has led to other design concepts based on more complex cycles, such as interrelated flow loops. In this new scenario, loops should be

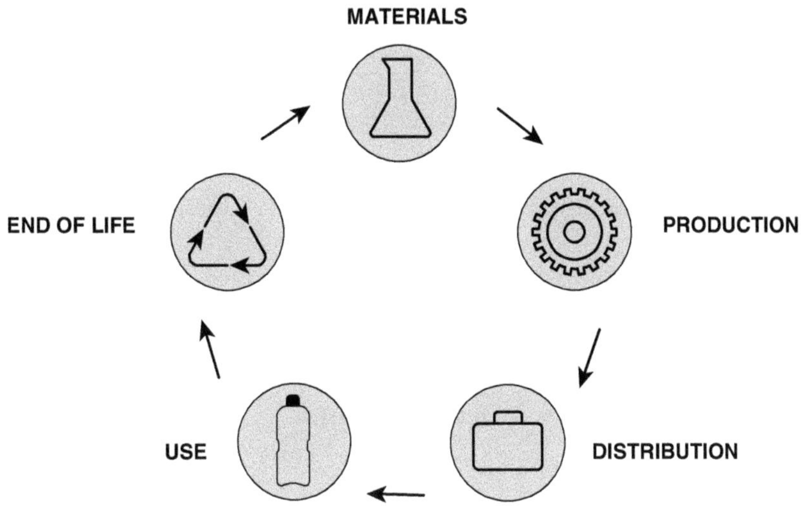

Figure 12.1 Lifecycle of products

considered for designing, as the lifecycle of a product could be extended by processes such as sharing, reusing, remanufacturing, repairing, and recycling.

Thus, design becomes a crucial factor in the circular economy strategy, where sustainability is also considered (Ceschin, 2016). As a preventative approach, designed to optimise the environmental performance of products or services while maintaining their functional qualities for as long as possible within the system, such a strategy provides genuine new opportunities for manufacturers, consumers, and citizens.

Advantages of ecodesign and circular design

Ecodesign as such is a method that does not modify the basic structure that is followed in the development of new products, but the design and development process is complemented by the inclusion of environmental criteria (CEGESTI, 1999). They are evaluated at the same level as other criteria such as economics, quality, functionality, ergonomics, aesthetics, or safety, among others (Figure 12.2).

Design under the scope of a circular economy (or circular design) also considers environmental impacts, but it puts the focus on keeping the value of the product for as long as possible within the system. Circular design comprises aspects like durability, repairability, compatibility, maintenance, refurbishing, recycling, etc. However, it also requires a new way of thinking: the best circular approach is, indeed, not to have the need for the product to exist (Figure 12.3).

The importance of considering environmental and circular criteria at the design stage relates to the fact that once the product has been manufactured and begins to be marketed, the possibility of reducing its environmental impacts is much smaller (or sometimes zero) and, complementarily, the space for increasing its circularity is also limited.

Environmental impacts also play a role within the circular economy, as they are interrelated. Considering the lifetime of a product as an example, let's assume it has a lifetime of one unit of

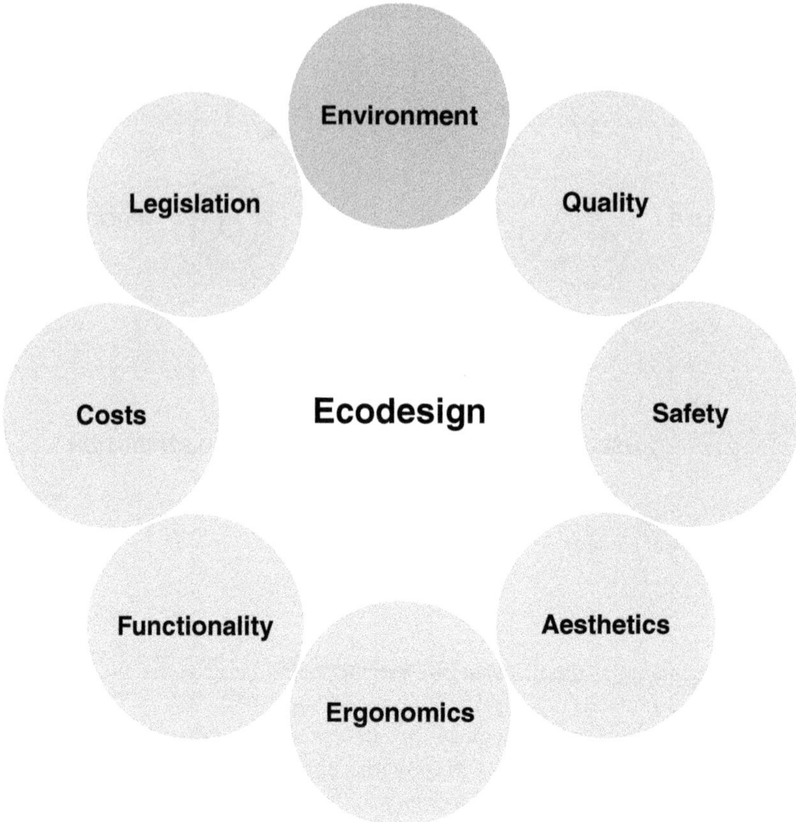

Figure 12.2 Ecodesign approach is not just about environment

time. When there exists the need to use it for two units of time, in a linear economy approach, two products would be needed. Even when ecodesigned to have the minimum environmental impacts, the total environmental impacts after the two units of time have passed correspond to the impacts of each product multiplied by two. However, if the lifespan is extended by circular design up to two units of time, thus increasing the circularity, the overall environmental impacts would be (presumably) lower than double, as the product was designed for it.

Ecodesign and circular design can also offer other advantages in addition to environmental improvements, such as:

- Cost reduction related to the consumption of raw materials and energy
- Quality improvement, as the product is designed based on the whole lifecycle
- Improvement of corporate image and prestige
- Accessibility to new markets, being able to serve as an ecological/circular marketing tool to attract new customers and consumers (creating new market niches)
- Stimulation of innovation, reaching products with better environmental performance and that are more innovative, with more efficient processes
- Advancement of new legal requirements, since these voluntary improvements make products ready for adaptation to mandatory requirements

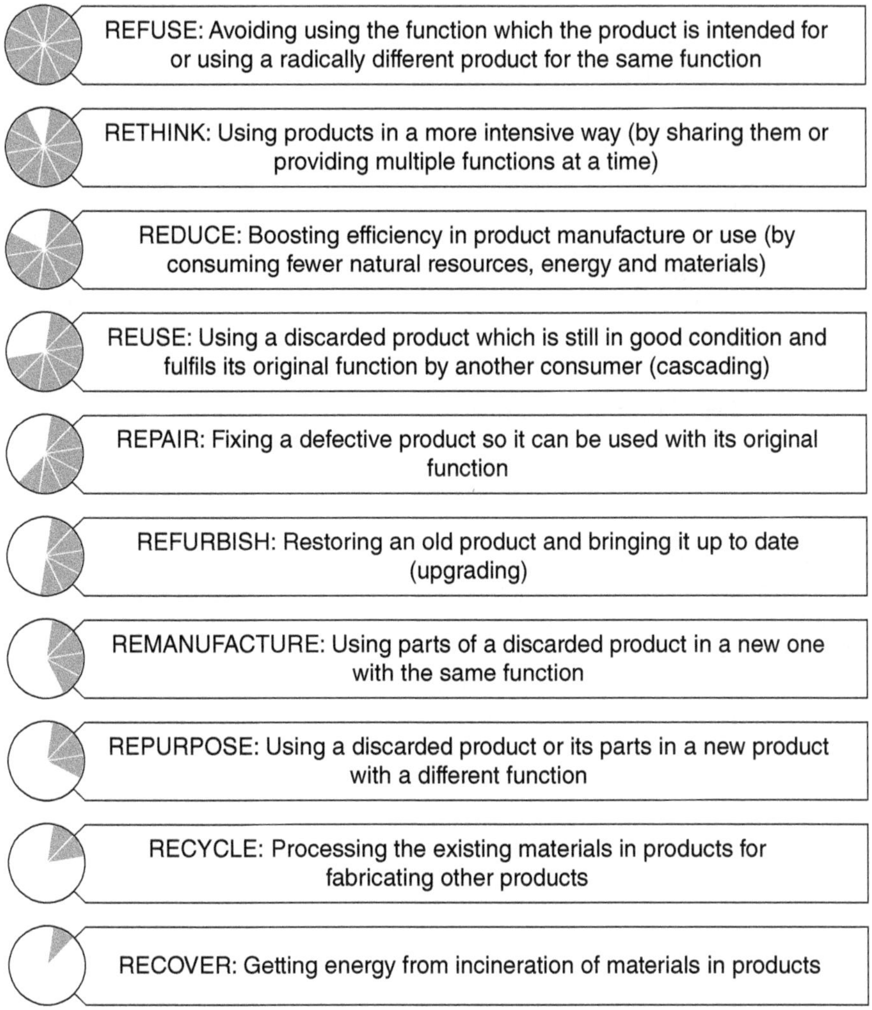

Figure 12.3 Ten circular approaches to be considered during circular design

Standardisation related to ecodesign and circular design

The standard ISO 14006 'Environmental Management Systems Guidelines for Incorporating Ecodesign' provides guidelines to assist organisations in establishing, documenting, implementing, maintaining, and continually improving their management of ecodesign as part of an environmental management system (EMS) (ISO, 2020). This standard is intended to be used by those organisations that have implemented an EMS in accordance with ISO 14001, but it can help in integrating ecodesign in other management systems. The guidelines are applicable to any organisation regardless of its size or activity (ISO, 2015). This standard applies to those product-related environmental aspects that the organisation can control and influence; it does not establish by itself specific environmental performance criteria and is not intended for certification purposes.

On the circular economy side, only the (2017) BS 8001 "Framework for Implementing the Principles of the Circular Economy in Organizations Guide" exists. In addition, the experimental French standard XP X30–901 (October 2018), "Circular Economy – Circular Economy Project Management System – Requirements and Guidelines," is also available. These two guides are not focused on products.

Nonetheless, different technical committees at both international and national levels, for example, ISO/TC 323, "Circular Economy," and CTN 323, "Economía Circular" (circular economy in Spanish), have been working on standards in this field focused on: (1) principles, framework, terminology, and management system standards; (2) guidelines for implementation and sectoral applications; (3) measuring circularity/supporting tools; and (4) specific issues for the circular economy.

Legislation related to ecodesign and circular design

There are pieces of European legislation related to ecodesign, focused on products that consume energy. The legislative framework in this field is constituted by:

- Ecodesign Directive (Directive 2009/125/EC) that provides consistent EU-wide rules for improving the environmental performance of products, such as household appliances, information and communication technologies, or engineering. The Directive sets out minimum mandatory requirements for the energy efficiency of these products. For energy-related products, in this Directive, the following main environmental aspects are considered: (1) energy consumption; (2) material consumption; (3) water consumption; (4) emissions to air, water, and soil; (5) hazardous content; and (6) waste aspects (waste generation; possibilities for reuse, recycling, and recovery) (EC, 2009).
- Energy Labelling Regulation (Regulation [EU] 2017/1369) may complement those ecodesign requirements with mandatory labelling requirements (EC, 2017).

Additionally, there are pieces of European legislation related to ecodesign and energy labelling for specific products, including computers, home appliances, lamps, pumps, and related devices consuming energy (EC, 2009).

Ecodesign is directly related to the European strategy for a circular economy. In December 2015, the European Commission adopted a Circular Economy Action Plan with the aim to set the European Union on the course of a transition towards a more sustainable model for economic development (EC, 2015). It includes, among other aspects, comprehensive commitments on ecodesign. The Plan highlighted the importance of design to reduce environmental impacts and the need to work on ecodesign of products (which will help with the reparability, upgradability, durability, and recyclability of products) by extending the scope of ecodesign requirements beyond energy efficiency.

The document states that:

- The Commission will promote the reparability, upgradability, durability, and recyclability of products by developing product requirements relevant to the circular economy in its future work under the Ecodesign Directive, as appropriate and taking into account the specificities of different product groups. The ecodesign working plan for 2015–2017 will elaborate on how this will be implemented. The Commission will shortly also propose ecodesign requirements for electronic display.

- The Commission will specifically consider proportionate requirements on durability and the availability of repair information and spare parts in its work on Ecodesign, as well as durability information in future energy labelling measures.

Procedure for designing a product

A design project that incorporates ecodesign or circular design can be broken down into four main steps (Figure 12.4):

1. **Project Preparation**. It is advisable to form a working team where participants have different knowledge of the product in question and the authority to make decisions on it. Therefore, it is preferable that the selected team has the following characteristics: multidisciplinary, small, organised, and with decision-making capacity
2. **Identification of Product Specifications**. The product must have enough degrees of freedom; that is, it must be capable of changing its design. It must also meet the motivating factors of the company for the application of design. Once the product to be designed or redesigned is selected, it is about identifying its environmental/circular profile and evaluating the potential circular/environmental aspects of the product able to be adopted. For this phase, specific analysis tools are needed
3. **Definition of Design Strategies**. There exist various established tasks, known as strategies, which help the team identify possible improvement actions in the product. Once selected, it is advisable to assess the degree of applicability related to the technical and economic viability
4. **Action Plan**. In this phase, the product is defined by the new requirements developed in the previous stages in terms of technical and circular/environmental requirements, such as components, dimensions, production processes, packaging, and providing the final sketch of the product.

The differences in the project design for ecodesign and circular design arise mainly from the way the product (or service) is analysed and assessed, that is, the scope of this assessment, but also the strategies to be applied to it. For this reason, these assessment tools and strategies will be defined in the following sections.

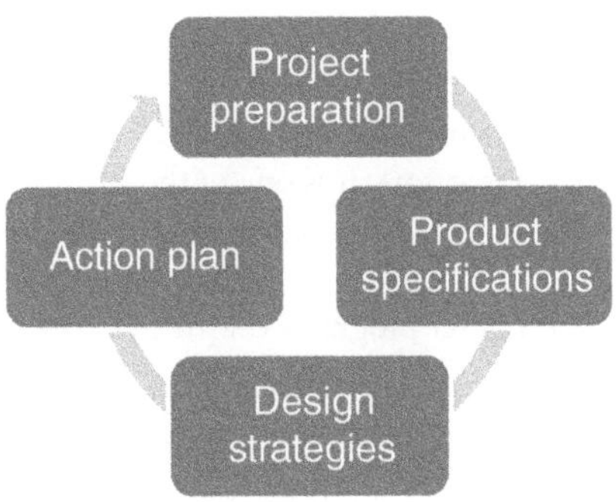

Figure 12.4 Steps for preparing a design project, both ecodesign and circular design

Tools for assessing the circular/environmental impacts of a product

Environmental assessment tools

There are a few environmental analysis tools which can be utilised to identify which aspects within a product have the greatest environmental impact. They serve to detect points of improvement or to compare products or design options. Most of these tools are based on the product lifecycle, which provides an overview of the environmental aspects of the product to avoid transferring the impacts from one stage of the cycle to another. The choice of one tool over another depends on the type of product, the objective of the study, and the availability of data, among other factors.

The following sections describe the environmental analysis tools that are considered most useful to introduce environmental criteria in the development of designs.

Materials use, energy use, and toxic materials matrix

The Materials use, Energy use and Toxic materials (MET) tool is based on a matrix of qualitative or semiquantitative metrics which offers an overview of the inputs (materials and energy) and outputs (emissions, spills, and toxic waste) of each phase of the product lifecycle. The main environmental aspects of the product and possible environmental improvement options are listed or quantified at each stage.

The main objective of the MET matrix is to identify environmental weaknesses or priorities on which to work for the product when very precise information is not available. It is used to detect in broad strokes the main environmental aspects of a product. These quantitative and/or qualitative data are arranged on a double entry matrix, where the main categories (**M**aterials use, **E**nergy use, and **T**oxic materials, emissions and waste) are shown in different columns in relation to the different rows where each stage of the product lifecycle is shown.

Once the MET matrix has been proposed, subjective conclusions can be drawn, allowing the identification of the main environmental weaknesses of the product and, therefore, showing at which stages the consideration of environmental criteria is a priority without the need to be an expert in the subject. When drawing conclusions from the matrix, it should be noted that the use of toxic and/or hazardous materials, as well as the consumption of energy or other auxiliary consumables of the product during use, will be aspects to be prioritised in the assessment scale of the improvement actions.

Lifecycle assessment

This tool includes computer programs based on the lifecycle assessment (LCA) methodology, which includes a database of eco-indicators. It is one of the key tools in the environmental analysis of products. It consists of an objective evaluation that prioritises the environmental aspects of a product and potential associated environmental impacts within a system using computer programs. There are different computer programs on the market that vary in extent and quality of data, economic cost, and so on.

The objective of this tool is to identify, quantify, interpret, and evaluate the environmental impacts of a product under a global vision and divided by each stage of its lifecycle. It allows visualisation of the main environmental aspects of the product in a graphical way, identifying improvement points, as well as comparing different product alternatives. It is standardised by

ISO 14040, 'Environmental Management – Life Cycle Assessment – Principles and Framework' (ISO, 2006). The steps to follow for applying this tool are mainly the following:

- Definition of objectives and scope of the study. In this phase, the purpose of the study is defined, and the limits of the system (what is to be included and what not) and the functional unit must be established
- Inventory analysis. This consists of obtaining detailed data so that all inputs and outputs of the product are identified and quantified, such as consumption of resources (raw materials, water, energy, etc.) and emissions to air, water, and soil. The data collection process is the most intensive step
- Impact assessment. The conversion of collected data into environmental impacts requires databases and computing software. There is no internationally accepted single database that evaluates the environmental impact of a product. The accuracy and veracity of the results depend on the quality of the data collected, the data source selected, and the robustness of the software
- Interpretation of the result. The LCA software provides numerical data and graphs that show what the main impacts of the product are and in which stages they are present. This stage makes use of them for detecting points of environmental improvements within the product.

Circular assessment tools

Widely accepted tools for assessing circularity do not exist yet. As the concept of circular economy encompasses a change in the way companies and products act, some of these existing tools are focused on companies or just materials, or they are based on the qualitative assessment of the circular approaches (as described in Figure 12.3).

As previously highlighted, ISO/TC 323 is working on tools for measuring circularity, among other issues, so there is not a common framework in which products/companies can work. British and French standards are not focused on products. Nonetheless, some tools have appeared for assessing circularity at the level at which they are created to do so, as described in Table 12.1.

Design strategies

Environmental strategies

In ecodesign projects, the definition of strategies is key. In general, there are eight ecodesign strategies applicable to any product when based on the lifecycle approach (InEDIC, 2011). Table 12.2 shows the relationship of these strategies to the stages of the lifecycle of a product.

For each of these strategies, several actions can be described, as follows.

New concept development

This strategy focuses on a stage prior to the product lifecycle. Specifically, it focuses on the conception of the product and its function. The strategy is based on optimising the function of the product or even replacing it with a service. Functional optimisation involves eliminating those superfluous components that do not fulfil any basic product function.

Table 12.1 Details for some circularity assessment tools

Tool	*Applies to*	*Description*	*Measures*	*Owner/Creator*	*Notes*
Material Circularity Indicator (MCI) (EMF, 2015)	Product	Measures the extent to which linear flow has been minimised and restorative flow maximised for its component materials and how long and intensively it is used compared to a similar industry-average product	From 0 (fully linear) to 1 (fully circular)	Ellen MacArthur Foundation	Focused only on materials
Circular Transition Indicators (CTIs) (EMF, 2020)	Business	They determine the ability of the company to ensure both resource extraction and material waste are reduced to a minimum	Some indicators include % circular inflow, % circular outflow, % water circularity, and % renewable energy	World Business Council for Sustainable Development	Under development
Circulytics	Business	It supports a company's transition towards the circular economy, regardless of industry, complexity, and size. Going beyond assessing products and material flows, this company-level measuring tool reveals the extent to which a company has achieved circularity across its entire operations	Overall single score from A+ to E	Ellen MacArthur Foundation	Based on a detailed survey. To be updated yearly

Table 12.2 Ecodesign strategies related to the lifecycle stages

Life-cycle stages	*Strategies*
–	New concept development
Extraction of raw materials	Selection of low-impact materials
	Reduction of material use
Production	Optimisation of production techniques
Distribution	Optimisation of distribution system
Use	Reduction of impact during use
	Optimisation of lifetime
End of life	Optimisation of end-of-life system

Another possibility is multifunction, which consists of integrating several functions in the same product, using the same amount of resources to provide it with more functions. This is the case with mobile phones, which have a wide variety of functions apart from their main function (e.g., calendar, alarm clock, camera, internet connection, etc.). Some actions for this new concept development are dematerialisation, product sharing, function integration, and functional optimisation of the product (components).

Selection of low-impact materials

A simple and effective way to select low-impact materials is through ecological labels, which indicate the environmental performance of raw materials and components. Another option is to work with suppliers that have an environmental management system or can demonstrate good environmental practices. The use of renewable, recycled, and recyclable materials contributes to reducing the environmental impact of the final product significantly. Recycling represents energy savings and the consequent reduction of greenhouse gas (GHG) emissions, since the extraction of virgin materials is avoided. Some actions for selecting lower-impact materials are using those that are cleaner in production and renewable, with a low use of energy and high recycled or recyclable content.

Reduction of material use

When designing a product, technical and quality specifications must be defined according to its intended use. Sometimes, specifications are over-sized, resulting in higher environmental and economic costs. There is the possibility of reducing the thickness of pieces and compensating for their mechanical properties through nerves or reinforcements if they do not condition the aesthetic specifications of the product.

The incorporation of foamed materials is also a very interesting option for lightening parts, especially in high-volume parts. Lightening not only reduces the consumption of raw materials but also the impact of transport of the resulting products. Some actions for reducing material use are based on reducing its weight and/or volume.

Optimisation of production techniques

The optimisation of the productive process happens in the first place by reducing the number of productive stages, for example, by reducing the surface treatments of the product. Another option to reduce the environmental impact of production is to implement the best available techniques (BATs), which are technologies with lower energy consumption, greater use of raw materials, lower generation of emissions and waste, and so on. These technologies usually require significant investments, although they are amortised in the medium term due to the saving of materials and energy.

A key aspect is temperature control of the processes themselves and the use of motor drives to optimise energy consumption. There is also the possibility of opting for renewable energy sources or, failing that, selecting those non-renewable sources that have a lower carbon footprint (e.g., natural gas is preferable to other fossil fuels). It is important for example, to recover the waste of the process itself in the same product through internal operations in the company itself. Finally, it is possible to avoid the loss of materials and minimise the generation of waste through proper stock management and rotation.

Some actions for optimising production techniques are using alternative techniques, dedicating fewer steps to the process, consuming less energy, using cleaner energy, producing less waste, or using fewer consumables.

Optimisation of distribution system

Optimisation of the the distribution system consists of selecting more efficient forms of distribution at an environmental level, considering the type of transport, logistics, and product packaging. Good planning of distribution routes is key to minimising distances travelled. It is also crucial to have good logistics to avoid 'empty' trips. In addition, it should be highlighted that not all means of transport are equally efficient (e.g., rail transport has a lower environmental impact than road transport).

The design of stackable, removable, or foldable products is a good solution to reduce the volume of products during storage and transport so that it is possible to transport more products on each trip. Another crucial aspect is the optimisation of the packaging of the product with a view to reducing the weight or volume to be transported. The use of reusable containers can also be interesting. Another option to consider is the use of recyclable containers and/or containers made from recycled materials. Some actions for optimising the distribution system are making the packaging lower, cleaner, or reusable or using energy-efficient means of transport.

Reduction of impact during use

Active products are those that consume auxiliary resources during use, such as water or energy. Normally, 90% of the environmental impact during the lifecycle of these products originates in the use stage. In these cases, it is very important to focus efforts on reducing environmental impacts during use. For this, it is possible to install saving devices to minimise the need for consumables or that are reusable and to optimise product insulation.

In this sense, insulating materials are frequently used in the construction sector to minimise energy losses in buildings. The use of light materials such as composites in the automotive and aeronautical sectors to lighten the weight of vehicles, reducing their fuel consumption and polluting emissions, is also increasingly common. Some actions for reducing impacts during use are lowering power consumption, using cleaner sources of energy, decreasing the need for supplies, or using cleaner consumables.

Optimisation of lifetime

Optimisation of the lifetime also focuses on the stage of use, but efforts in this case are to increase the durability of the product. The design of reusable products is also a good alternative, being one of the solutions promoted by the circular economy. The circular economy also requires that the maintenance and repair of the products be facilitated through a modular design and the development of parts and standard joints that allow component replacement. Classic designs are very important in products of medium-long life, since they are timeless designs that allow products not to go out of style and not to be replaced long before their end of life. Some actions for optimising lifetime are increasing trust and durability, allowing easier maintenance and repair, or producing modular structures with classic designs.

Optimisation of end-of-life system

Optimisation of the end-of-life system consists of designing the product in a way that ensures its effective management once it becomes waste. For this strategy, the different possible end-of-life scenarios and their order of priority must be considered. These options favour the reuse of the product or some components, the recycling of materials and components, or their final disassembly and subsequent recovery. This approach requires minimising the variety of materials used in the product, using materials for which there are recovery channels, using joints between components that allow easy separation, and so on. In addition, it is very important to clearly identify the different materials of the product to facilitate its recovery.

In this sense, the development of products designed for recycling is important, where aspects that harm recycling or reduce the quality of the recycled material obtained are minimised. Finally, we must not forget that some materials in the product, like plastics, have a high calorific value, so their energy recovery can be a viable option when it is not possible to apply other recovery options. Some actions for optimising end-of-life are reuse, remanufacturing, recycling, energy recovery, or incineration.

Circularity strategies

Conceptually, circular design is not directly based on a traditional life-cycle approach. In that sense, strategies for circular design are slightly different from those of ecodesign. However, when comparing ecodesign strategies and circular design approaches (Figure 12.4), some similarities are found. While 'refuse/rethink' circular approaches are aligned with the "new concept development" ecodesign strategies, 'reuse/repair/refurbish' could be related to 'optimisation of lifetime' and 'remanufacture/repurpose/recycle/recover' to 'optimisation of end-of-life.' Despite the fact that this can be used as a reference, other strategies seem to be more useful for circular design by conception.

Long-lasting

Long-lasting strategies allow products or materials to be kept for as long as possible in the system. In that sense, design is focused on increasing the use of products by durability approaches but also by easing repair and maintenance. These strategies are like those for environmentally based strategies focused on lifetime optimisation. The most difficult issue under this topic is to design for emotional durability, where products should avoid obsolescence through user love, trust, and added value.

Loop-lasting

For the loop-lasting strategy, issues related to the fact that a product finishes its life for a user are considered. Then, the product enters a new loop: to be reused, repaired, and so on until it ends in recovering. Actions here rely on design for recyclability but also for refurbishing, repurposing, and cascading with other users. This strategy meets both the ecodesign strategies on lifetime and end-of-life optimisation.

Bioinspired/biobased

For the bioinspired/biobased the techno-cycle meets the bio-cycle by adopting biomimetic, bionic, and/or nature-based solutions or using resources coming from/ending in renewable loops. This circular strategy could be focused on selecting lower-impact materials but also on the end of life for the bio-cycle (EMF, n.d.).

Conclusions

Strategies dedicated to ecodesign are well known. Some circular design strategies can fit on those based on ecodesign. However, others seem contrary to environmental aspects and ecodesign. Improving the durability of a product could involve, for instance, making it thicker. On the other hand, ecodesign would suggest making it thinner to use fewer materials. In that sense, the aim of the design should be defined clearly to ensure that the assessment of the improvements will be aligned.

Environmental assessment tools are also widely utilised for ecodesign. Product circularity evaluation is still not fully defined. Some proposals are ongoing; for example, ISO-level standards are being developed. Meanwhile, partial or business-based approaches exist for measuring the circularity of products but not for the product itself.

References

CEGESTI (1999). *Manual para la implementación de ecodiseño en centroamérica.* San José: CEGESTI, Marcel Crul y Jan Carel Diehl, C.R. 183pp. ISBN:9968-9821-2-1.

Ceschin F. (2016). Evolution of design for sustainability: from product design to design for system innovations and transitions. *Design Studies.* 47: 118–163.

EC (2009). Directive 2009/125/EC of the European parliament and of the council of 21 October 2009 establishing a framework for the setting of ecodesign requirements for energy-related products. *OJL.* 285.

EC (2015). *Closing the loop – an EU action plan for the circular economy.* Communication from the Commission to the European, the Council, the European Economic and Social Committee and the Committee of the Regions, Brussels.

EC (2017). Regulation (EU) 2017/1369 of the European parliament and of the council of 4 July 2017 setting a framework for energy labelling and repealing directive 2010/30/EU. *OJL.* 198: 1–23.

EC (2019). *The circular economy action plan.* Report from the Commission to the European Parliament, the Council, the European Economic and Social Committee and the Committee of the Regions, Brussels, Belgium.

EMF (Ellen MacArthur Foundation) (2015). *Material circularity indicator – assessment tool for companies to improve product design and material procurement.* Cowes: Ellen Macarthur Foundation.

EMF (Ellen MacArthur Foundation) (2020). *Circulytics – measuring circularity.* Cowes: Ellen Macarthur Foundation.

EMF (Ellen MacArthur Foundation) (n.d.). *Circular economy system diagram.* Available at: www.ellenmacarthurfoundation.org/circular-economy/concept/infographic

InEDIC (2011). *Ecodesign manual.* InEDIC – Innovation and Ecodesign in the Ceramic Industry. https://www.scribd.com/document/267599466/Inedic-Manual-En-Innovation-and-Ecodesign-in-the-Ceramic-Industry

ISO (2006). 14040. *Environmental management – life cycle assessment – principles and framework*. Geneva, Switzerland: International Organization for Standardization.
ISO (2015). 14001. *Environmental management systems – requirements with guidance for use*. Geneva, Switzerland: International Organization for Standardization.
ISO (2020). 14006. *Environmental management systems: guidelines for incorporating ecodesign*. Geneva, Switzerland: International Organization for Standardization.

13

APPROACHES TO MONITORING AND EVALUATION OF RESOURCE RECOVERY FROM WASTE TOWARDS A CIRCULAR ECONOMY

Eleni Iacovidou and Elena Lovat

Introduction

Solid waste, the materials, components, and products generated and/or produced by human and animal activities and discarded when they are no longer wanted or have any use, is the aftermath of an increased consumption of resources. Economic growth and technological advancements have given rise to a great variety of materials, components, and products, offering access to goods and services to a large part of the ever-growing global population. This has the unintended consequence of using natural resources at a faster rate than regeneration, while finite resources are severely depleted. In addition, materials, components, and products are often designed to have shorter lifespans, for example, single-use and disposable products, while others are selected or tailor-made to fit the needs of the specific area or aesthetic qualities (e.g., construction components, food), screening out numerous others that are perfectly functional and/or edible (Facchini et al., 2018).

Overuse, misuse, and depletion of natural resources and of the materials, components, and products made of them has escalated the issues associated with solid waste generation and management, with waste pollution becoming the new 'tragedy of the commons' (Hardin, 1968). Wastes that are not properly collected and managed remain in the environment, and now, large amounts of them are ubiquitously present in all environmental compartments – soil, air, and marine and fresh waters, as well as in biota across the world, causing detrimental effects on the environment, human health, and wildlife. At the same time, more natural resources are needed to make the same or similar materials, components, and products to meet the needs of the global population, leading to a vast dissipation of valuable resources both upstream and downstream of the point where waste is generated in the value chain. In view of these shortcomings, a number of strategies have been formulated to achieve or maintain sustainability in resource recovery from waste. These involve improving recycling rates, banning single-use materials, promoting prevention, and stimulating innovation in the development of alternative materials, components, and products or developing technologies for dealing with them when they become wastes (Iacovidou et al., 2017a).

The challenge involved with the implementation of such strategies is that they have reinforced siloed thinking, for example, leading to sector-specific activities for tackling issues at a specific stage in the value chain or influencing a certain group of stakeholders, neglecting to account for the multidimensional implications arising from them. Realisation of the flaws in these strategies has led to the development of the all-embracing concept of the circular economy (CE) (European Commission, 2015), where materials, components, and products behave as either 'technical or biological nutrients' that can be returned to the economy via circular loops (McDonough et al., 2003; Ellen MacArthur Foundation, 2014). The CE concept places specific emphasis on the importance of limiting waste generation and instead promoting the remanufacture, repair, reuse, recycling, and recovery of waste resources via, and to, different processes, depending on the remaining properties and characteristics of the materials, components, and products.

Notwithstanding the importance of the CE concept, numerous authors have raised concerns that often the circularity of resources and wastes can be insufficient for achieving sustainability in the system and in a way that reduces reliance on, and consumption of, natural resources (Cullen, 2017; Zink and Geyer, 2017). This is owing to the fact that circularity focuses on partial improvements in the system rather than promoting the sustainable management of resources (Geyer et al., 2016). This has implications in achieving the CE, and it is now advocated that effective management of resources through the concept of CE must be linked to an effective monitoring and evaluation process that measures both the feasibility of closing the material component and product loops and the sustainability of doing so (Iacovidou et al., 2020).

There is no 'one size fits all' method of monitoring the diverse mixture of resources and wastes across a range of environmental compartments. It is therefore important to determine exactly why the monitoring is being undertaken, for example, to track the flow of resources and assess environmental impacts; to determine critical thresholds and targets; to identify potential sources, stocks, and transformation mechanisms; or to measure the effectiveness of policy or other implemented measures. Moreover, it is critical to evaluate progress based not only on environmental and economic aspects, as is usually the case, but also based on social, technical, and political considerations. Compiling information from a range of literature, this chapter will outline and discuss the monitoring and evaluation methods proposed for measuring progress in achieving the CE. It will place focus on the implementation of these approaches at various levels and highlight the gaps in governing circularity effectively and, most importantly, sustainably.

Monitoring and evaluation: importance and definitions

Solid waste is a highly heterogeneous mixture, spanning orders of magnitude in size (from obsolete construction structures to microplastics measured in μm in diameter), and it can be in a wide range of shapes and types, arising from an equally diverse range of sources, for example, residential, industrial and commercial, and agricultural. The fate of solid waste once it is disposed of in a receptacle can depend on a number of factors, including, but not limited to, the collection and transport processes, the facilities where it is further processed, the proximity and technological advancement of facilities, and the regulatory framework in the area(s) where it is managed, as well as the physical properties of the waste itself (e.g., size, shape, density). Almost limitless combinations of these factors make it difficult to monitor and evaluate the potential of returning wastes sustainably back to the economy as resources.

As a result, there are various approaches used to monitor progress towards circularity which often depend on a number of factors, such as type of materials, components, and products and the context within which progress is assessed. This chapter adopts a generic view of the monitoring and evaluation methods, focusing particularly on those performed to measure progress towards the transition to a circular economy. As a result, monitoring and evaluation approaches to specific materials, components, and products (e.g., plastics, food, metals) and area of generation and/or presence in environmental compartments (e.g., terrestrial, marine environments) are excluded from the present discussion.

Monitoring is the process of continued collection of data against a specific set of indicators (referred to herein as *metrics*) for measuring progress towards the achievement of defined goals, targets, and other purposes. It can provide information at the beginning of a development, during its implementation, or after completion. Thus monitoring is our means to establish how things perform in relation to a defined purpose, helping us identify where interventions might be needed in relation to achieving the defined purpose. Evaluation is the process of assessing the impact of the progress made in terms of its sustainability potential, using both quantitative and qualitative measures (Sharp et al., 2010). It is useful, as it provides insight into domains/areas where improvements can be made in order to achieve the desired goals; thus, it is causally linked to monitoring. Both monitoring and evaluation must be performed in order to get a full picture of the impact on sustainability of circularity processes.

Monitoring and evaluation of the transition to CE involves the regular tracking of resources and measurement and analysis of the impacts on the environment, economy, society, and engineering systems. It regularly measures progress towards achieving specific objectives, making sure that these are being effectively delivered (Sharp et al., 2010). It also delivers feedback to strategy and planning development (Alaerts et al., 2019).

In the following sections, we will describe the variants between existing monitoring and evaluation approaches.

Monitoring and evaluation approaches in achieving a circular economy

Since the proposal of the Circular Economy Action Plan in Europe (European Commission, 2015), a number of monitoring and evaluation frameworks have sprung up by research centres, academia, governmental agencies, and charities aiming to assess one or more aspects of CE. Monitoring and evaluation are important to estimate the extent to which efforts to achieve CE initiatives are delivering results (Potting et al., 2017; Avdiushchenko and Zajaç, 2019) and to capture critical aspects of CE-related activities, which might result in problem shifting rather than an overall benefit (Corona et al., 2019; Mayer et al., 2019). For example, reuse by repair, refurbishment, or recycling is often assumed to result in net greenhouse gas (GHG) emissions savings. While this is true for the recycling of energy-intensive materials such as steel, aluminium, or plastics, recycling of materials such as concrete, plasterboard, bricks, and paint may result in higher net GHG emissions compared to producing their primary counterparts (Turner et al., 2015, 2016). This implies that monitoring and evaluation methods to assess the transition to a CE need to be scrutinised over their ability to effectively measure progress towards achieving sustainable circularity.

Existing monitoring frameworks seek to measure and evaluate progress of meeting CE policy interventions at different implementation scales, namely the *macro*, *meso*, and *micro* scales. According to some authors, *micro* refers to a single material, component or product, company, or consumer; *meso* refers to eco-industrial parks, industries, and companies; and *macro* refers to a city, province, region, or nation, or it may take a global perspective (Kirchherr et al., 2017; Moraga et al., 2019). Huysman et al. (2017) and EASAC (2016) have also associated countries and regions with a macro-level analysis and micro-level analysis with products and companies, while reference to the meso level has not been ratified. Edgerton et al. (2018), differentiates the monitoring and evaluation implementation levels into: *nano*, micro, meso and macro, corresponding to products and components; companies and consumers; inter-industry networks; and cities, countries, and international agencies, respectively.

At the macro level, monitoring and evaluation are mainly performed on mass flow exchanges between the economy and the environment, on international trade, and on material accumulations in national economies (Vercalsteren et al., 2018; Mayer et al., 2019). Macro monitoring and evaluation methods towards achieving the CE have been carried out via an economy-wide perspective at the national or higher scale (Giljum et al., 2015; Haas et al., 2015; Schandl et al., 2018). Mayer et al. (2019) suggest that only at this level it is possible to capture system-wide effects such as displacement or rebound effects and to assess whether absolute reductions in resource use and waste flows are achieved. As a result, they developed a mass-based monitoring framework based on a system-wide assessment of a CE at the global and European level from an economy-wide material flow (ew-MFA), using data on material consumption, transformation and use. Material flow analysis (MFA) – a quantitative procedure for determining the flow of materials and energy through the economy on multiple scales (e.g., area-, process-, and/or product-specific) – has been commonly used for the systematic monitoring of resources and wastes through socioeconomic systems, assessing where in the system loop-closing processes occur (Brunner and Rechberger, 2016). Mayer et al. (2019) underlined that although the MFA is useful in tracking the flows of materials, components, and products, its use is often hampered by the non-homogeneity of data available and uncertainties around the material, component, and product stocks (Mayer et al., 2019). As a result, the ew-MFA, which includes macroeconomic aspects, can be better suited to evaluate material circularity (Mayer et al., 2019) and to assess the state and changes of a system over time (Avdiushchenko, 2018).

In line with this approach, the material circularity indicator framework was developed by the Ellen MacArthur Foundation (2017) to monitor and evaluate progress to achieving the CE at the meso level. Meso-level monitoring and evaluation processes focus on mapping the consumption activities of industries and/or companies in relation to specific material, component, or product consumption patterns (Edgerton et al., 2018). The purpose of this framework is to monitor the activities of an industry or company and identify wastage incidents, pollution sources, and opportunities for improving their efficiency yields in the economic, environmental, or social domains (Vercalsteren et al., 2018). Nonetheless, the environmental and/or economic assessment elements of this framework are optional and often performed by companies that have the appropriate tools for their performance (Ellen MacArthur Foundation, 2015). This hampers the wider utilisation of the MCI framework for evaluating the sustainability potential of material, component, and product circularity (Vercalsteren et al., 2018).

It can be argued that MFA and ew-MFA approaches, albeit useful in monitoring the flows of materials, components, and products, tend to place focus on the technical and/or socio-economic aspects of CE at the macro- and meso-level implementation, while considerations regarding specific materials, components, and products in other domains (e.g., environmental, social) are not properly addressed (Iacovidou et al., 2017a). Some authors associate this limitation with the lack of targets (Potting et al., 2017), as well as the fact that there has been significantly longer experience in solid waste management, while CE is a new and underdeveloped concept (Kristensen and Mosgaard, 2020). With regard to the former, the European Commission has set specific targets for waste and material management but did not set targets that could facilitate stakeholders in defining priorities and carrying out comparative research (Potting et al., 2017).

To address this gap, the circular economy monitoring framework (CEMF) was developed for, and applied by, the European Commission to evaluate circularity performance at the national and European levels. In this framework, CE is evaluated through the use of metrics adopted by resource efficiency scoreboards (RESs) and raw material scoreboards (RMSs) measuring progress towards a circular economy in a way that encompasses its various dimensions at all stages of the life cycle of resources, products, and services. As a result, metrics are classified into production and consumption, waste management, secondary raw materials, and competitiveness and innovation (European Commission, 2019). Among the limitations of this framework is the lack of focus on product design towards the development of better components and products, while the export of recyclable materials to non-EU countries (exporting waste to third-world countries does not guarantee environmentally or socially friendly utilisation) has not been accounted for, raising issues with regard to our ability to evaluate the sustainability of transitions to CE (Moraga et al., 2019).

Potting et al. (2017) suggested broadening the scope of the CEMF by combining all three levels of implementation, that is, macro, meso, and micro. Micro-level monitoring and evaluation refers to the assessment of the activities performed for specific materials, components, and products in relation to, for example, their extraction, manufacturing, and end-of-life management practices. As a result, they proposed expanding the list of the ten metrics used in the CEMF to highlight linkages with the micro level and suggested associating the implementation levels of CE (i.e., macro, meso, micro) with metrics as a constructive and useful approach to enabling the monitoring and evaluation of the transition towards a CE. Classification of metrics to the macro, meso, and micro scales of implementation can make their selection best suited to the issues of concern and the questions being addressed, as methods and tools for 'calculating' indicators can be very divergent (Vercalsteren et al., 2018).

Macro-, meso-, and micro-level metrics used for monitoring and evaluation of achieving the circular economy

Metrics are fundamentally needed for evaluating, monitoring, and improving various policies and programs, and, as a result, they are closely associated with the implementation scales of the monitoring and evaluation frameworks towards the CE, the micro, meso, and macro. At the macro level, mass-based metrics summarise the progress at (supra)national level, while meso and micro metrics are tailored towards assessing circularity at the level of companies and/or materials, components, and products, respectively (Alaerts et al., 2019; Kristensen and Mosgaard, 2020).

Vercalsteren et al. (2018) describe macro-level metrics as those that can be used to support decisions in areas such as economic, trade, and environmental policy integration; sustainable development strategies and action plans; and national waste management and resource conservation policies. Meso-level metrics are those that can measure the economic, environmental, or social performance of a region (domain specific) or a product group or industry (process specific) and enable a more detailed analysis of material flows within the economy. Micro-level metrics provide detailed information for specific decision processes at the business or local level or concerning specific substances or individual products. Table 13.1 outlines three different types of metrics used in monitoring and evaluation methods for measuring progress towards achieving the CE based on macro, meso, and micro implementation scales. Domain refers to environmental, economic, social, and technical considerations that metrics should be able to measure for evaluating the sustainability of CE transitions.

Table 13.1 The scalability of metrics used in monitoring and evaluation frameworks proposed for measuring progress towards achieving the circular economy

Source	*Macro*	*Meso*	*Micro*
(Potting et al., 2017)	*Area specific – Lead (generic)*	*Domain specific – dashboard (generic)*	*Product specific*
(Alaerts et al., 2019)	*Area specific (supranational or regional)*	*Function specific (societal needs)*	*Product/service specific*
(Vercalsteren et al., 2018)	*Macroeconomic – area specific*	*Domain and process specific*	*Product specific*

According to Potting et al. (2017) in the macro category, the lead metrics for monitoring (transition-related metrics) and evaluation (effect-related metrics) are mass-based metrics and circularity strategies, respectively. Circularity strategies refer to circularity assessment tools that can either provide single or aggregate metrics or multiple metrics for specific case studies (CE assessment frameworks) (Corona et al., 2019). Moraga et al. (2019) classify the circularity metrics into three types, according to focus: (1) on a specific CE strategy, (2) on more than one CE strategy (e.g., cumulative exergy extraction from the natural environment [CEENE] indicators used to assess performances for post-industrial plastic waste recycling (Huysman et al., 2017), and (3) extrapolated from other assessment methods not directly related to CE assessment (e.g., resource efficiency scoreboard; European Commission, 2019). Kristensen and Mosgaard (2020) highlight that monitoring and evaluation of CE through single macro metrics might preclude a deeper understanding of CE achievements, and industry- and/or material-, component-, and product-specific indicators should be developed to promote utilisation by companies at the meso and micro levels to enable better evaluation processes.

In line with this, Potting et al. (2017) expanded their metrics list to include both meso- and micro-level metrics which are relevant to specific materials, components, and products. In the meso level, metrics are domain specific, that is, focusing on the environmental and economic domains (effect-related metrics) and process specific (transition-related metrics), while at the micro level, product-specific metrics are proposed for the transition process and its effects. However, macro-, meso-, and micro-level implementation of the CEMF seem to be divorced in time, as it is not clear how the metrics at the different implementation levels will be concurrently used to monitor and evaluate progress towards CE.

To address this shortcoming, Alaerts et al. (2019) refined the meso-level metrics as a way to link macro- and micro-level metrics to measure CE achievements and effects. They proposed that these metrics should be function specific, measuring the satisfaction of societal needs. The metrics proposed should be able to measure the functional aspects of material, component, and product circularity (Alaerts et al., 2019). In contrast, Vercalsteren et al. (2018) explained that lack of data and the time and effort investment required to perform such analyses are often key barriers to linking metrics on a macro and meso level with those on a micro level.

At the micro level, where the focus is on the material, component, or product, there are a large number of metrics that can be used for assessing aspects of CE (Kristensen and Mosgaard, 2020). Corona et al. (2019) indicate that such variety could be attributed to the lack of a common definition of CE, which inconsistently accounts for the environmental, economic, and social dimensions (Kirchherr et al., 2017). This has led to the development of a large number of metrics that measure progress and evaluate efforts to achieve the CE that can often be disproportional to the processes and scale of measurement and misleading (Moraga et al., 2019).

Discussion

Monitoring the transition to a circular economy and evaluating its sustainability and potential to align with new policies and trends necessitates a multiscalar analysis which requires a set of multilevel metrics. The classification of metrics into the macro, meso and micro scales of CE implementation, although useful for monitoring at higher level, can lead to a distorted depiction of the system's ability to become more circular and disaggregation of efforts to monitor and evaluate the transition to a sustainable CE. As suggested by Alaerts et al. (2019), the transition to CE is cross-sectorial and is thus expected to have multidimensional impacts.

In line with this, mass-based metrics may not encompass all aspects that should be monitored and evaluated for making a sustainable transition to CE but is essential in providing the baseline analysis of the system. This baseline analysis can lead to the development and management of metrics that measure the scale of input and output flows and which can be further developed to include environmental, economic, social, and technical indicators across all scales (meso level) and for specific materials, components, and products (micro level) (Iacovidou et al., 2017a; Mayer et al., 2019). Integrating the macro-, meso-, and micro-level metrics with the corresponding implementation scales of the CE can be succeeded by a view of linking the production, consumption, and management (PCM) stages of resources and wastes flows in a specific geographical context. In this way, the monitoring and evaluation method used should be able to account for the pertinent processes and stakeholders involved in that particular system, critical elements for enabling transitions to CE.

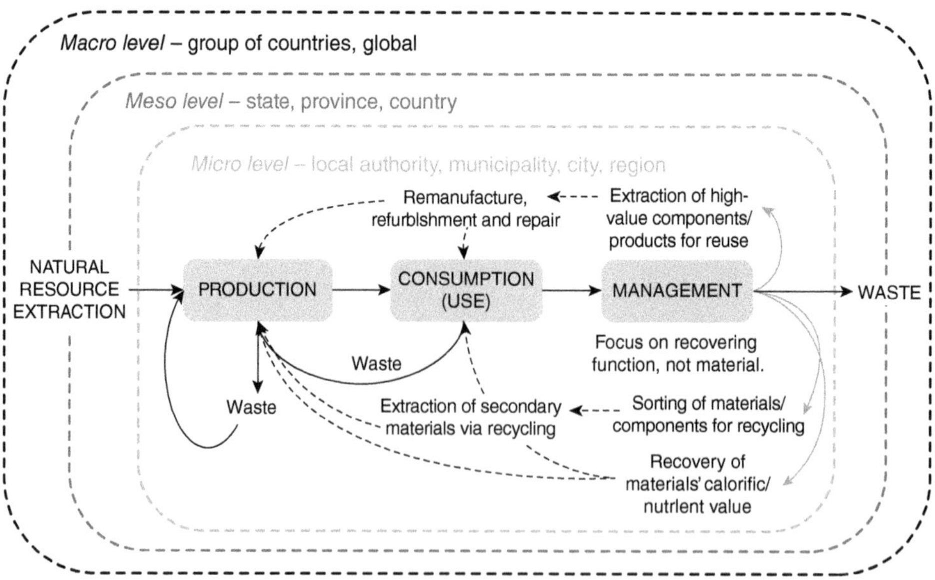

Figure 13.1 Depiction of the macro, meso, and micro scales as system boundaries within which we have production-consumption-management. The processes and stakeholders involved in PCM stages can be inside or outside the micro-level boundaries, depending on the material, component, and product system(s) under analysis.

We suggest that the monitoring and evaluation processes for achieving a CE should be context specific, with the macro, meso, and micro scales of implementation referring to geographical context, thus representing system boundaries, and that the material, component, and product be the central force at all scales. Figure 13.1 presents our approach to monitoring and evaluation of sustainable CE transitions.

As depicted in Figure 13.1, at the micro scale, we have the PCM activities implemented at the local authority, municipality, or regional level. The analysis on such activities can then be expanded at the meso scale, referring to province, sub-national, or national levels, while the macro scale describes PCM activities implemented at multinational or global level. The scale is thus defined by the grade of complexity attributed to the PCM processes involved within each system boundary. In each scale, all sustainability aspects (i.e., environment, social, economic, and technical considerations) are casually taken into account in order to analyse the achievements of CE from a holistic perspective. Thus, the metrics selected and/or developed to monitor and evaluate progress in each scale are multidimensional and evaluated simultaneously, without prioritisation of any of the processes involved in the PCM system. The MFA is a prerequisite in our approach, as it facilitates the mapping of all processes involved within the system and supports the identification of the stakeholders directly involved in the system (Iacovidou et al., 2017b). The implementation of CE activities depend on the stakeholders' values, interests, and behaviour; thus, they represent a key element in the monitoring and evaluation processes.

Based on our approach, the classification of metrics in the macro, meso, and micro scales is independent of the material, components, products, and functionality. The selection and/or development of metrics depends on the targets and objectives for achieving CE at each scale and can be used for monitoring and assessing progress in meeting the targets. This way, metrics remain simple, transparent, and easy to measure, and can be both system and stakeholder specific, which is key in assessing the sustainability potential of resource recovery from waste. This approach follows the rationale of evaluating the 'complex value' of materials, components, and products, that is, the holistic sum of their environmental, economic, social, and technical benefits and impacts across the system, from their production to their end-of-life management. This is further described in the newly developed complex value optimisation for resource recovery (CVORR) approach (Iacovidou, Millward-Hopkins et al., 2017; Iacovidou, Velis et al., 2017). CVORR supports understanding of how the various policy interventions, current and planned, need to be coordinated to deliver the desired outcomes, key in monitoring and evaluating progress towards achieving the CE.

Conclusions

The circularity of materials, components, and products in resource recovery systems is not necessarily sustainable. Understanding the processes and stakeholders involved in production, consumption, and management of materials, components, and products and is therefore necessary to assessing and monitoring progress towards transitions to circular economies. Up until now, progress has been slow due to the fragmented nature of the resource and waste management systems and the lack of coordination between all stakeholders involved. Adopting a whole-systems approach to selecting and/or developing the metrics needed for assessing and monitoring progress in the resource recovery system's performance and sustainability can help us rationalise and select the targets, measures, and interventions that are best suited to achieving the sustainable circularity goal.

References

Alaerts L, Van Acker K, Rousseau S, De Jaeger S, Moraga G, Dewulf J, De Meester S, Van Passel S, Compernolle T, Bachus K, Vrancken K, Eyckmans J. (2019). Towards a more direct policy feedback in circular economy monitoring via a societal needs perspective. *Resources, Conservation and Recycling*. 149: 363–371.

Avdiushchenko A. (2018). Toward a circular economy regional monitoring framework for European regions: conceptual approach. *Sustainability (Switzerland)*. 10 (12). doi:10.3390/su10124398.

Avdiushchenko A, Zajaç P. (2019). Circular economy indicators as a supporting tool for European regional development policies. *Sustainability (Switzerland)*. 11 (11): 1–22. doi:10.3390/su11113025.

Brunner P, Rechberger H. (2016). *Handbook of material flow analysis, for environmental, resource, and waste engineers*. Howick Place and London: CRC Press.

Corona B, Shen L, Reike D, Carreón JR, Worrell E. (2019). Towards sustainable development through the circular economy – a review and critical assessment on current circularity metrics. *Resources, Conservation and Recycling*. 151 (September): 104498. doi:10.1016/j.resconrec.2019.104498.

Cullen JM. (2017). Circular economy: theoretical benchmark or perpetual motion machine? *Journal of Industrial Ecology*. 21 (3): 483–486.

EASAC (2016). *Indicators for a circular economy*. Policy report by the European Academies' Science Advisory Council. German National Academy of Sciences Leopoldina 2016. ISBN: 978-3-8047-3680-1.

Edgerton B. et al. (2018) *Circular metrics: Landscape analysis*. May 2018. p. 35. Available at: https://www.wbcsd.org/Programs/Circular-Economy/Factor-10/Metrics-Measurement/Resources/Landscape-analysis

Ellen MacArthur Foundation (2014). *Towards the circular economy: accelerating the scale-up across global supply chains*. Davos-Klosters, Switzerland: Ellen MacArthur Foundation, McKinsey & World Economic Forum. World Economic Forum.

Ellen MacArthur Foundation (2015). Circularity indicators – an approach to measuring circularity. *Methodology*. doi:10.1016/j.giq.2006.07.001.

Ellen MacArthur Foundation (2017). *Measuring your company's circularity*. Available at: https://www.ellenmacarthurfoundation.org/resources/apply/circularity-indicators (Accessed 22/09/2019).

European Commission (2015). *Closing the loop – an EU action plan for the circular economy, COM/2015/0614. Communication from the Commission to the European Parliament, the Council, the European Economic and Social Committee and the Committee of the Regions*. Brussels, Belgium: European Commission.

European Commission (2019). *Circular economy monitoring framework*. Available at: https://ec.europa.eu/eurostat/web/circular-economy/indicators/monitoring-framework (Accessed 22/09/2019).

Facchini E, Iacovidou E, Gronow J, Voulvoulis N. (2018). Food flows in the United Kingdom: the potential of surplus food redistribution to reduce waste. *Journal of the Air & Waste Management Association*. 68 (9): 887–899.

Geyer R, Kuczenski B, Zink T, Henderson A. (2016). Common misconceptions about recycling. *Journal of Industrial Ecology*. 20 (5): 1010–1017.

Giljum S, Bruckner M, Martinez A. (2015). Material footprint assessment in a global input-output framework. *Journal of Industrial Ecology*. 19 (5): 792–804.

Haas W, Krausmann F, Wiedenhofer D, Heinz M. (2015). How circular is the global economy? An assessment of material flows. Waste production, and recycling in the European Union and the world in 2005. *Journal of Industrial Ecology*. 19 (5): 765–777.

Hardin G. (1968). The tragedy of the commons. *Science*. 162 (3859): 1243–1248.

Huysman S, De Schaepmeester J, Ragaert K, Dewulf J. (2017). Performance indicators for a circular economy: a case study on post-industrial plastic waste. *Resources, Conservation and Recycling*. 120: 46–54. doi:10.1016/j.resconrec.2017.01.013.

Iacovidou E, Ebner N, Orsi B, Brown A. (2020). *Plastic packaging – how do we get to where we want to be?* Report No. 14777, Department of Environment, Food and Rural Affairs, London: Defra.

Iacovidou E, Millward-Hopkins J, Busch J, Purnell P, Velis CA, Hahladakis JN, Zwirner O, Brown A. (2017). A pathway to circular economy: developing a conceptual framework for complex value assessment of resources recovered from waste. *Journal of Cleaner Production*. 168: 1279–1288.

Iacovidou E, Velis CA, Purnell P, Zwirner O, Brown A, Hahladakis J, Millward-Hopkins J, Williams PT. (2017). Metrics for optimising the multi-dimensional value of resources recovered from waste in a circular economy: a critical review. *Journal of Cleaner Production*. 166: 910–938.

Kirchherr J, Reike D, Hekkert M. (2017). Conceptualizing the circular economy: an analysis of 114 definitions. *Resources, Conservation and Recycling*. 127: 221–232.

Kristensen HS, Mosgaard MA. (2020). A review of micro level indicators for a circular economy – moving away from the three dimensions of sustainability? *Journal of Cleaner Production*. 243: 118531.

Mayer A, Haas W, Wiedenhofer D, Krausmann F, Nuss P, Blengini GA. (2019). Measuring progress towards a circular economy: a monitoring framework for economy-wide material loop closing in the EU28. *Journal of Industrial Ecology*. 23 (1): 62–76.

McDonough W, Braungart M, Anastas PT, Zimmerman ZB. (2003). Peer reviewed: applying the principles of green engineering to cradle-to-cradle design. *Environmental Science & Technology*. 37 (23): 434A–441A.

Moraga G, Huysveld S, Mathieux F, Blengini GA, Alaerts L, Van Acker K, de Meester S, Dewulf J. (2019). Circular economy indicators: what do they measure? *Resources, Conservation and Recycling*. 146: 452–461.

Potting J, Nierhoff N, Montevecchi F, Antikainen R, Colgan S, Hauser A, Günther J, Wuttke J, Kjær BJ, Hanemaaijer A. (2017). *Input to the European Commission from European EPAs about monitoring progress of the transition towards a circular economy in the European Union*. PBL Netherlands Environmental Assessment Agency.

Schandl H, Fischer-Kowalski M, West J, Giljum S, Dittrich M, Eisenmenger N, Geschke A, Lieber M, Wieland H, Schaffartzik A, Krausmann F, Gierlinger S, Hosking K, Lenzen M, Tanikawa H, Miatto A, Fishman T. (2018). Global material flows and resource productivity: forty years of evidence. *Journal of Industrial Ecology*. 22 (4): 827–838.

Sharp V, Giorgi S, Wilson DC. (2010). Methods to monitor and evaluate household waste prevention. *Waste Management & Research*. 28 (3): 269–280.

Turner DA, Williams ID, Kemp S. (2015). Greenhouse gas emission factors for recycling of source-segregated waste materials. *Resources, Conservation and Recycling*. 105: 186–197.

Turner DA, Williams ID, Kemp S. (2016). Combined material flow analysis and life cycle assessment as a support tool for solid waste management decision making. *Journal of Cleaner Production*. 129: 234–248.

Vercalsteren A, Maarten C, Van Hoof V. (2018). *Indicators for a circular economy*. Available at: https://circulareconomy.europa.eu/platform/sites/default/files/summa_-_indicators_for_a_circular_economy.pdf. SUMMA Policy Research Centre.

Zink T, Geyer R. (2017). Circular economy rebound. *Journal of Industrial Ecology*. 21 (3): 593–602.

14

COMPLEXITY AND THE CIRCULAR ECONOMY

Systems approaches for change

Martha Bicket

Introduction

Current trends in resource use are unsustainable. Resource consumption, waste, and pollution levels exceed several global biophysical limits (Rockström et al., 2009; IPCC, 2014), and their growing levels pose a threat to human well-being through impacts on health, resource security, development, and economic and political stability (European Commission, 2011; IPCC, 2014; McCarthy et al., 2017; OECD, 2019). The move to living within sustainable limits requires a fundamental transformation of our production and consumption systems (EEA, 2015). The circular economy is a proposed reimagining and restructuring of our current production and consumption systems to improve and optimise for sustainability. In contrast to our current, largely linear 'take-make-use-dispose' economy (Bicket et al., 2014), the circular economy seeks to optimise the use of products and materials in the economy by maintaining them at their highest value for as long as possible and by maximising their recovery and reintegration back into the economy and natural systems, avoiding waste (Webster, 2017; de Jesus and Mendonça, 2018).

Delivering this circular economy involves change across multiple systems – social, technical, natural, and economic. For example, the transition to the circular economy involves changes and innovation in recycling, repair, and remanufacture (McCarthy et al., 2017); material efficiency (Allwood, 2014); waste management (Greyson, 2007); business models (such as industrial symbiosis and product-service systems; Stahel, 1997; Eckelman and Chertow, 2013; Tukker, 2015); skills (Schröder, 2020); cultures, preferences; and behaviours (Shove, 2014; Kirchherr et al., 2018; Schröder, 2020) and across many different sectors (Köhler et al., 2019). Furthermore, the elements in these systems are deeply interlinked, with multiple interactions taking place between the individuals, businesses, governments, institutions, technology, and natural environment involved (Bastein et al., 2013; Kemp and Dijk, 2013; McCarthy et al., 2017; Domenech and Bahn-Walkowiak, 2019).

These interlinkages and dependencies mean that transitions to sustainability such as the circular economy "cannot be addressed by incremental improvements and technological fixes" (Köhler et al., 2019) nor by single policies (Hartley et al., 2020). Instead, the shift to a circular economy and greater resource efficiency calls for a holistic view of the systems involved and the interactions between them (Bastein et al., 2013; de Jesus and Mendonça, 2018;

Preiser et al., 2018) and for combinations of policies that address these interactions which lie at the heart of the transition to the circular economy (Hughes and Ekins, 2018; Milios, 2018; Hartley et al., 2020).

The concepts of *complexity* and *complex adaptive systems* can be particularly useful for those seeking to bring about change towards the circular economy (Davies et al., 2012; Levin et al., 2013; Reyers et al., 2018). Complex adaptive systems are made up of many diverse and interconnected components whose adaptive behaviour and nature of interactions makes the system's overall behaviour particularly hard to predict. Managing the transition to the circular economy can be described as a complex or 'wicked' issue (Carley and Christie, 2000); the circular economy is highly interconnected – indeed, it seeks to achieve greater interconnectedness between sectors through industrial symbiosis, between individuals through sharing, and between different stages of the value chain through leasing business models, for example. At the same time, the natural environment, individuals, and organisations involved will adapt their behaviour over time in response to the conditions, drivers, and barriers they meet along the way.

This chapter explores the concepts of complexity and complex systems, how an appreciation of these is important for those seeking to steer change towards the circular economy, and a range of promising tools and approaches that can help. We begin by examining in greater depth what complexity is and its relevance to – and implications for – transitions to the circular economy.

Complexity, systems thinking, and managing transitions to the circular economy

Complexity and complex systems

Closely related to the fields of cybernetics (Wiener, 1948) and systems theory (Boulding, 1956), complexity science is the study of systems with a particular structure and certain characteristic behaviours. Although a variety of descriptions and definitions exist (Crutchfield and Machta, 2011; Guijt et al., 2011), they usually share the following features in common: complex systems are made up of many diverse and interconnected components, which interact with each other in non-linear and non-proportional ways (Bicket et al., 2020). Additionally, in complex *adaptive* systems (Cilliers, 1998) – the type of system on which we focus in this chapter – these components can also learn and adapt their behaviour over time.

The size, interconnectedness, and non-linear and adaptive nature of these systems can give rise to a range of characteristic features and behaviours, for example:

- Self-organisation – the ability of components to self-arrange with each other to better interact with their environment (Briassoulis, 2004)
- Emergence – the interactions between components can give rise to new, higher-level behaviours which are said to be emergent if they "cannot easily be predicted from the properties of the lower-level components" (Bicket et al., 2020)
- Feedback and feedback loops – a circular process where an action has an effect on its actor (Bar-Yam, 2009); feedback can accelerate change (positive feedback) or suppress or dampen it (negative feedback)
- Attractors and tipping points – complex systems may gravitate towards relatively stable states of behaviour or equilibria, called attractors (Gunderson and Holling, 2002); a tipping

point is a threshold beyond which the behaviour of a system gravitates towards a different attractor state

- Path dependency – when the current state of a system depends on its past and not just its present conditions; that is, when placed under the same present conditions, a system with a different history might not behave in the same way
- Unexpected indirect effects – where change in one part of the system can lead to unexpected change in another seemingly remote part (Rittel and Webber, 1973)
- Openness – where the observational boundaries drawn around a system need to be continually redrawn and expanded due to its highly interconnected nature and unexpected indirect effects (Rittel and Webber, 1973: 159)
- Change over time and unpredictability – the previous characteristics of complex adaptive systems mean that their behaviour is dynamic and continuously liable to change over time; as a result, the behaviour of complex adaptive systems is difficult to predict with any precision (Haynes, 2015).

Complex adaptive systems may also be described as 'wicked' problems for policy (Rittel and Webber, 1973) – wicked in the sense that they are difficult to define, have many interdependencies, are often unstable, rarely sit conveniently within the responsibility of any one organisation, involve a range of stakeholders, usually have no clear solution, and attempts to address them often fail and/or lead to unforeseen consequences (Australian Public Service Commission, 2007). Examples of wicked policy problems include climate change, obesity, and natural resource management.

The circular economy as a complex system

Managing the transition to the circular economy is a complex, wicked issue. The components involved in the change to the circular economy are many and diverse, ranging from organisms and natural resources to technologies, sectors, markets, individuals, and multinational organisations (Köhler et al., 2019). In terms of sectors alone, for example, not only will energy, waste, food, transport, and primary industries such as raw material extraction affect and be affected by resource use and transitions to sustainability, but also health, poverty, international development, education, migration, and the financial sector. The actors involved in transitions range from individuals, households, organisations, and governments to groups from academia, industry, and civil society. They each have their own differing roles, influence, perspectives, beliefs, preferences, and goals (Köhler et al., 2019) and the ability to adapt and learn over time.

These system elements are highly interconnected and are organised across different scales and within multiple nested and overlapping hierarchies (Briassoulis, 2004; Bastein et al., 2013; Kemp and Dijk, 2013; McCarthy et al., 2017; Domenech and Bahn-Walkowiak, 2019). For example, "the use of natural resources through one type of human activity affects ecological processes in ways that interact with other people's uses of natural resources, both across geographic and ecosystem boundaries and time scales" (Ison, 2010). Similarly, the performance of local, national, and global economies depends on the interactions between resource use, institutions, policies, and systems of innovation (Domenech and Bahn-Walkowiak, 2019).

Characteristics of complex adaptive systems such as tipping points and feedback loops are frequently used to describe and explore the behaviour of socioecological systems (Preiser et al., 2018; Colding and Barthel, 2019), while attractors and path dependency are used to study behavioural and technological lock-in preventing the widespread adoption of green innovations such as electric vehicles and carbon-saving technologies (Unruh, 2000). Nonlinearity can be seen in the technological, political, and sociocultural processes that will be involved in circular economy transitions, for example, in political agenda U-turns, volatile oil prices, and the processes by which innovations fail or rapidly take off (Köhler et al., 2019).

This complexity in the systems involved in transitions to the circular economy is further exacerbated by the long timeframes over which transitions are likely to take place. It is therefore particularly important for those seeking to govern sustainability transitions and bring about change towards the circular economy to have an awareness of complexity and its implications for managing transitions (Davies et al., 2012; Levin et al., 2013; Reyers et al., 2018).

The implications of complexity for managing transitions to the circular economy

The high degree of interconnectedness and interactions between all of the components involved in moving to the circular economy demands a shift from studying individual parts of the system (Rosen, 1991) to an approach which takes account of the interactions between them and looks at the behaviour of the system as a whole (Bastein et al., 2013; Patterson et al., 2017; de Jesus and Mendonça, 2018; Preiser et al., 2018). A number of contributions to the field of sustainability transition studies have embraced this notion by moving away from looking at single sectors in isolation (such as energy or transport) to analysing combinations of sectors and the interactions between them (Raven and Verbong, 2007; Konrad et al., 2008; Papachristos et al., 2013).

Another consequence of the interconnected, interdependent, and non-linear nature of the circular economy as a complex system is that a circular economy transition will not resemble a linear process or be brought about by small isolated changes. The transition to the circular economy is instead likely to be made up of multiple interdependent developments, and these themselves may interact with each other in unpredictable ways (Scoones et al., 2015: 21; Köhler et al., 2019). We should therefore not seek to manage the change to the circular economy with single policies but use combinations of policies that address these interdependencies and systemic behaviours instead (Kivimaa and Kern, 2016; Hughes and Ekins, 2018; Milios, 2018; Hartley et al., 2020).

A key consequence of the system we are seeking to transition to the circular economy being complex is that it is impossible to explain comprehensively and accurately, and both its behaviour and the effects of proposed solutions are therefore largely unpredictable. There is uncertainty in both our current understanding of the system as well as about its future possible states – for example, due to future changes in social demands (Ison, 2010) and the existence of multiple promising transition pathways (Geels and Schot, 2007; Rosenbloom, 2017). Those looking to manage transitions to the circular economy should anticipate uncertainty (Preiser et al., 2018) and embrace it by adopting an adaptive management approach (Patton, 2011) which recognises that our understanding, the system's behaviour, and appropriate interventions may change over the course of a transition (Köhler et al., 2019).

Within this adaptive management mindset, experiments may play a useful role in niche development and regime change (Hoogma et al., 2002; Jurgilevich et al., 2016; Matschoss and Repo, 2018). The governance of wicked problems in sustainable resource management may also benefit from the application of a set of guiding principles rather than "management by strict rules" (Endl, 2017). Indeed, complexity and uncertainty urge us to aim to steer change rather than seeking to control it; "the transition to a circular economy can be compared to trying to sail across the ocean to another continent" (European Environment Agency, 2017). The phrase 'to govern' itself is thought to be derived from the Greek *kybernan*, meaning to pilot or steer (Levi-Faur, 2012).

Seeking to understand the system – and to update this understanding over time – is crucial. Transformations towards sustainability need to consider the dynamic behaviour of the system arising from its interacting parts and its propensity to change over time (Patterson et al., 2017). This calls for reflexivity in governance (Hendriks and Grin, 2007; Voß and Bornemann, 2011) and the continuous evaluation and adaptation of practice (Endl, 2017; Patterson et al., 2017). Evaluation can be described in lay terms as the science of finding out 'what works.' It may seek to establish of an intervention 'did it work' and 'how well' but also more formatively 'what worked, how and why' as a means to refine the intervention and inform future decision-making (Patton, 2008; Byrne, 2013). Evaluation is key to effective governance in general, but especially so to the governance of complex systems (Bicket et al., 2020), such as those involved in the transition to the circular economy. It plays a central role in learning and is critical to the success of interventions (Chelimsky, 2006): "understanding the key mechanisms shaping the design, production, use and end-of-life treatment of products in a linear economy is a prerequisite for identifying effective measures that can alter system dynamics and drive the shift to circular material use" (European Environment Agency, 2017). Despite being more important, evaluation is also made more challenging by the presence of complexity. For example, the complexity of the supply chain in the food processing sector has hindered the progress of understanding around important issues such as waste, reuse of resources, and greenhouse gas emissions (Genovese et al., 2017). How to conduct complexity-appropriate evaluation is a well-recognised challenge, and there is a growing base of literature and resources to support knowledge and capacity-building in this area (e.g., Patton, 2011; Bicket et al., 2020). Crucially, the monitoring and evaluation of complex systems and interventions towards the circular economy must take a systems perspective which is attuned to their context (European Environment Agency, 2017).

Another consequence of complexity is disagreement and controversy between actors involved in the transitions to the circular economy over the origin and understanding of problems and appropriate solutions (Ison, 2010; Köhler et al., 2019). Participatory approaches to governance and evaluation are key to help build a holistic and nuanced understanding of the complex system and its interacting elements (Hammond, 2005; Ison, 2018), which can help to reduce uncertainty about the system and identify promising ways forward (Patterson et al., 2017). The transition to the circular economy will require upskilling workers and communities, changes in cultures and traditions, and facilitating shifts in consumer preferences and behaviours; the engagement and participation of actors such as countries, communities, workers, and consumers is also key, therefore, to facilitate effective change and a just transition to the circular economy (de Jesus and Mendonça, 2018; Schröder, 2020).

In this section, we have explored what complexity means for how we manage change to a circular economy and how we seek to understand and influence the relevant social-technical and socioecological systems involved. The key points for those managing change towards the circular economy are to look at the system as a whole, anticipate uncertainty and use an adaptive approach to managing change, and aim to steer change rather than seeking to control it. Evaluation is a particularly important tool to help learn and gain an understanding of: the complex interactions between, and behaviour arising from, the many actors, components, and processes involved in the transition to the circular economy; how these may be changing over time; and whether and how policies and other governance strategies are helping to support the management of circular economy transitions. Last but not least, participatory approaches to both governance and evaluation are of key importance to help make sense of complexity, decide between multiple ways forward, and ensure a just transition and facilitate change to a more circular economy.

Tools and approaches for managing change to the circular economy

Top-down and bottom-up processes for change

Despite the uncertain and emergent nature of transitions, Berkhout et al. (2004) warn against overstating the role of niches as the main arena for driving system transformation. Small, bottom-up niche changes need to be nurtured by public policy (Kemp et al., 1998). Transitions to the circular economy need to be developed instead from both the bottom up and top down simultaneously (Jurgilevich et al., 2016; Patterson et al., 2017), for example, by considering the changes that arise from the interactions of components at the local, niche level (and how these might be scaled up) at the same time as how top-down measures can help to steer the transition by modifying the system's barriers and drivers and seeking to create optimal conditions for change. Indeed, the complexity of the circular economy transition and the adaptive management mindset require us to appreciate the interactions between all parts of the system and therefore the resulting interplay between top-down and bottom-up processes for change. The rest of this section looks at some of the approaches and tools available to those seeking to manage or steer change towards the circular economy, keeping this shared importance of both top-down and bottom-up approaches in mind.

High-level, goal-oriented governance approaches

Many governments have taken initial, explicit steps to promote change towards a circular economy through national and international strategies and policy frameworks, for example: China's 2008 Law for the Promotion of the Circular Economy, the EU's Roadmap to a Resource Efficient Europe and 2015 and 2020 Circular Economy Action Plans, Japan's Fundamental Law for Establishing a Sound Material-Cycle Society, and the United States' EPA Sustainable Materials Management Program (McCarthy et al., 2017; Schröder, 2020). These set out the rationale for moving towards more circular economy practices and establish targets for specific sectors and waste streams. They are complemented by cross-national sustainability agendas such as the UN Sustainable Development Goals and research initiatives such as Future Earth. These high-level, goal-oriented governance approaches are one important component of the transition to the circular economy; the setting of targets and guiding principles exemplifies the importance of seeking to steer, rather than control, complex systems, discussed earlier in this chapter.

Policy tools and approaches

A wide variety of barriers stand in the way of change to the circular economy. These include insufficient investment (e.g., in infrastructure, innovation, and technologies), resource prices which fail to internalise true costs or incentivise efficient use, limited consumer and business acceptance of alternative business models, shortfalls in consumer awareness and practice (e.g., regarding waste separation), lack of information and know-how (e.g., about the detailed composition of products for recycling or about circular product design)(Bicket et al., 2014), and technological lock-in (Preston, 2012). The policies needed to enable change towards the circular economy should draw from and seek to respond to these barriers, for example: policies which encourage manufacturers to design products with asset recovery in mind and support changes in consumption patterns and ownership models (Bicket et al., 2014).

The importance of the traditional policy toolbox in helping to address these barriers and enable the transition to the circular economy shouldn't be underestimated (Köhler et al., 2019). Relevant traditional policy instruments include:

- 'Hard,' legally binding regulatory instruments, such as: targets, product and process requirements, and regulatory instruments for extended producer responsibility, including guarantees and take-back requirements.
- 'Soft' instruments, such as: voluntary agreements and standards, recommendations, and green public procurement.
- Education and information instruments, such as: awareness campaigns, best practice platforms, and advisory services (e.g., for companies to find alternative uses for by-products or alternative sources of inputs).
- Market-based instruments, such as: taxes, charges, subsidies, and pay-as-you-throw schemes.
- Public investment, for example in R&D, skills, and infrastructure (Bicket et al., 2014).

However, as discussed in the section on complexity and systems thinking, the interdependencies and complexity involved in the transition to the circular economy mean that we should not seek to manage change using single policies, but with combinations or 'mixes' of them. Unfortunately, designing effective policy mixes is more challenging than simply assembling a loose bundle of individual instruments; designing a policy mix needs to consider the consistency (lack of contradictions), coherence (synergies between instruments and across different levels of governance), and appropriate sequencing of instruments in the mix (Ekvall et al., 2016). With the right mix, policy instruments can be used in combination to enhance one another's implementation, to target multiple market failures, and to address multiple sectors or steps in the supply chain in a coordinated manner across different groups, activities, and levels of governance (OECD, 2007).

Tools and approaches for evaluation and participation

Despite the importance of evaluation for understanding and managing the complex system associated with the transition to the circular economy, the complexity of the system also makes analysis, data collection, and evaluation more challenging. For example, the complexity of the supply chain in the food processing sector has hindered the progress of understanding around important issues such as waste, reuse of resources, and greenhouse gas emissions (Genovese et al., 2017), and a review of studies investigating the macroeconomic consequences of a transition to the circular economy concluded that current methods to quantify benefits were "not robust enough to serve as trustworthy policy tools" (Dubois, 2015).

In the United Kingdom, the Centre for the Evaluation of Complexity Across the Food-Energy-Environment-Water Nexus (CECAN) was tasked by UK government and research councils with developing evaluation methods that can be applied to complex environmental policy challenges, such as those associated with the circular economy. As part of this remit, CECAN worked with policymakers to develop and use methods to understand the interactions and interdependencies within their systems and to identify effective points of intervention for steering change. Drawing on the fields of complexity science and complex evaluation, a range of methods and approaches can used to better understand and navigate the challenges of working with complex systems. Examples include qualitative comparative analysis, agent-based modelling, theory-based approaches, and participatory systems mapping, which brings together stakeholders in complex policy domains to build a richer understanding of the interactions and interlinkages within them and uses network and causal loop analysis to investigate and explore ways to exploit these to bring about change towards desired objectives. CECAN's insights for the successful evaluation and management of complex systems have been summarised in a supplement to the United Kingdom's central government guidance on evaluation (see Bicket et al., 2020).

Conclusions

The transition to the circular economy involves managing a complex system of actors and sectors constantly adapting and interacting with each other within and across a diverse yet deeply interconnected and overlapping array of social, ecological, technical, economic, and political systems. Change towards the circular economy demands a style of governance that is sensitive to this complexity, exploiting approaches like systems thinking and adaptive management to understand and anticipate complexity and uncertainty in the system. This complexity, constant change, and resulting uncertainty underline the key roles of participation and continuous evaluation in understanding and supporting transitions to the circular economy.

Those seeking to steer change towards the circular economy can combine these overarching management principles and considerations with the well-known, wide range of policy instruments at their disposal, from 'hard' regulatory instruments to voluntary approaches and public investment. The lion's share of the work lies in choosing the appropriate mixes and sequencing of policies that can steer the system towards guiding goals and objectives over time. To assist with this, there are a range of innovative new evaluation approaches and methods that offer to help uncover and explore what works.

References

Allwood JM. (2014). Squaring the circular economy: the role of recycling within a hierarchy of material management strategies. In: Worrell E, Reuter MA. (Eds.), Handbook of recycling: state-of-the-art for practitioners, analysts, and scientists. Boston. 445–477. doi:10.1016/B978-0-12-396459-5.00030-1.

Australian Public Service Commission (2007). *Tackling wicked problems: a public policy perspective.* doi:10.4324/9781849776530.

Bar-Yam Y. (2009). General features of complex systems. In: Kiel LD. (Ed.), *Knowledge management, organizational intelligence and learning, and complexity. encyclopedia of life support systems (EOLSS).* Oxford: EOLSS Publishers, UNESCO.

Bastein T et al. (2013). *Opportunities for a circular economy in the Netherlands.* TNO. Available at: https://www.tno.nl/media/8551/tno-circular-economy-for-ienm.pdf

Berkhout F, Smith A, Stirling A. (2004). Socio-technological regimes and transition contexts. System innovation and the transition to sustainability: theory, evidence and policy. *Edward Elgar Cheltenham.* 44 (106): 48–75.

Bicket M et al. (2014). *Scoping study to identify potential circular economy actions, priority sectors, material flows and value chains.* Brussels, Belgium: Publications Office of the European Union.

Bicket M et al. (2020). *Magenta book 2020 supplementary guide: handling complexity in policy evaluation.* HM Treasury. Available at: https://assets.publishing.service.gov.uk/government/uploads/system/uploads/attachment_data/file/879437/Magenta_Book_supplementary_guide._Handling_Complexity_in_policy_evaluation.pdf.

Boulding KE. (1956). General systems theory – the skeleton of science, *Management Science.* 2 (3): 197–208.

Briassoulis H. (2004). Policy integration for complex policy problems: what, why and how. In: *Greening of policies.* Berlin: Interlinkages and Policy Integration.

Byrne D. (2013). Evaluating complex social interventions in a complex world. *Evaluation. SAGE Publications Ltd.* 19 (3): 217–228.

Carley M, Christie I. (2000). *Managing sustainable development.* Oxfordshire: Earthscan.

Chelimsky E. (2006). *The purposes of evaluation in a democratic society: the Sage handbook of evaluation*. London: Sage Publications. 33–55.

Cilliers P. (1998). *Complexity and post-modernism: understanding complex systems*. London: Routledge.

Colding J, Barthel S. (2019). Exploring the social-ecological systems discourse 20 years later. ecology and society. *The Resilience Alliance*. 24 (1). doi:10.5751/ES-10598-240102.

Crutchfield JP, Machta J. (2011). Introduction to focus issue on randomness, structure, and causality: measures of complexity from theory to applications. *Chaos: An Interdisciplinary Journal of Nonlinear Science. American Institute of Physics*. 21 (3): 37101.

Davies A et al. (2012). *Systemic innovation*. Available at: https://www.siceurope.eu/sites/default/files/field/attachment/SIE%20Systemic%20Innovation%20Report%20-%20December%202012_1.pdf

de Jesus A, Mendonça S. (2018). Lost in transition? Drivers and barriers in the eco-innovation road to the circular economy. *Ecological Economics*. 145(2016): 75–89. doi:10.1016/j.ecolecon.2017.08.001.

Domenech T, Bahn-Walkowiak B. (2019). Transition towards a resource efficient circular economy in Europe: policy lessons from the EU and the member states. *Ecological Economics*. 155 (2017): 7–19.

Dubois M. (2015). *Economic benefits of the circular economy: a critical review of the literature and opportunities for CGE modelling*. Paris. Available at: https://one.oecd.org/document/ENV/EPOC(2015)14/en/pdf

Eckelman MJ, Chertow MR. (2013). Life cycle energy and environmental benefits of a US industrial symbiosis. *The International Journal of Life Cycle Assessment*. 18 (8): 1524–1532. doi:10.1007/s11367-013-0601-5.

EEA (2015). *The European environment – state and outlook 2015: synthesis report*. Copenhagen: European Environment Agency. Available at: www.eea.europa.eu/soer-2015/synthesis/report.

Ekvall T et al. (2016). A systemic and systematic approach to the development of a policy mix for material resource efficiency. *Sustainability*. 8 (4): 373.

Endl A. (2017). Addressing "wicked problems" through governance for sustainable development – a comparative analysis of national mineral policy approaches in the European Union. *Sustainability (Switzerland)*. 9 (10). doi:10.3390/su9101830.

European Commission (2011). *Roadmap to a resource efficient Europe*. Communication from the Commision to the European Parliament, the Council and the European Economic and Social Committee and the Committee of the Regions.

European Environment Agency (2017). Circular by design: products in the circular economy. *European Environment Agency*. doi:10.1002/14651858.CD004884.pub3.

Geels FW, Schot J. (2007). Typology of sociotechnical transition pathways. *Research Policy*. 36 (3): 399–417. doi:10.1016/j.respol.2007.01.003.

Genovese A et al. (2017). Sustainable supply chain management and the transition towards a circular economy: evidence and some applications. *Omega (UK)*. 66: 344–357. doi:10.1016/j.omega.2015.05.015.

Greyson J. (2007). An economic instrument for zero waste, economic growth and sustainability. *Journal of Cleaner Production*. 15 (13): 1382–1390. doi:10.1016/j.jclepro.2006.07.019.

Guijt I et al. (2011). *Evaluation revisited: improving the quality of evaluative practice by embracing complexity: conference report*. Available at: http://www.fapesp.br/avaliacao/manuais/evaluation_revisited_2010.pdf

Gunderson LH, Holling CS. (2002). *Panarchy: understanding transformations in human and natural systems*. Island Press (Environmental management). Available at: https://books.google.co.uk/books?id=DHcjtSM5TogC.

Hammond D. (2005). Philosophical and ethical foundations of systems thinking. *Triplec: Communication, Capitalism & Critique*. 3 (2): 20–27.

Hartley K, van Santen R, Kirchherr J. (2020). Policies for transitioning towards a circular economy: expectations from the European Union (EU). *Resources, Conservation and Recycling*. 55: 104634. doi:10.1016/J.RESCONREC.2019.104634.

Haynes P. (2015). *Managing complexity in the public services*. Abingdon: Routledge.

Hendriks CM, Grin J. (2007). Contextualizing reflexive governance: the politics of Dutch transitions to sustainability. *Journal of Environmental Policy & Planning*. 9 (3–4): 333–350. doi:10.1080/15239080701622790.

Hoogma R et al. (2002). *Experimenting for sustainable transport: the approach of strategic niche management*. London: Routledge. doi:10.4324/9780203994061.

Hughes N, Ekins P. (2018). The role of policy in unlocking the potential of resource efficiency investments BT – investing in resource efficiency: the economics and politics of financing the resource transition. In: Flachenecker F, Rentschler J. (Eds.). Cham: Springer International Publishing. 247–281. doi:10.1007/978-3-319-78867-8_12.

IPCC (2014). Climate change 2014: impacts, adaptation, and vulnerability. Part A: global and sectoral aspects. In: Field CB et al. (Ed.), *Contribution of Working Group II to the fifth assessment report of the intergovernmental panel on climate change*. Cambridge and New York: Cambridge University Press.

Ison R. (2010). Systems practice: how to act in a climate-change world. 1–340. doi:10.1007/978-1-84996-125-7.

Ison R. (2018). Governing the human – environment relationship: systemic practice. *Current Opinion in Environmental Sustainability*. 33: 114–123. doi:10.1016/j.cosust.2018.05.009.

Jurgilevich A et al. (2016). Transition towards circular economy in the food system. *Sustainability (Switzerland)*. 8 (1): 1–14. doi:10.3390/su8010069.

Kemp R, Dijk M. (2013). *Analytical framework of drivers and barriers to resource efficiency (POLFREE – policy options for a resource efficient economy) D1.1*. Available at: https://www.ucl.ac.uk/polfree/publications/publications-2014/Deliverable_1.1_Reformatted_May2014.pdf

Kemp R, Schot J, Hoogma R. (1998). Regime shifts to sustainability through processes of niche formation: the approach of strategic niche management. *Technology Analysis & Strategic Management*. 10 (2): 175–198. doi:10.1080/09537329808524310.

Kirchherr J et al. (2018). Barriers to the circular economy: evidence from the European Union (EU). *Ecological Economics*. 150: 264–272.

Kivimaa P, Kern F. (2016). Creative destruction or mere niche support? Innovation policy mixes for sustainability transitions. *Research Policy*. 45 (1): 205–217. doi:10.1016/j.respol.2015.09.008.

Köhler J et al. (2019). An agenda for sustainability transitions research: state of the art and future directions. *Environmental Innovation and Societal Transitions*. 31: 1–32. doi:10.1016/j.eist.2019.01.004.

Konrad K, Truffer B, Voß JP. (2008). Multi-regime dynamics in the analysis of sectoral transformation potentials: evidence from German utility sectors. *Journal of Cleaner Production*. 16 (11): 1190–1202. doi:10.1016/j.jclepro.2007.08.014.

Levi-Faur D. (2012). From "big government" to "big governance"?'In: Levi-Faur D. (Ed.), *The Oxford handbook of governance*. Oxford: Oxford University Press. 1–17. doi:10.1093/oxfordhb/9780199560530.013.0001.

Levin S et al. (2013). Social-ecological systems as complex adaptive systems: modeling and policy implications. *Environment and Development Economics*. 18 (2): 111–132. doi:10.1017/S1355770X12000460.

Matschoss K, Repo P. (2018). Governance experiments in climate action: empirical findings from the 28 European Union countries. *Environmental Politics*. 27 (4): 598–620. doi:10.1080/09644016.2018.1443743.

McCarthy A, Dellink R, Bibas R. (2017). *The macroeconomics of the circular economy transition*. Paris. doi:10.1787/af983f9a-en.

Milios L. (2018). Advancing to a circular economy: three essential ingredients for a comprehensive policy mix. *Sustainability Science*. 13 (3): 861–878. doi:10.1007/s11625-017-0502-9.

OECD (2007). *Instrument mixes for environmental policy*. Paris: OECD Publishing.

OECD (2019). *Global material resources outlook to 2060*. Paris: OECD. doi:10.1787/9789264307452-en.

Papachristos G, Sofianos A, Adamides E. (2013). System interactions in socio-technical transitions: extending the multi-level perspective. *Environmental Innovation and Societal Transitions*. 7: 53–69. doi:10.1016/j.eist.2013.03.002.

Patterson J et al. (2017). Exploring the governance and politics of transformations towards sustainability. *Environmental Innovation and Societal Transitions*. 24: 1–16. doi:10.1016/j.eist.2016.09.001.

Patton MQ. (2008). *Utilization-focused evaluation*. Minneapolis, MN: Sage Publications.

Patton MQ. (2011). *Developmental evaluation: applying complexity concepts to enhance innovation and use*. New York: The Guilford Press.

Preiser R et al. (2018). Social-ecological systems as complex adaptive systems: organizing principles for advancing research methods and approaches. *Ecology and Society*. 23 (4). doi:10.5751/ES-10558-230446.

Preston F. (2012). *A global redesign? Shaping the circular economy*. London: Chatham House.

Raven R, Verbong G. (2007). Multi-regime interactions in the Dutch energy sector: the case of combined heat and power technologies in the Netherlands 1970–2000. *Technology Analysis & Strategic Management*. 19 (4): 491–507. doi:10.1080/09537320701403441.

Reyers B et al. (2018). Social-ecological systems insights for navigating the dynamics of the anthropocene. *Annual Review of Environment and Resources. Annual Reviews*. 43 (1): 267–289. doi:10.1146/annurev-environ-110615-085349.

Rittel HWJ, Webber MM. (1973). Dilemmas in a general theory of planning. *Policy Sciences*. 4 (2): 155–169. doi:10.1007/BF01405730.

Rockström J et al. (2009). A safe operating space for humanity. *Nature*. 461 (7263): 472–475. doi:10.1038/461472a.

Rosen R. (1991). *Life itself : a comprehensive inquiry into the nature, origin, and fabrication of life*. New York: Columbia University Press. doi:10.1177/027046769301300389.

Rosenbloom D. (2017). Pathways: an emerging concept for the theory and governance of low-carbon transitions. *Global Environmental Change*. 43: 37–50. doi:10.1016/j.gloenvcha.2016.12.011.
Schröder P. (2020). *Promoting a just transition to an inclusive circular economy*. Available at: https://www.chathamhouse.org/publication/promoting-just-transition-inclusive-circular-economy
Scoones I, Newell P, Leach M. (2015). *The politics of green transformations*. New York: Routledge.
Shove E. (2014). Putting practice into policy: reconfiguring questions of consumption and climate change. *Contemporary Social Science*. 9 (4): 415–429. doi:10.1080/21582041.2012.692484.
Stahel WR. (1997). The functional economy: cultural and organizational change. In: Richards D. (Ed.), *The industrial green game: implications for environmental design and management*. Washington, DC: National Academy Press. 91–100.
Tukker A. (2015). Product services for a resource-efficient and circular economy – a review. *Journal of Cleaner Production*. 97: 76–91. doi:10.1016/j.jclepro.2013.11.049.
Unruh GC. (2000). Understanding carbon lock-in. *Energy Policy*. 28 (12): 817–830. doi:10.1016/S0301-4215(00)00070-7.
Voß JP, Bornemann B. (2011). The politics of reflexive governance: challenges for designing adaptive management and transition management. *Ecology and Society*. 16 (2).
Webster K. (2017). *The circular economy: a wealth of flows*. Cowes: Ellen MacArthur Foundation Publishing.
Wiener N. (1948). *Cybernetics, or control and communication in the animal and the machine*. Oxford. England: John Wiley & Sons.

15

CIRCULAR ECONOMY MESO-LEVEL PLANNING

An approach with 'distributed economies'

Mario Augusto P. Monteiro and Cleber JC. Dutra

Introduction

The recent and increasing introduction of circular economy (CE) principles into policy and legislative instruments issued by state and supranational authorities aligns with a general view that economic advances are (and must be) compatible with environmental and social benefits. Thus, the pursuit of sound systems of production and consumption relying on standards able to simultaneously benefit a greater number of people and preserve the natural systems is increasing.

The unsustainable pattern of the traditional linear 'extract-produce-use-dump' material and energy flow model of the modern economic system – with its corresponding growth in demand for raw materials worldwide and the accumulation of waste – has accrued persistent criticism (Frosch and Gallopoulos, 1989; Meadows et al., 1972; Stahel and Reday-Mulvey, 1976; Robèrt, 1991). From an economic viewpoint: (1) production-consumption systems based on depletion, rather than on the restorative use of resources, entail significant losses all along the value chain, and (2) linear models of resource consumption increase the exposure of firms (and therefore of consumers) to risks of higher and more volatile production costs (Ellen MacArthur Foundation, 2013).

Hence, the concept of sustainability arises as a proxy for expectations and requirements of social agents engaged in/pursuing social and environmental outcomes provided by economic growth. Sustainability is the process of realising human development in an inclusive, equitable, prudent, and safe way (Gladwin et al., 1995). Furthermore, a 'sustainable enterprise' should seek to produce positive results not only in economic terms but also in social and environmental terms (Hart and Milstein, 2003).

Despite clear insights into the entangled aspects of the economic, environmental, and social dimensions of sustainable production-consumption systems, particularly at a micro-level view of operations, a challenge to policymakers and stakeholders charged with designing programmes aiming to enforce CE policies is a lack of guidance for the economic meso-level planning of such systems. This chapter adds theoretical elements for reflection on arrangements as potential models for organising regional and local production and consumption practices. Using a brief review of the central elements in the criticism of the neoclassical paradigm, we examine the fundamentals of economics related to sustainability issues as a way of analysing how the model of distributed economies (DE) might serve as a guide for the implementation of CE principles.

Grounds for expecting support from distributed economies to circular economy

Considering the process of adopting CE principles as a significant cultural change, the planning and implementing of programmes with this purpose requires not only sound models in combining sustainability dimensions but also innovative approaches, particularly in the social dimension. Inspired by the remarkable accomplishment of industrial symbiosis practices in the Scandinavian region, the model of distributed economies (DE) corresponds to the economic perspective that evolved from the examination of those highly complex closed-loop operations. These operations involved material and energy flows from industrial, as well as commercial and residential, production-consumption activities. Arising from social relationships among agents engaged in improving the efficient use of resources, the symbiotic concepts embody vital elements that constitute CE. Therefore, social innovations were present from the start in such programs in Scandinavia.

Those spontaneous (and innovative) cultural changes are translated by Johansson et al. (2005) as a concept of a "third culture" which aims to unite the technical-scientific culture and the culture of humanistic tradition and is characterised by a greater accountability of humanity in relation to the environment. Johansson et al. (2005) proposed the DE model as an alternative structure for managing the processes of production and consumption. It is based on smaller production units, which are flexible and integrated into regional networks, generating the innovations required to achieve a CE ideal.

Therefore, an analysis of the DE model from the perspective of the CE is important in terms of its contribution to overcoming the present environmental challenges, as well as the extent of its criticism of the neoclassical paradigm.

Theoretical foundations for the analyses of the distributed economies model

Underlying the DE model is the recognition of the limitations of production systems designed in the neoclassical paradigm, which are oriented towards maximising allocative and productive efficiency through the economies of scale offered by large production units. The model notes the need for a new paradigm based on a new economic thinking and on a broader view of supply chains to the extent that technological advances, though relevant, are not by themselves sufficient to solve the contemporary environmental dilemmas.

Fundamentals of economic analysis of sustainability: the neoclassical paradigm

The dominance of neoclassical paradigms on the formulation of economic policies and derived managerial practices, observed throughout the 19th and 20th centuries, can be attributed to the strength of two theorems related to welfare economics, namely: (1) Pareto efficiency, provided by maximisation of preferences under a budget restriction and profit maximisation given a certain technology, and (2) the efficient allocation of resources in a competitive economy, given appropriate initial conditions. Underlying these theorems, two fundamental premises to neoclassical economic thought were paramount, namely the strict rationality of economic agents and the prevalence of competitive market structures.

It is important to mention, however, that scholars who follow the neoclassical paradigm have, in recent years, made many attempts to incorporate new elements into their formulations in order to deal with the oversimplifying assumptions that reduce the ability of the theory to explain complex social phenomena (Gowdy and Erickson, 2004, 2005; Arnsperger and Varoufakis, 2006). Issues such as market power and imperfect competition, asymmetric information, the presence of externalities, and the existence of public goods and common resources are becoming increasingly important in economic literature (Salanié, 2000; Stavins, 2000; Kreps, 2004).

Despite efforts to add new dimensions to its analytical models, the main purpose of the neoclassical economic approach to the issues and challenges faced by society, including, for instance, global warming and loss of biodiversity, remains the creation of conditions for a competitive economy in order to achieve Pareto efficiency. In this context, the emphasis on determining the "correct values" for maximising social welfare reduces the relevance of social and environmental variables within the neoclassical paradigm. Moreover, the fragility of strict assumptions of rationality of economic agents and of the prevalence of competitive market structures restricts the capacity of neoclassical economics to deal with the current environmental and social issues, including those concerned with climate change, the destruction of biodiversity, and growing inequalities.

Criticism of the neoclassical economic paradigm is focused on the validity of some of these assumptions, concepts, and theories (Arnsperger and Varoufakis, 2006), notably in terms of their ability to represent and explain adequately the complex contemporary reality. The increasing relevance of alternative economic approaches, characterised by ecological concerns, points out the emergent strength of those alternatives.

Some limits and pitfalls of the neoclassical paradigm applied to sustainability analysis

The mechanistic-reductionist and deterministic nature of the scientific method associated with the neoclassical paradigm reinforces the inadequacies of its application for proposing solutions to current environmental challenges. Such inadequacies are discussed by Morin (1977; 1990), for whom the scientific method of discussion of environmental issues must break the methodological limitations of deterministic and reductionist paradigms, especially in the context of the opportunity of change represented by the contemporary environmental crisis.

Restrictions in the adoption of neoclassical models, grounded in a mechanistic and reductionist worldview, whose main assumptions are the existence of independent utility functions, private ownership of economic assets, and perfectly competitive markets, to deal with environmental issues are evidenced by different authors (Swaney, 1987; Söllner, 1997; Venkatachalam, 2007).

Individual consumers and firms are at the centre of neoclassical analysis, which strongly highlights the concept of *homo oeconomicus* (which seeks to maximise its utility guided by individual preferences, using opportunistic behaviour). The suitability of this concept to sustainability issues is questioned by Faber et al. (2002), who argue that only occasionally will individuals orient their preferences in favour of sustainability, because the gains from such behaviour are not clear for them. Similarly, Siebenhuner (2000), Gowdy and Erickson (2005), and Venkatachalam (2007) link the existing restrictions on the concept of economic man to the explanation of decisions on consumption of goods and natural resources.

Other critical aspect of the neoclassical approach to environmental issues is the denial of the existence of concrete limits to economic growth, based on the argument that the introduction of technological innovations reduces potential growth constraints resulting from scarcity of natural resources. Based on the second law of thermodynamics (entropy) applied to environmental issues, as proposed by Georgescu-Roegen (1971), the possibility of return through technological innovations to the situation existing prior to consumption/degradation of the productive resources available in the natural environment is questioned (Venkatachalam, 2007).

Besides the aspects previously mentioned, the following elements of neoclassical economic paradigm applied to sustainability issues are also questioned: (1) role and nature of externalities, (2) valuation of the social and natural environments, and (3) relevance of institutional arrangements (Swaney, 1987; Söderbaum, 1992; Spash and Hanley, 1995; Gowdy and Erickson, 2005).

Fundamentals of economic analysis of sustainability: the institutional/ecological paradigm

In opposition to assumptions/postulates of neoclassical economics, the institutionalist approach rejects the assumption of omniscience and efficiency of free markets, coupled with the reductionism mechanism, increasingly abstract, inherent to neoclassical paradigm. In this sense, the institutionalist approach reinforces its nature as more holistic, evolutionary, and inclusivist.

In methodological terms, the institutional paradigm presents three clear differences when compared to the neoclassical paradigm. The first difference is the emphasis on the institutionalist approach in the interaction between the various fields of human knowledge (inter)related to the search of solutions for specific issues, such as those related to the environment. The second difference refers to the rejection by institutionalists of the neutrality of the researcher before the research object, with the argument that subjective elements are inherent to the research process. Finally, the institutional approach rejects the neoclassical assumption that all assets and rights in a society, including those that are environment related, can be represented monetarily. As a consequence, when analysing environmental issues, the institutionalist method is oriented towards the incorporation of social, ethical, and ideological dimensions.

Söderbaum (1992), considering the heterogeneity of individuals and their diverse categories, argues that the institutionalist paradigm assigns a central role to the networking of the different categories of social actors during the process of social change. Adopting the institutionalist paradigm as reference, Gowdy and Erickson (2005) characterise the ecological economy as a source of a theoretical basis and public policy solutions for environmental issues, once it encompasses production and consumption in the broadest sense, including their environmental, social, and ethical dimensions, as well as their effects on the market.

Ecological economics incorporates the concept of entropy as the basis for the concepts of absolute scarcity of natural resources and strong sustainability (Turner et al., 1995). Moreover, ecological economists note current development patterns based on excessive use of energy, increasingly stimulated by the production and consumption of new energy-intensive products (Venkatachalam, 2007). The maintenance of such development patterns can lead irrevocably to resource depletion, which shall impose, from the ecological perspective, a change from these current patterns to another one that uses fewer resources and brings about less environmental degradation. Ecological economics advocates for the concept of *homo politicus* (Faber et al., 2002), understood as the individual whose actions are guided by the common good rather than the maximisation of individual welfare.

Therefore, the major conceptual issues of ecological economics can be summarised as follows: (1) the individual as social actor in opposition to a rational economic agent; (2) criticism of the applicability of the marginalist analysis; (3) treatment of uncertainty in the context of environmental issues; (4) emphasis on the concepts of fairness, stability, and resiliency of social and environmental systems (as opposed to efficiency) as drivers of public policy; and (5) the concept of production as a social and physical process.

Paradigmatic alignment of the distributed economies model with ecological economics' support of the circular economy

The backdrop of the discussions about how to reconcile economic development and environmental sustainability often fails to take account of constraints to the supply of natural resources and approaches to reduce or eliminate these constraints. More specifically, polarisation is noticeable in the two distinct approaches to the economic analysis of environmental issues which are represented by environmental economy and CE.

Generally, the main difference between these two approaches is their views on the relationship between economic growth and environment. Environmental economists state that environmental limits to economic growth are relative, given the inventive capacity of mankind (represented by its technology). However, those who advocate for CE argue that the environment presents absolute limits to economic growth. Underlying this divergence is the degree of comprehension and acceptance regarding the irreversibility of the effects on the environment caused by existing production and consumption practices.

According to the neoclassical paradigm, the quest for wealth creation implies an increasing scale of production units. Johansson et al. (2005) presented the DE model as a new industrial development strategy which is compatible with sustainability requirements. It addresses the inadequacy of the neoclassical paradigm of being grounded in achieving economies of scale through the concentration of the production process in large geographically concentrated production units. Such growth of productive economic units, made possible by acquisitions and industrial diversification policies, has some negative effects, including (1) environmental problems, related to increasing demands for natural resources; (2) inflexibility of production; (3) exclusion of consumers from production decisions; (4) subordination of quality to the requirements of continuous reduction of costs; (5) centralisation of the process of value creation, resulting in mental and cultural impoverishment of remote regions affected by this process; and (6) growth of inequality in the distribution of wealth generated.

In the process of formulating strategies that go beyond the efficiency postulates of neoclassical production theory, the structure and objectives of production-consumption systems should be challenged and changed, respecting the limits to economic growth imposed by nature. Concerning these limits, the continuity of the process of sustainable development depends on the maintenance of the energy supply, which requires simultaneously a continuous increase in productive efficiency and a change in consumption patterns to reduce the pressure on the stock of natural resources (Elzen and Wieczorek, 2005; Alcott, 2005). Due to the difficulty in providing solutions to the present environmental contradictions based solely on purely technical innovations (Elzen and Wieczorek, 2005; Alcott, 2005), proposals for innovative ways of organising production-consumption systems should be based on an alternative set of values. They should meet the demands of a new paradigm associated with the DE concept of a 'third culture,' to the extent that they: (1) refer to technical development as well as to economic development, (2) have an inclusive nature, and (3) emphasise the concept of global responsibility based on common ethical standards.

The DE model can serve as the basis for innovative strategies for organising production and consumption towards sustainability by grouping similar or complementary systems into networks to provide scale, without the disadvantages of inflexibility. Söderbaum (1992) categorises economic strategies for contemporary societies into two alternatives. The first strategy focuses on unconditioned economic growth (in terms of use of natural resources), resulting in the concentration of decision-making in large multinational corporations and homogenised consumption patterns. The second strategy focuses on sustainability and selective internationalisation.

It incorporates the premise that each region (or country) should take responsibility for the environmental degradation caused by its patterns of production and consumption. Thus, the DE model, by proposing a decentralisation of production and consequently including local communities in the decision-making process related to production and consumption, takes up a clear option for the second strategic alternative.

According to the DE model, production in smaller units creates the opportunity to focus on product quality, providing a competitive advantage to the firms within such an arrangement. Any disadvantages of smaller scale production can be offset by greater integration and coordination of such productive units, also taking account of the knowledge from universities and research centres located in those regions. The DE model seeks to bring many key issues related to sustainable development (even ethical considerations) closer to the individual agent, either as a consumer or producer. Progress, wealth, and quality of life assume a more tangible sense when incorporated into decisions about everyday life.

Regarding the environment, the DE model emphasises society's roles in the management and preservation of natural resources. In this sense, the model considers the possibilities of greater individual participation in production and consumption decisions, thereby giving these individuals the power to demand and implement actions aimed at environmental sustainability and which would increase their quality of life. Indeed, the structure of the production system proposed by the DE model is largely dependent on self-organised cooperative (non-competitive) processes involving a diversity of actors and regarding production as both a physical and a social process (Gowdy and Erickson, 2005).

Thus, it is possible to establish a clear identification between the DE model and ecological economics concepts. The conclusion which can be drawn, therefore, is that the system of production and consumption proposed by this model represents a sustainable alternative to economic development, particularly related to the environmental and social dimensions. By extension, it also provides a coherent foundation for the design of CE operational structures and practices.

Conclusion

CE practices are still far from being widespread, and scalability remains a major issue (Ghisellini et al., 2016; Linder and Williander, 2017). The DE model can address these issues. The recommendations by different authors of adopting a closed cycle of materials as a framework for a sustainable economy (e.g., Stahel and Reday-Mulvey, 1976; Pearce and Turner, 1990; Su et al., 2013) and its benefits for minimising waste and emissions by slowing, closing, and narrowing loops (Geissdoerfer et al., 2017; Stahel, 2016) reflect the original concepts that inspired the DE approach.

The DE model facilitates the pursuit of innovative economic opportunities (Ellen MacArthur Foundation, 2015; Stahel, 2014) and optimisation of natural and human resource use (Missemer, 2018) using symbiotic practices. Considering that lack of eliciting innovation could explain the slowness of circular economy adoption (Pietro-Sandoval et al., 2018), especially related to responses expected as social changes at societal sectors (Huesemann, 2004; Lozano, 2008), the social innovations encompassed by the description of the DE approach constitute an ideal guidance to promote CE practices.

The model is fully compatible with the design of radically new alternative solution possibilities of using multiple levels of application of the principles and practices of CE (Ghisellini et al., 2016; Lieder and Rashid, 2016; Sauvé et al., 2016; Blomsma and Brennan, 2017; Murray et al., 2017; Geissdoerfer et al., 2017). Such designs comprise the entire life cycle of any process, as well as all interactions between the process, the environment, and the economy in which it is embedded, implying clear progress of the entire living and economic system when compared with the prevailing linear model of resource consumption.

Furthermore, its objection to the neoclassical economic paradigm, particularly the 'technological fixes' (Solow, 1974) by proposing organisational and social changes in favour of economic advances, is fully suitable for the CE perspective. The DE model's paradigmatic choice aligns with the view that the economic subsystem belongs to a larger system in which it is articulated, integrated, and limited by the environmental and social subsystems (Boron and Murray, 2004; Mitchell, 2000). This perspective assumes that a system is composed of interrelated elements that constitute an entity or global unity, thus implying the complexity of its nature (Morin, 1977).

Considering that managerial practices mainly evolve from concepts and principles established by economic theories and paradigms, we contend that an in-depth examination of the fundamentals of economics related to sustainability issues can contribute to the development of CE management approaches. A deeper understanding of economic perspectives involving planning and implementation of CE propositions is essential for their successful performance, especially in times of great uncertainty about the global structures of the production-consumption system. Indeed, there are significant doubts about our dependence on supply global chains, as evidenced by the impacts of the COVID-19 pandemic. There is a need to encourage shifts towards 'localisation,' which is an underlying proposition promoted by the DE philosophy that we examined in this chapter.

References

Alcott B. (2005). Jevons' paradox. *Ecological Economics*. 54: 9–21.

Arnsperger C, Varoufakis Y. (2006). What is neoclassical economics? The three axioms responsible for its theoretical oeuvre, practical irrelevance and, thus, discursive power. *Panoeconomicus*. 1: 5–18.

Blomsma F, Brennan G. (2017). The emergence of circular economy: a new framing around prolonging resource productivity. *Journal of Industrial Ecology*. 21 (3): 603–614.

Boron S, Murray K. (2004). Bridging the unsustainability gap: a framework for sustainable development. *Sustainable Development*. 12 (2): 65–73.

Ellen MacArthur Foundation (2013). *The circular model: a brief history and school of thought*. Available at: www.ellenmacarthurfoundation.org/circulareconomy/circular-economy/the-circular-model-brief-history-and-schools-ofthought.

Ellen MacArthur Foundation (2015). *Towards the circular economy: economic and business rationale for an accelerated transition*. Available at: www.ellenmacarthurfoundation.org/assets/downloads/TCE_Ellen-MacArthur-Foundation_9-Dec-2015.pdf.

Elzen B, Wieczorek A. (2005). Transitions towards sustainability through system innovation. *Technological Forecasting & Social Change*. 72: 651–661.

Faber M, Petersen T, Schiller J. (2002). *Homo oeconomicus* and *Homo politicus* in ecological economics. *Ecological Economics*. 40 (3): 323–333.

Frosch RA, Gallopoulos NE. (1989). Strategies for manufacturing. *Scientific American*. 261 (3): 144–153.

Geissdoerfer M, Savaget P, Bocken NM, Hultink EJ. (2017). The circular economy – a new sustainability paradigm? *Journal of Cleaner Production*. 143: 757–768.

Georgescu-Roegen N. (1971). *The entropy law and the economic process*. Cambridge: Harvard University Press.

Ghisellini P, Cialani C, Ulgiati S. (2016). A review on circular economy: the expected transition to a balanced interplay of environmental and economic systems. *Journal of Cleaner Production*. 114: 11–32.

Gladwin TN, Kennelly JJ, Krause TS. (1995). Shifting paradigms for sustainable development: implications for management theory and research. *Academy of Management Review*. 20 (4): 874–907.

Gowdy J, Erickson J. (2004). *Ecological economics at a crossroads*. Rensselaer Working Papers in Economics n. 0417. Troy, NY: Rensselaer Polytechnic Institute.

Gowdy J, Erickson J. (2005). The approach of ecological economics. *Cambridge Journal of Economics*. 29 (2): 207–222.

Hart SL, Milstein MB. (2003). Creating sustainable value. *Academy of Management Perspectives*. 17 (2): 56–67.

Huesemann MH. (2004). The failure of eco-efficiency to guarantee sustainability: future challenges for industrial ecology. *Environmental Progress*. 23 (4): 264–270.

Johansson A, Kisch P, Mirata M. (2005). Distributed economies – a new engine for innovation. *Journal of Cleaner Production*. 13: 971–979.
Kreps DM. (2004). *Microeconomics for managers*. New York: Norton.
Lieder M, Rashid A. (2016). Towards circular economy implementation: a comprehensive review in context of manufacturing industry. *Journal of Cleaner Production*. 115: 36–51.
Linder M, Williander M. (2017). Circular business model innovation: inherent uncertainties. *Business Strategy and the Environment*. 26 (2): 182–196.
Lozano R. (2008). Envisioning sustainability three-dimensionally. *Journal of Cleaner Production*. 16 (17): 1838–1846.
Meadows D, Randers J, Behrens WW, Meadows DH. (1972). *The limits to growth: a report for the Club of Rome's project on the predicament of mankind*. New York: Universe Books.
Mitchell CA. (2000). Integrating sustainability in chemical engineering practice and education: concentricity and its consequences. *Process Safety and Environmental Protection*. 78 (4): 237–242.
Morin E. (1977). *The method, I: the nature of nature*. Paris: Seuil.
Morin E. (1990). *Science avec conscience*. Paris: Fayard.
Missemer, A. (2018). Natural capital as an economic concept, history, and contemporary issues. *Ecological Economics*. 143: 90–96.
Murray A, Skene K, Haynes K. (2017). The circular economy: an interdisciplinary exploration of the concept and application in a global context. *Journal of Business Ethics*. 140 (3): 369–380.
Pearce DW, Turner RK. (1990). *Economics of natural resources and the environment*. Baltimore, MA: John Hopkins University Press.
Prieto-Sandoval V, Jaca C, Ormazabal M. (2018). Towards a consensus on the circular economy. *Journal of Cleaner Production*. 179: 605–615.
Robèrt KH. (1991). The physician and the environment. *Reviews in Oncology: European Organisation for Research and Treatment of Cancer*. 4 (2): 1–3.
Salanié B. (2000). *Microeconomics of market failures*. Cambridge, MA: MIT Press.
Sauvé S, Bernard S, Sloan P. (2016). Environmental sciences, sustainable development, and circular economy: alternative concepts for trans-disciplinary research. *Environmental Development*. 17: 48–56.
Siebenhuner B. (2000). *Homo sustinens* – towards a new conception of humans for the science of sustainability. *Ecological Economics*. 32: 15–25.
Söderbaum P. (1992). Neoclassical and institutional approaches to development and the environment. *Ecological Economics*. 5: 127–144.
Söllner F. (1997). A reexamination of the role of thermodynamics for environmental economics. *Ecological Economics*. 22: 175–201.
Solow R. (1974). The economics of resources or the resources of economics. *The American Economic Review*. 64 (2): 1–14.
Spash C, Hanley N. (1995). Preferences, information, and biodiversity preservation. *Ecological Economics*. 12: 191–208.
Stahel WR. (2014). *Reuse is the key to the circular economy*. Brussels: European Commission.
Stahel WR. (2016). The circular economy. *Nature*. 531 (7595): 435–438.
Stahel WR, Reday-Mulvey G. (1976). *Jobs for tomorrow: the potential for substituting manpower for energy*. Report to the Commission of the European Communities (now the European Commission), Brussels.
Stavins RN. (2000). *Economics of the environment: selected readings*. New York: Norton. 4th Ed.
Su B, Heshmati A, Geng Y, Yu X. (2013). A review of the circular economy in China: moving from rhetoric to implementation. *Journal of Cleaner Production*. 42: 215–227.
Swaney J. (1987). Elements of a neoinstitutional environmental economics. *Journal of Economic Issues*. 21 (4): 1739–1779.
Turner K, Perrings C, Folke C. (1995). *Ecological economics: paradigm or perspective*. CSERGE Working Paper GEC 95–17. Available at: www.cserge.ac.uk/sites/default/files/gec_1995_17.pdf (Accessed 7/4/2020).
Venkatachalam L. (2007). Environmental economics and ecological economics: where they can converge? *Ecological Economics*. 61: 550–558.

Further reading

Kirchherr J, Reike, D, Hekkert M. (2017). Conceptualizing the circular economy: an analysis of 114 definitions. *Resources, Conservation and Recycling*. 127: 221–232. (An overview of CE conceptualisations and its aspects and principle convergence.)

Oltra V. (2008). Environmental innovation and industrial dynamics: the contributions of evolutionary economics. *Cahiers du GREThA*. 28 (27): 77–89. (A better understanding of determinants and features of environmental innovations, which are factors that favour the transition towards CE.)

Prieto-Sandoval V, Jaca C, Ormazabal M. (2018). Towards a consensus on the circular economy. *Journal of Cleaner Production*. 179: 605–615. (A proposal for a unifying concept of CE, including principles, structure, and determinants, highlighting the role of eco-innovations as determinants to CE implementations.)

Roscoe S, Cousins PD, Lamming RC. (2016). Developing eco-innovations: a three-stage typology of supply networks. *Journal of Cleaner Production*. 112: 1948–1959. (An elaborate conception of eco-innovation and a proposed typology of networks to develop it.)

Sehnem S, Vazquez-Brust D, Pereira SCF, Campos LM. (2019). Circular economy: benefits, impacts and overlapping. *Supply Chain Management*. 24 (6): 784–804. (Analyses of overlaps, complementarities, and divergences in the literature on CE and non-linear production models and frameworks.)

PART III

Policy and legislative considerations

16

THE ROLE OF POLICY IN CREATING A MORE CIRCULAR ECONOMY

Patrick J. Mahon

Introduction

This chapter discusses the role of policy – and, in particular, public policy, as generally created by governments – in making the economy more circular. If circularity is the objective, policy intervention will be necessary, because the current state of most, if not all, countries' economies is largely linear.

So, what do we mean by 'circular'? To quote one definition,

> A circular economy is an alternative to a traditional linear economy (make, use, dispose) in which we keep resources in use for as long as possible, extract the maximum value from them whilst in use, then recover and regenerate products and materials at the end of each service life.
>
> (WRAP, n.d.)

Government policy to drive circularity in the economy is a development of resources and waste policy, because its historical antecedents were generally concerned with controlling what happens to products and materials when they become waste. It may therefore be useful to briefly summarise the history of resources and waste policy in the United Kingdom as an example of the way in which the focus of government policy has evolved over time.

The first legal instrument to control what happened to waste in the United Kingdom was the Public Health Act 1848. This legislation was introduced in response to the negative health impacts of the Poor Law 1834 and was intended to improve public health through, amongst other actions, 'the removal of all refuse from houses, streets and roads' (UK Parliament, n.d.). This was followed a decade later by the Local Government Act 1858, which extended the powers of the Local Boards of Health set up by the previous Act to contracting for the removal of household waste. That this was an issue of great public interest can be seen from the fact that the pre-eminent British author of the day, Charles Dickens, focused the plot of his last completed novel, *Our Mutual Friend*, on the emerging business of waste (then called 'dust') management (Dickens, 1865).

This chapter is written in a personal capacity and should not be taken to represent WRAP's views.

The primary purpose of waste management – the protection of public health – remained broadly unchanged for the following century. However, in the 1960s, as the manufacturers of consumer goods in developed countries such as the United Kingdom and United States vied with each other to develop new products and thus gain market share, public attention started to shift towards the increasing wastefulness of consumer societies; see, for example, Packard (1960). Concerns around the efficiency of resource use were reinforced by the emergence of the environmental movement in the late 1960s, partly prompted by the famous 'Earthrise' picture taken during the Apollo 8 mission that circled the Moon in December 1968 (see, e.g., Drake, 2018) and the oil crisis in 1973, which led to urgent programmes of energy efficiency (see Darmstadter, 2014).

By the 1990s, these environmental concerns led to the development of policy measures focused less on protecting human health and more on resource efficiency. One key development was the introduction of the waste hierarchy. This was a preferential ordering of waste management options originally created by the Dutch politician Ad Lansink in 1979 (Lansink, 2017). It was formally adopted into European Union policy in 1989 in 'A Community Strategy for Waste Management' (European Commission, 1989). It became a legally binding five-step ordering (prevention, preparing for reuse, recycling, recovery, disposal) in the 2008 revision to the EU Waste Framework Directive (European Commission, 2008). Although some limited policy measures were adopted at EU level to promote waste prevention and reuse, the two elements at the top of the waste hierarchy, most attention was focused on legally binding targets to increase recycling rates as a concrete way of increasing the efficiency of resource use.

In the first decade of the 21st century, the focus on resource efficiency was supplemented by the introduction of the concept of a circular economy (CE); see, for example, Braungart and McDonough (2008), originally published in 2002. The objective here moved beyond both end-of-pipe solutions intended to protect human health and the environment and approaches that increased the efficiency of resource usage towards a holistic and collaborative redesign of the products and services that satisfy society's needs and wants. Enabling such systemic solutions through public policy has proven more challenging than at previous stages, when legislative measures to reduce negative outcomes were easier to design.

As we enter the third decade of the 21st century, we can look back on nearly two centuries of UK policy responses to concerns around waste, resources, and the circular economy as a specific example of the more general evolution of resources and waste policy worldwide. How we use policy to drive greater circularity in future is the focus of the rest of this chapter.

The policy process

Before starting to discuss specific policy measures that might help to create a more circular economy, we need to answer some introductory questions. What *is* policy? *Why* is it needed? *Why* should policymakers be interested? And *how* can policy drive progress towards a more circular economy?

What is policy?

'Policy' is defined as 'a set of ideas or a plan of what to do in particular situations that has been agreed to officially by a group of people, a business organization, a government, or a political party' (Cambridge Dictionary, n.d.).

I will focus in this chapter on public policy – that is, policy decisions made by the public sector and by central governments in particular.

Why is public policy ever needed?

Public policy measures are the mechanism through which a public body, such as a central government ministry or a local municipality, enacts its programme of action and encourages or requires activities to take place that it thinks are socially beneficial and would otherwise not take place, whilst also restricting or stopping other activities that are seen as socially undesirable. Some examples from the United Kingdom might include the provision of tax relief on private pension contributions to encourage saving for retirement or the 2007 ban on smoking in enclosed workplaces to reduce health risks from passive smoking. Economists (see, for example, OECD, 2006) often refer to policy being a response to a 'market failure,' the implication being that policy measures should only ever be needed if the free market does not lead to a socially desirable outcome spontaneously. This means that policy is politically determined and will vary over time and space in line with changing political and public priorities.

Why should policymakers be interested in creating a circular economy?

The world faces increasing challenges in ensuring access to the material resources that are needed by society, due to a number of demographic and other pressures:

- Demographics (United Nations, 2019):
  - The global population will grow from 7.3 billion in 2015 to 8.5 billion in 2030 and 9.7 billion in 2050;
  - The global middle class will reach 4.9 billion in 2030, up 145% from 2 billion in 2012 (Pezzini, 2012);
  - Increasing urbanisation: 50% of the world's population lived in cities by 2010. This will increase further, to 58% by 2025 and 67% by 2050.
- The impact of demographic and economic growth (WRAP, 2015: 8–9):
  - Global energy demand up 40% by 2030;
  - Global food demand up 50% from 2012 to 2030;
  - Severe water stress will affect 48% of the world's population by 2030 (up from 21% in 2012).
- The consequences:
  - Greater competition for material and energy resources, leading to higher prices and price volatility;
  - Countries will therefore benefit from greater resource efficiency and moves towards a circular economy, keeping resources in use for longer and reducing their demand for new, virgin materials.

At the same time, creating a circular economy can also have beneficial social effects in terms of job creation and the stimulation of economic growth – see, for example, WRAP and Green Alliance (2015), WRAP (2020).

How can policy drive a circular economy?

Most countries have developed a largely linear economy, as discussed in the Introduction. If a more circular economy is to be achieved, the market failures that led to this state of affairs must be addressed. These market failures principally include (Defra, 2011: 7–10):

- Environmental externalities, where an individual or firm does not bear the costs of the harm it imposes (e.g., most producers do not have to pay the waste management costs of dealing with their product at end of life);
- Incomplete or asymmetric information, where the two parties to an exchange do not have perfect information or have different levels of information (e.g., a waste reprocessor will often know less about the quality of a load she is about to buy than the waste manager who is about to sell it); and
- Public goods, where some goods can be accessed by multiple audiences simultaneously, and it is difficult or impossible to prevent anyone from accessing them, whether they have paid to do so or not (e.g., the illegal dumping of waste in public spaces is a negative example of this).

Public policy can be introduced to address these market failures – for example:

- Regulating to ensure that polluters are required to pay for the pollution they create to be abated (e.g., introducing extended producer responsibility legislation);
- Requiring the publication of information so that purchasers of products understand the environmental impacts of different choices (e.g., mandatory recyclability labelling of packaging); or
- Introducing taxes or permits to assign prices to unpriced public goods, ensuring that those who use them pay the environmental costs (e.g., landfill tax).

More generally, public policy can encourage more collaborative action across supply chains (e.g., through voluntary agreements), taking a more holistic view of the environmental costs and benefits of systemic action than may be taken by the individual supply chain actors working solely in their own individual interests.

The policy options

If the decision is taken that a policy intervention is needed to drive greater circularity, what is the range of policy measures that might be used? There is a wide variety of instruments available to policymakers, including government strategies, regulation, taxes, procurement requirements, standards, and voluntary agreements.

The Ellen MacArthur Foundation has divided them into six broad categories (2015: 47):

- Regulatory frameworks: legislation which requires certain actors to behave in certain ways. Examples include eco-design regulations, legally binding targets for resource productivity, or environmental permits for waste management facilities.
- Fiscal frameworks: changes in the government's taxation and subsidy regime to benefit circularity. Examples include VAT reductions for circular products or services, taxes on the extraction of virgin materials, and policies that shift taxation from labour to resources.

- Public procurement and infrastructure: the public sector has significant buying power and can use this to create or stimulate markets for environmental goods and services. Examples include public procurement rules requiring life-cycle costings (so that circular service contracts can compete with linear product sales) and co-investment in circular infrastructure projects that are not yet cost competitive with linear alternatives.
- Business support schemes: technical or financial support to circular businesses, enabling them to start up or expand. Examples include support in writing business plans, financial guarantees, and advice services.
- Collaboration platforms: the public sector can collaborate with the private sector to research circular economy opportunities, or the public sector can encourage collaboration across private sector supply chains. Examples include publicly funded research programmes and publicly supported voluntary industry agreements.
- Education, information, and awareness: awareness campaigns, provision of product information, and educational initiatives at schools and universities. Examples include CE modules in university design courses, public information leaflets, and publicly funded consumer behaviour change campaigns.

There are advantages and disadvantages to each category of policy instrument, and there is no 'one size fits all' solution to choosing the right policy option. Policymakers will need to go through a structured process (one such process is considered in detail in Ellen MacArthur Foundation, 2015) to arrive at their preferred solutions.

Literature review

We will briefly summarise here four documents of direct relevance to the question of how policy can help create a more circular economy:

- *The Waste Makers*, Vance Packard (1960);
- *Cradle to Cradle*, Braungart and McDonough (2008);
- *Delivering the Circular Economy: A Toolkit for Policymakers*, the Ellen MacArthur Foundation (2015); and
- *Circular Economy Policies in China and Europe*, McDowall et al. (2017).

The Waste Makers (Packard, 1960) is a classic early text on the industrial drivers in favour of a highly wasteful, linear economy and the methods used to generate unnecessary excess consumer demand. Packard discussed how continual increases in productivity, aided by increasing levels of mechanisation, led during the 1950s to a position where supply would massively outstrip demand. In order to sustain economic output, and the jobs and livelihoods that went with it, American industry came up with nine strategies of consumerism, the chief of which was planned obsolescence. Six decades on, it is worth reading Packard to understand the business motivations behind creating a throwaway society, as these may need to be reversed through policy measures if a circular economy is to be achieved.

Cradle to Cradle: Re-Making the Way We Make Things (Braungart and McDonough, 2008). Michael Braungart (a chemist) and William McDonough (an architect) coined the idea of the 'cradle to cradle' design philosophy which underpins the concept of the circular economy. They focus particularly on the idea that there are two different types of materials in our industrial economies, requiring distinct and parallel treatments at end of life. The first is a biological cycle of organic materials, which should be returned to the natural environment at the end of their lives to be composted or otherwise reabsorbed, in exactly the same way as occurs in nature. The second is a technical cycle of inorganic materials. These will not break down harmlessly if disposed of in nature and need to be collected and reused or recycled. Braungart and McDonough reject the concept of eco-efficiency, which they characterise as trying to do the wrong thing slightly less badly, and favour instead the idea of eco-effectiveness, where we design our products, services, and systems to echo nature's focus on building biodiversity through abundance.

Delivering the Circular Economy: A Toolkit for Policymakers (Ellen MacArthur Foundation, 2015). This is a practical guide to the thought processes that policymakers may wish to follow as they develop policies to deliver a circular economy. The toolkit is particularly useful, as it provides a detailed three step methodology:

- Step 1: map the system, define a realistic level of ambition, and decide on the sectoral scope;
- Step 2: assess and prioritise circular economy opportunities for each focus sector and identify sector-specific policy options to address these;
- Step 3: aggregate the sectoral analyses and use these to develop economy-wide policy measures. Decide on the prioritisation and sequencing of sectoral and economy-wide policies. Implement.

The report concludes with a detailed pilot study of how the toolkit was applied in Denmark.

Circular Economy Policies in China and Europe (McDowall et al., 2017) provides a valuable analysis of the similarities and differences between the policy responses to the circular economy concept in China and Europe. The paper notes that although the concept has come to the fore in both territories on a similar timescale, the interpretation of the idea is subtly different, reflecting the state of development and economic focus of each place. While the core in both cases is the treatment of resources and waste, in China, this is supplemented by a focus on pollution abatement, while in Europe, the additional focus area is economic growth and employment opportunities. Unsurprisingly, given the very different economic systems operative in Europe and China, the policy approaches that have been adopted to drive circularity are quite distinct. While Europe's focus is on product design and helping consumers to make better decisions, China's primary interest is in reducing pollution and improving the resource efficiency of their manufacturing industry.

Case studies

We highlight here four examples of policy measures intended to create a more circular economy, each focused on different aspects of the problem, each developed in a different context, and each using a different kind of policy measure.

Regulatory framework – packaging extended producer responsibility (United Kingdom)

Under the EU Circular Economy Package (European Commission, 2015), EU member states are required by 2025 to introduce extended producer responsibility (EPR) legislation covering four waste streams: packaging, waste electrical and electronic equipment (WEEE), end-of-life vehicles (ELVs), and batteries.

The UK government indicated its intention to introduce EPR legislation for packaging by 2023 in the Resources and Waste Strategy for England (Defra, 2018: 31–35). Key elements of the approach include:

- Packaging producers will bear the full net cost of managing their products at end of life;
- When producers put packaging on the UK market, they will have to pay a modulated fee per tonne of packaging, with the fee being lower for more recyclable and higher for less recyclable materials and formats;
- All packaging will have to show a mandatory label indicating whether it is practically recyclable, to aid consumer choice; and
- The income from producers will be used to help all local authorities to collect a more consistent set of materials, reducing consumer confusion as to what can and can't be recycled wherever they live in the country.

The use of legislation will ensure a level playing field between all packaging producers supplying the UK market, avoiding the problem of free riders. The adoption of modulated fees should have a transformational effect on packaging design, driving the least recyclable materials and formats off the market completely.

Business support scheme – the REBus project (European Union)

REBus was a €3.1-million EU LIFE+ funded research project to pilot Resource Efficient Business models which ran from 2013 to 2017. The project was a collaboration between organisations in the United Kingdom and the Netherlands: WRAP, the Rijkswaterstaat (Dutch Ministry of Infrastructure & Environment), the Aldersgate Group, the Knowledge Transfer Network (KTN), and the University of Northampton. The objective was to pioneer and test a methodology for helping companies transform their business models to be more profitable, resilient, and resource efficient. REBus trialled 30 pilot projects which delivered a total of £4.89 million in financial benefit, allowed for 62,619 tonnes in material savings, and reduced greenhouse gas emissions by 1,953 tonnes.

REBus is an example of a business support scheme, using public sector resources and support to research and pilot circular economy opportunities, with the learning being made widely available at the conclusion of the project (REBus, 2017).

Collaboration platforms – the Plastics Pacts (United Kingdom and global)

The UK Plastics Pact was launched in April 2018 jointly by WRAP and the Ellen MacArthur Foundation (WRAP, 2018). The Pact is a collaboration platform whose objective is to eliminate plastic pollution by creating a circular economy for plastics in the United Kingdom.

The Pact is a voluntary agreement with (at November 2019) 76 business members, responsible between them for 85% of the plastic packaging sold through UK supermarkets. They have signed up to four targets, to be achieved by 2025:

- Take actions to eliminate problematic or unnecessary single-use packaging items through redesign, innovation, or alternative (reuse) delivery models;
- 100% of plastic packaging to be reusable, recyclable, or compostable;
- 70% of plastic packaging to be effectively recycled or composted; and
- 30% average recycled content across all plastic packaging.

The UK Plastics Pact is the first in a planned series of global plastics pacts led by the Ellen MacArthur Foundation under its New Plastics Economy programme. As of late 2019, Plastics Pacts had also been initiated in France and Chile, with more to follow in 2020 (Ellen MacArthur Foundation, n.d.).

The choice of a collaboration platform is intended to encourage collaboration between organisations across the entire plastic packaging supply chain, enabling rapid but workable business-led change.

Public procurement and infrastructure – Chinese circular economy demonstrator projects

In the Literature Review, we discussed some of the differences between the implementation of the CE concept in China and the EU. One example of how this translates into a policy measure is the Chinese approach of designating specific geographical areas (e.g., provinces, cities or industrial parks) as CE pilots or demonstrators (McDowall et al., 2017: 655). Successful applicants for such a designation receive funding from the National Development and Reform Commission in return for a commitment to be assessed against CE performance targets, and lessons learned from these pilots are used to inform future policymaking.

Challenges

Given the holistic and systemic nature of circular economy solutions, designing measures to drive circularity may be a challenge to policymakers used to the more focused interventions often applied within a linear economy. We will highlight here six of the challenges that CE policymakers will have to overcome.

Swimming against the tide

Most countries' economies are overwhelmingly linear in nature, and that is what we are all used to as consumers, businesses, or policymakers. It can be difficult to disconnect from the dominant paradigm, with the result that many policies labelled 'circular economy' are, in fact, just minor evolutions of existing measures. An obvious example here is recycling. Many so-called circular economy policy documents – see, for example, European Commission (2015) and Defra (2018) – contain as a key legislative element mandatory recycling rate targets (a familiar and well-understood waste policy measure), while placing much less emphasis on novel policy approaches that could drive more radical solutions, such as circular economy business models (CEBMs).

Legislating is hard

Many policymakers, particularly in the EU, see legislation as the primary weapon in their armoury. However, as moving to a fully CE would involve major changes in the way that both businesses and consumers operate, it can be difficult to legislate for. Regulation enforcing radical change is likely to encounter significant resistance from incumbent businesses that will face high compliance costs or may even find their current mode of business no longer viable. The new, circular businesses that would be aided by such legislation are unlikely to exist yet and so cannot lobby in support of the change. The result is that radical legislative change is vetoed.

Where's the evidence?

It is difficult to design effective policy measures without a sound evidence base. Unfortunately, a lack of robust data and evidence is characteristic of the resources and waste sector in the United Kingdom and many other countries (Defra, 2018: 134). This problem has plagued traditional policymaking for decades. If more radical policy measures are under consideration, the need for a sound evidence base is much higher, to convince the sceptics, yet the potential to produce one is likely to be much lower. The result is an inability to convince stakeholders of a policy's merits.

Making the perfect the enemy of the good

A constant challenge for policymakers is to find the right balance between, at one end, designing a policy measure that would, in theory, deliver precisely the objective intended but is, in practice, impossible to implement and, at the other, a measure that is easy to introduce but whose practical effects bear but a poor relation to the original policy intention.

This general problem is of particular relevance here because of the complex nature of the problem to be addressed. To take a concrete example, the first case study described earlier concerns EPR for packaging in the United Kingdom. In designing the modulated fees that producers will pay when putting their packaging on the UK market, which should act as a key driver for future ecodesign efforts by those producers, the policymaker is faced with exactly this problem. Should the fees be modulated by reference to something that is well understood and easy to validate, such as the extent to which the particular packaging material and format are recycled in practice? Or should they reflect a more holistic judgement of the 'circularity' of the packaging over its entire life cycle, referencing many other issues beyond recyclability, even if such a metric is contested, hard to measure, and subject to data gaps? The former decision will enable early implementation but may have unintended consequences (see below for details); the latter may provide a more circular policy measure but may delay implementation until a robust version of the intended metric is available.

The risk of unintended consequences

This is another widespread difficulty in policymaking. Attempts to solve one problem through policy can inadvertently lead to the creation of other problems.

To continue with the example of packaging EPR discussed immediately above, one potential unintended consequence of implementing packaging EPR using the simple and workable metric of recyclability to determine the size of modulated fees can be seen in the following: in recent years, several instant coffee brands have introduced plastic refill pouches which enable consumers to reduce their waste production by up to 97% per unit of coffee bought, in terms of the weight of the plastic refill pouch compared to the traditional glass jar it replaces. Although the glass jar is recyclable, the non-recyclable plastic pouch is significantly more resource efficient, given that both glass and plastic are carbon-intensive materials to produce, and most environmental impacts are broadly proportional to weight. However, if packaging EPR uses recyclability as its sole metric for setting modulated fees, the heavier glass jar will survive, while the more resource-efficient plastic pouch is likely to be discontinued.

What do we want?

One further challenge to CE policymaking is the response of consumers. There are many examples in the literature of well-meaning environmental policy measures – particularly those focused on providing information and raising awareness – which failed to have any substantive impact on the behaviour of mainstream consumers. Multiple examples are discussed in Phillips and Scott (2012), for example. This is particularly relevant to those trying to promote a circular economy. More than six decades of marketing efforts focused on persuading us all of the necessity of owning the newest, latest, shiniest gizmo (see Packard, 1960, for an early analysis) will not be easily reversed, particularly when the vast majority of today's marketing budgets remain dedicated to promoting the products of a largely linear economy.

Future opportunities

The use of policy measures to drive greater circularity is in its relative infancy, and much experience to date has represented a limited evolution beyond existing resources and waste policy. It is to be hoped that the CE policy landscape will become more revolutionary over time.

To give just one example, circular economy business models are one of the most promising approaches for driving real circularity, yet to date, it has proved deeply challenging to find an effective way of supporting them through policy. In future, perhaps CEBMs will be supported through an aggressive combination of green public procurement (encouraging the public sector to procure their goods and services preferentially from businesses using CEBMs) allied to fiscal measures (e.g., a reduction in VAT or corporation tax for qualifying businesses).

Conclusion

We conclude with three observations.

Policy measures can have a genuine impact on the level of circularity of the economy, and it is therefore worth considering in detail how policymaking can be tuned to the rather particular requirements of a circular economy.

However, the use of policy to drive circularity is far from trivial. Policy measures are easiest to apply when they are focused on a single problem, a single actor, or a confined geography. Creating a circular economy requires complex, holistic solutions integrated across supply chains. Developing policy measures to deliver such solutions is complicated.

To date, most 'circular economy' policy measures have actually represented a very limited evolution of more traditional waste management policies. If we are to drive real progress towards a truly circular economy, as the scale of the environmental challenges facing us would suggest we must, then more radical policy approaches will be needed.

References

Braungart M, McDonough W. (2008). *Cradle to cradle: re-making the way we make things*. London: Jonathan Cape.

Cambridge Dictionary (n.d.). *Policy*. Available at: https://dictionary.cambridge.org/dictionary/english/policy (Accessed 31/10/2019).

Darmstadter J. (2014). *Reflections on the oil shock of 40 years ago*. Resources. Available at: www.resourcesmag.org/archives/reflections-on-the-oil-shock-of-40-years-ago/ (Accessed 1/11/2019).

Defra (2011). *The economics of waste and waste policy*. Available at: https://assets.publishing.service.gov.uk/government/uploads/system/uploads/attachment_data/file/69500/pb13548-economic-principles-wr110613.pdf (Accessed 8/11/2019).

Defra (2018). *Our waste, our resources: a strategy for England*. London: HM Government.

Dickens C. (1865). *Our mutual friend*. London: Chapman & Hall.

Drake N. (2018). *We saw Earth rise over the Moon in 1968. It changed everything*. National Geographic. Available at: www.nationalgeographic.com/science/2018/12/earthrise-apollo-8-photo-at-50-how-it-changed-the-world/ (Accessed 1/11/2019).

Ellen MacArthur Foundation (n.d.). *Plastics Pact.* Available at: www.newplasticseconomy.org/projects/plastics-pact (Accessed 8/11/2019).
Ellen MacArthur Foundation (2015). *Delivering the circular economy: a toolkit for policymakers.* London: Ellen MacArthur Foundation.
European Commission (1989). *A community strategy for waste management. SEC (89) 934 final.* Brussels: European Commission.
European Commission (2008). *Directive 2008/98/EC on waste and repealing certain directives. OJEU L 312.* Brussels: European Commission.
European Commission (2015). *Closing the loop – an EU action plan for the circular economy.* COM(2015) 614 final. Brussels: European Commission.
Lansink, A. (2017). *Challenging changes – connecting waste hierarchy and circular economy.* The Hague, The Netherlands: LEA Nijmegen.
McDowall W et al. (2017). Circular economy policies in China and Europe. *Journal of Industrial Ecology.* 21 (3): 651–661.
OECD (2006). *Glossary of statistical terms: market failure.* Available at: https://stats.oecd.org/glossary/detail.asp?ID=3254 (Accessed 3/11/2019).
Packard V. (1960). *The waste makers.* London: Longmans, Green and Co Ltd.
Pezzini M. (2012). *An emerging middle class.* OECD Observer. Available at: http://oecdobserver.org/news/fullstory.php/aid/3681/An_emerging_middle_class.html (Accessed 8/11/2019).
Phillips R, Scott F. (2012). *Neither sermons nor silence: the case for national communications on energy use.* London: Green Alliance.
REBus (2017). *Summary of learnings.* Banbury: WRAP. Available at: www.rebus.eu.com/wp-content/uploads/2017/10/REBus-project-Most-common-learning-points-for-organisations.pdf (Accessed 8/11/2019).
UK Parliament (n.d.). *The 1848 public health act.* Available at: www.parliament.uk/about/living-heritage/transformingsociety/towncountry/towns/tyne-and-wear-case-study/about-the-group/public-administration/the-1848-public-health-act/ (Accessed 31/10/2019).
United Nations (2019). *World population prospects 2019.* New York: United Nations.
WRAP (n.d.). *WRAP and the circular economy.* Available at: www.wrap.org.uk/about-us/about/wrap-and-circular-economy (Accessed 31/10/2019).
WRAP (2015). *Resource revolution: creating the future.* Available at: www.wrap.org.uk/about-us/our-plan (Accessed 8/11/2019).
WRAP (2018). *UK businesses make world-leading pact to tackle plastic pollution.* Available at: www.wrap.org.uk/content/uk-businesses-make-world-leading-pact-tackle-plastic-pollution (Accessed 8/11/2019).
WRAP (2020). *How moving to a circular economy can help the UK to build back better.* Available at: www.wrap.org.uk/buildbackbetter (Accessed 23/9/2020).
WRAP and Green Alliance (2015). *Employment and the circular economy: job creation in a more resource efficient Britain.* Available at: www.wrap.org.uk/content/employment-and-circular-economy (Accessed 3/11/2019).

Further reading

Braungart M, McDonough W. (2008) *Cradle to cradle: re-making the way we make things.* London: Jonathan Cape. (A guide to implementing circular economy principles through design.)
Hill J. (2011). *The secret life of stuff: a manual for a new material world.* London: Vintage. (An investigation of where the materials underpinning modern life come from.)
Lansink A. (2017). *Challenging changes: connecting waste hierarchy and circular economy.* The Hague, The Netherlands: LEA Nijmegen. (An examination of how the concept of the waste hierarchy can be used to drive progress towards a circular economy, from the inventor of the hierarchy.)
Packard V. (1960). *The waste makers.* New York: David McKay. (An early exploration of the corporate need for over-consumption to absorb ever-growing industrial supply, and the many creative ways in which American business created excess demand.)
Stahel WR. (2019). *The circular economy – a user's guide.* London: Routledge. (A recent overview of circular economy principles from one of the originators of the concept.)
WRAP (2020). *How moving to a circular economy can help the UK to build back better.* Available at: www.wrap.org.uk/buildbackbetter (Accessed 23/9/2020). (A recent discussion of how circular economy principles could enable a green recovery following the COVID-19 pandemic.)

17
LEGAL CONSIDERATIONS FOR A CIRCULAR ECONOMY

Sean Thomas

Introduction

The circular economy (CE) has developed in the context of environmentalism (Hu et al., 2018: 990) and against the background of scientific (and, more recently, social scientific and normative) identification, analysis, and critique of the (polluting) impact of linear consumption (see generally Ellen MacArthur Foundation, 2013). Nevertheless, moves towards CE are likely to be highly contested. It arguably involves "nothing short of a wholesale transformation of the basis of contemporary capitalism and consumption" (Gregson et al., 2015: 224), and it certainly "must be understood as a fundamental systemic change instead of a bit of twisting of the status quo to ensure its impact" (Kirchher et al., 2017: 229). However, the extent to which this systemic change is actually occurring remains unclear. Explorations of how this change could take place, and indeed how such changes are actually happening, are often absent in CE debates (Hobson, 2016). On this broad point, the importance of the role of ownership and property rights within law contrasts heavily with the rather glib way by which it has been analysed in CE literature generally, as well as in the context of the (limited) literature on law and CE (Thomas, 2018, 2019a). Additionally, the best that can be said about the seemingly ambitious and wide-ranging CE package developed in the European Union is that it merely focuses on reintegrating waste: "The wrong signal is therefore sent to the private and public market: the focus on getting scarce materials back into the loop instead of a systemic change of production processes themselves" (Mélon, 2019). Generally speaking, CE entails a radically different understanding and approaches to those which have so far dominated legal discourse and the structuring of relevant legal frameworks.

CE cannot be constrained within arbitrary geo-political, and thus also jurisdictional, boundaries. On one level, CE will involve (just as is the case with a linear economy) a necessarily extensive level of imbrication between domestic, regional, international, and supranational jurisdictional structures. At another level, the drivers towards CE are the global impacts of pollution and resource-consumption, and the actual practices within a CE will invariably move from relatively small-scale examples to more complex commercial arrangements which will involve multiple jurisdictions. This may not be easily recognisable by the participants within CE transactions. It may also be problematic for regulators and other such governing institutions to acknowledge that the transactions concerned will involve multiple parties across different boundaries: the recent EU Directive on Waste (European Commission, 2018a) presents itself as being a foundation for CE, but it does so clearly within the confines of the EU itself. This inevitably generates further tension as to the value, or even validity, of legal responses to CE that rest in one particular jurisdiction or in one particular legal family.

This chapter thus can only offer a brief glimpse of the possible legal considerations of CE, and it certainly cannot examine in depth whether the identified problems may be jurisdictionally specific or whether they are common to different jurisdictions and/or legal families. In common-law jurisdictions such as England and Wales, law is developed as much by judicial responses to disputes (between private parties or between private parties and the state) as it is by legislation. Civilian legal systems such as France or Germany (and also socialist-inspired systems such as China) place much less emphasis on judicial decision-making as a source of law compared to legislative action (and, to different degrees, academic works). This is a very crude comparison (for useful introductions to the comparison of law, see, e.g., Zweigert and Kötz, [1977] 1988; Reimann and Zimmerman, 2006; Örücü and Nelken, 2007). The focus here is not to provide a comprehensive study but to give a brief outline of legal issues of CE from the general perspective of the law of England and Wales (including EU law: the Brexit issue shall necessarily be put to one side).

Selected legal issues

Meaning and enforcement of circular economy

The legal meaning and status of CE is unclear. This is to be expected, given the relative novelty of legislation in this area. The first legislative step is commonly regarded as the 2008 Circular Economy Promotion Law of the People's Republic of China, which was considered "an integrated state strategy rather than simply an environmental policy" (Hu et al., 2018: 991). Since then, there have been a variety of domestic and international legislative actions, such as the Directive (European Commission, 2018a) amending the 2008 Waste Directive to more explicitly engage with CE aims and practices. Accompanying such actions have been various governmental reports, plans, strategies, and similar such official outputs which have identified CE thinking as being at least valuable, if not of central importance, to dealing with the various substantive material problems of overconsumption, waste, and pollution (Department of Environment, Food and Rural Affairs, 2018; European Commission, 2019b). The need for economy prevents a comprehensive detailed exposition and critique of all such examples (see generally Thomas, 2019a; Hu et al., 2018), but examination of the various legislative responses clearly shows up the difficult issue of enforcement. Thus, with regard to the Chinese example, the lack of clarity as to the meaning of CE makes it "very difficult to establish a behaviour system that conforms to the legal norms" (Hu et al., 2018: 1007). There is clear evidence as to the multiplicity of meanings attached to CE, with the possible effect that CE will eventually become rather meaningless (Kirchher et al., 2017; Murray et al., 2017). Certainly, there is no specific legal definition in the EU's CE package, which seems to scupper any attempt to realistically enforce a CE. Instead of attempting to legislate for a CE at a general level, though, it may be more valuable to concentrate on how circularity can be built into commercial practices by considering how different areas of law can be used, manipulated even, to generate, protect, and enhance CE practices.

Circular economy literature: whither law?

Whilst there is growth in the legal literature on CE, alongside tentative legislative steps, it is clear that the non-legal CE literature has so far taken only the very briefest account of the legal position. It may just be that the majority of CE writers were interested in other aspects, such as technical and scientific issues of biomimicry and resource exploitation or the vast array of non-legal normative and positive implications of turns towards CE. It may also simply be a reflection of the lack of legal writers in the field of CE. The relative novelty, and until relatively recently the relative obscurity, of CE itself may well have had a restrictive effect on legal analysis: we lawyers simply did not know much if anything at all about it. Certainly this author has experienced such reactions over the past few years when presenting and discussing this topic. To recast this complexity: the lack of legal analysis of CE and of legal writers within the CE literature has resulted in much CE literature presenting as if the necessary changes to shift from linear to CE practices can occur without consideration of the legal impact or, more likely, without consideration of how law may restrict or alternatively enhance such moves.

Given the limited implementation of meaningful and/or enforceable legislation (or other such doctrine, that is, law) and the relatively limited engagement with law on the part of CE proponents and literature of the same, it is perhaps unsurprising that there has until very recently been very limited academic legal interest. To illustrate, the author undertook a search (on 5 November 2019) of Westlaw, one of the leading databases for legal research. The search was for the term "circular economy," and 124 items were returned. Space precludes a detailed overview, but there was a variety of different types of sources ranging from EU directives, to official publications, to short comments and notes. However, there were only six substantive articles (defined here as being a single focused piece [as opposed to an update] of greater than five pages in a journal) which had CE either as a primary focus or as a substantive element (Thomas, 2019a; Dawson, 2019 [an eight-page analysis of DEFRA, 2018]; Maitre-Ekern and Dalhammar, 2019; Mak and Lujinovic, 2019; Turunen, 2017; and de Römph, 2016). The other substantive articles only very briefly referred to CE (Cavoski, 2019: 112 [referring once to the EU's CE package]; Moutsipai and Geiger, 2019: 233 [referring once to the EU CE Action Plan]; Bostoen and Devroe, 2018: 412 [referring once to the shift away from ownership as part of the CE]; Adekilekun et al., 2018: 46 fn 75 [referring once to the EU's CE Action Plan]; Ijaiya et al., 2018: 282 [referring to the Ellen MacArthur Foundation]; Adhikari et al., 2016 [referring once to the PRC's approach]; Hillyear, 2015: 146; Tromans, 2010: 183 [referring once to the Chinese 2008 CE law]). It is also worth very briefly noting that a similar search on HeinOnline, another extensive repository for legal work, generates 337 results. Economy prevents a detailed systematic review of this, but that search, whilst repeating much of the Westlaw results, does indicate some other substantive articles (such as Porcelli and Martinez, 2018; Bradshaw, 2018; Thomas, 2018; Nair and Kraus, 2019).

Thus, with some limited exceptions (Bonciu, 2014; Dalhammar, 2015; de Römph, 2016; Steenmans et al., 2016; Cassotta, 2016; Turunen, 2017), the first substantive legal analysis was probably Backes (2017), with a focus on the EU (and to a lesser extent, the Netherlands). The first substantive analyses of English law appear only at the end of 2018 and the start of 2019 (Thomas, 2018, 2019a). Whilst there is growth both in the number of pieces published which substantively engage with the issues of CE (as well as academic conferences and calls for papers) and in the range of legal areas being brought under legal academic analysis (with examples such as real property [Ploeger et al., 2018] to consumer law [Mak and Lujinovic, 2019; Maitre-Ekern and Dalhammar, 2019], as well as specific responses to problematic issues such as plastics [de Römph and Van Calster, 2018; Thomas, 2019b; European Commission, 2019a]), there are numerous areas of law which remain untouched. Thus, there remains no substantive work to this author's knowledge (again, with the further caveat that this author's field of knowledge is both restricted in its subject matter and in its jurisdictional scope) concerning company law and CE, though there is some closely connected work on sustainability (such as Sjafjell, 2018). Thus, as Mélon (2019) correctly writes:

> the private sector further encounters impediments to circularity in current legislation on . . . competition law and the general corporate law favouring and prioritising linear business practices, lacking clear guidance on circularity. These are all examples of policy incoherence, some representing a direct example of incoherence . . ., others an indirect example.

Another area of considerable importance with only limited examination is the field of intellectual property, where the only substantive work appears to be Thomas's critique of the approach taken to so-called intelligent assets by the Ellen MacArthur Foundation (2016) (Thomas, 2018).

The obvious problem with the limited level of academic engagement with the legal issues concerning CE is that sometimes writers may inadvertently engage with legal concepts, which in turn demonstrates the potential difficulties facing CE practices. Thomas (2018, 2019a, 2020) has pointed out that CE literature has tended strongly towards making considerable questionable assumptions about the nature of property and ownership. The ease with which such foundational conceptual structures (both of the specific English legal framework per se and of the more general market-capitalist social structures at the heart of Western and modern life) are elided over or brutally overturned in CE literature raises questions about the normative or theoretical basis of law in a CE, as well as more pragmatic questions about the specific form of the doctrines attending to CE transactions. It is not uncommon to see brief references to areas of law concerning ownership, property interests and rights, commercial transactions, forms of agreement, and so on, without any explanation or analysis as to how complex legal matters can be effectively, let alone efficiently, brought within the CE aura. Even when there are works (such as Hieminga, 2015; European Investment Bank, 2015) on the need to address the often unique financial structures and transactional forms attending to circular economic practices, they can project insufficiently precise and thus potentially misleading representations of the role of property, title, and ownership within financing (certainly under English law) (Thomas, 2019a, 2020).

Procurement and ownership

Where there has been some work is in the context of public procurement. The initial steps came from the EU Parliament, which in June 2017 published a report connecting "green public procurement" and CE. This was swiftly followed by a brief document from the European Commission's Directorate General for the Environment on best practices in circular economy and public procurement (EC, 2017). At the time, Sanchez-Graells, a leading expert on public procurement, perceptively referred to this as "fluffy guidance" at a "high level of generality" which does not provide much in the way of legal guidance (2017). He also noted the connection between public procurement, circular economy, and general environmentalism, and he rightly questioned whether this was properly coordinated within the EU. For the purposes of this chapter, it is worth noting the underlying "greening" aspect here:

> Circular public procurement is an approach to greening procurement which recognises the role that public authorities can play in supporting the transition towards a circular economy. Circular procurement can be defined as the process by which public authorities purchase works, goods or services that seek to contribute to closed energy

> and material loops within supply chains, whilst minimising, and in the best case avoiding, negative environmental impacts and waste creation across their whole life-cycle.
>
> (EC, 2017)

It is suggested that this focus on the "green" aspect is part of both an attempt to greenwash the more profound implications of moving towards a CE, as well as an attempt to provide an otherwise reasonable justification for the policies themselves. Thus, whilst the EU Directive on Public Procurement (Directive 2014/24/EU) "may help incorporate environmental criteria," the conditional "may" is telling:

> there is no evidence on how well these [sustainable and green public procurement] criteria actually capture the circular elements. Using several predefined environmental criteria in the procurement process does not necessarily make the outcome of the procurement more sustainable, green, or circular.
>
> (Alhola et al., 2019: 97)

Overall, public procurement and CE in general remain a rather flawed and inchoate area of law: limited illustrative examples provide the extent of the evidential basis, and loose guidance and an absence of detailed, specific legislative provision essentially prevent the generation and implementation of CE public procurement. A substantial problem with loose legal analysis of CE situations is a consistent failure to acknowledge the role of property and ownership within society and in particular a rather glib approach to critically examining the role of property and ownership within the legal architecture of (1) currently existing systemic institutions and structures and (2) whether such roles would be the same or, more importantly, how and to what extent they would differ in the context of moves towards CE. This has been analysed in depth by Thomas (2018, 2019a, 2019b, 2020), but it is worth adding to the critique by considering how this problem appears in the context of public procurement.

Thus, for example, it has been suggested that moves towards CE would include "leasing, hiring, and sharing of products, as opposed to direct ownership" (Alhola et al., 2019: 97). This controversial statement, baldly put, is highly problematic. At one level, there is some evidence of potential commercial problems with such simplistic approaches towards ownership structures. This can be seen in the context of the WEEE Forum of the International Association of Electronic Waste Producer Responsibility Organisations (a not-for-profit organisation encompassing various national bodies concerning waste electronics), which at the 2019 Rome Conference of the EU Network for the Implementation and Enforcement of Environmental Law stated that

> it was noted that the leasing of products had not come out as strongly in the survey responses as designing for ease of recycling and that the project faced some challenges in convincing the sector of the merits of the leasing model, one of the facets of the circular economy it will be looking at.
>
> (WEEE, 2019)

This replicates earlier findings of the European Investment Bank (2015: 27–31).

Furthermore, the notion of "direct ownership" is rather jurisdictionally loaded. Depending on whether you are coming from a common-law or continental tradition, such a phrase may make more or less sense (Thomas, 2019a). Secondly, there are also jurisdictionally relevant issues with, for example, the conflation of leasing, hiring, and sharing: these three concepts entail different proprietary structures, and considerable debates have arisen over the proprietary status of a lease (Swadling, 1998; Watt, 2003; MacFarlane, 2003; McMeel, 2003). Thus, whether a hire agreement has an option or an obligation to purchase at the end of the term can have substantive commercial implications (*Helby v Matthews* [1895]), whilst at the same time, proprietary consequences can clearly follow from merely possessory dealings with personal rights (*The Manchester Ship Canal Company Ltd v Vauxhall Motors Ltd* [2019]). Certainly there is a fundamental logical question here: if there is a move towards leasing or hiring or sharing, then that still presupposes a party with a more powerful ownership interest above that of the lessee, hirer, or those coming to share the asset. CE will thus entail *enhancement* of direct ownership in the hands of a limited number of initiators (Thomas, 2019a: 63) of CE transactions, a position that raises substantial normative problems (Thomas, 2018, 2020).

If we look to some of the best-practice guidance published, it becomes clear that there is a lack of examination of underlying legal structures. Thus, consider the Aalborg procurement of new school furniture (European Commission, 2018b). Whilst it is interesting to note that participants in the exercise demonstrated pre-existing circular practices, there was also some note of reservation about what CE meant in contrast to more general environmental aspects. For our purposes, it is also worth noting that there was no clear outline of legal issues: the most the example gives are references to a beguiling all-encompassing "framework agreement." This is problematic given how a similar approach (that is, the use of a "framework agreement") can be discerned in practical examples in England (York, North Yorkshire, East Riding Enterprise Partnership, 2019). Given that there is a strong possibility in English law that some of the products covered might become part of the land (and thus not remain as goods), with complicated shifts in ownership and new proprietary relationships forming as a result (Thomas, 2015), this lack of clarity makes it difficult to assess what legal risks, and what financial risks tied to the legal allocation of property rights and interests, may have arisen in this context. This issue of the land/goods divide becomes even starker when we consider the potentially viable project of concrete recycling (EC, 2019). Some more detail on this issue is found, however, in the Malmö furniture procurement project (EC, 2019). Whilst there was reference to the bland "framework agreement," there were some indications that this would entail some sort of buy-back obligation on the seller, as well as recognition of the potential issues concerning the power of the municipality to sell unneeded goods. Recognising these issues is a good first step, but there clearly needs to be acknowledgement that the obligations and duties regarding ownership (or whatever proprietary or non-proprietary interests in goods are identified) need broader examination.

Overcoming contradictions?

Consider this definition of circular public procurement: "procurement of competitively priced products, services, or systems that lead to extended life spans, value retention, and/or remarkably improved and non-risky cycling of biological or technical materials, making use of and supporting the circular business models and related networks" (Alhola et al., 2019: 105). What exactly is a standard of "remarkably improved and non-risky"? This appears to raise considerable questions concerning the application of precautionary principles regarding technologies: can something ever be "non-risky," or is that term merely a reference to a limited level of risk (to which we can further ask: What exactly is the tipping point for the relevant level of risk?). This may be a key issue in the implementation of CE practices, as recently recognised by the European Court of Justice (Second Chamber) in the context of a dispute over the capacity to alter the end-of-waste status of sewage sludge.

In *AS Tallinna Vesi v Keskkonnaamet* [2019], there was a need to deal with the absence of defined criteria of "end-of-waste." This is of key importance for operating within the Waste

Directive regulatory framework (Directive 2008/98/EU). The Estonian Environmental Board refused to recognise end-of-waste status for sludge that had undergone treatment with the aim of marketing the resulting product as "greening soil." The Court held that Article 6(4) of the Directive allowed member states to take the approach that some waste could not cease to be waste and that that would be allowed even if it would effectively negate the achievement of the objectives of the Directive, such as the recovery of waste. Furthermore, the Court held that the recovery of sewage sludge entailed certain environmental and health risks, and this meant that the discretion awarded to member states could be operational here. The Court further held ([2019] Env L R 30 [27]) that whilst the Advocate General's observation that member states' abstention from adopting legislation concerning end-of-waste status was acceptable, they

> must ensure that such abstention does not amount to an obstacle to the attainment of the objectives set by Directive 2008/98, such as . . . encouraging the recovery of waste and the use of recovered material in order to preserve natural resources and to enable the development of a circular economy.

This was the sole reference to circular economy. It suggests that the EU's juridical approach to the policy role of circular economy is that whilst it may be a factor, it ultimately is subsidiary to the more explicit legislative provisions of the Waste Directive. The waste-holder could not demand that the relevant regulatory body determine the end-of-waste status of products such as sludge, that is, products which have risky aspects to their material nature. This illustrates a paradox: one arm of law will prevent what another arm may require. This may be another instance of the lack of coordination; it may be something more substantive in that there will need to be a radical change to the nature of environmental law, at least in the context of the Waste Directive and its use of the waste hierarchy, in order to properly account for the necessarily broader concept of CE.

Suggested future directions

To a degree, it could be said that this radical change has begun. In 2018, the EU amended the 2008 Waste Directive (European Commission, 2008). The preamble's reference to the need to "take additional measures . . . by focusing on the whole life cycle . . . in a way that preserves resources and closes the loop" (European Commission, 2018a) [1]) suggests that new legislative measures will be required. Certainly Preamble (17) lists a variety of measures member states could use to ensure that material recovered from waste ceases to be waste and appears to clearly deal with the core issue at the heart of the dispute in *AS Tallinna Vesi v Keskkonnaamet*. Furthermore, Article 1 of the Waste Directive now states that the Directive

> lays down measures to protect the environment and human health by preventing or reducing the generation of waste, the adverse impacts of the generation and management of waste and by reducing overall impacts of resource use and improving the efficiency of such use, which are crucial for the transition to a circular economy and for guaranteeing the Union's long-term competitiveness.

The reference to *transition* to CE is important. However, it is still hemmed in by both the notion of dealing with waste and the need to guarantee "competitiveness." Whether this reference to competitiveness will generate a conflict between CE and other economic goals is unclear.

Moreover, there is no clarity as to the role (if any) that changes to or new forms of ownership and property structures will play in this. Preamble (29) of the 2018 Directive implies that there needs to be facilitation of business models that promote reuse, refurbishment, repurposing, and sharing of products. Such facilitation may well take the form of manipulating property interests, but the Directive avoids detailing this explicitly. Similarly, preamble (61) refers to the dissemination of best practices as to the meaning of "waste" and "discard," accounting for CE where "a substance or object is transferred from one holder to another holder without the intention to discard." This is clearly an issue relating to ownership and property rights, though how precisely this should be dealt with is unclear. Ultimately, the potential for use of property and ownership structures to incentivise CE practices is not explicitly mentioned in the Annex of examples of economic instruments and other measures added by the 2018 directive to the 2008 Waste Directive (European Commission, 2018a) Annex, inserting Annex IVa), and it is rather difficult (though not impossible) to imply into those measures that are specifically mentioned. At best, this Directive is merely a "prelude" to a legal definition of CE (Lingl and Marcó, 2019) and the beginning to a coherent and non-contradictory legal framework for CE transactions.

The legal aspects of CE need to be subject to much more detailed and critical analysis. At the outset, the problems of comparative legal analysis were briefly noted: this chapter must thus stand as an invitation to others to undertake such work. In addition to such issues, efforts are required to develop coherent theoretical frameworks, connecting legal doctrine with the policies and practices of CE. This should not merely be undertaken as a means of catalysing doctrinal change to help develop, support, and enhance CE practices. Rather, given the potentially problematic consequences of CE practices (see, e.g., Thomas, 2018), explication of the policy grounds allows for critical examination. This deeper understanding of CE may help with the problems of enforcement. As noted previously, there is a lack of consensus over what CE actually is. However, rather than fretting over the meaning of CE, it may be more useful to consider what options for enforcement can be drawn from commercial practice.

There could be an attempt to present obligations to undertake practices that at least accord with some sort of common framework of CE. This could take the form of some sort of industry-wide code similar to examples such as that concerning mortgages (Pre-Action Protocol, 2008). Alternatively, there could be some form of enforceable soft law encoded into transactions through essentially invariable contracting practice, such as that found in the law concerning documentary credit (International Chamber of Commerce, 2006). What is worth noting is neither is specifically enforceable (for criticism of the Pre-Action Protocol, see Dixon, 2008). Without effective regulation, a soft-law code essentially relies on mere market-influenced enforcement, with participants dealing with marginal transactions less likely to follow. Sanchez-Graell's perceptive identification that CE in Europe involves "weak policy intervention in the form of best practice dissemination (i.e., even weaker than soft law guidelines)" (2019) thus raises serious questions about the need for substantive regulatory engagement with CE.

Conclusion

This brief chapter has provided an outline of legal issues likely to affect moves to a CE. It has identified a tendency towards exhortatory legislative responses, along with a tendency towards engaging with specific legal areas in the literature (such as the focus on property law or procurement law). Analysing law's response as a whole, rather than the ad hoc nature of the current approach, would clearly be a massive endeavour. What may therefore be more palatable is simply a call for greater volumes of work in general in the context of legal analysis of CE practices and policies. There needs to be a shift in perception among academic lawyers as to the nature of what CE involves: it needs to be understood through something more multifaceted than the current, rather myopic, (re)presentation of the issues through the ultimately flat lens of sustainability and waste (Mélon, 2019). Issues of sustainability, and issues concerning waste, are of course hugely important to CE (and they of course raise complex substantive issues of law by themselves). However, they are both merely part of a broader idea: an idea which necessarily will involve multiple other areas of law by virtue of the transactionalism than inheres within CE practices. In particular, there will need to be broader and deeper examinations of the doctrinal issues that will arise when CE transactions collapse, including collapses away from the start or end-points of supply chains, and thus generate problems other than those concerning sustainability or waste. The minimal attention paid to property and ownership structures will be hugely problematic given the importance of such concepts to commercial transactions.

References

Adekilekun MT, Gan CC, Fuguo C. (2018). International legislative frameworks for public-private partnerships: an evaluation. *Public Procurement Law Review*. 1: 33–50.

Adhikari DR, Gautam DK, Chaudhari MK. (2016). Corporate social responsibility domains and related activities in Nepalese companies. *International Journal of Law and Management*. 58 (6): 673–684.

Alhola K, Ryding SO, Salmenperä H, Busch NJ. (2019). Exploiting the potential of public procurement: opportunities for circular economy. *Journal of Industrial Ecology*. 23 (1): 96–109.

AS Tallinna Vesi v Keskkonnaamet (2019). *Environmental Law Review*. 30.

Backes C. (2017). *Law for a circular economy*. The Hauge: Eleven International Publishing. Available at: www.uu.nl/sites/default/files/rgl-ucowsl-backes-law_for_a_circular_economy.pdf.

Bonciu F. (2014). The European economy: from a linear to a circular economy. *Romanian Journal of European Affairs*. 14 (4): 78–91.

Bostoen F, Devroe W. (2018). From sales to subscriptions in the car sector: competition law implications of servitisation and the refusal to sell to customers. *European Competition Law Review*. 39 (9): 411–419.

Bradshaw C. (2018). Waste law and the value of food. *Journal of Environmental Law*. 30 (2): 311–332.

Cassotta S. (2016). The Paris Agreement in logic of multi-regulatory governance: a step forward to a new concept of 'global progressive adaptive-mitigation. *European Energy and Environmental Law Review*. 25 (6): 196–212.

Cavoski A. (2019). EU environmental compliance assurance. *Environmental Law Review*. 21 (2): 111–118.

Dalhammar C. (2015). The application of 'life cycle thinking' in European environmental law: theory and practice. *Journal of European Environmental & Planning Law*. 12: 97–127.

Dawson L. (2019). Our waste, our resources: a strategy for England – switching to a circular economy through the use of extended producer responsibility. *Environmental Law Review*. 21 (3): 210–218.

de Römph TJ. (2016). Terminological challenges to the incorporation of landfill mining in EU waste law in view of the circular economy. *European Energy and Environmental Law Review*. 25 (4): 106–119.

de Römph TJ, Van Calster G. (2018). REACH in a circular economy: the obstacles for plastics recyclers and regulators. *Review of European, Comparative & International Environmental Law*. 27 (3): 267–277.

Department of Environment, Food and Rural Affairs (2018). *Our waste, our resources: a strategy for England*. Available at: www.gov.uk/government/publications/resources-and-waste-strategy-for-england.

Dixon M. (2008). Editor's notebook. *Conveyancer and Property Lawyer*. 474.

Ellen MacArthur Foundation (2013). *Towards the circular economy: economic and business rationale for an accelerated transition*. Available at: www.ellenmacarthurfoundation.org/publications.

Ellen MacArthur Foundation (2016). *Intelligent assets: unlocking the circular economy potential*. Available at: www.ellenmacarthurfoundation.org/publications.

European Commission (2008). *Directive 2008/98/EC of the European Parliament and of the council of 19 November 2008 on waste and repealing certain directives*. Brussels, Belgium.

European Commission (2017). *Public procurement for a circular economy: good practice and guidance*. Available at: https://ec.europa.eu/environment/gpp/pdf/Public_procurement_circular_economy_brochure.pdf.

European Commission (2018a). *Directive (EU) 2018/851 of the European parliament and of the council of 30 May 2018 amending Directive 2008/98/EC on waste*.

European Commission (2018b). Circular procurement for a sustainable learning environment. *Green Public Procurement Practice*. 79. Available at: https://ec.europa.eu/environment/gpp/pdf/news_alert/Issue79_Case_Study_155_Aalborg.pdf.

European Commission (2019a). *A circular economy for plastics: insights from research and innovation to inform policy and funding decisions*. De Smet M, Linder M. (Eds.). Available at: https://publications.europa.eu/s/mTES.

European Commission (2019b). A low carbon, circular economy approach to concrete procurement. *Green Public Procurement in Practice*. 88. Available at: https://ec.europa.eu/environment/gpp/pdf/news_alert/Issue_88_Case_Study_168_Zurich.pdf; https://ec.europa.eu/environment/gpp/pdf/news_alert/Issue_88_Case_Study_169_Malmo.pdf.

European Investment Bank (2015). *Access-to-finance conditions for projects supporting circular economy*. Available at: www.eib.org/attachments/pj/access_to_finance_study_on_circular_economy_en.pdf.

Gregson N, Crang M, Fuller S, Holmes H. (2015). Interrogating the circular economy: the moral economy of resource recovery in the EU. *Economy and Society*. 44: 218–243.

Helby v Matthews [1895] AC 471.

Hieminga G. (2015). *Rethinking finance in a circular economy: financial implications of circular business models*. Available at: www.ing.com/About-us/Ourstories/Features/Circular-economy-challenges-financial-business-models.htm.

Hillyear S. (2015). Supply chain opportunities in the UK offshore oil and gas decommissioning. *Construction Law Journal*. 31 (3): 139–147.

Hobson K. (2016). Closing the loop or squaring the circle? Locating generative spaces for the circular economy. *Progress in Human Geography*. 40 (1): 88–104.

Hu Y, He X, Poustie M. (2018). Can legilsation promote a circular economy? A material flow-based evaluation of the circular degree of the Chinese economy. *Sustainability*. 10 (4): 990–1012.

Ijaiya H, Abbas WI, Wuraola OT. (2018). Re-examining hazardous waste in Nigeria: practical possibilities within the United Nations system. *African Journal of International and Comparative Law*. 26 (2): 264–282.

International Chamber of Commerce (2006). *Uniform customs and practice for documentary credits: UCP600*. London: International Chamber of Commerce.

Kirchher J, Reike D, Hekkert M. (2017). Conceptualizing the circular economy: an analysis of 114 definitions. *Resources, Conservation and Recycling*. 127: 221–232.

Lingl CM, Marcó RP. (2019). First EU attempt of a legal definition of 'circular economy'. *Lexology.com*. Available at: www.lexology.com/library/detail.aspx?g=6aa8a7d7-abb1-4511-b1e1-c07f05c767be.

MacFarlane B. (2003). Identifying proprietary right: a reply to Mr Watt. *Conveyancer and Property Lawyer*. 67: 473.

McMeel G. (2003). On the redundancy of the concept of bailment. *Lloyds Maritime and Commercial Law Quarterly*. 169.

Maitre-Ekern E, Dalhammar C. (2019). Towards a hierarchy of consumption behaviour in the circular economy. *Maastricht Journal of European and Comparative Law*. 26 (3): 394–420.

Mak V, Lujinovic E. (2019). Towards a circular economy in EU consumer markets – legal possibilities and legal challenges and the Dutch example. *Journal of European Consumer and Market Law*. 8 (1): 4–13.

Mélon L. (2019). *Is circular economy a move towards or away from sustainability? A short piece on the (ab)use of the concept of circularity*. Available at: www.howtocrackanut.com/blog/lela-melon-circular-economy-sustainability.

Moutsipai E, Geiger A. (2019). Lobbying in EU competition law. *European Competition Law Review*. 40 (5): 229–237.

Murray A, Skene K, Haynes K. (2017). The circular economy: an interdisciplinary exploration of the concept and application in a global context. *Journal of Business Ethics*. 140: 369–380.

Nair PS, Kraus JH. (2019). Food industry trends towards sustainability and the circular economy and the mixed effect of laws. *SciTech Lawyer*. 15 (4): 10–25.

Örücü E, Nelken E. (Eds.) (2007). *Comparative law: a handbook*. Oxford: Hart Publishing.

Ploeger H, Prins M, Straub A, Van den Brink R. (2018). Circular economy and real estate: the legal (im)possibilities of operational lease. *Facilities*. 37 (9/10): 663–668.

Porcelli AM, Martinez AN. (2018). Legislative analysis of the circular paradigm economy. *DIREITO GV Law Review*. 14 (3): 1067–1105.

Pre-action protocol for possession claims based on mortgage or home-purchase plan arrears in respect of residential property (November 2008). Available at: www.justice.gov.uk/guidance/courts-and-tribunals/courts/procedure-rules/civil/contents/protocols/prot_mha.htm.

Reimann M, Zimmermann R. (Eds.) (2006). *The Oxford handbook of comparative law*. Oxford: Oxford University Press.

Sanchez-Graells A. (2019). *A quick, non-comprehensive update on circular economy and public procurement*. Available at: www.howtocrackanut.com/blog/update-circular-economy-procurement.

Sanchez-Graells A. (2017). *Public procurement for a circular economy: some thoughts on policy coordination and fluffy guidance*. Available at: www.howtocrackanut.com/blog/2017/10/26/public-procurement-for-a-circular-economy-some-thoughts.

Sjafjell B. (2018). Redefining the corporation for a sustainable new economy. *Journal of Law and Society*. 45 (1): 29–45.

Steenmans K, Marriott J, Malcolm R. (2017). *Commodification of waste: legal and theoretical approaches to industrial symbiosis as part of a circular economy*. Working Paper presented at the SMART conference Life-cycle based management and reporting for sustainable business. Available at: https://ssrn.com/abstract=2983631 (Accessed 29–30/11/2016).

Swadling W. (1998). The proprietary effect of a hire of goods. In: Palmer N, McKendrick E. (Eds.), *Interests in goods*. London: LLP. 2nd Ed. 491.

The Manchester Ship Canal Company Ltd v Vauxhall Motors Ltd [2019] UKSC 46.

Thomas S. (2015). Mortgages, fixtures, fittings and security over personal property. *Northern Ireland Legal Quarterly*. 66 (4): 343–365.

Thomas S. (2018). Law, smart technology, and circular economy: all watched over by machines of loving grace? *Law, Innovation & Technology*. 10 (2): 230–265.

Thomas S. (2019a). Law and the circular economy. *Journal of Business Law*. 62–83.

Thomas S. (2019b). Personal property law for a zero-waste circular economy: using retention of title clauses to reduce plastics waste. *Law, Environment and Development Journal*. 15 (2): 176–207.

Thomas S. (2020). Circular economy, title, and harmonisation of commercial law. In: Akseli ON, Linarelli J. (Eds.), *The future of commercial law: ways forward for harmonisation*. Oxford: Hart Publishing.

Tromans S. (2010). China's environmental law – a changing landscape. *International Energy Law Review*. 6: 182–195.

Turunen T. (2017). Deconstructing the bottlenecks caused by waste legislation: end-of-waste regulation. *Journal for European Environmental & Planning Law*. 14 (2): 186–207.

Watt G. (2003). The proprietary effect of a chattel lease. *Conveyancer and Property Lawyer*. 67: 61.

WEEE (2019). *WEEE forums*. Available at: https://weee-forum.org/ws_news/connecting-policy-law-and-practice-with-the-circular-economy/.

York, North Yorkshire, East Riding Enterprise Partnership (2019). *Make your product law forever*. Available at: www.businessinspiredgrowth.com/wp-content/uploads/2019/10/Warp-It-MP.pdf.

Zweigert K, Kötz H, Weir, T. (Trans.) ([1977] 1998). *An introduction to comparative law*. Oxford: Oxford University Press. 3rd Ed.

Further reading

Backes C. (2017). *Law for a circular economy*. The Hauge: Eleven International Publishing. Available at: www.uu.nl/sites/default/files/rgl-ucowsl-backes-law_for_a_circular_economy.pdf. (A detailed examination of the legal implications for moves to a CE. Focused primarily on EU and Dutch law, with a concentration on the impacts on environmental law.)

Gregson N, Crang M, Fuller S, Holmes H. (2015). Interrogating the circular economy: the moral economy of resource recovery in the EU. *Economy and Society*. 44: 218–243. (Valuable critical evaluation of the normative basis of arguments in favour of the CE. Questions the value of the EU-focus to the CE literature.)

Mak V, Lujinovic E. (2019). Towards a circular economy in EU consumer markets – legal possibilities and legal challenges and the Dutch example. *Journal of European Consumer and Market Law*. 8 (1): 4–13. (Brief overview of legal problems concerning EU consumer law and CE.)

Maitre-Ekern E, Dalhammar, C. (2019). Towards a hierarchy of consumption behaviour in the circular economy. *Maastricht Journal of European and Comparative Law*. 26 (3): 394–420. (Deep theoretical overview of implications for consumers of moves to CE.)

de Römph TJ, Van Calster G. (2018). REACH in a circular economy: The obstacles for plastics recyclers and regulators. *Review of European, Comparative & International Environmental Law*. 27 (3): 267–277. (Brief, enlightening overview of the implications of transitions to CE, in light of current legal-technical standards governing a specific type of goods [plastics].)

Thomas S. (2019). Law and the circular economy. *Journal of Business Law*. 62–83. (Outlines the possible implications for English law of CE practices. Considers the relationship between CE and the sharing economy.)

Thomas S. (2018). Law, smart technology, and circular economy: all watched over by machines of loving grace? *Law, Innovation & Technology*. 10 (2): 230–265. (Examines the value of intellectual property law as a means of moving to CE. Detailed critique of notion of "intelligent assets." Analyses possible dangers of moves towards tracking and controlling goods. Questions the normative foundations for CE.)

18

ECONOMIC AND TRADE CONSIDERATIONS OF CIRCULAR ECONOMY APPROACHES

Paul Sheeran

Introduction

The circular economy and the international trade of goods present a number of problems and challenges. The economics of closed trade loops operating at an optimum circularity would theoretically eliminate the need for international trade. Local production and markets, 3D printing, and a radical consumption boycott of high-carbon products, for example, may reduce material use, waste, and scarce resources, but it is unrealistic to believe radical transformations in circularity internationally will be made in the short-term. The global economic consequences of COVID-19, secular stagnation, slumps in international trade, and shifts in consumer attitudes are, however, animating economic agents to review efficiencies, which may accelerate the adoption of circularity.

The ongoing commitment to gross domestic product (GDP) and short-term equity performance is related to a linear trade model that prioritises the continued expansion of production, consumption, and international exchange. There is little evidence that the absolute decoupling of resource use and carbon emission from GDP will be successful (Hickel and Kallis 2020: 470). Therefore, the incremental alignment of trade to circularity will be largely related to efficiency gains in trade. The balance between superabundant profits and natural capital needs to be reset. The following section examines the impact of trade on externalities. It is clear that a more informed understanding of the benefits of the circular economy should be related to science-based evidence. Digital and data analysis and metrics offer a range of tools improving the gathering and monitoring of trade efficiencies and losses. Information gleaned from the tools monitoring trade activities needs to be considered, along with a range of political, economic, and social influences that have diverging positions on the circular economy. The need for substantive evidence illustrating the value of the circular economy to complement sustainable growth is crucial in influencing decision-making among producers and consumers, international organisations, and governments. Institutional economics as a discipline is providing insight into the impediments to institutional change, which is useful in assessing economic and trade relations.

Efforts by urban organisations such as the London Waste and Recycling Board have been instrumental in highlighting practical initiatives that have wide circularity benefits. Similar initiatives are being introduced in cities the world over. The global 24/7 news media and the blogosphere are repeatedly highlighting the perilous nature of climate challenge, which is impacting governments and companies concerned, respectively, with an informed electorate and reputational risk. Groups such as Plan B Earth and Extinction Rebellion are raising the awareness of climate breakdown and the catastrophic impact of a linear economy model on natural capital. In the United Kingdom, the 'State of Nature' report produced by the National Biodiversity Network provides insightful material that highlights the deterioration of natural capital.

This short chapter is essentially concerned with the transportation of goods across borders: road haulage, rail, airfreight, and shipping, with container shipping being the most pervasive. In Europe, the marine industry is making efforts to adapt to circularity. This shift can be noted in industry bodies such as Seawork International (2019), with exhibitions and conferences illustrating the importance the sector is placing on green restructuring. Whilst promising, the international transition from trade transport based on the linear model requires a substantive reappraisal to ensure the full benefit from a circular economy is realised.

The following material considers three key related themes: (1) circularity gains in the transportation of international goods, (2) circularity clauses in international trade agreements, and (3) the wider transition from a carbon-intensive model of international trade to a circular trade model.

Circularity and international trade

Estimates suggest that the circular economy is currently 9% of the world economy (Circle Economy, 2019). Whilst useful, the full capture of the circular economy is currently ambiguous and inconclusive. The circular economy in academic literature can be traced back to Kenneth Boulding. Boulding viewed circularity from a holistic standpoint, where he presented the earth as a 'spaceship,' a closed circular system with the economy and the environment understood as an integrated set of complex technical and natural activities.

The circular economy is 'restorative and regenerative by design and aims to keep products, components and materials at their highest utility and value at all times distinguishing between technical and biological cycles' (Ellen MacArthur Foundation, 2015: 19). The report toolkit provided a default basis for industry and academia to scope projects and research programmes in the circular economy. Elsewhere, Brown (2018: 24) suggests "in contrast to our growth-orientated, waste-producing, business-as-usual means, a circular economy may not only safeguard the integrity of ecosystem services essential for humanity's survival but also help rebuild natural capital." Deviating slightly to include a commercial impression, Deloitte's (2018) advice to clients is that the circular economy is one that is restorative and regenerative by design – it is about optimisation of value circulation, not prevention of waste generation. It is clear that the circular economy includes a range of approaches that are set out subsequently. These categories are not necessarily exhaustive, but they do cover the core practices associated with the circular economy. These are listed with a brief alignment to international trade.

- Cradle to cradle: continuing the recapture of nutrients in product design to climate waste; energy and fuel
- Performance economy: economies in closed loops and increased taxes on non-sustainable activities; integrated logistics and transportation
- Biomimicry: innovation inspired by nature; natural methods in protecting harbours, ports, and coastal areas
- Industrial ecology; design function reducing and eliminating waste; fuel derivatives; marine, aviation, and freight transport pollution

- Natural capitalism: increasing productivity with inputs from natural capital
- Blue economy, sustainable ports and transportation, containerisation and the related sustainable performance of economic activity on and around the sea, rivers and lakes
- Regenerative design; innovation, creativity, and imagination in solving, transforming, or mitigating existing impediments not compliant with the circular economy
- Externalities: accounting for the value of nature and the full cost of its use in economic activities
- Gift economies: gift without the expectation of future reward (bi-lateral networks in small island states and near-neighbour trade exchanges)
- Sharing economy: collaborative and sharing economies: sharing assets, materials, and intellectual resources to realise a shared objective; trade diplomacy and exchange

The challenges facing international trade and circularity

Is cross-border trade circularity impossible? From a European perspective, the 'Stern Review on the Economics of Climate Change' released by the UK Treasury on the 30 October 2006 made a definitive statement on the negative impact on the environment and climate change from business-as-usual production and consumption practices. The Paris Agreement 2015 facilitated through the UN Framework Convention on Climate Change recognised the United Nations General Assembly Resolution A/RES/70/1 'Transforming Our World: the 2030 Agenda for Sustainable Development,' which included the 17 Sustainable Development Goals (SDGs) and promoted a common cause in limiting a global temperature change to below 2 degrees Celsius. The United States has announced its intention to leave the agreement. The Conference of the Parties (COP23) held in Bonn in 2017 continued the development of a 'rulebook' that would bind parties to a climate agreement. In 2018, the sixth assessment report from the Intergovernmental Panel on Climate Change (IPCC) presented a target figure of 1.5 degrees C to reduce the impact on climate change on the planet. The International Trade Centre (ITC) Sustainability Map and the World Economic Forum New York Environmental Stewardship Conference (September 2018) made clear the relationship between international trade, the environment, and the Sustainable Development Goals. The impact from global issues such as climate change, resource scarcity, population growth, expanding markets, and digital globalisation complicates the operation of trade theories such as gravity theory, which is based on the size of the economy and distance. Within the international business community, firms such as EY are advising clients on the main drivers of the circular economy and potential benefits to business models. Programmes and projects such as the OCED in its Re-Cycle series are promoting resource efficiency within a market economy model. Therefore, the potential for approaches within the circular economy to encourage sustainable international trade and services is increasingly being acknowledged. Nevertheless, progress is uneven due to a lack of commitment in terms of recognition and ratification throughout the international system. Despite limited adoption of the circular economy, key problems persist, for example, in the export of hazardous waste, which is 'regulated' by the Basel Agreement to limit the worst manifestations of it but continues to be an unacceptable example of the linear trade catastrophe.

The first international meeting of the World Social Forum held in Barcelona in 2019, pre-empting the World Social Forum of Transformative Economies 2020, illustrates the movement to embed a circular economy ethos in discussions based around the sharing economy and participatory economics including trade. Parente et al. (2018: 56) makes clear the global benefits of the sharing economy between firms operating internationally. Embedding the circular economy in developing countries is reliant on trade agreements at the regional and international level that overcome existing barriers to the cross-border exchange in inputs and on ideas in the circular economy (Preston et al., 2019). Concerns that circularity is, however, a form of protectionism continue to be an impediment to implementing circularity in the Global South. The Global South must be an equal partner in shaping policies and adopting practices consistent with a global approach to the circular economy.

The challenge to influence the processors related to international trade with policies promoting circularity can be shaped by regulation such as the Climate Change levy or incentives. The work by Nobel Prize winner Richard Thaler on 'nudge theory' (subtle signal encouraging a change in behaviour) potentially offers wider techniques in encouraging an engagement with the circular economy. In a complex open economy, trade flows are not without a multitude of impediments: trade between countries is conditional on legislation, permits, custom requirements, patents, import and export documentation, insurance, classification, registration, and reporting. Sustainability strategies, policies, and regulation that successfully limit the exploitation and destruction of natural capital are all conditional on establishing well-considered trade agreements that look at the full impacts of international trade.

Knowledge transfer in the circular economy in the United Kingdom is being driven by key organisations such as the Ellen MacArthur Foundation; the Department for Environment, Food and Rural Affairs (DEFRA); and the nexus of public and private organisations, associations and think-tanks, and stakeholders who recognise the gains from circularity techniques and methods.

Within the UK public sector, governance of circular economy initiatives are framed by DEFRA. DEFRA has been the central UK public body promoting sustainable policies and practice. The department has championed sustainable policies for a number of years and promotes three key goals in its strategic literature: (1) a just society, (2) a healthy environment, and (3) a productive economy. Scotland has been a particularly creative agent in introducing sustainable policies. Wales and Northern Ireland are also addressing challenges raised by a range of environmental challenges from climate change, waste management, and sustainability. The United Kingdom does not have a dedicated public body responsible for coordinating, implementing, and monitoring a circular economy. DEFRA does, however, promote better environmental decision-making and policies that address waste, inefficiency, and related sustainable practices. It also works with numerous agencies to promote good environmental practice through programmes and projects. Whilst DEFRA is an important organisation influencing environmental protection in the United Kingdom, it does not explicitly provide strategic leadership on the circular economy. It is likely, however, that DEFRA will continue to be the government's representative agency shaping and influencing sustainable policy. The degree to which it will initiate fully the circular economy, as in a full-scale environment protection policy, is uncertain.

Government policy on procurement might be used to encourage an alignment with the circular economy through incentives in public contracts. This may encourage organisations unable or unwilling to adopt a circular economy model to reconsider over-reliance on familiar practices and change behaviour. The EU Commission has been actively pursuing and promoting circular economy standards and frameworks.

The potential to include the full cost accounting to economic activity to facilitate transparency in real cost of production, consumption, and transportation is increasingly available through big data, analytics, and accounting. Input on the circular economy from Wales, Scotland, and Northern Ireland is recognised as a necessity in realising a cohesive UK public position. At an organisational level, the introduction of BS 8001 from the BSI Group is an important step in the standardisation of the circular economy and resource stewardship. Providing a usable, credible, and transparent circular economy standard is an important anchoring mechanism in shaping sustainable future free trade agreements and international trade relations.

Growth and circularity

The relationship between output and the full cost of growth on natural capital and waste requires a forensic reassessment. Gross domestic product provides a guide to the expansion or contraction of an economy, but it is not, nor is it intended to be, a reliable measure of the full costs of growth. Further development of accounting, evaluation, and reporting methods capturing the full cost of production, consumption, and exchange potentially offers an enlightened model that will improve decision-making and policy formation in commercial organisations and statutory bodies. Whilst a consensus has not been reached on the system for calculating green GDP across the globe, the linear model measure of progress is being questioned, and alternative methods are being proposed (Bleys, 2012; Fioramenti, 2013).

Whilst other measures such as Human Development Index are useful in measuring indicators such as a long and healthy life and a decent standard of living, the full-cost capture of the production and consumption circle, including externalities, remains elusive. Kate Raworth's work on doughnut economics (Raworth, 2018) and the UN's Gapframe go some way in challenging the orthodox linear model, but more might be revealed from the rapid take up of digital and data analysis, new productive methodologies, and consumption shifts from purchasing to renting or doing without to inform the practical applications and solutions derived from the circular economy.

A range of surveys on millennial and post-millennial attitudes to the environment have heightened concerns about sustainability in all parts of life. Generational expectations concerning corporate social responsibility (CSR) and responsible consumption are increasingly influencing policy in public and private organisations. The shift from the purchase of goods model to renting is reshaping orders, inventories, and sales. Sustainable prosperity for the future will need to engage with a generation questioning the suitability of a linear model in meeting contemporary and future growth challenges. This is likely to involve new kinds of public debate based on a diversity of voices based on movements such as the Citizen's Assembly on Climate Change.

In the private sector, trading facilities such as FTSE for Good promote responsible business and encourage reputational risk protection and sustainability. The linking of finance to a range of environmental, social, and governance (ESG) issues is expanding to other areas of sustainable finance. Insurers, underwriters, and risk assessors are increasingly attributing high risk to organisations that are environmentally reckless. The success of the exchange has encouraged good corporate governance and smart growth. Internationally, the move to a global circular economy is still in its early stages, but global policies amongst international governmental organisations and agencies within the United Nations system are initiating structural changes to require multinational companies to improve procurement and global supply chain strategies to meet sustainability indicators and policies. This is influencing risk models in trade finance. Corporate entities such as Unilever and Patagonia have been proactive in embedding sustainable policies in business models, strategically and operationally.

The leading UK market research agency Ipsos Mori has established a Sustainable Development Research Centre to capture data on the economy and environmental impacts. Generational change and a range of targeted education initiatives in higher education are shaping a professional awareness of sustainability, SDGs, and the principles of responsible management (PRME). This is being noted by personnel departments and human resource management agencies. The growing impact on decision-making that stems from intelligence gleaned from multiple surveys and data analysis on a range of environmental impacts appears to be influencing consideration of models related to the circular economy, responsible trade, and sustainable growth. On the operations side, the circular economy provides the potential for intellectual engagement with innovators, engineers, and inventors in dealing with a range of technical possibilities and global challenges.

International trade and growth: trends and anomalies

Trade restrictions, protectionism, tariffs, non-tariffs, custom unions, and trade impediments continue to operate throughout the international trading system. Whilst the World Trade Organization seeks to promote free trade, trade in goods and services is not free of political interferences, technical impediments, reciprocal disincentives, complex documentation, and uneven representation.

Economists such as John Galbraith and David Pearce have previously warned of the depletion of natural resources and the impact on human society from non-sustainable trade and consumption. Economics and the environment cannot be separated in the long term. The academic literature covering the circular economy repeatedly illustrates the relationship between trade, the environment, and legislation. For example, in vivo studies on the impact of transportation on biological entities through forensic analysis are increasing environmental awareness. The empirical focus that underpinned the United Kingdom's ship recycling strategy is an example of how biological studies and industry analysis are changing business models. Of the closed-loop archetypes (closed regional and local loops, closed global supply loops, geographically open cascades, etc.), the ones that are organised locally rather than globally should, in theory, exhibit superior economies. Typically, the greater the distance, the more the transport and indirect costs. But this is not always the case. Global trade volumes are increasingly containerised, and empty containers need filling to offset the structural imbalance of trade flows (WEF, 2014). Logistics closes loops, but it fails to do so fully despite technological innovations. This is due to the lack of market actors that are willing to entirely embrace the full remit of circular economy principles (Fennemann et al., 2018).

A deep understanding of the construction of markets and a dedicated resource base shed light on the use of forensic analysis in undertaking investigations and implementing sustainable solutions as a market device to the economisation of recycling (Gregson et al., 2013).

Emphasis within the literature is placed on approaches that engage with the sustainability of global value chains (GVCs) and global production networks (GPNs), which potentially allow for higher levels of long-term returned investment based on stable growth. Several mechanisms are available to trade delegations aiming to reduce trade barriers created by regulations: harmonisation, mutual recognition and equivalency, and regulatory cooperation are among the main ones (Couvreur, 2015). New trade agreements may facilitate stronger circular economies of scale (Dhingra et al., 2018), which is particularly relevant in navigating the labyrinth of custom papers, completing compliance regulations, and pervasive non-tariff measures that complicate trading activity (Owen et al., 2017).

Where disputes in the terms of trade appear intractable, alternatives drawn from the circular economy may provide solutions to issues of resource scarcity, sustainable trade, and environmental protection.

Comprehensive Economic and Trade Agreement: the future of a United Kingdom–Canada circular economy

This section includes a brief review of an Economic Social Research Council (ESRC, UK) – Social Sciences and Humanities Research Council (SSHRC, Canada) project titled the Future of UK-Canada Circular Economy Trade (De Lange et al., 2019). The full report included meta-analysis and synthesis of the circular economy literature and considered in detail the Comprehensive Economic and Trade Agreement (CETA) between Canada and the European Union. This section relates specifically to material collated from a UK perspective. It is included in the chapter to highlight the potential for approaches in the circular economy to be aligned to future international free trade agreements, which the United Kingdom is currently examining in response to its international trade position post-Brexit.

The Comprehensive Economic and Free Trade Agreement between Canada and the European Union was signed in 2016. Scrutiny in several EU countries has delayed the full implementation of the agreement, approval in the United Kingdom having been examined by the Regulatory Policy Committee, Parliament, and the Department of International Trade.

The academic literature on circular economy, commercial surveys, and corporate reports inferred that a range of practical factors could be incorporated into a post-CETA trade agreement. This is based on several key circular theories: cradle to cradle, performance economy, biomimicry, industrial ecology, natural economy, and the blue economy, each offering solutions to key trade and environmental challenges mentioned earlier.

The opportunity to re-draft a bilateral trade agreement that includes a fully worked appraisal of the circular economy could be developed from the existing CETA documentation, specifically Chapters 22–24. These chapters cover sustainability and the environment in a general sense. The specific contribution of a circular economy to an international trade agreement would be a shift from a linear model (take, make, waste) to a circular model (make, use, return).

CETA was referred to in the literature review, but an extensive in-depth analysis of the agreement in academic and practical terms will not become available until the agreement is fully implemented in all EU countries. The CETA agreement is not without criticism. It should be recognised that existing trade agreements within the EU do not operate without elements of technical and practical impediments. Green activists and members of the European Parliament (MEPs) representing various political parties are critical of the CETA agreement on the basis that it favours the interests of big business to the detriment of the environment. In an early review considering the economic and sociopolitical dimensions of transatantlic trade integration and policy, Deblock and Rioux (2010) considered the potential for Canada and Europe to place environmental, public health, and human rights concerns in a more balanced position with the interests of corporations Nonetheless, despite pressures and successes from green lobbyists to see sustainable practices within the CETA documentation, an explicit engagement with the circular economy is absent.

A future trade agreement between Canada and the United Kingdom could revise sections to explicitly include circularity. There is an opportunity to establish a circular trade gold standard, which might be the basis for other trade agreements either with the commonwealth or international trade partners. It is clear that common understanding of circularity must be consistently communicated and applied across industries and countries. Crucially, the circular economy must not be a vehicle to generate impediments to trade through non-tariffs and other subtle forms of mischievous interference, which may include voluntary sustainability standards (VSS). Trade policy should ensure that transparency is promoted and protected to secure consistent efficiency and accountability.

The UK capture of the literature and subsequent interviews with trade and environmental specialists identified several related issues, gaps, and opportunities to inform decision-making, policy, and practice in formulating the future of a Canada–UK Trade relationship. Key points are summarised in the following:

- The circular economy literature is wide and frequently repetitious – terms originally identified in the work of the UK economist David Pearce and the Ellen MacArthur Foundation being prominent

- More science-based research on the full environmental impacts of international trade, including the unintended consequences of trade, is required on a case-by-case basis to provide a measurable understanding of international trade and circularity
- The benefits gleaned from digital technology in the area of communication networks (recognised as an important element in the sharing economy), 3D printing, reciprocal arrangements, and bar code monitoring need to be aligned with research in the circular economy
- The circular economy perspective in international trade requires additional research in the areas of intellectual property, knowledge transfer, and design mechanisms, which are invariably, but not equally, shaped by innovation, legislation, regulation, and consumer demand
- More research on circular economy incentives in trade agreements and government procurement contracts requires further consideration – for example, circularity alignment in preferential tariffs agreements, tender applications, and sustainability clauses would make choices clear
- Incentives in free trade agreements (FTAs) and government procurement contracts require further consideration. Circularity in procurement policies needs to be transparent and should be supported with clear evidence of the full benefits of sustainable performance. This effort might overcome resistance in the form of restrictions or incentives on trade being assumed to be anti-competitive or protectionist
- Circularity commitments in FTAs potentially promote and facilitate best sustainable practice and could be aligned to multilateral trade agreements
- Increased science-based evidence on climate change and marine and aviation transportation (marine litter, fuel pollution, and damage to ecosystems and atmosphere; the global commons and the blue economy) is generating a range of local, sub-national, regional, and international legislation, which may trigger prohibitive environmental tariffs, non-tariffs, and related restrictions
- Circularity is being used in industry – however, it may not necessarily be counted or understood as an alignment to the circular economy. For example, circularity exists in the energy sector in decommissioning platforms and land-based facilities – this is undertaken on a practical reuse basis, which naturally addresses resource scarcity, cost logic, and efficiency
- Research is required to identify key areas of international trade and FTAs that would benefit from alignment with circularity.

Conclusion

The economic and trade considerations stemming from the transition from a linear economy to a circular economy need to be aligned with the identification and implementation of circular efficiencies thoroughly undertaken. This is particularly relevant to the international trade in commodities, intermediaries, and finished goods.

The advances in monitoring, surveillance, and interpretation from a global digital infrastructure offer extensive opportunities to apply approaches common to the circular economy to trade efficiencies and agreements. The following points highlight the potential for the data gathering, monitoring, and analysis of international trade circularity to identify gaps and potential solutions to mitigate the limitations noted in the linear trade model. Key recommendations are:

- Increased evidence-based understanding of the full cost of international trade activities and impacts on externalities using metrics and digital metrics to measure specific waste and design functions to re-envisage the re-design, reuse, and recapture sector

- Use of the data and metrics in international trade hubs, including ports, harbours, rail, and road freight to identify and moderate inefficient gaps
- Recognising and recording the value of the circular economy and the digital economy on transaction costs
- Policy co-ordination of international trade and circular economy through trade and commercial diplomacy to benchmark circularity in free trade agreements.

The circular economy transition requires in-depth evidence-based research to support ongoing technical specialisation in measuring circularity. The availability of transparent evidence on the accurate performance of the circular economy would influence trade policy and trade conduct. It is crucial that the emerging trade framework be transparent and representative of interests in the Global North and Global South. This will provide opportunities in the form of design, technical application, and cross-border collaboration in the framing of international trade agreements and trade transportation to promote sustainable trade relations.

Anchoring circularity to production, consumption, and exchange offers an incremental solution to trade challenges in the Global North and Global South. Spaceship Earth requires prosperity and sustainability. Trade between countries will continue to evolve despite trade wars and trade disputes. The objective should not be to eliminate differences of opinion in trade relations but to understand the full impact of externalities on the planet. The ability to gather, inform, shape, and negotiate the transition from a linear to a circular international trade economy is being expanded through science-based evidence. Failure to accurately recognise negative trade anomalies, inefficiencies, and errors that extend resource scarcity and fail to address domestic and global waste will compromise sustainable international trade in physical goods. Circularity offers a solution to future shocks in the linear model and potentially rebalances wealth and the health of nations.

References

Bleys B. (2012). Beyond GDP: classifying alternative measures for progress. *Social Indicators Research*. 109 (3): 335–376.

Boulding KE. (1966). The economics of the coming spaceship earth. In: Jarrett H. (Ed.) *Environmental quality in a growing economy, resources for the future*. Baltimore, MD: Johns Hopkins University Press. 3–14.

Brown H. (2018). Towards a circular energy economy. Cross-sector successes in Brazil and India. *Journal of Sustainable Development*. 20 (1): 23–42.

Circle Economy (2019). *Circularity gap report*. Available at: https://circulareconomy.europa.eu/platform/sites/default/files/circularity_gap_report_2019.pdf

Couvreur A. (2015). New generation regional trade agreements and the precautionary principle: focus on the CETA agreement between Canada and the European Union. *Asper Review*. 265–292.

Deblock C, Rioux M. (2010). From economic dialogue to CETA: Canada's trade relations with the European Union. In: The other transatlantic relationship: Canada, the EU and 21st Century Challenges. *International Journal*. 66 (1): 39–56.

De Lange D, Sheeran P, Welch P. (2019). *The future of UK–Canada circular economy trade*. SSHRC-ESRC research paper. Available at: www.ryerson.ca/content/dam/tedrogersschool/trsm-news/General_Public/documents/UK-Canada-Trade-Report-Final.pdf

Deloitte (2018). *Circular economy: from theory to practice*. Available at: https://www2.deloitte.com/content/dam/Deloitte/fi/Documents/risk/Circular%20economy%20FINAL%20web.pdf

Dhingra S, Ottaviano G, Rappoport V, Sampson T, Thomas C. (2018). UK trade and FDI: a post-Brexit perspective. *Papers in Regional Science*. 97 (1): 9–24.

Ellen MacArthur Foundation (2015). *Delivering the circular economy: a toolkit for policymakers*. Available at: https://www.ellenmacarthurfoundation.org/publications/delivering-the-circular-economy-a-toolkit-for-policymakers

Fennemann V, Hohaus C, Kopka J. (2018). *Moving in circles: logistics as key enabler for a circular economy*. Fraunhofer Institute for Material Flow and Logistics. Available at: http://p:ublica.fraunhofer.de/documents/N-502288.html

Fioramenti L. (2013). *Gross domestic problem*. London: Zed Books.

Gregson N, Watkins H, Calestani M. (2013). Political markets: recycling, economization and marketization. *Economy and Society*. 42 (1): 1–25.

Hickel J, Kallis G. (2020). Is green growth possible? *Journal of New Political Economy*. 25 (4): 469–486.

Owen J et al. (2017). *Brexit report*. London: Institute of Government. Available at: www.instituteforgovernment.org.uk/sites/default/files/publications/IFGJ5896-Brexit-Report171214-final_0.pdf

Parente RC, Geleilate JMG, Rong K. (2018). The sharing economy globalization phenomenon: a research agenda. *Journal of International Management*. 24 (1): 52–64.

Preston F, Lenhe J, Wellesley L. (2019). *An inclusive circular economy: priorities for developing countries*. Chatham House and the Department for Energy, Environment and Resources Department. Available at: https://www.chathamhouse.org/publication/inclusive-circular-economy-priorities-developing-countries/2019-05-22-Circular%20Economy.pdf

Raworth K. (2018). *Doughnut economics: seven ways to think like a 21st century economist*. London: Random House.

Seawork International (2019). It's all go for green. *Maritime Journal*. 378: 35.

London Waste and Recycling Board (2015). *London: the circular economy capital*. White Paper, London.

World Economic Forum (WEF) (2014). *Towards the circular economy: accelerating the scale-up across global supply-chains*. Available at: http://www3.weforum.org/docs/WEF_ENV_TowardsCircularEconomy_Report_2014.pdf

19
MANAGING WASTE AT THE NATIONAL AND LOCAL LEVELS

Christian Zurbrügg

Managing solid waste at the national and local level

Stakeholders and their roles in solid waste governance

The complexity of managing solid waste becomes clear when considering its wide range of stakeholders. Stakeholders can affect or be affected by waste management system elements, its objectives, or related policies and regulations.

The concept of "Integrated Sustainable Waste Management (ISWM)" (Van de Klundert and Anschütz, 2001) describes how most successes in solid waste management will depend on how a range of stakeholders has been integrated into the process, capitalising on their strengths to shape a strong team with clearly distributed roles and responsibilities. This involves addressing practices, attitudes, and behaviours of waste generators and skills, work performance, and quality of service provided by the staff of the municipality, private enterprises, or informal sector such as waste pickers, waste dealers, and recyclers. The current state of the art in solid waste management thus involves all stakeholders in a participatory process to foster positive perceptions and attitudes (Bolaane, 2006). In this regard, methodologies for stakeholder inclusivity and respective techniques play a crucial role and are required as key skill sets in local government bodies managing solid waste.

Not all stakeholders are equal in terms of attitudes, interests, influence, roles, or responsibilities. Any change of the solid waste management system will affect each stakeholder either negatively or positively. This implies that whatever is done to change and hopefully improve or redirect solid waste management towards a circular economy must consider the needs and perceptions of all stakeholders and see how these can be taken into account (Alamgir et al., 2012; Bao et al., 2012). Examples for stakeholders in solid waste management and their roles are shown in Table 19.1. It shows that to a large degree, the local government (municipality) retains a major role in most activities of solid waste management. In certain functions of solid waste management, however, such as treatment, recycling, or collection, the private sector (formal or informal) plays a significant role and can be integrated more strongly as a service provider. This shows in the global trend of municipal authorities increasingly contracting the private sector for certain waste activities or even attempting to integrate the informal sector into the overall management system (Rouse and Ali, 2008; Furedy, 1986). While the formal private sector is typically mandated for collection, treatment, or disposal through a contract or license,

Table 19.1 Typical stakeholder roles (XX: dominant role and X: occasional role) in the municipal solid waste management system.

Stakeholder	*Awareness raising*	*Waste Generation*	*Primary collection*	*Sweeping*	*Secondary collection*	*Treatment Recycling*	*Disposal*	*Monitoring*	*Policy*
Residents as waste producers		XX				XX		X	
Commerce and industrial waste producers		XX				XX		X	
Informal private sector			XX	X	X	XX	X		
Formal private sector			X	X	XX	XX	XX		
Non-governmental organisations	XX							XX	X
Local government and municipal authorities	XX		X	XX	XX	XX	XX	XX	XX
Provincial and national government authorities	X							XX	XX
International support agencies	X							X	X

Source: Adapted from Rouse and Ali (2008)

the informal sector typically engages in recycling activities and, given its informal status, remains without contractual security. In low- and middle-income countries, the informal sector – the sector which includes all livelihood opportunities that are not recognised as normal income sources and on which taxes are not paid – plays a significant role in solid waste management and recycling activities (Ali and Snell, 1999; Henry et al., 2006).

It is the interaction and collaboration between these stakeholders which is crucial to success, whatever the type of solid waste project. Waste collection service needs the cooperation of households, such as placing their waste at the curb on defined days. It also needs their collaboration when payment is due. A few principles of successful collaboration are (UN-Habitat, 2001):

- Early involvement of all participants and sufficient time
- Willingness to participate (valuing the benefits)
- Transparency
- Cultured conflict: This does not mean the absence of disagreement but should ensure that all views are represented, participants listen to one another, take each participant's perspectives seriously, and attempt to address the concerns voiced
- Sustained dialogue which seeks consensus
- Capacity of facilitation, monitoring, evaluation

Participatory methods and techniques are central tools in community development, and there is a continuously growing set of tools to foster participation, transparency, and accountability (UN-Habitat, 2001).

An organisation in charge of solid waste management, independent of its organisational status (e.g., community-based, cooperative, informal private sector, formal private sector, NGO, or governmental) must have the ability to communicate, coordinate, and collaborate with this variety of stakeholders. This is a challenge for most. Municipal officers generally feel that this is not their job description, and they receive no training in communication, outreach, leadership, and collaborative problem solving. They frequently lack the capacity to fulfil the role of facilitator and mediator. The lack of integration of the informal recycling sector into the overall waste management strategy is the most prominent example of this challenge to integrate and cooperate (Scheinberg, 2001; Scheinberg et al., 2010). Although the informal sector is involved in the various stages of solid waste management, questions remain as to whether its inclusion is by participation or only by information.

The green versus the brown agenda

Wilson (2007) describes how in what are now high-income countries, the drivers for waste management historically developed from resource use to public health and then environmental protection, which lasts to date (however underscored by a public awareness and the drive to 'close the loop' and achieve 'sustainable consumption and production'). On the other hand, in low-income countries, the main driver remains public health. Environmental protection as a driver remains relatively low on the public and local political agenda. However, given the global impact of solid waste management, it is not only these local stakeholders that influence and act upon a city's solid waste management system. Increasingly, it is also the global community and its international representatives who are pushing local governments to prioritise the "green agenda" of waste management.

This green agenda directs priority towards issues of environmental and ecosystem protection and the mitigation of detrimental effects from human activity on the environment at the regional and global scale. Typical examples of this green agenda are the greenhouse gas (methane) emissions at disposal sites or uncontrolled landfills. When organic waste decomposes in landfills and uncontrolled dumps, it produces methane (CH_4), one of the major greenhouse gases that contribute to climate change. In developed countries, landfill CH_4 emissions have been stabilised due to widespread initiatives of landfill gas-capturing or incineration plants. On the other hand, in developing countries with continuing improvements towards controlled (anaerobic) landfilling practices, CH_4 emissions could even increase (Bogner et al., 2008). Therefore, from a waste management perspective, opportunities to reduce greenhouse gas emissions lay in finding better solutions for organic waste treatment, thereby diverting biowaste from landfills and thus reducing uncontrolled anaerobic decomposition and methane emissions (UN-Habitat, 1996).

The global pressure to mitigate these emissions is reflected in the establishment of the Paris Agreement of 2016, which has brought all nations into a common cause to undertake ambitious efforts to combat climate change and adapt to its effects, with enhanced support to assist developing countries to do so. Another example is the global efforts to combat plastic waste in the oceans. Again, the impact is global, but measures are necessary at the local or national level. Such a green agenda may not be of immediate priority to the community and local government, which is more concerned with the "brown agenda" (McGrahanan and Satterthwaite, 2000). The brown agenda is the focus upon direct human health and well-being, targeting immediate problems at the local level, especially those suffered by low-income communities. Protecting human health is the main reason cities collect solid waste. Thus, from a local-level perspective, it is often not landfill gas emissions which are ranked high in the list of priorities but rather ensuring regular and reliable collection services to all of the urban population, especially the poor in urban slums. Waste which remains uncollected in the neighbourhoods reflects an immediate and direct health threat for the population.

Private, public, or merit good?

From an economic perspective, solid waste management lies in a complex space between private, public, and merit good. Public and merit goods benefit society as a whole, while private goods or "services" have the characteristics of: (1) excludability: where customers of the good can be excluded from consuming them if they do not pay for them; (2) rivalry: where the consumption of the good by one customer reduces the amount for other customers to consume; and (3) rejectability: where goods and services can be declined should the tastes and preferences of the customer change (Riley, 2019b).

Certain services in the solid waste system show typical characteristics of a private good. A good example for such a private good in solid waste collection is a better or higher level of "collection service" – in other words, an upgrade of the basic service. This upgrade can take place, for instance, in terms of where waste is collected (door collection services rather than a community collection point) and when waste is collected (daily instead of weekly collection frequency). Such a service can be provided only if paid for (excludability), the number of possible customers may be limited (rivalry) depending on the capacity of the collection system, and service can be stopped if the customer has other preferences (rejectability). Similarly, collection of recyclable materials at the household level can also be considered a private service. Such private good services are, however, always an upgrade to existing baseline services.

Baseline services – also called basic services – are under the responsibility of the local government and are not a private good, as they should be available to all irrespective of payment. The aspect of excludability, therefore, does not apply. It is of public interest that waste collection service be available to all, and when payment for services is defined as a requirement, the service must nevertheless be included for "free-riders" to avoid littering and indiscriminate disposal in and around the neighbourhood. Similarly, street sweeping, secondary collection from community bins, and transfer stations or safe disposal at sanitary landfills are clearly not private goods, as all residents, paying or non-paying, benefit from the service of a cleaner environment. Hygiene, public health, and protection of the environment are the drivers to provide waste management services to all independently of whether they pay. Such aspects are typical of merit goods, where it is judged that an individual or society should have this good on the basis of some concept of need and public interest rather than the ability, willingness, or affordability of paying (Riley, 2019a). It can simply be seen as a good that has positive externalities associated with it.

How does this discussion of private, public, and merit goods affect the provision of solid waste management? Better understanding the elements and the inherent reasons for service provision helps to clarify aspects of responsibility and accountability. For instance, if a higher standard of collection services, for example, more frequent collection or household collection instead of community bins, is requested by some households, this relates to a private good for which service may be provided by a private entity at a price. However, the overall resulting negative externalities of no or deficient waste management services are the main reasons local government, that is, a municipality, must retain overall responsibility for the service. Even if certain critical tasks are delegated to the private sector, in light of public interest, the service providers are accountable to local government, and control of services is in municipal hands. Also, the far-reaching global impacts of deficient waste management clearly set waste management in the realm of a public or merit good. This underlines the significant needs of strategic steering and operational planning by national and local government. However, weak institutional capacity, limited financial means, and also the lack of political will hinder such planning and are typical for low- and middle-income economies (UNEP and UNITAR, 2013).

Defining a glocalised strategy for solid waste management?

There is a general consensus in most high-income countries on the importance of the "waste management hierarchy" (Figure 19.1) as a leading strategic concept that defines and underlines the following policy principles in their order of decreasing priority.

Figure 19.1 National and local factors (circular boxes) influencing the core concepts of the waste management hierarchy (triangle) in which priority of solid waste elements diminishes from top to bottom

Implementation and prioritisation of this hierarchy at the national and local level is, however, less evident and easy for different reasons. To ensure "avoidance and minimisation," for instance, action would be needed to control production and packaging, thus targeting designers, engineers, and managers in industry who decide what is manufactured, constructed, and/or processed and packaged. At national level, however, policies around waste management are typically designed and pushed forward by the environmental sector without much interaction and exchange with the sector of production and trade, which, given globalisation trends, may also be dominated by international trade decisions and agreements. Also, at the local operational level, stakeholders in solid waste management (typically the municipal solid waste department) have little influence on avoidance or minimisation of waste. Furthermore, implementing certain activities of avoidance, minimisation, or reuse may contradict social (lifestyle) choices or economic feasibility without significantly adding value to the environmental sustainability objectives (Kijak and Moy, 2004). Also, such initiatives towards avoidance might even be misguided, considering that the average per capita waste amount produced in low-income countries is significantly lower than in more affluent countries (Hoornweg and Bhada-Tata, 2012). Countries of the Organisation for Economic Co-operation and Development (OECD) generate almost half of the global waste amount, while Africa and South Asia are the regions which produce the least waste. Thus, global pressure to put a high priority on waste "avoidance" does not necessarily respect local conditions and priority targets, as low-income economies already "achieve" low waste generation. Also, given a growing threat of climate change and increasing restrictions on supply of resources, a global effort towards resource-use efficiency and the concept of doing more with less is being pushed forward. By this, a strategy of circular economy and low carbon footprint is establishing itself worldwide as an essential element of municipal solid waste management and sustainable development with the priorities placed on it by the World

Summit on Sustainable Development (WSSD) and one of its key outcome documents – the "10 Year Framework Programme on Sustainable Consumption and Production" (Barber, 2003). Responsible governments, civil society, and other stakeholders are encouraged to build strategic alliances and develop a vision for the medium- and longer-term future in this regard, although urgent local needs might be to improve waste collection and the financial stability of the solid waste management system. Policies and strategies must therefore try to balance considerations at both local and global levels, described by the term "glocalisation."

National-level guidance on local waste management

Although waste management is a local matter, there is a strong case to argue for national-level guidance, policies, and strategic direction. Addressing solid waste issues at the national level brings the topic to a higher political and public visibility, sets a framework to reconcile inconsistent local practices, and can ensure that resources are used to reflect these national interests, be this in terms of capacity and skills development or earmarked funding for research or infrastructure. Policies and strategies are given flesh through their translation into legislation. But laws and regulations and their enforcement are just one way to support implementation of strategies and policies. Other instruments that can also help support implementation are instruments of social mobilisation or economic instruments such as incentives or disincentives. Depending on the objective, a mix of these instruments can be used (Figure 19.2).

For instance, if the goal is to ensure that waste management activities do not pollute the environment, it can be made mandatory that facilities need a license to operate, and as precondition for the license, they would have to install a certain technical infrastructure. If the goal, however, is to increase waste segregation at the source, various social instruments would probably be more beneficial than only regulations and enforcement. Typical social instruments are awareness-raising campaigns; these are often, for instance, used to combat littering and in general to change perceptions and behaviour. Also, economic instruments can be used to support achievement of certain policy goals.

In Bangladesh, to implement the national 3R strategy and promote recycling, the government (through the Ministry of Finance and its National Board of Revenue and Board of Investment) developed three economic instruments: (1) tax holidays for five to ten years for all formal

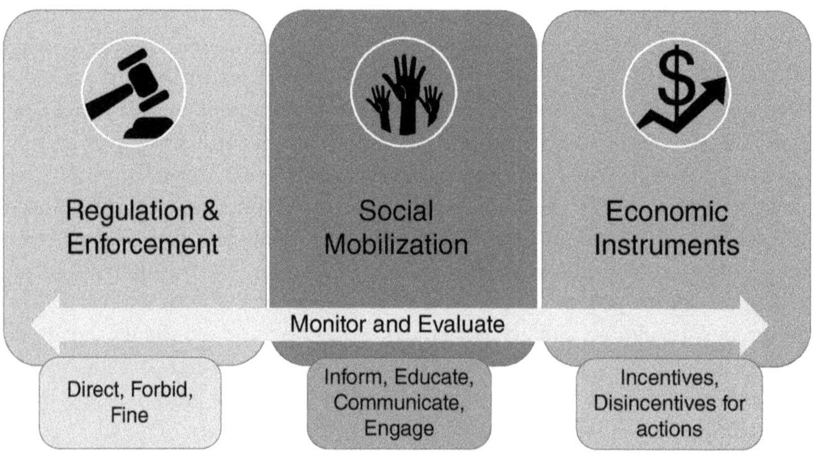

Figure 19.2 Policy implementation tools

waste treatment and recycling plants, (2) less import duties on relevant waste management equipment, and (3) no VAT/sales tax on sales of compost (Matter et al., 2013). All such policy instruments are prepared or implemented by various governmental agencies, depending on the institutional setup and the specific mandates. An inclusive governance system, however, also negotiates policies and seeks agreement to planned processes and practices with all stakeholders, in other words, the inclusion of organisations and people in a participatory decision-making process.

In India, the composting project of Tikri in Delhi, which processes market waste, was strongly supported by the municipal council as well as the Agriculture Produce Market Committee (APMC), and land was provided to the project proponent on a lease-rent basis (CDM, 2009). Another example is from the city of Mumbai, where the municipal authority proactively supports citizens' waste management initiatives within the framework of a program called Advanced Locality Management (ALM). The city offers these initiatives a contact person for assistance as well as a regular forum with specialists and authorities. Here problems can be addressed, and potential solutions are jointly formulated (Zurbrügg et al., 2004). The national government of India also identified 100% sanitation as a priority goal for the Swachh Bharat Mission. This provoked national support to local city governments with financial assistance for state-level strategies and city-level plans, awareness generation, and capacity building, as well as legal support for initiating public–private partnerships (CPHEEO, 2016).

In Brazil, at the municipal, state, and national level, "catador de material reciclável" (collector of recyclables) has been formally included as a profession in the Brazilian Occupation Classification (CBO). The local government of Belo Horizonte furthermore introduced legislation that made recycling, social inclusion, job creation, and income generation the four main pillars of solid waste management. This paved the way for an integration of recycler cooperatives and recycling enterprises in the overall solid waste management plans through the lead of the Public Cleansing Authority (SLU) (Dias, 2011).

In the Philippines, the local government of Iloilo City recognised the potential of existing waste pickers and assisted in forming a Waste Workers Association, which was then registered as a formal business enterprise. This set the stage for an effective recycling program well integrated into the formal solid waste management system (Paul et al., 2012).

In Mexico, the federal government declared the "crusade for a clean Mexico" in 2001, recognising the current inefficient operations and assigning a clear role to SEMANAT, the national ministry of environment, to intervene with support for local authorities in capacity development with targeted training programs (Wehenpohl, 2007).

In Pakistan, the case of a successful community-based sanitation project in the slum of Orangi-Karachi was possible thanks to a major shift in policy by the municipal government in the 1970s to legalise the slum settlement. This enabled people to buy titles to their homes, giving them a sense of permanency and the incentive to invest in improvements (Khan, 1997).

Biowaste management as a priority

Biowaste is organic and biodegradable waste consisting of food waste, kitchen waste, green garden waste, paper, and cardboard. In low- and middle-income countries, biowaste is typically 50–70% of all municipal solid waste (Wilson et al., 2015). If waste is not managed appropriately, this biowaste fraction will attract various animals that transmit diseases, such as flies, rats, or other animals scavenging the garbage. If collected and disposed at uncontrolled dumpsites, the biowaste fraction undergoes anaerobic degradation and generates methane, a greenhouse gas that is 25 times more potent than carbon dioxide (CO_2), thus severely contributing to global warming. In neighbourhoods as well as dumpsites, soil and water in contact with waste become rapidly contaminated, threatening soil quality and food safety, as well as surface and groundwater resource quality. Solid waste littered in drainage channels will cause blockage, flooding, or stagnant ponds. This can propagate the breeding of mosquitoes that transmit malaria, dengue, Zika virus, and yellow fever. Finally, yet importantly, indiscriminately dumped solid waste in a settlement area is unappealing. It lowers the attractiveness for economic activity (e.g., tourism) and lowers the resilience and self-esteem of communities. In high-income countries biowaste constitutes around 30% of the total waste generated. This is predominantly food waste, in other words, food resources wasted during food distribution from retail and at the consumer level – although overproduction at farms, incentivised through agricultural subsidies, may also be a source.

If biowaste can be safely managed and recycled, the impact will not only be significant on the whole solid waste management system but will also be a major step towards a more circular economy. The carbon, nutrient, and energy content of biowaste can be recycled as organic fertiliser, soil improver, or growing media or transformed into a range of different bio-based products or renewable energy (Moreno et al., 2016).

However, when mixed with other waste, biowaste becomes contaminated and more difficult to valorise. Mixed waste also lowers the value of "dry recyclable" waste materials such as metals, paper, glass, and textiles. Therefore, the success of all recycling depends critically on materials being kept separate and clean after waste is generated. The solution is "household or commercial segregation" into at least two fractions: organic "wet" waste and non-organic "dry" waste. Fostering waste segregation at the level of waste generator requires a dynamic and vigorous interaction with community and/or private-sector members to incentivise their participation and change of behaviour (Mosler, 2012). Regardless of what biowaste treatment technology is used, it is essential to evaluate and understand which key factors are crucial for durability and sustainability of this operation.

Overall, three different feasibility domains can be distinguished: (1) *technical feasibility:* includes the space and materials required for construction and operation; the technical skills and capacity to build and operate the facility; and the suitability and accessibility of biowaste type, quantity, and quality; (2) *economic feasibility:* comprises the expected capital and operational costs of a facility, as well as the possible revenue streams based on the value and demand for the product derived from biowaste recycling; and (3) *social feasibility:* includes all aspects of community and stakeholder social acceptance and support for the specific biowaste recycling facility and/or its derived products. Careful evaluation of these factors will help decide what type of biowaste treatment technology is most feasible for implementation (Zabaleta et al., 2019).

As safe management of biowaste can contribute significantly to an overall improved solid waste management system and a circular economy, support with policy instruments and legislation are important prerequisites. Local and national governmental authorities must ensure that this issue is listed as a high-priority strategic goal and is followed up on the one hand with policies, policy instruments, regulations, and legislation, and on the other with technology selection support, as well as financing structure. Direct regulation includes legislation and its enforcement. It serves to protect common interests in a society, such as public health and the environment. For instance, waste generators can be given the duty of and made responsible for waste segregation, or local authorities can be made responsible for facilitating implementation of any appropriate processing of biodegradable wastes. A 'direct regulation' approach needs monitoring, inspection, and enforcement; therefore, a commitment to good and continuous data management is essential. Economic policy instruments, on the other hand, help direct stakeholder behaviours and practices towards biowaste management using market-based incentives and disincentives, for instance, by subsidising compost to increase its competitiveness with regard to the already subsidised chemical fertilizers or fiscal benefits for companies engaging in biowaste recycling. Finally, 'social' policy instruments can be supported and implemented to strengthen the biowaste management approach. These are based on communication and interaction with stakeholders, such as awareness-raising campaigns to affect people's waste attitudes and behaviours or leading by example (e.g., using compost in public spaces).

Although various biowaste management initiatives and projects have been implemented at different scales and with different outcomes, these are seldom documented comprehensively. Available data are rarely analysed and are often not even saved in a form which allows good analysis. Especially transparent cost-revenue information is key to understanding financial feasibility of the case for future replication and scaling.

Implementation at the local level

Any solid waste management activity will cause an impact on the socioeconomic and natural environment, be it positive or negative. Such impacts will alter the perceptions and interactions among stakeholders, and this change in societal conditions can influence the "enabling environment," which either fosters or hinders the solid waste activity. This interaction is expressed well by the "pressure-state-response framework," as shown in Figure 19.3 (OECD, 1993). Waste management exerts pressure (impacts) on the environment, thereby changing its state. This change and the perception of its importance and severity trigger societal processes at different societal levels (households, administrations, etc.) which then again result in societal responses (actions, policies, etc.). The responses affect the enabling environment for the original solid waste activity or future similar projects.

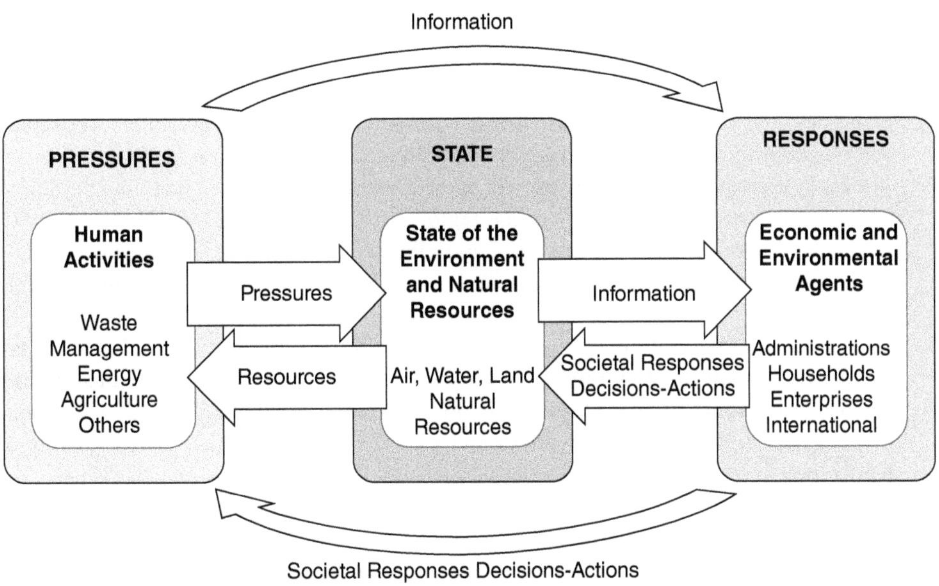

Figure 19.3 Elements of the pressure-state-response framework

The best example in solid waste management of such a pressure-state-response situation is the "not in my back yard" (NIMBY) effect. NIMBY characterises opposition by residents to a new development, for instance, a landfill or a waste treatment facility (e.g., incineration) because it is close to them (Zurbrügg, 2003). The NIMBY example also shows that societal responses are not necessarily linked to the specific impact of the proposed project but rather to the societal perception of impact based either on previous experience or obtained external information. The public resistance to a sanitary landfill is a result of past experience or information on how open dumps have negatively impacted neighbourhoods. Building of public opinion can thus be based on information from other cities, regions, or even nations. The success of any solid waste management action needs support by a favourable "external" environment. This is called the enabling environment. Without it, the resources committed to bring about change will hardly be effective (Lüthi et al., 2011). "Enabling environment" is used as a term to describe a group of interrelated features and conditions which, in a specific context, will set the basis for change and improvement. Furthermore, these features of an enabling environment will support and sustain this improved state over time. The enabling environment can, therefore, also be described as a set of factors impacting the solid waste management project or activity of concern during the whole project cycle. This set of enabling conditions is dynamic and will change more or less rapidly with time. Economic crisis, research and innovation, and the associated mutual learning process or political change are typical examples of a changing environment which is influenced from outside the local solid waste management domain.

1. *Supported by government and legislation*: Represents the extent to which government endorses and supports the project and how it coincides with national laws, regulations, standards, and codes.
2. *Enabled through an effective organisational structure:* Which is clearly defined in its goals and objectives; has a strong forward-looking leadership; and operates under the principles of quality control, accountability, transparency, and equity. Here, a sound partnership with other solid waste stakeholders and networking are considered decisive to build on strengths and opportunities. In-house capacity to fulfil the quality of service envisaged is reflected by the skilled, motivated, and continuously trained staff.
3. *Embedded in a financially sound setup:* Involves a viable business model and business plan, access, and the capacity to mobilise investment capital and mechanisms to recover capital and operational costs through reliable revenue sources.
4. *Technically appropriate:* Where the project operates with locally proven technologies suited to the local context, ideally built in the region with local materials and skills, and with a reliable service chain to ensure rapid and effective maintenance and repair. Flexibility of the technology to cope with changing conditions is another feature of a suitable technology.
5. *Environmentally sound:* Where project activities monitor emissions and comply with environmental regulations. Environmentally sound operations also strive to reduce energy and natural resource consumption; minimise emissions to water, air, and soil; avoid other nuisances; and safeguard workers and adjacent resident health – independently of legislation.
6. *Socioculturally accepted and beneficial:* Involves the endorsement and support of the project by the community as well as their motivation and willingness to participate and contribute to the process and objectives of the project. This also comprises recognised and valued benefits for the community, not only in terms of improved cleanliness but also with regard to employment opportunities and local social and economic development.

References

Alamgir M, Bidlingmaier W, Cossu R. (2012). Editorial: successful waste management strategies in developing countries require meaningful involvement of the concerned stakeholders. *Waste Management*. 32: 2007–2008.

Ali M, Snell M. (1999). *Lessons from community-based initiatives in solid waste*. Loughborough: WEDC.

Bao PN, Aramaki T, Hanaki K. (2012). Assessment of stakeholders' preferences towards sustainable sanitation scenarios. *Water and Environment Journal*. 27 (1): 58–70.

Barber J. (2003). Production, consumption, and the world summit on sustainable development. *Environment, Development and Sustainability*. 5: 63–93.

Bogner J, et al. (2008). Mitigation of global greenhouse gas emissions from waste: conclusions and strategies from the Intergovernmental Panel on Climate Change (IPCC) fourth assessment report. Working group III (mitigation). *Waste Management & Research*. 26 (1): 11–32.

Bolaane B. (2006). Constraints to promoting people centred approaches in recycling. *Habitat International*. 30: 731–740.

CDM (2009). *Project design document – version 03. Upgradation and expansion of A.P.M.C. compost plant at Tikri*. Delhi: CDM – Executive Board.

Central Public Health and Environmental Engineering Organisation (2016). *Municipal solid waste management manual – part 1: overview*. Delhi: Ministry of Urban Development, Government of India.

Dias SM. (2011). *Recycling in Belo Horizonte, Brazil – an overview of inclusive programming*. WIEGO Policy Brief (Urban Policies) No. 3. Manchester, UK.

Furedy C. (1986). The people who get in the way: changing values in urban waste management. *Water Science and Technology*. 18: 121–128.

Henry RK, Yongsheng Z, Jun D. (2006). Municipal solid waste management challenges in developing countries – Kenyan case study. *Waste Management*. 26: 92–100.

Hoornweg D, Bhada-Tata P. (2012). *What a waste – a global review of solid waste management*. Washington, DC: Urban Development and Local Government Unit, World Bank.

Khan AH. (1997). *The sanitation gap: development's deadly menace*. New York, NY: UNICEF, Progress of Nations.

Kijak R, Moy D. (2004). A decision support framework for sustainable waste management. *Journal of Industrial Ecology*. 8: 33–50.

Lüthi C, Morel A, Tilley E, Ulrich L. (2011). *Community-led urban environmental sanitation planning (CLUES)*. Dübendorf: Swiss Federal Institute of Aquatic Science and Technology (Eawag).

Matter A, Dietschi M, Zurbrügg C. (2013). Improving the informal recycling sector through segregation of waste in the household – the case of Dhaka Bangladesh. *Habitat International*. 38: 150–156. https://doi.org/10.1016/j.habitatint.2012.06.001.

McGrahanan G, Satterthwaite D. (2000). Environmental health or ecological sustainability? Reconciling the brown and green agendas in urban development. In: Pugh C. (Ed.), *Sustainable cities in developing countries*. London: Earthscan.

Moreno M, De los Rios C, Rowe Z, Charnley F. (2016). A conceptual framework for circular design. *Sustainability*. 8 (9): 937.

Mosler HJ. (2012). A systematic approach to behavior change interventions for the water and sanitation sector in developing countries: a conceptual model, a review, and a guideline. *International Journal of Environmental Health Research*. 22: 431–449.

OECD (1993). *OECD core set of indicators for environmental performance reviews – a synthesis report by the group on the state of the environment*. Paris: Organisation for Economic Co-operation and Development, OECD.

Paul JG, Arce-Jaque J, Ravena N, Villamor SP. (2012). Integration of the informal sector into municipal solid waste management in the Philippines – what does it need? *Waste Management*. 32: 2018–2028.

Riley G. (2019a). *Merit goods and services*. Economics Resources – AS & A2 Economics Revision Workshops. Available at: www.tutor2u.net/economics/revision-notes/as-marketfailure-merit-goods.html (Accessed 12/11/2019).

Riley G. (2019b). *Public goods and private goods*. Economics Resources – AS & A2 Economics Revision Workshops. Available at: www.tutor2u.net/economics/revision-notes/as-marketfailure-public-private-goods.html (Accessed 12/11/2019).

Rouse J, Ali M. (2008). *Planning for sustainable municipal solid waste management*. Warwickshire: Practical Action.

Scheinberg A. (2001). Micro- and small enterprises in integrated sustainable waste management. In: Scheinberg A. (Ed.), *Tools for decision-makers – experiences from the Urban Waste expertise programme (1995–2001)*. Nieuwehaven: WASTE.

Scheinberg A, Wilson DC, Rodic L. (2010). *Solid waste management in the world's cities. UN-Habitat's third global report on the state of water and sanitation in the world's cities.* London: Earthscan.
UNEP, UNITAR (2013). *Guidelines for national waste management strategies – moving from challenges to opportunities.* United Nations Environment Programme. ISBN:978-92-807-3333-4.
UN-Habitat (1996). *An urbanizing world: global report on human settlements.* Oxford: United Nations Centre for Human Settlements, Oxford University Press.
UN-Habitat (2001). *Tools to support participatory urban decision making.* Nairobi, Kenya: UN-Habitat, Urban Governance Toolkit Series.
Van de Klundert A, Anschütz J. (2001). *Integrated sustainable waste management – the concept; tools for decision-makers – experiences from the urban waste expertise programme (1995–2001).* Gouda: WASTE.
Wehenpohl G. (2007). Training and building awareness: network of environmental promoters for SWM in Mexico. In: Diaz LF, Eggerth LL, Savage GM. (Eds.), *Management of solid waste in developing countries.* Padova: CISA. 245–251.
Wilson DC. (2007). Development drivers for waste management. *Waste Management & Research.* 25: 198–207.
Wilson DC, Rodic L, Modak PSR, Carpintero A, Velis CA, Iyer M, Simonett O. (2015). *Global waste management outlook.* UNEP. http://web.unep.org/ourplanet/september-2015/unep-publications/global-waste-management-outlook.
Zabaleta I, Mertenat A, Scholten L, Zurbrügg C. (2019). *Selecting organic waste treatment technologies.* Dübendorf, Switzerland: SOWATT. Eawag: Swiss Federal Institute of Aquatic Science and Technology.
Zurbrügg C. (2003). *Solid waste management in developing countries.* Available at: www.sandec.ch/SolidWaste/Documents/04-SW-Management/Basics_of_SWM.pdf (Accessed 10/11/2019).
Zurbrügg C, Drescher S, Patel A, Sharatchandra HC. (2004). Decentralised composting of urban waste – an overview of community and private initiatives in Indian cities. *Waste Management.* 24: 655–662.

PART IV

Sharing economies and capacity building

20
MAKING SUSTAINABLE MARKETS AND THE FORMING OF A CIRCULAR ECONOMY

Katy Mason and Thomas Jalili Tanha

Markets and the work of concerning market actors in order to bring about change

For many years, the discipline of marketing has taken the concept of markets for granted – not questioning what they are or how they work and instead focusing on the tools needed by managers to 'enter' them (cf. Ansoff, 1965). Only recently, since the turn of the century, have market studies started to change that. Drawing heavily on sociological studies of markets (Fligstein and Dauter, 2007) and the French tradition of economic sociology (Callon, 1998, 1999; Callon et al., 2002; Callon and Muniesa, 2005), the market studies approach is developing deeper understandings of what markets are and how they work: who forms them, how, and with what materials. In other words, the proponents of market studies pay attention to the unfolding practices and materialities that constitute markets as complex sociotechnical and political systems of action (see Araujo, 2007; Kjellberg et al., 2015).

An increasingly prominent debate in market studies raises questions about the political nature of markets and specifically how markets are unsettled, made dynamic, and reformed to take account of social concerns. Geiger et al. (2014) refer to markets experiencing such turbulence as "concerned markets." For them, concerned markets are a specific market form where sociopolitical unease is invoked by multiple market actors to contest extant market practices, images, competences, and ideas (also see Cochoy, 2014; Fernandes et al., 2019; Mallard, 2016). The notion of concerned markets is useful in thinking about circular economies for two reasons. First, the work of concerning markets, politicising them, making people care about change, and mobilising different actors to bring about such change is the key to taking into account important elements that make markets sustainable. Stigzelius et al. (2018), for example, show how concerned housewives in wartime Sweden had to address 'resource scarcity' in order for both markets and consumption to be sustained (also see Phipps and Ozanne, 2017). Stigzelius et al. (2018) follow 'two matters of concern' in the Swedish housewives magazine Husmodern (1938–1958) – 'the scarcity of resources in food markets' and 'the scarcity of time in consumption' – and show how published narratives that explain how housewives should be acting in and out of the kitchen as they shop, cook, and clean lead to the changing practices of *thrift* and *convenience* in the kitchen, forming a *market-consumption junction* (Cowan, 1987). The mechanisms of *concerning* (Mallard, 2016), whereby housewives are shown in the magazine how to save food, make food go further, and use up leftovers, and *agencing* (Cochoy et al., 2016), whereby new market objects (meat bouillon and meat grinders, for example) are introduced to support the better use of leftovers, drive changes at the boundary between markets and consumption. The point here, as Stigzelius et al. (2018: 347) so clearly point out, is that as time goes by, markets and consumption change "New spaces and practices emerge as markets bridge, connect with, transform, and are transformed by consumer cultures." This puts the *concerning work* that actors do centre stage in the making of sustainable markets, whether the concerns are managers, entrepreneurs (Hopkinson and Aman, 2017), consumers (Harrison and Kjellberg, 2016), pressure groups (Rao, 2009), media (Hopkinson, 2017), and/or governments (Mason and Araujo, 2020).

Second, the work of *concerning markets* often reaches across discrete market categories. This claim positions the market intervention as a purposive activity that sets out to disrupt and destabilise entangled market practices and change them in some way that produces a 'better' or perhaps more sustainable market. That is to say that such disruptive practices necessarily cross multiple organisational and market boundaries. In their discussion on the multiple and overlapping version of markets, Kjellberg and Helgesson (2006: 849) illustrate this point:

> the combined activities of Procter & Gamble, its intermediaries and customers, including market surveillance and segmentation practices, goods distribution, advertising, shelf management, media consumption and coupon handling, create a multi-segment shampoo market. This, however, does not preclude that parallel practices may perform an undifferentiated mass-market version of this market. In short, different exchange, normalizing and representational practices may simultaneously perform different versions of the same market. Further, these versions need not be at odds with each other, but may more or less peacefully co-exist.

Through this simple illustration, we begin to see how a single intervention, in a single market category, may well have consequences for many other interdependent markets and market actors. In this Proctor & Gamble illustration, we begin to see beyond the market through the entangled complex of practices, to what might be called *the shampoo economy*.

In a circular economy, the initial market intervention is often concerned with the sustainability of a market. In exploring the transformation of the unsustainable coffee market in Uganda, Onyas and Ryan (2015) show what market actors do to map out how their market works and work out whom they need to engage and enrol in order to draw attention to the effects and problems with existing market configurations. This study foregrounds the important implication for market actors in accounting for concerns as a form of market-making work that brings about co-ordinated, collective action for change. The point here is that by making something into a concern for a group, the collective can be mobilised to address it (Chakrabarti and Mason, 2014). As Cochoy (2014) explains, the work of mobilising concerns amongst a collective is also the work of questioning what markets are and what they do. This very process of collective inquiry opens up efforts to imagine what those markets should be, what the different actors should do, and how they – the collective – could act differently to deliver such value(s) (Mason et al., 2017). This view foregrounds the need to understand how notions of sustainability are formed and by what individual and collective efforts. It is to these issues we now turn.

Making up sustainable markets as moral markets

In general, it is recognised that the achievement of sustainability goals requires changes to market systems (Bruntland, 1987; Capra, 1983; Shove and Walker, 2010). For Onyas and Ryan, sustainability means understanding the emergent market being designed to enable Ugandan coffee growers and exporters to advance ecological and social sustainability goals: developing fair-trade and organic certification programs and brands aimed at enhancing the livelihoods of small-scale farmers in subsistence marketplaces. These small-scale farmers are carers of the land but live at near-subsistence levels, despite being important actors in the coffee value chain, connecting southern producers to northern consumers. As the indigenous Ugandan coffee exporter Good African Coffee (GAC) promotes organic farming and offers superior prices for the higher-quality coffee that its farmers produce, it enables farmers to save for the future, and a new sustainable market begins to emerge. GAC emphasises the 'trade not aid' motto, promoting practices which empower farmers to become entrepreneurial and autonomous as well as a sustainability market model which makes environmental conservation and social responsibility begin to emerge. Here the sustainable market is judged as moral, as a better market, by the multiple actors that perform it: the farmers, exporters, and consumers. The devices that are made up and inserted into this world – fair-trade and organic certification – equip actors to make such judgements of morality. We argue that the role of calculation (working things out), valuation (knowing what is worth doing), and evaluation (knowing if things worked well) is central to how actors judge what is moral and worth doing (cf. Anteby, 2010; Callon and Muniesa, 2005; Lamont, 2012).

Doganova and Karnøe (2015: 22) illustrate the valuation process required to create a sustainable market for slurry acidification products, revealing how a new market for "clean technology" that reduces ammonia emissions from farms emerged as the result of an EU directive that calculates market actions differently. The EU Integrated Pollution and Prevention Control (IPPC) directive required polluting industries, including livestock farms, to implement the "best available techniques" in order to reduce their emissions. The driving morality behind the directive was to create markets for clean technologies that are both environmentally and economically valuable. In this sense, regulators set out to use market mechanisms to trigger a virtuous cycle of demand, supply, and technological innovation, which would allow emissions to be reduced at a reasonable cost thanks to modern equipment. Doganova and Karnøe (2015: 30) explain, as the Danish government and entrepreneurs put the directive into practice, a key device begins to play a significant role in what was judged as valuable and worth doing,

> In the Danish regulation, a peculiar device instrumented the definition of best available techniques: the Technology list. This device not only lists which technologies are approved for complying with the law but also describes these technologies in terms of the emissions reductions that they enable, and the costs incurred to farmers.

Doganova and Karnøe's (2015) analysis reveals how the Technology list intervenes in multiple market practices to create a market in which products are valued based on the environmental benefit of curbing emissions *and* the economic cost of investing in a new technology. But this is not a smooth process. As market actors attempt to get recognised by "the list," controversies are triggered in which political, scientific, and economic issues become increasingly entangled. Ultimately, the construction of the new, sustainable market hinges upon the composition of a complex network of actors with divergent, and sometimes conflicting, interests, in which the market devices (in this case, the technology list) calculate what forms of action should be taken by farmers; "the list" plays a pivotal role as a market-making device. Critical market devices that calculate value,

> appear not as something that individual market participants 'have,' but as material devices that are collectively constructed and negotiated. Overflows [elements that

> emerge that have not yet been taken into account] are not challenges that lie beyond the frame or obstacles that impede actors' ability to comply with it but concerns that are triggered by the very process of framing as controversies develop over the translations and boundaries drawn by the frame.
>
> (Doganova and Karnøe, 2015: 30)

This puts not only the market device but the *valuation practices* that construct it centre stage in the development of sustainable markets and reveals the experimental nature of market emergence (Callon, 2009). An issue raised by the 'clean technologies' case is the role that market devices play in accounting for moral actions.

While not paying particular attention to market devices, Anteby (2010) foregrounds the notion of moral markets to show that markets are more than categories of goods and geographies. Rather, they are bundles of entangled practices that take place in space and time to hold 'good trades' in place as a recursive, ongoing way of performing a market. Anteby's analysis provides rich descriptions of the judgements that market actors (medical schools, doctors, councils, mortuaries, and families), continuously make in the trading of cadavers for the progression of medical science. His argument is that the judgements of these professions are based on the multiplicity of value systems they have in place and as such provide the basis for moral markets to be conceptualised and continuously performed across organisational boundaries. We go further, arguing that these value systems (at least in part) are what become materialised in market devices such as Doganova and Karnøe's (2015) "Technology list." The sociology of valuation and evaluation (SVE) literature offers valuable insight here.

In the SVE literature, valuation and evaluation are understood to be practically inseparable (Lamont, 2012). Valuation is understood as the processes and practices that make things valuable, that is, the process of producing 'valuable' advice for entrepreneurs or product developers, and the valuation practices as the engaging, analysing, and presenting of advice as critical components of that process. Evaluation is understood as the means by which such productive process and practices are themselves valued and collectively understood as valuable. This approach recognises that multiple valuation practices and their conventions can be at play in a single setting (Boltanski and Thévenot, 2006; Stark, 2009). We argue that understanding which valuation practices are at play in a sustainable market and, further, in a circular economy setting (where the nexus of practices combines the valuation practices of regulators, government agencies, blue chip businesses, entrepreneurs, and users) is likely to be critical to developing useful evaluative frames that help actors work out what actions to take (cf. Espeland and Sauder, 2007). Exploring how the resultant valuations can be made commensurable across the different value systems of the distributed actors and across organisational boundaries will be equally critical. Commensuration, the transformation of different qualities into a common metric, allows the comparison of different entities – particularly a sustainable market setting, which is designed to bring multiple valuation norms together: from regulators, government agencies, blue chip businesses, entrepreneurs, and users. Hence, commensurability equips actors to make judgements about the world – about what is most valuable and what is worth engaging with and investing in and why.

> Whether it takes the form of rankings, ratios, or elusive prices, whether it is used to inform consumers and judge competitors, assuage a guilty conscience, or represent disparate forms of value, commensuration is crucial to how we categorise and make sense of the world.
>
> (Espeland and Stevens, 1998: 314)

Thus, we argue that how values are materialised into market devices and how devices are used in the performance of markets are likely to be important in co-ordinating collective action and holding the different actors to account for their role and performance of the moral market of which they hope to become part.

As Mason et al. (2017; Mason, 2019) point out, the very process of bringing new concerns into the market questions what value(s) are at stake, what market 'solutions' should/could be offered, and how these solutions might make life 'better' for some social group in need. In this way, the work of raising concerns seems to play a critical role in shaping what kinds of valuation practices are performed. Thus, valuation practices often focus simultaneously on what work should be done to make an emergent technology or innovation valuable, as well as on how the market should work to hold the new 'valued' market object in place and available for exchange (cf. Whitham et al., 2019). This dualism is sometimes referred to as marketisation work (Mason et al., 2017) and, we argue, provides us with a critical bridge from the making of sustainable markets to the making of circular economies.

From sustainable markets to circular economies

In the transformation of a market to a sustainable market, we have argued that concerns must be raised across a collective, which brings into question the value of current market objects and market practices. These practices, we have argued, form part of a larger *marketisation* process, which, broadly understood, is the work done to enable the conceptualisation, production, and exchange of goods (cf. Araujo and Pels, 2015). As we have seen, many studies have cited efforts to transform market structures, introduce market devices, alter market behaviour, and reconstitute market agents as the outcomes of coordinated efforts of actor networks (Doganova and Karnøe, 2015; Kjellberg et al., 2015; Onyas and Ryan, 2015). These studies are beginning to report on how actors work to create new valuation methods and practices through the introduction, presentation, and circulation of new forms of scientific, technical, and market knowledge (Çalışkan and Callon, 2010). Traditionally, the marketing literature has focused on enabling exchange between buyers and suppliers, explaining how managers align product characteristics with customer demands (Baker and Sinkula, 2002), and persuading unknowing potential market actors to value innovative offerings (Jaworski et al., 2000). But exchange is only a part of the marketisation process. A focus on exchange neglects both the conceptualisation and production work needed to constitute, innovate, and reconfigure market systems. The point here is that the broader system of provision configures connections *between* markets (Fine, 2002). Taking into account this broader system of provision, what might also be described as chains of markets, creates a critical insight for our concern with the circular economy and the generation and co-ordination of market-making activity across multiple markets.

In a bioscience incubator setting, where technologies advance and innovations emerge, firms collaborate to generate new, sustainable health markets to accommodate them (Aarikka-Stenroos and Sandberg, 2012). Mason et al. (2017) study *marketisation work* in this setting to reveal the importance of connections made across markets in the marketisation process. They conceptualise marketisation work as the strategic and deliberate practices performed by market actors to shape market futures (also see Cochoy and Dubuisson-Quellier, 2013). By studying the *practices* that constitute this work, they pay particular attention to the "routinized way in which bodies are moved, objects are handled, subjects are treated, things are described and the world is understood" (Reckwitz, 2002: 250) and the valuation practices that produce new conceptualisations of future markets and how they might work. Their findings show how networks, intentionally formed for a specific purpose (also see Adner and Kapoor, 2010; Möller and

Halinen, 1999; Möller et al., 2005), collectively imagine how future health markets might work *together* in a connected system of provision. Thus, managers imagine and start to put in place practices that begin to bridge multiple *communities of practice* (Lave and Wenger, 1991) – connecting specialist forms of technical and professional knowledge in new ways – to generate new and more holistic forms of knowing and acting. Three specific forms of marketisation work are identified: (1) the conceptualisation of actors' roles, (2) the conceptualisations of markets, and (3) the conceptualisation of goods. These findings show the strategic net as a well-positioned and well-equipped institutional form for engaging in marketisation and more specifically a particular form of marketisation work that Mason et al. (2017) label *conceptualisation work*.

Findings show that the routinised, recursive nature of conceptualisation practices becomes part of the market architectures that reproduce social bonds (Latour, 2005) and bridge different communities of practice (Möller and Svahn, 2006). In particular, it is through these specific forms of conceptualisation work that strategic networks act to continuously generate new market devices that can calculate, both *within* and *across* these communities of practice, organisational and market boundaries and the unfolding value of scientific discoveries in the broader system of provision for healthcare. While this system of provision is not conceptualised as circular by Mason et al. (2017), it is not difficult to imagine how such circularities might be generated through such practices. That is, if circularities – the capture of waste and the reuse and revaluing of waste into new socioeconomic objects of value – are what we value in markets and what we see as performing the morality of the market, then it is not a big stretch to see how understanding the construction of sustainable markets through the circularities created might be possible and valuable.

Thus, we put the marketisation work that crosses market boundaries at the core of what it means to create a circular economy. In our recent work, we have been studying the making of a sustainable market for energy storage: crossing the boundaries of scientific knowledge markets, insurance markets, car battery design markets, and beyond. We study and to some extent have engaged in the making of a sustainable market for electrical energy storage, in particular, through the reuse of decommissioned electric vehicle (EV) batteries. EV batteries typically end their life of powering an EV propulsion system when they reach ~30% degradation due to limited and unreliable mileage concerns of EV users and producers. In our study, we have sought to understand how these batteries can be given a 'second life' by transforming them into valuable stationary, industrial energy storage units. The market-making work to make 'second life' EV batteries valuable has connected multiple, unforeseen market concerns (e.g., air pollution, battery life assurance, and waste), actors (battery scientists, insurance underwriters, carbon emissions measurement specialists, and industry managers), and markets (for energy storage, grid services, EVs, extended warranty insurance, and carbon accounting). In the course of our study, it became clear that it was the marketisation work that crossed these multiple markets that are starting to form a circular economy.

The authors worked with these market actors to develop a material, carbon-neutral conceptualisation of a circular economy: a conceptual ideal that could be shared and would be collectively worked towards. The circular economy concept began to connect and was being shaped by the multiple actors identified here, with each having a sustainable market versions of their very own, as well as versions of what a circular economy for electric vehicle batteries might be and understandings of the actions that need to be engaged in to make particular stages in the circular biography of the batteries and their materials valuable to markets.

By following a second-life EV battery energy storage start-up from an "embryonic stage," we studied and gave feedback on their efforts to attach social and economic value to their products for potential customers, distributors, and investors alike. Our intervention began when we realised that potential investors and customers needed to be equipped to calculate that EV batteries had not become 'waste' after ~30% degradation of the battery. Rather, the work of converting degraded batteries into reliable, resilient storage systems was a technological and economic activity worth investing in: to lower carbon emissions and to retain or increase their productivity levels. We worked with the start-up and others to calculate the value of the storage system in terms of environmental, social, and economic value. The second-life EV battery storage unit had to become a feasible, viable, and desirable economic object.

As potential investors and customers were concerned and uncertain about the remaining lifetime of the batteries, we needed to show how battery life in storage could be ensured and understood. As industrial-scale energy storage is a high capital expenditure project, there had to be assurance that the lifetime of the second-life batteries would at least equal the time it would take for the storage to pay back the initial expenditure and investments in it. To make these calculations, the start-up made new connections with insurance markets (drawing on underwriters' expertise) to see if it would be possible to make provision for an extended warranty product for storage units, attached to expected payback periods for customers taking out financing arrangements to buy the storage products. In the process of understanding how this might be calculated, potential customers voiced a moral issue, and the carbon advantage of second-life batteries became a pressing concern. It was evident to customers that the reuse of a battery product was strongly associated with carbon savings. Carbon savings stem recycling activities. However, it was not clear what the quantitative extent of the carbon savings were or how they might be usefully calculated to show the unique value of this second-life battery product. As an increasing number of organisations (potential customers and investors) enrolled in sustainability schemes, they wanted to be provided with data and knowledge showing *how* "second-life" could help them to achieve their carbon emissions target measures. Such accounting required the start-up to make connections with carbon accounting market specialists, as carbon accounting requires specific forms of calculative "know how." The start-up's engagement with carbon calculators opened up new questions about practically conceptualising a circular economy to account for the full circle of supply chain production and the recycling activities to be avoided with their second-life solutions being created. Our study therefore reveals a circular economy in the making.

In problematising the different carbon emitters in the supply chain, it became clear to the market actors involved in developing and conceptualising the sustainable market for second-life EV batteries that what they were doing was necessary, though not sufficient, to create their ideal of a circular economy. The actors, including ourselves, developed a new understanding and agenda for making a circular economy for second-life batteries. Using 'circular economy' as a perspective of inquiry, we realised and have started developing a strategic network of action across multiple markets to attach value to enrolling global supply chain members into optimising their carbon footprint.

One final observation here is the entrepreneurial effort it takes from start-ups and established market actors (such as the EV car manufacturers and battery designers) to concern markets, question values, and challenge the moral obligations and practices of market actors. In mobilising these concerns and collective actions, we also find value in Raghu Garud's inspirational work that explains how individual actors can drive institutional change (Garud et al., 2007, 2011, 2014, 2018). We see this as an additional future strand of work to explore the creation of circular economies.

References

Aarikka-Stenroos L, Sandberg B. (2012). From new-product development to commercialization through networks. *Journal of Business Research*. 65: 198–206.

Adner R, Kapoor R. (2010). Value creation in innovation ecosystems: how the structure of technological interdependence affects firm performance in new technology generations. *Strategic Management Journal*. 31: 306–333.

Ansoff HI. (1965). *Corporate strategy*. London: Penguin Books.

Anteby M. (2010). Markets, morals, and practices of trade: jurisdictional disputes in the US commerce in cadavers. *Administrative Science Quarterly*. 55: 606–638.

Araujo L. (2007). Markets, market-making and marketing. *Marketing Theory*. 7: 221–226.

Araujo L, Pels J. (2015). Marketization and its limits. *Decision*. 42: 451–456.

Baker WE, Sinkula JM. (2002). Market orientation, learning orientation and product innovation: delving into the organization's black box. *Journal of Market – Focused Management*. 5: 5.

Boltanski L, Thévenot L. (2006). *On justification: economies of worth*. Princeton, NJ and Oxford: Princeton University Press.

Bruntland G. (1987). *The Bruntland report. Our common future. World commission on environment and development*. Oxford: Oxford University Press.

Çalışkan K, Callon M. (2010). Economization, part 2: a research programme for the study of markets. *Economy and Society*. 39: 1–32.

Callon M. (1998). *The laws of markets*. Oxford: Basil Blackwell.

Callon M. (1999). Actor-network theory – the market test. *The Sociological Review*. 47: 181–195.

Callon M. (2009). Civilizing markets: carbon trading between in vitro and in vivo experiments. *Accounting, Organizations and Society*. 34: 535–548.

Callon M, Meadel C, Rabeharisoa V. (2002). The economy of qualities. *Economy of Society*. 31: 194–217.

Callon M, Muniesa F. (2005). Peripheral vision: economic markets as calculative collective devices. *Organisation Studies*. 26: 1229–1250.

Capra F. (1983). *The turning point: science, society, and the rising culture*. London: Flamengo.

Chakrabarti R, Mason K. (2014). Designing better markets for people at the bottom of the pyramid: bottom-up market design. In: Geiger S, Harrison D, Kjellberg H, Mallard A. (Eds.), *Concerned markets: economic ordering for multiple values*. Cheltenham, MA: Edward Elgar Publishing Inc. 153–177.

Cochoy F. (2014). *A theory of 'agencing': on Michel Callon's contribution to organizational knowledge and practice*. Oxford: Oxford University Press.

Cochoy F, Dubuisson-Quellier S. (2013). The sociology of market work. *Economic Sociology – The European Electronic Newsletter*. 15: 4–11.

Cochoy F, Trompette P, Araujo L. (2016). From market agencements to market agencing: an introduction. *Consumption Markets & Culture*. 19: 1–14.

Cowan RS. (1987). *The consumption junction: a proposal for research strategies in the sociology of technology. The social construction of technological systems: new directions in the sociology and history of technology*. Cambridge, MA: MIT Press.

Doganova L, Karnøe P. (2015). Building markets for clean technologies: controversies, environmental concerns and economic worth. *Industrial Marketing Management*. 44: 22–31.

Espeland WN, Sauder ML. (2007). Rankings and reactivity: how public measures recreate social worlds. *American Journal of Sociology*. 113: 1–40.

Espeland WN, Stevens ML. (1998). Commensuration as a social process. *Annual Review of Sociology*. 24: 313–343.

Fernandes J, Mason K, Chakrabarti R. (2019). Managing to make market agencements: the temporally bound elements of stigma in favelas. *Journal of Business Research*. 95: 128–142.

Fine B. (2002). *The world of consumption: the material and cultural revisited*. Oxford: Routledge.

Fligstein N, Dauter L. (2007). The sociology of markets. *Annual Review of Sociololgy*. 33: 105–128.

Garud R, Gehman J, Giuliani AP. (2014). Contextualizing entrepreneurial innovation: a narrative perspective. *Research Policy*. 43: 1177–1188.

Garud R, Gehman J, Kumaraswamy A. (2011). Complexity arrangements for sustained innovation: lessons from 3M corporation. *Organization Studies*. 32: 737–767.

Garud R, Gehman J, Tharchen T. (2018). Performativity as ongoing journeys: implications for strategy, entrepreneurship, and innovation. *Long Range Planning*. 51: 500–509.

Garud R, Hardy C, Maguire S. (2007). *Institutional entrepreneurship as embedded agency: an introduction to the special issue*. London: Sage Publications Sage.

Geiger S, Harrison D, Kjellberg H, Mallard A. (2014). *Concerned markets: economic ordering for multiple values*. Northampton and Cheltenham: Edward Elgar Publishing.

Harrison D, Kjellberg H. (2016). How users shape markets. *Marketing Theory*. 16 (4): 445–468.

Hopkinson GC. (2017). Making a market for male dairy calves: alternative and mainstream relationality. *Journal of Marketing Management*. 33: 556–579.

Hopkinson GC, Aman A. (2017). Women entrepreneurs: how power operates in bottom of the pyramid-marketing discourse. *Marketing Theory*. 17: 305–321.

Jaworski B, Kohli AK, Sahay A. (2000). Market-driven versus driving markets. *Academy of Marketing Science Journal*. 28: 45.
Kjellberg H, Azimont F, Reid E. (2015). Market innovation processes: balancing stability and change. *Industrial Marketing Management*. 44: 4–12.
Kjellberg H, Helgesson CF. (2006). Multiple versions of markets: multiplicity and performativity in market practice. *Industrial Marketing Management*. 35: 839–855.
Lamont M. (2012). Toward a comparative sociology of valuation and evaluation. *Annual Review of Sociology*: 38.
Latour B. (2005). *Reassembling the social: an introduction to actor-network-theory*. Oxford: Oxford University Press.
Lave J, Wenger E. (1991). *Situated learning: legitimate peripheral participation*. Cambridge: Cambridge University Press.
Mallard A. (2016). Concerning urban consumption: on the construction of market agencements for retail trade. *Consumption Markets & Culture*. 19: 56–70.
Mason K. (2019). Markets under the microscope: making scientific discoveries valuable through choreographed contestations. *Journal of Management Studies*. 56: 966–999.
Mason K, Araujo L. (2020). Implementing marketization in public healthcare systems: performing reform in the English National Health Service. *British Journal of Management*. In Press.
Mason K, Friesl M, Ford C. (2017). Managing to make markets: marketization and the conceptualization work of strategic nets in the life science sector. *Industrial Marketing Management*. 67: 52–69.
Möller K, Halinen A. (1999). Business relationships and networks: managerial challenge of network era. *Industrial Marketing Management*. 28: 413–427.
Möller K, Rajala A, Svahn S. (2005). Strategic business nets – their type and management. *Journal of Business Research*. 58: 1274–1284.
Möller K, Svahn S. (2006). Role of knowledge in value creation in business nets. *Journal of Management Studies*. 43 (5): 985–1007.
Onyas WI, Ryan A. (2015). Agencing markets: actualizing ongoing market innovation. *Industrial Marketing Management*. 44: 13–21.
Phipps M, Ozanne JL. (2017). Routines disrupted: reestablishing security through practice alignment. *Journal of Consumer Research*. 44: 361–380.
Rao H. (2009). *Market rebels: how activists make or break radical innovation*. Princeton, NJ: Princeton University Press.
Reckwitz A. (2002). Toward a theory of social practices: a development in culturalist theorizing. *European Journal of Social Theory*. 5: 243–263.
Shove E, Walker G. (2010). Governing transitions in the sustainability of everyday life. *Research Policy*. 39: 471–476.
Stark D. (2009). *The sense of dissonance: accounts of worth in economic life*. Princeton, NJ: Princeton University Press.
Stigzelius I, Araujo L, Mason K, Murto R, Palo T. (2018). Kitchen concerns at the boundary between markets and consumption: agencing practice change in times of scarcity (Husmodern, Sweden 1938–1958). *Consumption Markets & Culture*. 21: 347–372.
Whitham R, Pérez D, Mason K, Ford C. (2019). Realising the value of open innovation in policy making: equipping entrepreneurs for valuation work. *The Design Journal*. 22: 189–201.

21

BECOMING ECO-LITERATE THROUGH EXPERIENTIAL ENCOUNTERS WITH FOOD

Gia Daprano

Introduction

Environmental education has been tasked with the imperative of equipping us with a working knowledge of the complexities of ecological systems so that humanity's social existence is commensurate with the integrity of the bio-physical world. Yet our individual and collective eco-literacy falls far short of what is necessary, so our behaviours and actions remain detrimental to and at odds with the integrity of the ecology of the world. A transformative learning experience that could facilitate greater ecological awareness is muted, situated as it is within a modernist, commodified, and capitalist paradigm of education. Learning as a transformative experience is promoted as the means to inculcate ecologically literate citizens, with an ethical perspective as a necessary component. Nowhere is this state of affairs more relevant to debates regarding the need for eco-literacy than in our most primary of relationships with the bio-physical world – that of our necessity to consume food. Food, and its production and consumption, is essentially the primary connection to the web of life, and it is through an ecologically literate agenda that we can rewrite the manual for an epistemological shift towards a shared commonality with all life. This shift allows for 'living as knowing'[1] in a way that facilitates sustainable behaviours around food driven by an ecologically literate sensibility that utilises other ways of knowing the world through our lived experience of it. This chapter addresses a range of questions that strike at the heart of this debate to include, first, how and why the environmental educational project is considered to have failed in teaching us about food and, second, what is eco-literacy and in what ways can it help address this failure? Ultimately, these considerations ask of us how it is we apprehend and experience what it is to be human in a more-than-human world. The suggestion from an eco-literate perspective is that this is achievable through facilitating a more embodied and experiential learning experience that embeds the sense of interconnectivity of all life. In this way, the scene is set for a subsequent discussion of research into community food-growing projects that offers practical examples of embodied and experiential learning as a means to engender a connection to life through food as showcasing ecological literacy in action.

Learning for living: a failure of environmental education?

> When the call goes out that you can have it all . . . your world should know no limits . . . the co-literate citizen will recognise this for what it is . . . the stuff of epitaphs.
>
> *(Orr, 1992: 93)*

The scenario encapsulated in the previous quote is part of the apocalyptic projection that environmental education has been tasked with resisting. However, environmental education is critiqued from various angles in the literature, accused of being dominated by a Westernised developed world praxis of modernity, techno-centricity, or bio-colonialism (Shiva and Moser, 1995; Maxey, 2007; Orr, 1994). Various assaults on its assumed integrity towards learning about the environment have also come from eco-feminism (Plumwood, 2002) and a deep ecological antithesis towards environmental education's emphasis on an anthropocentric conservational and resource management agenda (Naess and Jicking, 2000). These wide-ranging debates pivot around accusations from a political ecological focus on the detrimental effects of environmental education that operates within a neo-liberal commodification perspective (Maxey, 2007) and an overt managerial approach to education in general (Sterling, 2001; Orr, 1992). Institutionally sanctioned interventions from a top-down approach[2] are efforts to gear environmental education towards informing modernist-oriented societies of the need for societies to practice sustainable actions with regard to the environment and the bio-physical world. However, without an underpinning dimension of ethical responsibilities entailed in this, such a requirement remains unfulfilled. This ethical dimension is also one that goes beyond the overriding influence of a historical position of Judeo-Christian stewardship (McShane, 2009) towards the incorporation of a radical *relational* ontology as promoted by adoption of bio-centric world views (Sessions, 1995; Fox, 1990; Roszak, 2001).[3] The suggestion is that such a relational ontology needs to be delivered through an environmental education that honours the embodied, subjective, and affective experiences of being in the world.

The paradigm shift that is increasingly required from education concerning learning *about* the environment to learning *with* the environment is given short shrift when it has to adhere to a type of educational value system which not only replicates the resource-based attitudes that are part of the problem but also leaves unquestioned the schism between human and nature, mind and body, self and other that underpins the ontological and epistemological basis of our problematic relationship with the bio-physical world. It is a truncated experience that undermines a role for affective and embodied ways of encountering the environment and knowledge creation. An environmental education that is culpable for teaching out of a commodified and exclusively *cognitive* approach is diametrically opposed to the facilitation of an *embodied relationality* as an essential marker of ecologically literate citizens (Sterling, 2009; Gruen, 2009; Clarke, 2012; O'Sullivan et al., 2002).

Recent developments within environmental education have been heavily influenced by Orr's seminal narration of the failure of environmental education (Orr, 1992, 1994). The ecological crisis he identified over 30 years ago is still with us and is one that, as he stated then, not only reflects the previous issues within education but is also central to the crisis within it. His anthropological understanding of learning with and from the environment prior to formalised education points to learning in a manner that was and could still be integrated with embodied, everyday encounters that maintain a relational awareness between ourselves and the environment we so rely upon. Even more radically, though, for Orr, an eco-literate knowledge is only achievable if we as a society insist that ecological awareness should be at the very heart of *all* education, not merely a marginalised subject vying for attention amongst others. Orr's fundamental critique of environmental education as a marginalised subject within education, reflects for him the disconnect and fragmentation of the very social structures education lies within. The dominance of an industrialised neo-liberal approach of global food commodification creates for us an 'epistemological distancing' (Carolan, 2007). The environmental effects of how our food is produced become part of an invisible chain of provenance of which we have little understanding or experience, so that the associated detrimental social and environmental effects are disguised.

Both Sterling (2001) and before him Bowers (1996) outline that an environmental education that views nature from a resource conservational mentality alone results in the replication of discourses around nature that deprive it of its own intrinsic worth as well as its worth to more-than-human frames of reference. The spaces of capitalism and commodification of nature are so prevalent in forming our experiences of nature and the environment, the resultant paucity of landscape encounters with the more-than-human is commensurate with a paucity of mind-scapes unable to apprehend the lessons deriving from the natural world (Gruenewald, 2003a). The tendency of education to treat environmental issues as ones that rely on technological solutions and a scientific positivist attitude towards ecological knowledges discounts the relevance of social knowledges as part of the solution to ecological challenges. Social learning is a key concept to act out of for eco-literate citizens. Its learning lessons are those of acceptance of diversity and inclusivity and the valorisation of subjectivity as expressed through engagement with others and are transferable to more-than-human others (Frisk and Larson, 2011; Sterling, 2009).

The pedagogic landscape of eco-literacy: what does it mean to be 'eco-literate'?

Within this broader framework of environmental education, several authors have contributed theoretical underpinnings in order to define more specifically what an eco-literate sensibility could consist of and in what ways it can engender a holistic environmental learning experience. The term 'eco-literacy' has undergone a reconceptualisation and definition within the research field in recent years in order to give a deeper appreciation of how we as a species acquire the necessary behaviours to live within our ecosystem means. At the forefront of this movement are educational models that are open to more holistic learning whereby 'literacy' about the environment must engage 'heads, hands and heart' and not just the 'head' alone (Miller, 2008). The term 'eco-literacy' may also be experiencing a renaissance given that 'sustainability' has suffered from a greenwashing effect and overuse by capitalist agendas. How eco-literacy as a concept triumphs over that of sustainability is in its insistence in taking into account not only our own multifaceted ways of learning but also the fact that environmental problems cannot be neatly packaged into technological or scientific responses alone. Indeed, the persistence of a systems theory approach outlines the prevalence of so-called 'wicked'[4] problems, a term that alludes to complexity theories born out of a full appreciation of how ecosystems actually function (Sterling, 2001).

McBride (2011) charts the first use of the term 'eco-literacy' by Roth in 1986 as well as highlighting the major contribution by both Capra (1999) and Orr (1994) in developing different nuances around the concept of 'eco-literacy.' Both, however, are unified in drawing the link between becoming ecologically literate and the ability of societies to redesign themselves as sustainable endeavours. Both are also in agreement as to the cause of a lack of eco-literacy in an American educational setting in that it is reduced to a range of classroom-based skills rather than what we as species really require, which is 'the ability to observe nature with insight' (Orr, 1992: 86). Orr's criteria for eco-literacy include that citizens have a healthy affective connection to nature created through a sense of wonder as well as being able to comment on the soil, geology, weather patterns, and common flora and fauna of their locale. An important addition to this range of knowledge includes for Orr some understanding of where their food originates, how it is grown, and how waste is dealt with. The overriding theme for Orr is that having a comprehensive knowledge of the facts needs to be complemented by a feeling of kinship with all life, 'the sort that cannot be entirely put into words' (Orr, 1992: 87). This tension within environmental education as to what sorts of knowledge are needed to meet our need for developing eco-literately based behaviours persists. A synthesis of science and affective modes of apprehending the natural world is not seen to be the norm within environmental education, given that the overwhelming view is for an enhanced role for ecology and the natural sciences to communicate all we need for the development of sustainable societies (McBride, 2011). The tension between competing requirements in teaching eco-literacy causes problems for the application of the concept pedagogically. Macbride et al. (2013) contend that its vagueness when trying to accommodate an ethical, moral, and emotional framework as part of its remit renders it open to a lack of criticality, with accusations of eco-literacy as yet another form of educational environmentalism masquerading as Western bio-imperialism (Boehnert, 2012). In light of these debates as to what constitutes eco-literacy and how to teach for it, environmental education tends to stick rigidly to a natural science approach in the belief that this renders it immune from the complexities around ethical discussions that muddy the water to its detriment. Whilst pluralism has its advantages, Nicholls (2010) also believes that diversity of definitions is problematic for researchers' and educators' attempts to communicate eco-literacy with clarity and has increased the difficulties in education standardising its teaching methods. In response to these concerns and to enhance its utility to educators, Capra in 2006 took up the challenge of defining eco-literacy by promoting core principles designed to anchor the concept in behaviours that acted in accordance with an understanding of the systemic equilibrium as a marker of environmental health. This would include learning through Capra's insistence on other ways of knowing about the world that lead to an ethical dimension, as well as an emphasis on the interconnectivity between human health with more-than-human health. Capra's core principles put forward as essential components of teaching for eco-literacy include imparting knowledge of living systems through a systems thinking perception of the environment and transitioning towards sustainable practices through design inspired by nature in order to build in a form of citizenship based on a collaborative and community-building ethos.

Barnes's (2013) awareness to action continuum model aligns with Capra's suggestions in that it incorporates problem-solving strategies derived from a knowledge of ecosystems, whilst Barnes's other levels of learning in order to facilitate eco-literacy are focused on a subjective experience of responsibility towards the environment and an awareness of one's attitudes and values around it prompted by the development of emotional encounters with it. Many of the food projects discussed in the third section of this chapter adhere to this type of learning as the means to engender eco-literacy around food based on an emotional connection and a sense of an ethics of care for both humans and non-humans involved in food pathways. Nicholls (2010) undertakes a 'concept analysis' of the term 'eco-literacy' so as to provide standardised meanings and allow educators to move beyond a basic knowledge of scientific ecology and incorporate these ethical and emotive responses to the lifeworld. Nicholls favours four principles of eco-literacy as consistent throughout the range of literature, namely: (1) conceptual knowledge, (2) a sense of place through emotional identification, (3) morals and ethical dimensions as respect for the other and for diversity, and (4) acquisition of skills or competencies provided by a range of cognitive and practical skills. The frame of mind that Nicholls promotes for eco-literacy is one that shifts from an anthropocentric and cognitive framework alone towards a post-humanist education that incorporates experiential learning through a connection with the more-than-human world as exemplified in our food encounters.

This 'more-than-human' approach within a contemporary environmental educational remit for greater ecological awareness is also touched on by the work of Sterling (2001), who, in agreement with Frisk and Larson (2011), promotes the utilisation of a transformative educational model in order to achieve the previously stated aims for learners of eco-literate behaviours. Sterling outlines how, at present, environmental education is held within what he terms 'transmissive learning.' Within this level of learning, knowledge is replicated without an opportunity for a reflexive, subjective, and critical approach. These particular assets in learning are important for Sterling because they allow for the capacity to expose unchallenged assumptions and values around the environment and other beings. Within Sterling's pedagogic schemata, there are three levels of learning that gradually build upon each other so as to enable fully reflexive learners to be transformed through a relationship with the environment. This approach again necessitates the incorporation of an experiential learning so as to deconstruct the overly anthropocentric understanding of being in the world in favour of one based more securely in a relational experience with other beings, many of whom are not human. O'Sullivan et al. (2002) also point to the need for a pedagogy that facilitates a transformative learning experience in order to acquire knowledges derived from holistic educational practices that can account for learning through a 'head (cognitive), hands (practical) and heart (affective)' approach (see also Sipos et al., 2009). Other recommendations for experiential pedagogies also include the place-based work within environmental education of Gruenewald (2003a) and Selby (2002), with the insistence that an environmental education must teach for knowledges that are not separate from 'the land and its lessons' in order to deliver an effective level of eco-literacy (Gruenewald, 2003b: 33). The reasoning for this is based on an understanding that mastery of a subject alone is not enough; what is required is a meaningful connection between these three facets of our body-mind-spirit – a holistic learning for the whole person (see also Orr, 1994, 1992; Miller, 2008). This transformative identification with nature through our knowledge (critical and reflexive), our care for it (as relational), and our materially deepened connections is a true form of eco-literacy which demands nothing less (Singleton, 2015).

The pedagogic landscape of food: a role for experiential learning as ecological literacy?

> Eating is a social act, a political act, a cultural act, an economic act and an environmental act. Above all it is a pedagogical act.
>
> *(Sumner, 2008: 23)*

Of the various themes that are discussed concerning what eco-literacy is and how it can be taught for, two central ideas emerge which now form the basis of further exploration in the development of a food eco-literacy. First, there is the consideration of the negative effects of working within a neo-liberal cultural framework upon food encounters and how teaching for an ecological food literacy can overcome these, and second, there is the understanding of how an experiential embodied approach to eco-literacy can aid in developing an ethical relationship to food through a recognition of the interconnectivity between human and the more-than-human. Given that eating is closely tied to our ecological systems, it can become a powerful vehicle for facilitating eco-literacy through opportunities for transformative learning experiences. These sorts of learning experiences as essential components of becoming eco-literate are generated through encounters that engage us experientially through the sensing body but also through a politically engaged critical pedagogy.[5] This latter educational approach requires us to ask questions of systemic dynamics within society that maintain inequalities of food access. In this manner, a critical *and* an experiential pedagogy become part of an eco-literate gestalt around food, derived as they are from meanings related to one's own subjective experiences. We are all food consumers affected by the neo-liberal politics of the day, as well as embodied sensing beings within the lifeworld. Further development of the role of eco-literacy in facilitating a food ethic that is transformative rests on the assumption that it is not only 'the moment' of the experiential embodied encounter with non-humans as food but that it also allows for retaining the experience as a lifelong attitude to the environment through deep commitments to the environment based upon an ongoing identification with it (Hayden and Beck, 2012).

A critical political pedagogy as eco-literacy for food

Pollen (2008) states that ideally, food can be experienced as much more than just a biological necessity. It is about community, family, spirituality, pleasure, intergenerational connection, and a relational encounter with more-than-human aspects of our world. As a pedagogic tool, it can express our identity as an aspect of culture. It involves teaching and learning, sharing, inclusion and exclusion, celebrations, and alliances. However, this idealised role for learning for eco-literacy through food faces serious blocks given our current state of normalised industrialised patterns of behaviour around food. The problems for learning are legion when one considers the lack of transparency of food supply chains and environmental conditions associated with capitalist global food chains for the average consumer.[6] By becoming ecologically literate about our food, a gateway is opened that allows us to question and seek resolutions about what makes for a just society, being as it is inextricably linked to women's rights, human health, poverty, and an acknowledgment of the more-than-human that we share the planet with. Potential sites for learning an eco-literacy of food could include volunteering at the food bank, participating in local food chains, becoming part of the slow food movement, buying directly from farmers and markets, growing food, foraging for food, or joining a food co-op. Maxey (2007) and DeLind (2002) both highlight how community horticulture in the civic or urban realm can be instrumental in affording alternative economic systems through alternative food pathways (the now well-known phrase 'from field to fork' is a part of this discourse) that seek to re-address the disconnect to food from its global scale of operation. A critical pedagogy of food should also make much of the literature from a vast body of work around political ecology. Seminal debates would include development of Lefebre's 'right to the city' whereby urban horticulture and agriculture as represented by community, school and home gardens, allotments, and guerrilla gardening are all part of a political agency that not only transforms and re-signifies urban spaces but also that of citizen identification with urban nature inclusive of its more-than-human co-habitants (Purcell and Tyman, 2015; Certoma and Tornaghi, 2015; Tornaghi, 2010; Wekerle and Classens, 2015). Furthermore, the creation of an urban public food commons (whether this is the food forests of Seattle or the food co-ops of Detroit) is a form of political activism that lends itself neatly to fostering an eco-literacy around food that is politically aware. All these actions become part of an armoury of critical thinking based on actual lived, everyday experiences of food that seeks to build an alternative and sustainable future.

Sumner (2008) also supports this theme of the value of eco-literacy arising out of framing eating as part of emancipatory learning through a critical pedagogy. For Sumner, a way to approach the damaging effects of a lack of eco-literacy about our food as a consequence of being embedded within a neo-liberal political agenda means that a political economic perspective of eating needs to consider the relationship between politics and economy so as to raise questions of power and profit as detrimental to an eco-literate sensibility around food. The imbalance in power dynamics that favours growth and profit over provision of food as a universal right, as well as highlighting our dependency on export markets and economic growth over feeding people, builds upon a body of literature that critiques neo-liberalism's practices of flooding markets through free trade and undermining the livelihoods of local growers as part of the problem with our relationship to food. The growth of a local food movement within the United Kingdom, for example, exemplifies attempts to reconnect to our food as a means to question the effects of globalised food chains upon local places. Sumner promotes the idea of a civil commons that protects this vital connection to food and cites examples of a food civil commons as food eco-literacy in action to include community kitchens, allotments, farmers' markets, soup runs, local food systems, community gardens, and a burgeoning interest in edible landscapes across urban public spaces. Community as well as school gardens have tended to dominate the environmental educational literature because of their capacity to provide a valuable educational resource for greater eco-literacy and the complexity of achieving sustainability, as well as providing empirical evidence for a range of health and well-being effects (Corkery, 2004). Corkery makes a crucial distinction between using such community spaces as tools for eco-literacy in that they not only serve learning outcomes of education *about* sustainability (ecological and biological processes, for example) but also provide education *for* sustainability, with its equal emphasis on values and ethics derived from experiential encounters with growing food in community. Complimentary research into alternative spatial arrangements for food growing has been conducted by Clarke (2012), who, when encountering food-growing interventions in underused public space, found that the resultant eco-literacy renders communities 'naturally smart,' as it builds local resilience as well as spatial and intergenerational connections between schools and local communities. Smith (2007) also builds upon this and provides a detailed analysis of the intergenerational effects between school children and community elders as responsible for growing ecological consciousness as well as food.

The sensing body as a site for learning food eco-literacy

The examples now given focus on human–plant interactions as demonstrative of the development of a sensorial and tactile eco-literacy with regard to foodstuffs. The background to exploring an eco-literate food engagement as an embodied experience is partly derived from research theories that acknowledge the vitalism and agency of plants and the material encounters between humans and the more-than-human. Such work challenges representational theories within the social sciences and has given rise to a post-humanist agenda for research enquiry that regards material bodies, some of which are not human, as a site for knowledge generation (Bennett, 2010; Law, 2004; Lorimer, 2007).[7] Ways of investigating the 'everyday routines, fleeting encounters, embodied movements, precognitive triggers, practical skills, affective intensities . . . and sensuous dispositions' of the lifeworld (Lorimer, 2005: 8) form the backdrop to a wealth of research into food encounters as a pedagogic tool for a deeper eco-awareness. In acknowledging the precognitive sensorial aspects of everyday experiences around food, all that is vital and relational in the material encounters within it can give rise to a holistic eco-literacy that engages us in all our totality. Moreover, as described in the first section of this chapter, environmental education encourages a dualistic ontology of the world as 'out there,' one that can only ever be apprehended through a predominantly human cognitive explanation; the need for a return to 'things and materials' and our visceral experiences as an essential component of learning can rise to the challenges of teaching eco-literacy through food experiences. This turn earthwards is more than prosaic concern with matter and becomes part of the 'doing' of things that dissolves the illusion of the separateness between living beings – of which our food exclusively consists. The animism that underlies relational food encounters between human and more-than-human decentres us as the cause of such relational encounters, as well as expanding the membership of the social towards inclusivity of the more-than-human (Hitchings, 2003). From this, an ethic of care, responsibility, and wonder emanates and becomes part of a deeper eco-literacy than that on offer from purely a natural sciences input. This predominantly entangled material, embodied, sensory, and visceral engagement between humans as growers and plants as foodstuffs exemplifies the experiences within community, home, and school gardens. Opportunities present themselves that suggest, in such situations, we are actively learning with and from plants regarding our place in the world as interconnected beings and as such can claim a deeper eco-literate sensibility. Lorimer (2005), for example, highlights how the repetitive practice of skills needed in community gardens highlights how we are not disembodied minds apprehending the world through symbolic representations. Our bodily sensing and the affects born out of practical skills used in growing food reinforce the connections between our material bodies and all material bodies, inclusive of plants, animals, and the whole earth.[8] The 'doing' of growing food gives rise to the 'meaning' of growing food in this scenario. This scenario is referred to in the literature as being one replete with the entanglements of human as grower and plant as foodstuff, also incorporating the wider ecology and even social aspects of such a network. For example, Hawkes and Acott (2013) follow the role of plants in generating ecological identities on allotment sites, firmly establishing the social, ecological, and material bricolages that act as transformative elements towards a relational experience of food growing with plants and with other growers. Similarly, Taylor and Lovell's (2014) work makes the most of the porous boundaries experienced in food growing in home gardens as a 'performative' encounter. When viewed as 'actors' embroiled in this encounter, the space is given to learning that requires a reconnection of human with non-human as derived from the shared physical interactions between human and plant as food.[9] Power (2005) also speaks of the de-centring of the human mind and draws attention to the site of food growing as a relational achievement based on a continuous negotiated, contested, and collaborative endeavour in relationship with plants as food by demonstrating that the relationship consists of 'involving processes of collaboration, negotiation, challenge and competition' (39) with vital and agential plants. Nature's active presence in the food garden deconstructs the passive assumptions of its materiality that require human agency to fulfil it and improve it. This approach allows for the body and materiality to become sites of knowledge creation and activism towards an eco-literacy that attends to builds upon the visceral qualities of food spaces (Hayes-Conroy and Hayes-Conroy, 2010).

Learning food eco-literacy through sensorial engagement with plants is of social significance, as it also begins to ask questions of the degrees to which different culturally signified bodies can participate in developing eco-literacy (Turner, 2011). The richly descriptive effects of somatic and emotional encounters with plants as food upon such sites are often overlooked as sources for learning eco-literacy, but certain authors promote learning for eco-literacy that accounts for the everyday, tapping into the 'fundamental sensibility of the human condition' (Anderson and Smith, 2001: 9) as the only means to implant an eco-literacy that has real tangible outcomes in terms of more sustainable food behaviours. In terms of the holistic educative model promoted as the means to acquire eco-literacy, a hands-on experiential learning must incorporate the emotional impacts of growing food (whether plant or animal), because the emotions act as the bridge between our sensing bodies and thinking minds (Judson, 2010; Anderson and Smith, 2001; Wright and Hodge, 2012). As Littledyke (2008) maintains, teaching for a 'felt knowledge' – one that integrates both cognitive understanding *and* affective experiences – facilitates the sense of relationality as the crucial element in delivering environmental care and responsibility in food encounters of an ecologically literate kind.

Conclusions

By highlighting how environmental education discounts our own sensing and emotional connection to food, because of its entrenchment within a modernist framework of industrialised food pathways, eco-literacy is promoted as an educational tool for righting the disassociation with foodstuffs. In order to engender an environmental ethic towards the more-than-humans that comprise our food, with specific reference to work that examines human–plant interactions, the sorts of eco-literate knowledges that are required are presented as two interweaving themes derived from an overview of how various authors define the concept of eco-literacy, first as part of a critical and therefore political pedagogy and second as part of a holistic teaching framework that engages our thinking heads, emotionally responsive hearts, and tactile hands. These ways of learning for eco-literacy have the potential for transformative effects regarding our relationship to food, given that they both involve our own experiential, everyday encounters with food. In this way, we reconnect with not only our food but also a broader understanding of a relational social world that incorporates both humanity and the more-than-human with which we share the world.

Notes

1 Akin to the ancient Greek concept of 'Paidea.'
2 For example, see the Belgrade Charter. Available at: http://portal.unesco.org/education/en/file_download.php/47f146a29d047189d9b3ea7651a2b98The+Belgrade+Charter.pdf
3 However, critiques of deep ecology do exist within environmentalism. For example, see Bookchin M. (1987). *Social ecology versus deep ecology: a challenge for the environmental movement.* Available at: http://theanarchistlibrary.org/library/murray-bookchin-social-ecology-versus-deep-ecology-a-challenge-for-the-ecology-movement
4 'Wicked' problems within ecology management refers to the inherent complexities of ecosystems and the tendency for simple solutions to generate more problems for the system because of a lack of understanding regarding spatial and temporal relationships within a system.
5 The work of Paulo Freire is seminal in this respect; for example, see Freire P. (2005). *Education for critical consciousness.* Bloomsbury Academy
6 Ian Cook's 'follow the things' project uses creative methods to explore consumerism's blind spots. Available at: www.followthethings.com/
7 The work of such authors is referred to as 'New Materialisms.'
8 The eco-phenomenological work of authors such as Brown and Toadvine (2003) develops this line of thinking; for example, see Brown CS, Toadvine T. (2003). *Eco phenomenology: back to the earth itself.* Albany: State University of New York.
9 Hitching (2003) utilises actor network theory to view the relationship between human and food plants, concluding that this research lens highlights how an environmental ethic arises from a distribution of agency through the food-growing site in question.

References

Anderson K, Smith S. (2001). Emotional geographies. *Transaction of the Institute of British Geographers.* 26: 7–10. Wiley.

Barnes JC. (2013). Awareness to action: the journey toward deeper ecological literacy. *Journal of Sustainability Education.* Available at: http://www.susted.com/wordpress/content/awareness-to-action-the-journey-toward-a-deeper-ecological-literacy_2013_05/

Bennett J. (2010). *Vibrant matter: a political ecology of things.* Durham, NC: Duke University Press.

Boehnert J. (2012). The visual communication of ecological literacy: designing, learning and emergent ecological perception. PhD thesis. University of Brighton, Brighton.

Bowers C. (1996). The cultural dimensions of ecological literacy. *Journal of Environmental Education.* 27 (2): 5–10.

Brown CS, Toadvine T. (2003). *Eco phenomenology: back to the earth itself.* Albany, NY: State University of New York.

Capra F. (1999). *Eco-literacy: the challenge for education in the next century.* Liverpool Schumacher Lectures, Centre for Eco-literacy. Available at: www.ecoliteracy.org.

Capra F. (2006). *Sustainable living, ecological literacy, and the breath of life.* Available at: https://cjee.lakeheadu.ca/article/download/624/507 (Accessed 2/11/2016).

Carolan MS. (2007). Introducing the concept of tactile space: creating lasting social and environmental commitments. *GeoForum*. 39: 1264–1275.
Certoma C, Tornaghi C. (2015). Political gardening: transforming cities and political agency. (Editorial). *Local Environment*. 20 (10): 1123–1131.
Clarke P. (2012). *Education for sustainability: becoming naturally smart*. London: Routledge.
Corkery L. (2004). Community gardens as a platform for education for sustainability. *Australian Journal of Environmental Education*. 20 (1): 69–75.
DeLind LB. (2002). Place work and civic agriculture: common fields for cultivation. *Agricultural and Human Values*. 19: 217–224.
Fox W. (1990). *Towards a transpersonal ecology: developing new foundations for environmentalism*. Dartington: Resurgence Books.
Frisk E, Larson KL. (2011). Educating for sustainability: competencies and practices for transformative action. *Journal of Sustainability Education*. 2: 1–20.
Gruen L. (2009). Attending to nature: empathetic engagement with the more than human world. *Ethics and the Environment*. 14 (2): 23–38.
Gruenewald DA. (2003a). The best of both worlds: a critical pedagogy of place. *Educational Researcher*. 32 (4): 3–12.
Gruenewald DA. (2003b). At home with the other: reclaiming the ecological roots of development and literacy. *The Journal of Environmental Education*. 35 (13): 33–43.
Hawkes FM, Acott TG. (2013). People, environment and place: the function and significance of human hybrid relationships at an allotment in South East England. *Local Environment*. 18 (10): 1117–1133.
Hayden J, Beck D. (2012). Doing CSA: tactile space, affect and effects of membership. *Geoforum*. 43: 332–341.
Hayes-Conroy JH, Hayes-Conroy AH. (2010). Visceral geographies, mattering, relating, defying. *Geography Compass*. 4 (9): 1273–1283.
Hitchings R. (2003). People, plants and performance: on actor network theory and the material pleasures of the garden. *Social and Cultural Geography*. 4 (1): 99–114.
Judson G. (2010). *Imagination in mind: educating for ecological literacy*. The Centre for Strategic Education Seminar Series. Available at: www.academia.edu/30354333/Imagination_in_mind_Educating_for_ecological_literacy.
Law J. (2004). *After method: mess in social science research*. London: Routledge.
Littledyke M. (2008). Science education for environmental awareness: approaches to integrating cognitive and affective domains. *Environmental Education Research*. 14 (1): 1–17.
Lorimer H. (2005). Cultural geography: the busyness of being 'more-than-representational'. *Progress in Human Geography*. 2: 83–94.
Lorimer H. (2007). Nonhuman charisma. *Environment and Planning: Society and Space*. 25 (5): 911–932.
Maxey L. (2007). From 'alternative' to 'sustainable' food geographies. In: Maye D, Holloway L, Kneafsey M. (Eds.), *Alternative food geographies: representations and practices*. Oxford: Elsevier Science.
McBride BB. (2011). *Essential elements of ecological literacy and pathways to achieve it: perspectives of ecologists*. PhD thesis. University of Montana, Missoula.
McBride BB, Brewer CA, Berkowitz AR, Borrie WT. (2013). Environmental literacy, ecological literacy, eco-literacy: what do we mean and how did we get here? *Ecosphere*. 4 (5): 1–20.
McShane K. (2009). Environmental ethics: an overview. *Philosophy Compass*. 4 (3): 407–420.
Miller JP. (2008). Holistic education and environmental education. In: Gray D, Selby D. (Eds.), *Green frontiers: environmental educators dancing away from mechanism*. Rotterdam: Sense Publishers. 82–94.
Naess A, Jicking B. (2000). Deep ecology and education: a conversation with Arne Naess. *Canadian Journal of Environmental Education*. 5 (1): 48–62.
Nicholls BH. (2010). *Essential eco literacy or 'earth smarts': defining and validation a pragmatic educational construct based on quality of life*. Available at: www.jsedimensions.org/wordpress/content/essential-ecoliteracy-or-%e2%80%9cearth-smarts%e2%80%9d-defining-and-validating-a-pragmatic-educational-construct-based-on-quality-of-life_2010_05/.
Orr DW. (1992). *Ecological literacy: education and the transition to a post-modern world*. Albany: State University of New York Press.
Orr DW. (1994). *Earth in mind: on education, environment and the human prospect*. Washington, DC: Island Press.
O'Sullivan E, Morell A, O'Connor MA. (2002). *Expanding the boundaries of transformative learning: essays on theory and praxis*. New York: Palgrave Macmillan.
Plumwood V. (2002). *Environmental culture: the ecological crisis of reason*. London: Routledge.

Pollen M. (2008). *In defense of food: an eater's manifesto*. New York Penguin Press.

Power ER. (2005). Human-nature relations in suburban gardens. *Australian Geographer*. 36: 39–53.

Purcell M, Tyman SK. (2015). Cultivating food as a right to the city. *Local Environment: The International Journal of Justice and Sustainability*. 20 (10): 132–147.

Roszak T. (2001). *Voice of the earth*. Grand Rapids, MI: Phanes Publishers.

Selby D. (2002). The signature of the whole: radical interconnectedness and its implications for global and environmental education. In: O'Sullivan E, Morrell A, O'Connor MA. (Eds.), *Expanding the boundaries of transformative learning*. New York: Palgrave Macmillan.

Sessions G. (1995). *Deep ecology for the twenty first century*. Boston, MA: Shambala Publishers.

Shiva V, Moser I. (1995). *Biopolitics: a feminist and ecological reader on bio-technology*. London: Zed Books. Chapter 12.

Sipos Y, Batistti B, Grimm K. (2009). Achieving transformative sustainability learning: engaging head hands and heart. *International Journal of Sustainability in Higher Education*. 8 (1): 68–86.

Smith M. (2007). On 'being' moved by nature: geography, emotion and environmental ethics. In: Davidson J, Smith M, Bondi L. (Eds.), *Emotional geographies*. Hampshire: Ashgate Publications.

Sterling S. (2001). *Sustainable education: re-visioning learning and change*. Schumacher Briefings No. 6. Totnes: Green Books.

Sterling S. (2009). Ecological intelligence: viewing the world relationally. In: Stibbe A. (Ed.), *Handbook of sustainable literacy*. Totnes: Green Books.

Sumner J. (2008). Eating as a pedagogical act: food as a catalyst for adult education for sustainability. *Kursiv – Journal für politische Bildung*. 4: 23–37.

Taylor JR, Lovell ST. (2014). Urban home gardens in the Global North: research traditions and future directions. *Agriculture and Human Values*. 31 (2): 285–305.

Tornaghi C. (2010). Public space, urban agriculture and the grassroots creation of new commons: lessons and challenges for politic makers. In: Viljoen A, Wiskerke H. (Eds.), *Sustainable food planning: evolving theory and practice*. Wageningen: Academic Publishing. 349–363. Chapter 29.

Turner B. (2011). Embodied connection: sustainability, food systems and community gardens. *Local Environment*. 16 (6): 509–522.

Wekerle GR, Classens M. (2015). Food production in the city: (re)negotiating land, food and property. *Local Environment*. 20 (10): 1175–1193.

Wright S, Hodge P. (2012). To be transformed: emotions in cross-cultural, field-based learning in Northern Australia. *Journal of Geography in Higher Education*. 36 (3): 355–368.

22

IMPLEMENTING LOW-CARBON STRATEGIES – ANALYSIS OF BARRIERS

Johannes Fresner, Fabio Morea, Christina Krenn, Anton Kleshkov, and Fabio Tomasi

Spurring investment in low-carbon technology

Barriers for investment in energy efficiency

The report of the International Energy Agency, "World Energy Investment" (2017), found that after years of growth, combined global investment in renewables and energy efficiency declined by 3% in 2017. Investment growth has again weakened in 2018 (CNBC, 2018). This is observed in spite of various authors demonstrating that there is a serious potential for energy efficiency in many industries (Panayiotou et al., 2017) and significant potential for renewable energy (International Energy Agency, 2014).

Over two-thirds of efficiency potential is still untouched (OECD, 2017). In 2017, the world spent about EUR 200 billion on energy efficiency. This was invested largely in heating, cooling, and lighting improvements in buildings. Tapping the full potential of energy efficiency will require at least four times as much investment than is spent at present. The International Energy Agency warns these investments are actually slowing down, partly due to the slow implementation of energy efficiency policies.

The most relevant economic and financial barriers for investing in industrial energy efficiency are, according to the Department of Energy (2015):

- Long payback time
- Volatility of energy prices
- Competition for capital
- Failure to recognise non-monetary benefits of energy efficiency.

An additional reason is the generally low relevance of energy cost in production companies.

• *Payback time*

The payback period of an investment is defined as the period of time in which the initial capital expenditure is recovered. According to Walsh, several authors agree that decisions on energy efficiency or renewable energy investment are taken mostly evaluating payback of investment

(Walsh and Thornly, 2012). Calculation of simple payback is one of the most common methods to evaluate a capital investment, especially among engineers. Payback frequently is used as a first assessment of whether detailed planning is worthwhile (Lefley, 1996).

Methodologically, the payback period does not include risk assessment and ignores the time value of money (Mayes, 2003). As the payback becomes longer, however, the more likely it is that unexpected changes might render the project not profitable (Gallo, 2016). The maximum acceptable payback is usually determined by a management decision and not based on a specific economic or scientific foundation. Production companies frequently require short payback periods of one to three years (Solnordal and Foss, 2018; Anderson and Newell, 2004) for investment. Larger companies, or owner-managed companies, sometimes accept three-year paybacks. Required payback times also depend on the type of equipment. In any case, payback has to be shorter than the remaining useful life of the equipment.

- *Volatility of energy prices*

Volatile energy prices will make decisions to invest in energy efficiency more risky because savings out of the investment become less certain (Monitoringstelle, 2017). In Austria, energy prices show volatility: According to Eurostat, natural gas prices for medium-sized industry, not considering households or big industrial consumers, peaked in 2013 and declined since then to about 80% of the peak by 2017. The producer price index for investment goods in Austria has increased by 10% in the same period (Eurostat, 2018). Cost of labour has increased. As a result, payback times for investment in energy efficiency in general have become longer.

- *Competition for capital*

Industrial firms do not make investment decisions related to energy separately from other aspects of operations. Measures which result in loss of production are usually not accepted. They are considered too much of a risk to the general business. Energy savings are perceived as side effect of other investments rather than as an independent value-generating activity (Department of Energy, 2015).

In Austria, industrial investments in general have been sinking over the last ten years (Austrian Chamber of Labor, 2017). From this, it can be concluded that in times of general lower investment, investment into activities, which apparently are not directly linked to productivity, is also lower.

- *Failure to recognise non-monetary benefits*

Non-financial barriers include limited in-house skills (Trianni et al., 2013a, 2013b) and expertise to identify (Fleiter et al., 2011) and implement energy saving projects (Trianni et al., 2013a, 2013b), as well as difficulty in gathering external skills and lack of time (Cagno et al., 2013). Information issues (such as lack of information on cost and benefits, unclear information by technology providers, difficulties in assessing the risks associated with the intervention, no trust in the information sources) emerged as an important barrier in the findings of Rohdin et al. (2007) and Trianni et al. (2013a).

Industrial energy efficiency measures may deliver substantial benefits in addition to energy cost savings – enhancing competitiveness, profitability, production and product quality; reduced maintenance; improving the working environment; and environmental compliance. The multiple-benefits approach tries to identify and appreciate these.

A survey of 200 companies which had implemented energy efficiency measures showed, however, the following results (Rasmussen, 2011):

- The main drivers for the implementation of energy efficiency are reduction of cost and increase of productivity, while corporate reputation and government regulations remain low on the list of influencing factors.
- Cost control is over five times more likely to be a driver of energy efficiency initiatives than compliance with government regulations, according to the organisations surveyed.
- Seventy-six percent of respondents recognised that other benefits in addition to energy savings are achieved through energy efficiency projects.
- Low relevance of energy cost.

The relevance of energy cost for industry is generally low, as data for Germany reveal (Table 22.1):

Table 22.1 Average cost contribution in German processing industry (VDI, 2018)

Cost contribution	*Percent (%)*
Energy	2
Commodities	13
Labour	19
Material	42
Other costs (depreciation, capital cost, taxes, profit)	24

There is, of course, a wide variation in energy cost depending on the production sector. In minerals, metal smelting, and wet finishing in the textile industry, energy cost is more likely to be in the range of 10% or more of overall expenses (Energy Efficiency Consultants Network, 2019).

• *Lack of strategic relevance*

De Canio and Watkins (1998) asked the question whether energy efficiency investments depend on company characteristics rather than profitability of the investment and answered it affirmatively. Starting from this observation, Coormans (2012) developed an investment decision-making model showing four levels of barriers (Figure 22.1) and tested it empirically with 28 companies.

As shown in Figure 22.1, there are four levels of organisational barriers to energy efficiency investments. Coormans (2012) labelled the four barrier levels "Base," "Symptom," "Real," and "Hidden." The level "Real" barriers stands mainly for lack of strategic fit of a project. "Base" barriers concern the lack of knowledge regarding energy efficiency measures. Small and medium-sized enterprises (SMEs) in general might not be aware of their energy-related expenses and are not using state-of-the-art sensors and meters to monitor and control energy consumption. "Symptom" barriers actually express deeper problems: capital might be allocated to other investments; risk is said to be high. "Real" barriers are the real obstacle to energy efficiency investments, the low or nonexistent strategic character of the investment for companies "which consider energy or energy use neither as a contributor to their competitive advantage nor as a critical resource" (Coormans, 2012). It appears from the empirical research that strategy is more influential than profitability. Profitability apparently is a generally necessary but insufficient condition. Coormans (2012) suggested as a solution to the energy efficiency gap to strongly communicate the strategic relevance of energy efficiency investments.

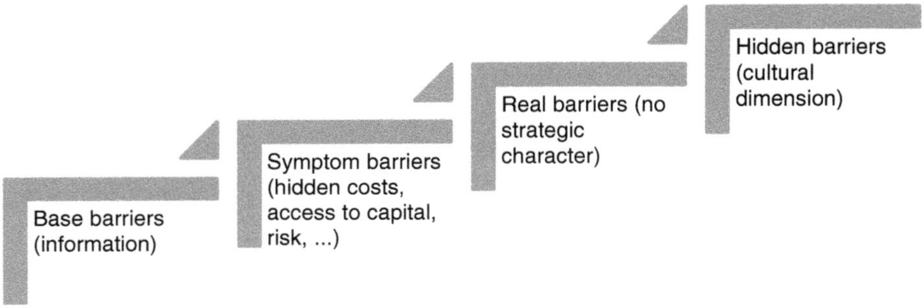

Figure 22.1 Barriers to energy efficiency investment

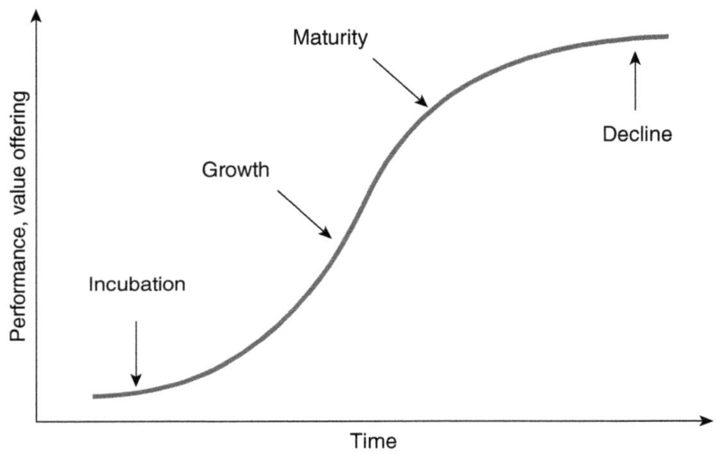

Figure 22.2 S-curve describing the development of performance and value over time

Basic enterprise strategies and relation to energy efficiency investment

The strategic focus of management of an enterprise will vary depending on the maturity level of the company. To explain this, the authors used the S-curve model to represent the development of enterprises. Performance of processes and infrastructure can be described by S-curves, which are logistic patterns consisting of four stages: incubation, growth, maturity, and decline (Figure 22.2) (Foster, 1986; Abernathy and Wayne, 1974). The four stages and their key characteristics are described in the following paragraphs (Röglinger et al., 2012; Miller and Friesen, 1985; Smith et al., 1985) and summarised in Table 22.2.

- *Incubation*

Most production processes start on a small scale. They are subject to frequent modifications and changes in this first phase of the life cycle. Quality is critical, because defective products can damage the reputation of the product or even the entire firm. In the incubation stage, profits are low because research and development, production, and marketing costs are high.

Table 22.2 Technical features of the life-cycle stages

	Life-cycle stage			
	Incubation	*Growth*	*Maturity*	*Decline*
Recommended strategy	"Make it work": Create product and provide it to the market, deliver products	"Grow and finance growth": Maximise market share	"Minimise cost, consolidate": Maximise profits while defending market share	"Prepare for next generation": Reduce expenditure in general, harvest
Product	Basic product	New features, services, extensions	Diversification of brand and models, reduce production cost	Phase out weak items
Characteristics of management	Owner-controlled, simple organisation, few employees, few formal systems	Functional managers, start of controlling system, delegation	Experienced management, stability in technology and market shares	Strong leadership, output related
Characteristics of cash flow	Focus on generation of cash to break even and replace wear off of capital assets	Probably high debt/equity rate	Stagnant margins, lowering revenue	Dropping margins, few investment opportunities
Corresponding barrier for energy efficiency investment	High priority of production increase, lack of people responsible and knowledgeable for energy utilities, resistance of management to change	Competition for capital, unfavourable to commit resources, banks perceive energy efficiency and renewable energy as risky	None specific identified	Technological change, conflicting priorities to convert mature industry and prepare new products and processes

Source: Modified from Burgelman et al. (2008)

The primary interest is to increase production while complying with safety, labour, and environmental regulations. Investments that are directly related to these aspects will be given the highest priority. Investments related to reduction of expenses are not considered as important at this stage. From an organisational point of view, the enterprise will be controlled by the owner with few formal management systems.

- *Growth*

During the growth phase, production increases with demand. While profits are high, it is important to stay ahead of competitors. During growth, the challenge is to increase production while maintaining quality and controlling cost. Ideally, capacity is added, gradually anticipating demand. In practice, this is difficult, as technical facilities can be up-scaled only in discrete steps (Quinn and Cameron, 1983). The organisation will be controlled by functional managers. A controlling system will be established.

- *Maturity*

During maturity, sales still increase. As a result of increasing competition, prices will lower. The product or service will be adapted to differentiate from competitors. At the same time, cost must be reduced to maintain profits. As earnings improve, mature companies will generate more cash from their products than they need for reinvestment. The question of how the company should return cash to stockholders becomes relevant at almost every mature company. Management is experienced and stable. Not all mature companies are large companies. Many small companies reach their growth ceiling quickly and essentially stay small, mature firms (Wendler, 2012; Howard and Hine, 1997).

- *Decline*

In the fourth stage of the S-curve, prices fall and sales drop. An appropriate strategy can be to reintroduce the product with a new feature. A refreshed marketing strategy is important to reach current and prospective customers. Consequently, processes will be scaled down and unprofitable products stopped. Investment will be directed towards new products and processes. Companies in this stage have few investment opportunities that generate value. These enterprises will have trouble in refinancing debt, since lenders will demand more stringent terms. The organisation will become mainly production output related. Strong leadership is needed to identify a strategy to sustain (Daft, 1998).

Experiences in implementing low-carbon strategies

Results of energy audits in 280 enterprises

The company data from the PINE project (Fresner et al., 2017) were analysed as a base of this study. The sample of 280 SMEs considered in the study represents a wide range of sectors and company sizes. The companies came from a variety of sectors, including mining, minerals processing, metal processing, engineering, wood processing, food processing, packaging, plastic processing, tyre manufacturing, and construction. In total, these energy audits resulted in almost 500 improvement suggestions. The calculated technical savings are about 60,000 MWh. The energy efficiency measures involve over EUR 10.7 million investments in sustainable energy within three years, to achieve 6,500 tons of oil equivalent (t_{oe}) primary energy savings and reduce GHG emissions by 13,500 tons of carbon dioxide (tCO_2). The project leads to annual savings of more than EUR 3 million.

Results related to life-cycle phase of the companies

For this analysis, the measures implemented in the 20 companies that participated in the full audits in Austria were analysed. Some 87 options were originally identified, out of which 77 were implemented. Ten of the options were dropped. Data were collected for these measures describing the nature of the measure, investment, savings, payback, and position of the process and the infrastructure (buildings and basic utilities like boilers) on the S-curve.

The participating companies were classified according to the four stages as they are used in literature describing the maturity of enterprises. The assessment of life-cycle stage of products, processes, and infrastructure was done through discussion with company representatives using the features presented in Table 22.2, without a quantitative evaluation of company data and market data.

A total of 46 measures were implemented in companies in which processes or buildings were rated as "mature." Some 18 measures were identified in companies which were rated as "growing." Some 13 were identified in companies which were rated as "declining" and none in a company which was classified as "incubating" (Table 22.3). Implementation was not always directly linked to short payback times. The share of measures with short payback (less than two years) was largest during the "growth" phase (69%), followed by measures implemented in the "mature" phase (63%), and lowest in the "decline" phase (54%). The latter measures were mostly building renovation measures to upgrade old buildings (windows, lights, ceilings) (Table 22.4). This observation goes together with the business strategies described.

Table 22.3 Analysis of the implemented energy efficiency and renewable energy measures from PINE in processes

Phase	*Implemented measures*	*Payback <2a*	*Payback 2–6 a*	*Payback >6 a*
Incubation	0	0	0	0
Growth	13	11	0	2
Maturity	36	25	3	8
Decline	7	4	1	2

Table 22.4 Analysis of the implemented energy efficiency and renewable energy measures from PINE in buildings

Phase	*Implemented measures*	*Payback <2a*	*Payback 2–6 a*	*Payback >6 a*
Incubation	0	0	0	0
Growth	5	0	3	2
Maturity	10	4	0	6
Decline	6	2	0	4

The analysis showed the following observations.

No company in the incubation phase was participating. In this phase, when investment into new equipment is done, the focus can be on energy-efficient equipment. From the interviews conducted with management of the participating companies, the authors conclude that if energy-efficient equipment is known at the incubation stage and economical, it might well be implemented.

During the incubation phase, the strategic focus is on product development and quality. During this phase, there is no explicit priority on energy efficiency. No company classified in this stage participated in the PINE project. Table 22.5 relates the generic strategy of the stage to energy efficiency and the use of renewable energy.

During the growth phase, the main strategic focus is on sustaining growth and growing market share. Three companies and three buildings of companies of the PINE project were classified as "growing." In these, 18 measures were implemented. In this category the share of measures with a payback of less than two years was the largest one. Table 22.6 relates the generic strategy of the stage to energy efficiency and the use of renewable energy.

During the maturity phase, the strategic focus is on consolidation and cost minimisation. A total of 14 companies and seven buildings were classified as "mature." Some 46 measures were implemented in companies or buildings of this category. In this phase, there was the biggest share of process-related measures. The share of measures with a payback time higher than two years was slightly higher than in "growth." Table 22.7 relates the generic strategy of the stage to energy efficiency and the use of renewable energy.

During the decline phase, the strategic focus is on the reduction of expenditure in general. A total of three companies and ten buildings were classified in this category. Some 13 measures were implemented in companies or buildings of this category. The share of measures with a payback time higher than two years is higher than in "growth" and "maturity." In this category, more than 50% of measures are implemented in buildings. Table 22.8 relates the generic strategy of the stage to energy efficiency and the use of renewable energy.

Table 22.5 Energy efficiency strategy during incubation

Business strategy	*Corresponding strategy for energy efficiency*
"Make it work": Make the product and supply it to the market, deliver products	Energy efficiency is not a priority at this stage

Table 22.6 Energy efficiency strategy during growth

Business strategy	*Corresponding strategy for energy efficiency*
"Grow and finance growth": Maximize market share	Implement very profitable energy efficiency measures to support the efficiency of processes

Table 22.7 Energy efficiency strategy during maturity

Business strategy	*Corresponding strategy for energy efficiency*
"Minimise cost, consolidate": Maximise profits while defending market share	Implement energy efficiency measures to systematically reduce operational cost

Table 22.8 Energy efficiency strategy during decline

Business strategy	*Corresponding strategy for energy efficiency*
"Prepare for next generation": Reduce all expenditure in general, collect profit	Implement energy efficiency measures to improve the infrastructure for the future

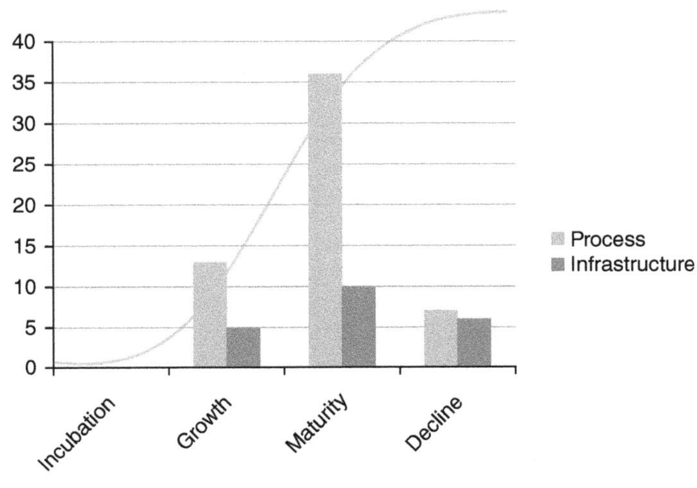

Figure 22.3 Implemented measures categorised by life-cycle phases of process and infrastructure

These results are from a group of 20 companies who volunteered for energy audits out of a general invitation to all the enterprises in the Austrian federal state of Styria. They participated following an open call in the region by the chamber of commerce, published in the weekly newsletter which is distributed to all the members of the chamber and announced in several public conferences. It can be concluded that the volunteering companies were already sensitive towards energy efficiency and renewable energy. As consequence, however, there is not equal distribution of life-cycle stages of processes and infrastructure (Figure 22.3). The results probably reflect the practice of companies already interested in energy efficiency and renewable energy.

Conclusions

Despite high attention to energy efficiency, it can be taken from the literature that a high number of available cost-efficient energy efficiency improvement measures are not implemented in European enterprises. Energy audits are one of the most widespread instruments used to overcome barriers to energy efficiency and promote energy efficiency in industry.

After identification of options, still there exist barriers to the implementation of energy efficiency and renewable energy. In Austria, the price of natural gas has significantly decreased over the period of 2013 to 2018. Prices for equipment and construction have increased. Increasing investment cost together with decreasing gas prices increase the payback time and reduce the motivation for investment into energy efficiency and renewable energy in industries.

This chapter proposes a new way of organising barriers to the implementation of energy efficiency measures along the stages of the life cycle of processes and buildings. In an exploratory

Table 22.9 Life-cycle phases of enterprises and corresponding investment strategies into energy efficiency

	Phase			
	Incubation	*Growth*	*Maturity*	*Decline*
Opportunity for energy efficiency and renewable energy	Basically high; however, as energy consumption is still low, priority is low	Moderate; investment focus on process and product optimisation	High; focus on (short-term) cost reduction	General low investment activity, focused on upgrading infrastructure and preparing for a change

study, the connection of implementation of energy efficiency measures and investment into renewable energy and life-cycle stage was tested on measures which were implemented as a result of 20 energy audits. Table 22.9 summarises the conclusions from this research.

Companies with new products and processes can benefit by including energy efficiency and renewable energy in the design of processes and equipment (Table 22.9). However, in this stage, cost needs to be considered carefully, as financial capital will probably be limited and energy efficiency will rarely be dealt with independently from general business. Therefore, proven energy-efficient options should be carefully integrated with process technology. If the measures are introduced early at the design stage, the cost is more economical, as there is no cost for demolishing and modifying existing equipment as in the case of retrofitting energy-efficient options.

Growing companies provide a moderate opportunity. Their strategy is focused on optimizing their product and production processes, not necessarily on cost reduction. In mature companies, the business strategy is cost reduction. Companies with mature processes might choose energy efficiency measures which pay back in less than two years. The required payback time of two to three years for energy efficiency investments results from the fact that the remaining potential time of use of the process is limited, and yet the investment should pay back before the main process becomes obsolete and renders connected equipment valueless. Therefore, in this phase, the implementation of good housekeeping measures (optimisation of control, optimisation of process parameters, optimised works planning) has priority.

Low-cost options should be identified, and their potential should be clearly demonstrated: A large share of energy is used for support processes such as lighting, heating, and cooling, which are similar across SMEs. This opens up possibilities for standardised solutions, which can be shared effectively.

Investment with a long payback time still might be implemented if the remaining useful lifetime of equipment justifies the investment. This is especially true for buildings, which, in Austrian SMEs participating in the PINE project, represent a significant share of overall electricity and heat consumption. At the time of writing, energy prices are not likely to accelerate energy efficiency investment at the required rate. There are "multiple benefits" on a macroeconomic level, but these are beyond the control and immediate priorities of production companies unless they match requirements from clients. Therefore, targeted energy efficiency policies are needed supporting the implementation of energy efficiency and renewable energy options.

The political strategy from our research should target the following interventions: Maturing processes require simple energy efficiency upgrades which reduce cost and do not interfere with the productivity of the main technology. Examples include optimising process parameters, upgrade of control equipment, retrofitting variable-speed drives on fans and pumps, and so on. Ideally, these solutions should be standardised and modular to reduce planning and transaction cost.

The focus of political support should be on technologies for which a pronounced learning curve can be expected. This includes the development of standardised designs and simplified technologies and use of cheaper materials. A strong learning curve can reduce uncertainty and make investment decisions less risky. Therefore, targeted grants can stimulate research on energy-efficient equipment and support demonstration projects to showcase energy-efficient solutions. Existing information channels, for example, via chambers, can be used to disseminate information on available options.

Companies with products and processes in the incubation phase require energy efficiency measures which are already integrated into the new equipment. Therefore, targeted support of suppliers of key energy-efficient process components is paramount, supporting research and development of innovative energy-efficient products and process equipment to provide options which are at the same time cost and energy efficient when innovative new processes are installed. Relevant information could be spread, for example, in incubators targeting start-up companies.

Companies in decline with old infrastructure are more likely to invest in buildings if these can be used for changed processes and new products. Therefore, for these cases, there should be financial support to lower investment barriers. The findings also suggest a targeted awareness-raising approach towards companies accounting for their varying motivation. Upfront information and support on energy efficiency and renewable energy are mostly needed for incubating companies.

Acknowledgements

The PINE project (IEE/11/885/SI2.615936) was co-funded by the Intelligent Energy Europe Programme of the European Union. The Austrian component was also cofunded by SFG (Steirische Wirtschaftsförderung). This is gratefully acknowledged.

References

Abernathy WJ, Wayne K. (1974). Limits of the learning curve. *Harvard Business Review*. 52 (2).

Anderson ST, Newell RG. (2004). Information programs for technology adoption: the case of energy-efficiency audits. *Resource and Energy Economics*. 26: 27–50.

Austrian Chamber of Labor (Arbeiterkammer) (2017). *Value added barometer (Wertschöp fungs barometer)*. Available at: https://ooe.arbeiterkammer.at/service/presse/PKU_2019_Wertschoepfungsbrometer2018_21.01.2019.pdf.

Burgelman R, Christensen C, Wheelwright S. (2008). *Strategic management of technology and innovation*. Boston: McGraw Hill.

Cagno E, Worrell E, Trianni A, Pugliese G. (2013). A novel approach for barriers to industrial energy efficiency. *Renewable and Sustainable Energy Reviews*. 19: 290–304.

CNBC (2019). *Investment in energy fell again in 2017*. Available at: www.cnbc.com/2018/07/16/investment-in-energy-fell-again-in-2017-international-energy-agency.html.

Coormans C. (2012). Investment in energy efficiency: do the characteristics of investments matter. *Energy Efficiency*. 5 (4): 497–518.

Daft RL. (1998). *Organization theory and design*. Cincinnati: South Western College Publishing.

DOE (Department of Energy) (2015). *Barriers to industrial energy efficiency*. Report to Congress. Washington, DC: United States Department of Energy.

Energy Efficiency Consultants Network (Energieberaternetzwerk) (2019). *Energy cost in different sectors (Energiekosten in verschiedenen Branchen)*. Available at: www.energieeffizienz-im-betrieb.net/energiekosten-unternehmen.html.

Eurostat (2018). *Prices for gas, steel, copper.* Available at: http://ec.europa.eu/eurostat/web/energy/data/main-tables.

Fleiter T, Worrell E, Eichhammer W. (2011). Barriers to energy efficiency in industrial bottom-up energy demand models – a review. *Renewable and Sustainable Energy Reviews.* 15: 3099–3111.

Foster R. (1986). Working the S-curve: assessing technological threats. *Research Management.* 29: 4.

Fresner J, Morea F, Krenn K, Aranda J, Tomasi F. (2017). Energy efficiency in small and medium enterprises: lessons learned from 280 energy audits across Europe. *Journal of Cleaner Production.* 142 (4, 20): 1650–1660.

Gallo A. (2016). A refresher on payback method. *Harvard Business Review.* 16 (2).

Howard D, Hine D. (1997). The population of organisations life cycle (POLC): implications for small business assistance programs. *The International Small Business Journal.* 15 (3): 30–41.

International Energy Agency (2014). *Capturing the multiple effects of energy efficiency.* Paris: IEA Publications. ISBN:9789264220720-612014321E1.

Lefley F. (1996). The payback method of investment appraisal: a review and synthesis. *International Journal of Production Economics.* 44 (3): 207–224.

Mayes TR. (2003). *Investment criteria.* Available at: http://clem.mscd.edu/~mayest/FIN3300/Files/ch9.ppt (Accessed February 2019).

Miller D, Friesen PH. (1985). A longitudinal study of the corporate life cycle. *Management Science.* 30 (10): 1161–1183.

Monitoringstelle Energieeffizenz (2017). *Development of energy efficiency measures, energy audits and other energy services reporting year 2017 (Marktentwicklungen von Energieeffizienzmaßnahmen, Energieaudits und anderen Energiedienstleistungen gemäß § 24 Abs. 2 Z 8 EEffG Berichtsjahr 2017).* Available at: https://www.monitoringstelle.at/index.php?id=795.

OECD (2017). *Renewable energy for industry – from green energy to green materials and fuels.* Paris: OECD.

Panayiotou GP, Bianchi G, Georgiou G, Aresti L, Argyrou M, Agathokleous R, Tsamos KM, Tassou SA, Florides G, Kalogirou SA, Christodoulides P. (2017). Preliminary assessment of waste heat potential in major European industries. *Energy Procedia.* 123: 335–345.

Quinn RE, Cameron K. (1983). Organizational life cycles and shifting criteria of effectiveness: some preliminary evidence. *Management Science.* 29 (1): 33–51.

Rasmussen J. (2011). Business models to improve industrial efficiency. In: *ECEEE industrial summer study proceedings energy efficiency first.* The Foundation of a Low-Carbon Society. Available at: https://www.eceee.org/library/conference_proceedings/eceee_Summer_Studies/2011/.

Röglinger M, Pöppebuß J, Becker J. (2012). Maturity models in business process management. *Business Process Management Journal.* 18.

Rohdin P, Thollander P, Solding P. (2007). Barriers to and drivers for energy efficiency in the Swedish foundry industry. *Energy Policy.* 35 (1): 672–677.

Smith KG, Mitchell RE, Summer CE. (1985). Top level management priorities in different stages of the organizational life cycle. *Academy of Management Journal.* 28 (4).

Solnordal MT, Foss L. (2018). Closing the energy efficiency gap – a systematic review of empirical articles on drivers to energy efficiency in manufacturing firms. *Energies.* 11 (3): 518.

Trianni A, Cagno E, Thollander P, Backlund S. (2013a). Barriers to industrial energy efficiency in foundries: a European comparison. *Journal of Cleaner Production.* 40: 161–176.

Trianni A, Cagno E, Worrell E, Pugliese G. (2013b). Empirical investigation of energy efficiency barriers in Italian manufacturing SMEs. *Energy.* 49 (1): 444–458.

Walsh C, Thornley P. (2012). Barriers to improving energy efficiency within the process industries with a focus on low grade heat utilization. *Journal of Cleaner Production.* 23: 138–146.

23

OVERCOMING FINANCIAL, SOCIAL, AND ENVIRONMENTAL CHALLENGES FACED BY COOPERATIVES

Case studies from the state of Rio Grande do Sul, Brazil

Tania Nunes da Silva and Eugenio Avila Pedrozo

Introduction

Like many other developing countries, Brazil faces issues around a rising population, growth of economic activity, and rapid urbanisation, leading to changing consumption patterns and the fast-growing generation of urban solid waste. Around 95% of all the waste generated in Brazil, which is one of the highest producers of waste globally, is sent to landfill (da Silva, 2018). Despite the implementation in 2010 of the National Solid Waste Policy (NSWP), Federal Law 12.305/10, most municipalities have not drafted or implemented their own municipal policy. Curitiba is one of the few municipalities to have developed a plan (da Silva, 2018). The National Solid Waste Policy established guidelines for integrated waste management, proposed sustainable consumption, and emphasised recycling, with landfill as a last resort (Ryen et al., 2018). However, 58.4% of urban solid wastes are disposed of in landfills, with the rest being deposited in inadequate or unsuitable places (ABRELPE, 2016). In addition, the municipal cost of waste management is the third-highest element in terms of expenses, equivalent to almost 6% of the entire budget of Brazilian capital cities (SNIS-RS, 2016). The NSWP also facilitates a shared responsibility approach for the product life cycle and establishes reverse logistics as an instrument of economic and social development (Azevedo, 2015; Demajorovich and Migliano, 2013). In 2004, a national policy to support local productive arrangements (LPAs, which are composed of companies focused on a similar product or service) was created with the intention of increasing competition, creating jobs, and increasing the capacity for cooperation to strengthen the country's industrial policy (Figal Garone et al., 2015).

There are only a few empirical studies in Brazil of the utilisation of circular economy (CE) approaches (e.g., Moreno et al., 2016; da Silva, 2018; de Oliveira et al., 2018; do Amaral et al.,

2018; Sehnem et al., 2019; Silva et al., 2019). Moreno et al. (2016) examine design strategies for a CE by considering knowledge about resource cycles and designs for longer product life. da Silva (2018) focuses on the development of a systems approach for the management of solid waste in the municipality of Curitiba. de Abreu and Ceglia (2018) found that 'interorganisational strategies and management' were key driving forces in bringing together a 'community' of actors to successfully implement the circular economy concept in a group of organisations. Similarly, Sehnem et al. (2019) showed that the support of senior managers, particularly those with formal sustainability education and an understanding of national sustainability strategy, was crucial for successful implementation. Silva et al. (2019) noted that there was a need for more cross-sectoral cooperation, as well as long-term approaches to implementation of the circular economy concept.

Brazil is considered one of the global centres for agribusiness (UNCTAD, 2017). The agrifood sector is one of the key areas of focus for a more circular approach due to the potential for recovery of value and the need to manage the impacts of increasingly intensified farming (e.g., waste, water scarcity, chemical contaminants, adaptation to climate change, etc.) (Piedrá-Muñoz et al., 2016; Godoy-Durán et al., 2017; Stock et al., 2017).

Cooperatives play a vital role in this more circular approach. Indeed, Clinton and Whisnant (2014) assert that they are innately a more sustainable form of business, one that is able to plan for the longer term and avoid the pressures that come from the demand for immediate profits. The International Cooperative Alliance defines a cooperative as "an autonomous association of persons united voluntarily to meet their common economic, social and cultural needs and aspirations through a jointly-owned and democratically-controlled enterprise" (ICA, 2014). They are composed of an elected board of members which hires professional staff to handle the day-to-day operations (Coren and Clamp, 2014).

Cooperatives play a particularly crucial role in the global agrifood sector and are responsible for over 60% of the collection, marketing, and processing of agricultural commodities at the source, generating in excess of US$290 billion (Campos-Climent and Sanchis-Palacio, 2015). For example, about 30% of the fresh vegetables produced in Spain, and 18% of those in Europe, are produced by small-scale family farms in Spain (Cajamar, 2016). Of all the sectors, it is in farming and agriculture where the cooperative business model is most widely utilised (cooperatives together have an estimated 32% of global market share). Cooperatives therefore play a key role in feeding the world's population.

Types of cooperatives

Dunn (1988: 85) described an agricultural cooperative as "[a] user-owned and controlled business from which benefits are derived and distributed on the basis of use." However, Emery et al. (2017) argue that this definition implies commercialisation and a degree of formality, which is not always the case. Instead, recent articles note a wider range of practices and more flexibility of interpretation (Emery et al., 2017; Gonzalez, 2017). For example, Fonte and Cucco (2017) state that cooperatives have a multifaceted nature. Using the example of Italian cooperatives, Fonte and Cucco (2017) contend that after World War II, agricultural cooperatives in the country moved towards a more productivist model and employed a 'competitive market paradigm' that went beyond merely economic issues and inspired movements for social and community economies. They have been shown to be important mechanisms for the economic and social empowerment of women and for facilitating gender equality (Majurin, 2012; World Bank, 2012; Leucoutere, 2017).

There are two main schools of thought about the role of cooperatives (Fonte and Cucco, 2017). One is that they are self-help groups, primarily concerned with their members' interests and controlling their social and economic welfare. The second views them as radical, alternative economic projects, with explicit societal and community-level objectives. This dichotomy can present tensions and challenges between producers and between producers and other stakeholders if cooperatives seek to expand in size or pursue wider territorial, societal, or environmental issues (Jaklin et al., 2015; Fonte and Cucco, 2017).

Various cooperative approaches have been employed. Kolk et al. (2008) note that the use of partnerships emerged in the 1990s as a means of facilitating capacity building and market access for small farmers. These partnerships have generally tended to focus on extension services, training, and technical assistance (Royer et al., 2017). For example, the use of vertical coordination was utilised in the 2000s in China (Jia et al., 2012). This coordination was achieved through the establishment of so-called 'dragon-head-driven' organisations. These organisations, for example, provided technical expertise and seeds and grain to farmers in exchange for subsidies and support from the government. There has also been the use of 'farmer professional economic cooperatives' (FPCs) as a means of facilitating access to markets by a large number of farmers and traders in the agrifood sector (Jia et al., 2012). Stock et al. (2017) take this concept further to describe shifts towards autonomy and neoliberalism (where the farmer is free to choose to protect, destroy, or enhance the natural environment) as serving to facilitate competition and the devolution of responsibility for environmental outcomes away from the state to the farm or individual farmer.

Shared-services cooperatives can provide either a single service or a range of services and thus can facilitate economies of scale and increased profits and reduce sole proprietors' need to be jacks of all trades (Clamp and Alhamis, 2006). Shared-services cooperatives are owned and controlled by their members and provide benefits to their members as users, not investors (Bhuyan, 1996). They are usually found in non-agricultural contexts. Agricultural cooperatives are producer led. They are often multipurpose and are created to address failures of the market (e.g., in purchasing and marketing). The goal of shared-services cooperatives is to meet an unmet need. Members pay a joining fee and any assessments that may arise and participate in governance and the cooperative's operation.

Partnerships between cooperatives and trading companies have also been utilised (Bitzer and Bijman, 2014). In addition, Royer et al. (2017) make mention of the use of cooperative unions, which market the products that primary cooperatives produce, and the increasing use of contract farming agreements (CFAs) with agribusiness companies to increase profits (Bellemare, 2010). Partnerships play a key role in farming meeting export requirements and gaining certification (Kersting and Wollni, 2012; Jaffee et al., 2011). The use of technological innovation has been shown to be of vital importance for cooperatives to scale or indeed to survive, particularly in developing countries (Garnevska et al., 2011; Luo et al., 2017). This innovation can take different forms, including: (1) technological innovation due to the acquisition of new technology or upgrading of existing technologies; (2) grassroots innovation whereby there is an incremental change to the technology, or there is a shift towards more sustainable consumption and production practices (Monaghan, 2009); and (3) social innovation, where the goal might be to build a better society and/or the use of social processes to develop new products and services in a more efficient manner (Luo et al., 2017; Lopez and Guerra, 2011).

Most of these studies have been undertaken within a Western context. There has been limited empirical research in other contexts. In addition, few have focused on community and the impact of social value factors. This study, therefore, aimed to address some of these issues.

Methods

Study context

In large part due to the use of subsidised rural credit from state-owned banks in the 1970s and 1980s, to utilise agricultural technologies, Brazil was able to transform itself from an importer to the third-largest food exporter in the world (Chaddad, 2016). This study was undertaken in the southern Brazilian state of Rio Grande do Sul during October 2017. Despite being one of the smallest states in Brazil, Rio Grande do Sul has a strong reputation as one of the highest producers of agri-commodities in the country, and it is one of the regions where cooperatives and contract farmers have played a key role (Chaddad, 2016). For example, it is the highest producer and exporter of wines in Brazil. Thus, it serves as an ideal case study region.

The research approach and data collection

Based on previous similar studies (e.g., Coren and Clamp, 2014; Royer et al., 2017), this study employed a case study approach. It utilised a descriptive, narrative approach of the experiences of representatives of the cooperatives, who were selected based on known contacts, and targeted a range of sectors and cooperative sizes. Semi-structured stakeholder interviews were conducted. These interviews were undertaken primarily in Portuguese, but in two cases, English was utilised. Notes were taken during the interviews, and then these were utilised to compile the narratives.

A total of eight interviews were undertaken with a large (1,100 members) producer of (sparkling and house) wines, juices, and brandy (Interviewee A); a small producer (100 members) of primarily rice, beef, and pork (Interviewee B); a small (80 members) producer of fruits and vegetables (Interviewee C); a small producer (35 members) of strawberries (Interviewee D); and a small producer (85 members) of citrus products, compost, and biogas (Interviewee E).

Based on Royer et al. (2017), an abductive narrative analytical approach was employed. Abductive research is a combination of inductive and deductive approaches (Eisenhardt and Graebner, 2007; Pratt, 2009). First, the literature provided an outline of the key issues to be examined. Second, working inductively, open questions were utilised to examine the challenges and key success factors. In addition, the findings of the initial interviews also served to shape the types of questions posed for the latter interviews as a means of identifying trends throughout all the interviews.

Results

General findings

All of the interviewees had sustainability certification from national certification agencies (e.g., IBO and Organico Brazil), as well as international certificates (e.g., Rainforest Alliance and USDA Organic). All employed the cooperatives as a way of not only delivering economic value but also social value for members and the community as a whole. The concept of circular economy played a fundamental underlying role in the practices and services of all (e.g., through the use of business models such as sharing and value recovery). All produced organic products that utilised no chemicals or limited quantities of chemicals.

Key findings from each interviewee

Interviewee A produced wines for local and international markets. It had certification for ISO 14001 and ISO 9001, which served to enhance the efficiency of its operations. The oak barrels which were used to store the wine were used only five times, following which they were used to mature the brandy. Wines and other products were exported outside of Brazil. From an environmental perspective, the main challenge faced related to having too much rain. There had been occasions where, due to severe frost (e.g., in 2016), harvests had been poor; however, annual disruptions due to changes in the weather were becoming more frequent.

Interviewee B had organised itself to form an 'Agrovila' (a micro agricity) that produced rice and also reared cows and pigs. The overarching aims revolved around provision of land for the members of the cooperatives, as well as improving the levels of education, health, and development of its members. For example, some of the rice and beef was used to feed children in the local schools. The key challenge was limitations in markets and financial resources. For example, the milk from the cows was sent to another cooperative to make milk powder. It was reinvesting some of its monies into building a new abattoir and looking at producing new products such as bread. However, it struggled to develop packaging that was fit for purpose (e.g., that enabled international export) and to find markets for the products. However, it was working with a federal agency to address some of its trading issues and had recently starting exporting to Venezuela. It also had limited access to scientific technical support. Indeed, in recent years, it had lost access to technical and financial support from the government. Another key challenge lay in encouraging the younger family members to stay and continue the traditions of the cooperatives and wider movement. Various incentives (e.g., use of ciranda – i.e., facilitation of schooling, for younger members up to 11 years old) had been developed to try to encourage younger members to remain. The concept of a circular economy was heavily utilised within everyday practices. For example, rice husks were used for fertiliser, and waste rice was mixed with corn and used as animal feed.

Interviewee C operated as an open market, opening to the public twice a week every week. There were around 70 producers that participated each week, and many sold organic produce (fresh fruits and vegetables). The key challenge was limitations in access to local markets. This lack of access meant that they had to sell their produce to intermediaries, which meant gaining a lower price. Despite this, there was the desire to maintain the weekly face-to-face market rather than, for example, setting up an online presence. The market gave both producers and customers the opportunity to meet and chat not only about the produce but also about what was happening generally in the community. The producers and the customers had become like a family. The concept was based on the traditions of the cooperative movement that had come out of, for example, the church and the community working together, as well as members of the community acting as unofficial doctors for others, whereby, for example, through their knowledge of the medicinal powers of plants, others in the community would visit them to be 'cured' of various ailments. Thus there was a hesitation to use an online platform, as it was feared that this would take away from the face-to-face interactions and community spirit that had been cultivated. They also faced difficulty in maintaining sustainability certification for products due to the associated costs. However, at the same time, certification served as a means of exchanging information and knowledge amongst producers within the cooperatives. Another challenge was that of the removal or relaxation of federal environmental legislation and policies, which meant, for example, that the use of pesticides was becoming more widespread. This therefore made it difficult for community groups, particularly small producers that grew and sold organic produce.

At the time of the study, Interviewee D had operated for over 15 years. Like Interviewee B, its biggest challenges were with regard to limitations in markets and access to technical scientific support. In addition, due to the nature of its main produce (strawberries), it faced challenges with control of pests. Changes in the climate also meant that the seasons were not as clearly defined as they had been. This had led to different varieties of strawberries having to be planted to ensure resilience. Similarly to Interviewee B, it had in recent years also lost access to technical and financial support from the government. As a result, it found it difficult to increase production levels, therefore becoming caught in a cycle. However, it had put in place a number of measures to increase productivity and revenue. For example, one approach that was being taken was to begin to process some of the strawberries, thus extending their life cycle and also expanding the product range. The 'waste' from the processing of the strawberries was also being sent to a company to make jam. To further increase revenues, the aim was to bring this jam making activity in-house in future.

Interviewee E had been in existence since 1990. It had a broad product range that included biogas generation, production of citrus fruits, and compost. It had received commercialisation support during the initial stages from the German organisation GTZ (now GIZ). It had experienced difficulty in raising capital during its initial years. However, through repaying initial loans on time, it had been able to demonstrate credibility to its creditors. It had been one of the first in Rio Grande do Sul to seek and gain sustainability certification. Certification served to document the processes and add value, particularly with respect to the sharing of best practices and knowledge (through national certification) and enhancing export potential (through international certification). It had grown and expanded its product and service range through strategic partnerships. In some cases, these partnerships had not been planned. For example, its move into composting had resulted through discussions with a company that had been brought in to assist with a technical issue. Expansion, however, brought with it challenges. It had difficulty keeping up with fulfilling its full market capacity and was also being impacted by strategic changes that were taking place within one of its partner organisations.

Discussion

As with other studies (e.g., Royer et al., 2017), linking the farmers and producers to high-value markets was a key challenge. Similarly to other studies (e.g., Navarro, 2014), access to support (e.g., financial support) is vital to enhancing the viability and productivity of the cooperatives. A combination of vertical and horizontal value chains proved crucial for Interviewee E to enhance the viability of its operation by facilitating strategic partnerships that opened up access to a wider range of services and markets. Indeed, the vertical value chain also enabled circularity in the manner in which it operated, in that different entities within the cooperatives were able to utilise the 'waste' products from other entities. Thus, strategic partnerships were employed to expand the operations and thus the cooperative's viability (de Abreu and Ceglia, 2018; Silva et al., 2019). Indeed, of all the cooperatives, Interviewee E operated as a shared-service cooperative, which meant that it could 'exploit' the expertise and skills of its strategic partners. It was therefore able to offer a wider range of services and products, not only to its members but also to its wider local and international communities (Coren and Clamp, 2014; Clamp and Alhamis, 2006). Thus, as outlined in the NSWP, all were using circular approaches (Azevedo, 2015; Demajorovich and Migliano, 2013).

All utilised sustainability certification as a way of increasing their potential to penetrate existing and new (overseas) markets. This was particularly the case for Interviewees A and E, who at the time of the study were exporting on a regular basis. In addition, however, certification also served as a way for members to share information and best practices (e.g., Interviewees C and E. Thus, all had both federal-level certification and international certification. However, both Interviewees C and E made mention of the high costs of sustainability certification. At the same time, like all of the others, they both also found that Brazilian certification was vital in sharing best practices and knowledge amongst producers. It therefore meant that there was a continual trade-off between the costs of the certification and the benefits that it brought.

Access to finance and capacity-building opportunities were key challenges for all of the cooperatives. This was especially so for the smaller cooperatives that generally required assistance with scaling up their commercialisation activities and accessing (local and international) markets. However, similarly to commercial entities, in the case of Interviewees B and D, who faced the biggest challenges of limitations in access to markets and technical support, these hurdles were resulting in innovation of the operations to develop new products. For example, the processing of the strawberries by Interviewee D served to preserve the fresh fruits and also to open up additional product ranges and markets. What was also interesting was that while environmental issues were in some cases perceived to be a challenge (e.g., for Interviewees C and D), the concepts of sustainability were being employed by all of the interviewees as a means not only of reducing the impact of their operations on the environment but also to enhance economic viability and productivity. Indeed, for Interviewee E, the concepts were utilised as a means of expanding the services of the cooperative, building strategic partners and increasing revenue. In addition, all produced organic products which sold at a higher premium as compared to other products and were produced using limited or no chemicals. As in Sehnem et al. (2019), the expertise and knowledge of the senior managers was a key driving factor.

All the cooperatives had gone beyond merely focusing on economic factors to also take account of social and community concerns (Fonte and Cucco, 2017). For example, for Interviewee C, their weekly market functioned not only as a commercial entity but also as a means of creating a 'family' network within the community. In this way, the cooperatives played a much wider social role. However, at the same time, the need to maintain this ethos was also hindering upscaling of its activities, as it was not prepared to, for example, develop an online presence even though this would provide it with a much wider customer base.

Conclusions

This study has demonstrated that there are some similarities in the challenges faced by the case study cooperatives and those to be found elsewhere. For example, there were issues around access to markets and funding and limitations in capacity-building opportunities. However, the manner in which the case studies sought to manage these challenges is instructive for others. For example, the utilisation of the circular economy principle, as well as the innovation in practice by Interviewees D and E and use of strategic partnerships by Interviewee E, all served to drive viability. These are approaches that would therefore be useful for others to adopt as a means of overcoming the inherent challenges that most cooperatives face globally.

References

ABRELPE (Associação Brasileira de Empresas de Limpeza Pública e Resíduos Especiais) (2016). *Panorama dos resíduos sólidos no Brasil 2016*. Available at: www.abrelpe.org.br/Panorama/paronama2016.pdf (Accessed 5/9/2019).

Azevedo JL. (2015). A economia circular aplicada no Brasil: uma análise a partir dos instrumentos legais existentes para a logística reversa. *Anais Congresso Nacional de Excelência em Gestão*. 11: 1–16.

Bellemare MF. (2010). Agricultural extension and imperfect supervision in contract farming: evidence from Madagascar. *Agricultural Economics*. 41 (6): 507–517.

Bhuyan, S. (1996). *Potential applications for shared-services cooperatives in North Dakota*. Fargo, ND: Department of Agricultural Economics, North Dakota State University.

Bitzer J, Bijman V. (2014). Old oranges in new boxes? Strategic partnerships between emerging farmers and agribusiness in South Africa. *Journal of Southern African Studies*. 40 (1): 167–183.
Cajamar (2016). *Análisis de la Campaña Hortofruticola de Almería, Campaña 2014/2015*. Almería, Spain: Cajamar.
Campos-Climent V, Sanchis-Palacio JRS. (2015). How much does size matter in agri-food firms? *Journal of Business Research*. 68: 1589–1891.
Chaddad F. (2016). Agriculture in Southern Brazil: cooperatives and contract farming. In: *The economics and organization of Brazilian agriculture: recent evolution and productivity gains*. Cambridge: Academic Press. 45–72.
Clamp C, Alhamis, I. (2006). *Carpet One: a case study of cooperative entrepreneurship. Carpet_one_case_study (final) – cooperative-curriculum*. Available at: http://cooperative-curriculum.wikispaces.com/file/view/Carpet_One_Case_Study+(final).pdf (Accessed 8/2/2018).
Clinton L, Whisnant R. (2014). Model behavior: 20 business model innovations for sustainability. *Sustainability*. Available at: http://sustainability.com/our-work/reports/model-behavior/ (Accessed 8/2/2019).
Coren C, Clamp C. (2014). The experience of Wisconsin's wine distribution co-operatives. *Journal of Co-operative Organization and Management*. 2: 6–13.
da Silva CL. (2018). Proposal of a dynamic model to evaluate public policies for the circular economy: scenarios applied to the municipality of Curitiba. *Waste Management*. 78: 456–466.
de Abreu MCS, Ceglia D. (2018). On the implementation of a circular economy: the role of institutional capacity-building through industrial symbiosis. *Resources, Conservation and Recycling*. 138: 99–109.
Demajorovic J, Migliano JEB. (2013). Política nacional de resíduos sólidos e suas implicações nacadeia da logística reversa de microcomputadores no Brasil. *Gestão and Regionalidade*. 29 (87): 64–80.
de Oliveira FB, França SLB, Rangel LAD. (2018). Challenges and opportunities in a circular economy for a local productive arrangement of furniture in Brazil. *Resources, Conservation and Recycling*. 135: 202–209.
do Amaral MC, Zonatti WF, da Silva KL, Junior KL, Neto JA, Baruque-Ramos J. (2018). Industrial textile recycling and reuse in Brazil: case study and considerations concerning the circular economy. *Gestão & Produção São Carlos*. 25 (3): 431–443.
Dunn J. (1988). Basic cooperative principles and their relationship to selected practices. *Journal of Agricultural Cooperation*. 3: 83–93.
Eisenhardt KM, Graebner ME. (2007). Theory building from cases: opportunities and challenges. *Academy of Management Journal*. 50 (1): 25–32.
Emery SB, Forney J, Wynne-Jones S. (2017). The more-than-economic dimensions of cooperation in food production. *Journal of Rural Studies*. 53: 229–235.
Figal Garone L, Maffioli A, de Negri JA, Rodriguez CM, Vázquez-Baré G. (2015). Cluster development policy, SME's performance, and spillovers: evidence from Brazil. *Small Business Economics*. 44 (4): 925–948.
Fonte M, Cucco I. (2017). Cooperatives and alternative food networks in Italy: the long road towards a social economy in agriculture. *Journal of Rural Studies*. 53: 291–303.
Garnevska E, Liu S, Shadbolt NM. (2011). Factors for successful development of farmer cooperatives in Northwest China. *International Food and Agribusiness Management Review*. 14: 69–84.
Godoy-Durán Á, Galdeano-Gómez E, Perez-Mesa JC, Piedra-Muñoz L. (2017). Assessing eco-efficiency and the determinants of horticultural family-farming of Southeast Spain. *Journal of Environmental Management*. 204: 594–604.
Gonzalez RA. (2017). Going back to go forwards? From multi-stakeholder cooperativeseratives to open cooperativeseratives in food and farming. *Journal of Rural Studies*. 53: 278–290.
ICA (The International Cooperativeserative Alliance) (2014). *Cooperativeserative identity, values and principles*. Available at: https://ica.coop/en/whats-cooperatives/cooperativeserative-identity-values-principles (Accessed 8/2/2019).
Jaffee S, Henson S, Díaz Rios L. (2011). *Making the grade: smallholder farmers, emerging standards, and development assistance programs in Africa*. A Research Program Synthesis. Washington, DC: The World Bank.
Jaklin IJ, Stummer S, Milestad R. (2015). Why do farmers collaborate with a food cooperative? Reasons for participants in a civic food network in Vienna, Austria. *International Journal of Sociology, Agriculture and Food*. 22 (1): 41–61.
Jia X, Huang J, Xu Z. (2012). Marketing of farmer professional cooperatives in the wave of transformed agrifood market in China. *The China Economic Review*. 23: 665–674.
Kersting S, Wollni M. (2012). New institutional arrangements and standard adoption: evidence from small-scale fruit and vegetable farmers in Thailand. *Food Policy*. 37: 452–462.

Kolk A, Van Tulder R, Kostwinder E. (2008). Business and partnerships for development. *European Management Journal*. 26: 262–273.

Leucoutere E. (2017). The impact of agricultural cooperatives on women's empowerment: evidence from Uganda. *Journal of Cooperatives Organization and Management*. 5: 14–27.

Lopez M, Guerra R. (2011). Strategic management for the internationalization and cooperative markets. *Procedia Social and Behavioral Sciences*. 24: 769–780.

Luo J, Guo H, Jia F. (2017). Technological innovation in agricultural cooperatives in China: implications for agro-food innovation policies. *Food Policy*. 76: 19–33.

Majurin E. (2012). *How women fare in East African cooperatives: the case of Kenya, Tanzania and Uganda*. Dar es Salaam: The International Labour Office.

Monaghan A. (2009). Conceptual niche marketing of grassroots innovation for sustainability: the case of body disposal practices in the UK. *Technological Forecasting and Social Change*. 76: 1026–1043.

Moreno M, De Los Rios C, Rowe Z, Charnley F. (2016). A conceptual framework for circular design. *Sustainability*. 8 (9): 937.

Navarro V. (2014). The case of Mondragon. *CounterPunch*. Available at: https://www.counterpunch.org/2014/04/30/the-case-of-mondragon/

Piedrá-Muñoz I, Galdeano-Gómez E, Pérez-Mesa JC. (2016). Is sustainability compatible with profitability? An empirical analysis of family farming activity. *Sustainability*. 8: 893.

Pratt MG. (2009). From the editors: for the lack of a boilerplate: tips on writing up (and reviewing) qualitative research. *Academy of Management Journal*. 52 (5): 856–862.

Royer A, Bijman J, Abebe GK. (2017). Cooperatives, partnerships and the challenges of quality upgrading: a case study from Ethiopia. *Journal of Cooperatives Organization and Management*. 5: 48–55.

Ryen EG, Gaustad G, Babbitt CW, Babbitt G. (2018). Ecological foraging models as inspiration for optimized recycling systems in the circular economy. *Resources, Conservation and Recycling*. 135: 48–57.

Sehnem S, Jabbourb CJC, Pereirac SCF, Jabbour ABLS. (2019). Improving sustainable supply chains performance through operational excellence: circular economy approach. *Resources, Conservation and Recycling*. 149: 236–248.

Silva FC, Shibao FY, Kruglianskas I, Barbieri JC, Sinisgalli PAA. (2019). Circular economy: analysis of the implementation of practices in the Brazilian network. *Revista de Gestão*. 26 (1): 39–60.

SNIS-RS (Sistema Nacional de Informações de Saneamento – Resíduos Sólidos) (2016). *Diagnóstico resíduos sólidos*. Available at: www.snis.gov.br/diagnostico-residuossolidos (Accessed 13/10/2019).

Stock PV, Forney J, Emery SB, Wittman H. (2017). Neoliberal natures on the farm: farmer autonomy and cooperation in comparative perspective. *Journal of Rural Studies*. 36: 411–422.

Unctad (2017). *Statistics*. Available at: http://unctad.org/en/Pages/statistics.aspx (Accessed 19/12/2019).

World Bank (2012). *World development report 2012: gender equality and development*. Washington, DC: The World Bank.

24

THE INFORMAL RECYCLING SECTOR – ENVIRONMENTAL CRIMINALS OR THE FUTURE OF THE CIRCULAR ECONOMY?

Mike Webster

> We have the Street-Finders, or those who, as I said before, literally "pick up" their living in the public thoroughfares. They are the "pure" pickers, or those who live by gathering dogs'-dung; the cigar-end finders, or "hard-ups," as they are called, who collect the refuse pieces of smoked cigars from the gutters, and having dried them, sell them as tobacco to the very poor; the dredgermen or coal-finders; the mud-larks, the bone-grubbers; and the sewer-hunters.
>
> *London labour and the London poor. (Mayhew, 1865)*

Introduction

In those places where there are no laws that regulate waste, where no-one collects the rubbish, let alone provides an organised recycling collection, there is always one group plying their trade. They could be described as 'the original recyclers,' and what they do constitutes a major source of income for the urban poor and economically marginalised. As indicated by the quote that opens this chapter, dating from the middle of the 19th century, they have been a part of the urban landscape since the industrial revolution and probably much before – the informal recycling sector.

The importance of the resource value of waste in developing countries has long been noted as a key driver for its collection, historically in economically developed countries and currently in the developing world (Wilson, 2007). As urban populations grow, bringing with them a surplus of labour, this value has attracted the urban poor into collection of such secondary materials for reprocessing.

This self-organised army has a significant impact on the world's cities. It provides income-earning opportunities for many of the world's poorest and provides environmental services in parts of urban areas where more formalised collections either struggle to physically reach or can't be afforded by the local population.

Meanwhile, though, there are tensions between the 'informal' sector, the focus of this chapter, and the 'formal' waste management sector, with the latter generally seeing the former as illegitimate competition. We examine some attempts to integrate the informal sector, with varying degrees of success. Such tensions need to be addressed to ensure that we maintain the benefits of the informal economy whilst ensuring all waste is collected and managed properly.

As humans become a predominantly urban species and the global population races towards 10 billion, the numbers engaged in informal recycling will only grow (Velis, 2017), with 40 megacities expected by 2030 (United Nations, 2016). With 27 of them in countries without comprehensive waste collection, and as traditional approaches to collection and disposal fail to keep pace, new models of waste collection, reprocessing, and revalorisation are clearly required.

There is growing understand within the global development community as to the importance of decent solid waste management in itself to global development. Spurred on by, amongst other things, the increased public interest in marine litter as well as recognition of the importance of public health to urban communities, there is renewed interest in *what works* in lower-income countries. Indeed, perhaps informal approaches point to a future of high-quality, comprehensive, and flexible collection systems that traditional municipal approaches sometimes struggle to deliver.

As international aid agencies focus more closely on pro-poor policies (for example, the UK Department for International Development's Leave No-One Behind agenda), there is renewed recognition of the importance but also the opportunity for an inclusive circular economy. On the one hand, it provides the poorest with a clean, healthy environment in which to thrive and, on the other, it is an opportunity to provide meaningful incomes for the poorest by engaging in recycling and repair.

What is informal recycling?

Characterised by Wilson et al. (2006) as "small-scale, labour intensive, largely unregulated and low-technology manufacturing or provision of services," the key driver of the sector is usually simple: sorting recyclable materials that have a high and clear secondary value from mixed waste for onward sale. There are a variety of broad categories of informal waste managers:

- Those paid by businesses and householders to collect and transport residual waste – either to a transfer site or municipal landfill or dumpsite (where they may pay to dispose) or to an illegal dumpsite – an example being the *pousse-pousseurs*[1] of Kinshasa
- Those focusing on the collection of specific materials for onward sale to middlemen or reprocessors – for example, metals, some plastics, textiles, or cardboard. They might buy or collect material door to door, sort through communal bins, or pick from the dumpsite where many live (Wilson et al., 2006). This practice is widespread in many middle- and lower middle-income countries, such as Mexico, Colombia, Thailand, and the Philippines.
- Those engaged in artisanal recycling or reprocessing – for example, aluminium drink cans into cooking pots, metal containers hammered into sieves or graters
- Municipal waste collection crews – individuals that may have a formal waste collection job and collect recyclable materials as a secondary source of income.

A flourishing informal recycling system is closely associated with inadequate formal solid waste management. There needs to be access to dumpsites or openly dumped waste, as well as a lax regulatory environment where individuals can collect materials directly from households or businesses without fear of prosecution. It can therefore be clearly seen that formalisation of waste collection, where this access is withheld, might be considered a threat by informal recyclers.

Who are informal recyclers?

The field is dominated by family and microenterprises composed of women, children, and elderly relatives (Kaza et al., 2018). These groups tend to sell the material onto micro and small enterprises, scrap buyers, who purchase or collect small quantities of recyclable waste materials from households. Above this level are, then, different levels of traders, many of whom are also informal and not registered as businesses. The materials recovered by this informal valorisation sector are sold on to the larger industrial value chain.

Waste picking, the activity at the bottom end of the hierarchy, is often practised by marginalised groups, perhaps regionally or internally displaced persons or those from a certain ethnic or caste group. The actual drivers vary from place to place, but it is often striking quite how consistent a group of informal recyclers is within a certain locale – there could be a stratified society in which a certain group is expected to work with waste (for instance, in Cairo, Egypt, with the Zabbaleen or in India with Dalit groups.) Elsewhere there may be internally displaced or marginalised groups, for instance, in Ecuador, where waste pickers are disproportionately represented by displaced indigenous groups or in northern Ghana, where waste pickers are over-represented by those fleeing intercommunal violence in Mali.[2] Essentially, informal recycling, and particularly waste picking, is something that can be done by those with little access to finance and without a local support network. Over time, a certain degree of organisation takes place, as can be seen with the Zabaleen (see Box 24.1). Previous authors have listed the various names and groups for informal recyclers around the world (Wilson et al., 2006, after Medina and Dows, 2000; Berthier, 2003):

> Zabbaleen in Egypt, Pepenadores, Catroneros and Buscabotes in Mexico, Basuriegos, Cartoneros, Traperos and Chatarreros in Colombia, Chamberos in Ecuador, Buzos in Costa Rica and Cirujas in Argentina; Kabadiwallas in India

Waste picking is practiced by the marginalised because, quite simply, it is one of the few occupations open to them. Spurned by local, established populations, it is attractive because of the low barriers to entry and the relatively high profit margins: no one else want to deal with waste and certainly not being seen to deal with waste by their peers. It is open to those individuals willing, or desperate, enough to bear being seen collecting, handling, and making a living from waste.

The scale and impact of informal recycling

Despite the small scale of individual operations, with individuals or small groups handling perhaps no more than a few kilogrammes of waste per day, what makes the informal sector significant in environmental, economic, and global development terms is the sheer number of people involved.

It is by its nature a tricky area to measure: a transient workforce made of marginalised groups, unregistered by definition and firmly existing in the grey economy is never going to be easy to count. Various estimates indicate that millions work within the field, that half a percent of urban populations are working in informal-sector recycling, some 15 million people worldwide (Velis, 2015), and in some areas, informal waste collection alone occupies 2% of the population (Gunsilius et al., 2011). In India alone, there are an estimated 1.5 million people involved in informal waste management (WIEGO, 2010) and somewhere between 500,000 and four million people in Latin America (Marello and Helwege, 2014).

Meanwhile, the recycling levels that can be achieved by the informal sector are looked on with envy by municipal systems. Recycling recovery rates as high as 80% have been achieved by the Zabbaleen in Cairo due to the labour-intensive approach to collection and sorting and their expertise at extracting value from waste and direct financial interest in extracting as much value from the waste as possible in a way that a collector working for a set wage, regardless of the quantity and quality of recycling they collect, might not be (Gunsilius et al., 2011).

This has led to the somewhat paradoxical situation of many cities in lower-income countries, without comprehensive waste collection and disposal, enjoying higher recycling rates than their higher-income counterparts. Examples include Nairobi recycling around 30% of its waste and Bamako 20% or Quezon City recycling almost 40% (Fargier, 2015). This compares favourably with Rotterdam at 23% and London at 33% (LWARB, 2017) and is delivered at little or no cost to the public.

Key challenges faced by informal recyclers

The informal sector is firmly part of the grey economy: they do not pay taxes, are generally unlicensed and hence working illegally, and are excluded from social welfare or government insurance schemes (Haan et al., 1998). This gives them a flexibility that the formal sector often lacks, as well as an ability to undercut regulated competitors and provide services at rates that are affordable for those that might not be able to afford licensed services.

Operating on the margins of society means that they generally lack access to legal protections and benefits such as social security coverage, health and unemployment insurance, pensions, and union representation. Working with waste on a daily basis exposes them to multiple occupational hazards, including pathogens, chemicals, and insanitary conditions. This is exacerbated if protective equipment and clothing are unavailable or, more likely, unaffordable (Wilson et al., 2006). This means people working on dumpsites that could contain any type of waste: municipal, hazardous, clinical, with pickers stepping across needle-strewn dumpsites wearing nothing more than rubber slippers as footwear and picking up items without gloves.

GIZ (2010) describes their plight:

> The majority of informal waste workers (dump and street waste pickers) constitute the lowest level of society. Working conditions are unimaginable and include permanent exposure to dangerous, toxic and contagious substances. Waste pickers are often bitten by dogs, subjected to harassment from officials, exploited by traders and have no legal protection. They live in inhuman and humiliating circumstances and generally lack of sanitary services, health care and social benefits. Child labour is the rule, not the exception, and life expectancy is low. Their task is carried out in the most dehumanising manner, with workers obliged to sort through putrefying garbage to retrieve the scraps of paper and plastic that will earn them a living.

The impact of this hard life on longevity can be dramatic. One study of waste pickers in Mexico City estimated life expectancy to be just 39 years, compared to 67 years among city residents overall (Wilson et al., 2006).

The health and personal risks are in addition to other kinds of confrontation. Organised crime has been well documented as active in several dumpsites around large cities, one example being Dandora dump in Nairobi (Muindi et al., 2016). Meanwhile, formal waste collectors – licensed companies or municipal services – see them at best as an irritant and worst as criminals. They often regard the sector as the cause of dumping and littering. In one case, the waste pickers of Mumbai were blamed for large and ongoing dump fires that led to smoke pollution throughout the city (Lakshmi, 2016). They are seen as an embarrassment (particularly those waste pickers living on dumpsites, notably children) and are seen as backward, unhygienic, and not suited to a modern, clean urban environment (Lerpiniere et al., 2014). They are also seen as unfair competition by the formal waste collectors, removing the higher-value materials within the waste stream and reducing profitability.

There also appears to be a lack of economic mobility, particularly for waste pickers. A lack of access to credit means they are unable to scale up and purchase equipment that might allow them to enjoy greater economies of scale such as balers, shredders, and vehicles (CWG and GIZ, 2011). This is a vicious circle – it weakens their bargaining power with buyers and in turn does not allow them to build up the resources to invest in scaling up (Fergutz et al., 2011). The less organised the informal recycling sector is, the less the people involved are capable of adding value to the secondary raw materials they collect and the more vulnerable they are to exploitation from intermediate dealers. Many who recognise the value of the recycling work carried out by waste pickers are often sceptical that this activity has the potential to become a desirable and permanent livelihood opportunity (Chikarmane, 2012).

Informal recyclers are also vulnerable to the vagaries of global recycling markets and are reliant on local buyers to access to world markets. This access to global markets requires a certain level of national economic development and stability – examples being the greater recovery of materials by the informal sector in relatively industrialised countries such as Pakistan, with a demand for secondary materials, as opposed to less activity in those places that are less industrialised (e.g., Somaliland).[3] One of the goals of those aiming to improve the economic agency of the informal sector is to provide more predictability in incomes (see Plastics for Change, Box 24.4).

Box 24.1 What happens if you try to get rid of the informal sector?[4]

The Zabbaleen, which translated means garbage people, are the traditional waste collectors of Cairo. They live in the settlement of Mokattam, with a population of around 20,000 to 30,000, and are over 90% Coptic Christians. Originally migrants from upper Egypt but long settled in Cairo, they have one of the world's most efficient recycling systems. Waste collection started with organic waste to be fed to their pigs in return for a small monthly fee paid by residents. Over time, this has moved to include dry recyclables as well, and the systems have mechanised, moving from donkey carts to trucks today. Their supporters claim they have greatly improved the capacity of Cairo to manage its waste at minimal cost or effort to the city administration. However, this system is under threat. Since 2003, the Cairo governorate has been implementing a policy of privatisation of waste collection and removing their central economic asset: municipal solid waste. While the Zabbaleen had previously recycled 80% of the waste they collected, the new contractors are required to recycle only 20%, the rest of which would go into landfill. The Zabbaleen were permitted to keep their jobs as wage workers with these companies, and they would also be responsible for street sweeping and placement of garbage bins. The Zabbaleen claim, however, that the salaries offered are less than what they used to make independently and that they used to earn 90% of their income from recycling rather than from the collection fee.

Citizens still prefer the traditional door-to-door collection method of the Zabbaleen. Furthermore, the large trucks of the companies cannot go into the narrow streets of Cairo, requiring the placement of bins in central collection points, to the dismay of residents and leading to large amounts of open dumping.

Recycling rates have collapsed, and the amount of waste sent to landfill has increased. Traditionally, the Zabbaleen used to feed the organic waste to their pigs, but even this system was ruined by the state when it culled all their pigs in 2009 to avoid swine flu. Pig rearing was more important to the Zabbaleen than initially thought, since they sell the meat to tourist facilities for good prices. With their main processor of organic waste gone, the Zabbaleen refused to collect organic waste from Cairo, leaving piles of garbage in the streets, replacing the threat of swine flu with the threat of typhus. Many Zabbaleen quit the business because without the rearing of pigs, the tedious work of sorting through garbage became economically unfeasible.

Their situation was made worse following the official policy of moving their activities, including sorting, recovery, trading, and recycling, outside of the city under the guise of making their neighbourhoods cleaner and healthier, increasing Zabbaleen's travel distance and consequently cost of services delivered to residents.

The citizens prefer the Zabbaleen system with its cheaper fees. They reject the government's plan to pay extra fees to private companies. As such, the Zabbaleen still collect municipal solid waste alongside companies and local municipalities.

An inclusive urban resource economy?

As urban areas develop economically, the way that waste and recycling collections are organised changes as policy makers come to grips with the problems caused by waste, regulation strengthens, and the capacity of the government to deliver and manage services increases.

Wilson (2007) lays this path out clearly: The initial driver of formal waste collection is public health – getting it off the streets, stopping open burning, and preventing openly dumped waste blocking drains and becoming a vector for disease. This is followed by steps to improve waste disposal, ensuring containment in sanitary landfills and protecting the local environmental around these sites. Ultimately, resource efficiency and global environmental issues direct policy, leading to a focus on segregating waste, producer responsibility, and novel approaches to incentivising positive behaviours, including financial and behaviour change strategies.

This formalisation process inevitably leads to conflict with the informal sector, who depend on a level of deregulation to access waste. However, given that the ultimate aim of this process is resource efficiency, of which recycling is a large part but, as outlined previously, leads to a period of reduction in recycling rates, there are obvious crossovers with the aims of the informal recycler, that is, to divert material for reprocessing from disposal.

Meanwhile, informal collectors are disproportionately composed of the poorest – an obvious priority for economic development. There is therefore a clear need to find some solution that leads to comprehensive waste collection and reduced open dumping whilst at the same time not harming the livelihoods of the informal sector. Surely there must be a way to take the best of what informal recyclers can deliver whilst working to ensure comprehensive collection of all materials, whether desirable to the informal sector or not.

Box 24.2 Getting it right? Integration of informal recyclers in Pune, India.[5]

In Pune, India, a cooperative of 3,000 waste pickers has successfully won better working conditions by joining forces with municipal authorities to collect waste door to door. In 1993 they organized themselves into the *Kagad Kach Patra Kashtakari Panchayat* (KKPKP) union. Thereafter, KKPKP campaigned for waste pickers, buyers, and collectors to be recognised as workers. They underlined the benefits of what their members delivered for the municipality and the city more broadly in terms of reducing waste disposal cost and jobs created, as well as improved public health and the environment. In 2007, they set up a wholly worker-owned cooperative of waste pickers, itinerant waste buyers, and waste collectors named SWaCH, an acronym for Solid Waste Collection Handling that also means 'clean' in the local language. SWaCH Seva Sahakari Sanstha Maryadit, as it is formally known, provides front-end waste management services to Pune City, with support from the Pune Municipal Corporation (PMC). Thousands of waste pickers walked the streets, proclaiming "*Kachra amchya malkicha . . . nahi kunachya bapacha*" (we own the waste; it does not belong to anyone's father!) and staking claim to the recyclables that they collected. KKPKP argued that collection of recyclable materials constituted work and sought recognition for the workers from the PMC. A series of collective actions that protested against abuse and discrimination of aggrieved waste pickers followed.

The circular economy and global development

Development practitioners are beginning to note the importance of the resource recovery sector in pro-poor development. Circular economy approaches are a classic win-win-win: A clean and healthy urban environment reduces the risk of chronic ill health linked with poor sanitation and helps local populations to thrive; the employment provided is disproportionately weighted towards the poor, providing a good return on donor investment; and capturing and revalorising materials also addresses the climate emergency and broader challenges such as marine plastics. However, the concept has until recently been almost entirely absent from development discourse (Tearfund and IDS, 2017). In 2012, only 0.3% of development aid went into solid waste management (Lerpiniere et al., 2014), although this has increased in recent years as some donors and international non-governmental organisations have begun to recognise the opportunities that valorisation of secondary materials offers.[6] This is still arguably a neglected area when compared to other thematic areas such as water, sanitation, and hygiene (WASH).

Informal workers in the contemporary economy: fourth industrial revolution

The fourth industrial revolution and globalisation throw up opportunities and challenges. On the one hand, the availability of cheap smartphones has allowed systems such as ScrapQ (see Box 24.3) to work more efficiently and engage with new customers whilst still allowing them the flexibility and independence that life as an informal worker previously offered. Meanwhile, automation (driverless cars, robotic collection and sensor-based sorting, big data and Internet of Things) provides new ways to recover resources but also threatens much of the work of the informal recycler, which is labour intensive by its nature (Velis, 2017).

Box 24.3 Uber for waste pickers: SCRAPQ

ScrapQ, based in Hyderabad, India, is a service provider that collects recyclables, enabling recycling, confidential data destruction, and scrap disposal services, working with local REOs (recycling officers). The philosophy at ScrapQ is that the solid waste service business is a local business managed by professionals living and working in the communities around Hyderabad, with ScrapQ acting as an aggregator to connect customers and the local Kabbadiwala (waste collectors).

Essentially, ScrapQ operates like an "Uber" for waste pickers with a network of 2,300 kabidawallas serving more than 12,000 customers across Hyderabad. They have set prices that fluctuate daily depending on world market prices. They service single or multifamily residences, commercial institutions, industrial locations, construction sites, or special events.

Whereas before, a kabadiwalla would walk down the street calling for householders to bring out their recyclables, ScrapQ allows them to wait in an office, arrange a collection with the householder, and collect the materials at the agreed-upon time. Payment goes via the app, avoiding the need to handle cash.

Meanwhile, other schemes such as Plastics for Change (see Box 24.4) link waste pickers into the global demand for recovered plastic at a fair and predictable price. It links up international companies (such as Body Shop) to source their plastics ultimately from the informal sector – such schemes could be an opportunity to set up such "fair trade" plastic schemes, ensuring that the informal sector gets a fair price and those involved can escape the poverty trap.

Box 24.4 Fair prices for waste pickers: Plastics for Change and The Body Shop

Plastics for Change is a for-profit organisation that partners with local NGOs Hasiru Dala and Hasiru Dala Innovation (HDI) to provide the Indian city of Bengaluru's waste pickers with a stable income and better opportunities.

PFC has developed an ethical sourcing platform to create sustainable livelihoods for the poor while transitioning the industry towards a circular economy. It catalyses the recovery and recycling of plastics waste in a profitable manner while helping brands meet and exceed their extended producer responsibility. It creates transparency and accountability from the base of the supply chain to the manufacturer, resulting in efficient shared-value chains and higher-quality recycled plastic.

The mobile platform helps wholesalers source plastic from waste pickers and also provides access to high-value markets with brands by addressing three problems: (1) reducing volatility in price by facilitating long-term relationships with buyers who guarantee a fair minimum rate in advance, (2) providing access to working capital finance to ensure prompt payment can occur at the point of exchange throughout the supply chain, and (3) preventing the exploitation of informal waste workers through a peer-to-peer rating system and audit system.

PFC is now partnering with the Body Shop, which has started using plastics sourced through PFC in 250-ml shampoo and conditioner bottles.

Conclusions

Informal recyclers provide a service where no one else does and deserve recognition for the vital public health and environmental services they provide where the public and formal private sector can't or won't go. The sector matters because of the sheer number of individuals engaged in it and, regardless of the small scale of individual operations, is significant globally. Furthermore, the example of Cairo (see Box 24.1) shows that we have to get integration between formal and informal right; otherwise, this can lead to less circularity and less just outcomes.

The existence of the informal sector shows that there is no "one size fits all" approach to the development of waste services. There have been approaches that consider integration, such as the example of Pune (see Box 24.2). With the increased focus of aid agencies on the poorest and most marginalised, there should be a focus on helping the informal sector collect more waste with a greater number of waste streams, more safely, and derive a more robust livelihood. The question remains what the best interventions are. Given the challenges they face, such a manifesto for inclusive recycling should include:

- Providing access to reduce the environmental health challenges faced by workers such as waste pickers. Measures as simple as ensuring access to personal protective equipment as well as health insurance could do much to increase the quality and length of the informal recycler's life.
- Ensuring an inclusive approach to the development of waste collection services: supporting waste pickers to set up cooperatives to help them understand their legal rights and protections and making sure that the policy environment means that they can be involved as waste services in a city develop.
- Providing credit and business support to help those on the lowest rungs of the collection ladder to scale up their operations and move out of poverty.

Development of new technology allows new approaches and opportunities for the informal recycling sector to reach out to new customers and improve profitability. Such *uberisation* of waste means that householders and companies can be incentivised to segregate their waste and arrange separate collections for separate materials, engaging with specialist collectors who know how to maximise the value of their specialist waste stream waste (e.g., the Zaballeen with organic waste and pigs). There are opportunities to leap-frog over rich world countries not only in terms of matching in waste collection coverage but in the quality and levels of recycling.

Indeed, it could be argued that the thriving repair and reuse economics seen in many of the megacities of the Global South offer a model that circular economy practitioners would do well to adopt in the North. Take-back schemes and specialisation of collection are key characteristics of circular economy systems, offering improved models for repair and refurbishment of otherwise end-of-life items. These tend to be a vibrant ecosystem of small, bespoke, specialist enterprises, each filling a niche, rather than the top-down paternalistic municipal systems we see in richer countries. For instance, the vibrant e-waste repair, reuse, and upgrade markets of West Africa are a great example of the informal sector thriving in a low-regulation environment. An enabling environment would allow public, private, and third-sector involvement in the resource sector – "let a thousand flowers bloom."

Notes

1 Named for the *pousse-pousse* (push-push) carts that are pushed along the streets of the city.
2 These are examples taken from the personal experience of the author.
3 As witnessed by the author.

4 Adapted from the Environmental Justice Atlas (2019).
5 Adapted from WIEGO (2010), Organizing Informal Waste Pickers: A Case Study of Bengaluru, India.
6 For example, UK NGO Tearfund has started a programme of developing solid waste sector community waste programmes, and there are indications of increasing donor interest within the field (as of 2019).

References

Berthier HC. (2003). Garbage, work and society. *Resources, Conservation and Recycling*. 39 (3): 193–210.

Chikarmane P. (2012). *Integrating waste pickers into municipal solid waste management in Pune, India*. WIEGO Policy Brief (Urban Policies) No. 8. Boston, MA: WIEGO.

CWG and GIZ (2011). *The economics of the informal sector in solid waste management*. Eschborn: Collaborative Working Group on Solid Management in Low- and Middle-Income Countries (CWG); Deutsche Gesellschaft für Internationale Zusammenarbeit (GIZ).

Environmental Justice Atlas. (2019). Multinational takeover threatens the livelihood of the Zabbaleen. *Egypt EJAtlas*. Available at: https://ejatlas.org/conflict/zabbaleen-cairo (Accessed 4/11/2019).

Fargier M. (2015). *Imperial college UG4 thesis/research paper*. London: Imperial College.

Fergutz O, Dias S, Mitlin D. (2011). Developing urban waste management in Brazil with waste picker organizations. *Environment and Urbanization*. 23 (2): 597–608.

GIZ (2010). *The waste experts: enabling conditions for informal sector integration in solid waste management. Lessons learned from Brazil, Egypt and India*. Available at: https://www.giz.de/en/downloads/gtz2010-waste-experts-conditions-is-integration.pdf

Gunsilius E, Spies S, García-Cortés S, Medina M, Dias S, Scheinberg A, Sabry W, Abdel-Hady N, dos Santos A, Ruiz S. (2011). *Recovering resources, creating opportunities: integrating the informal sector into solid waste management*. Eschborn: GIZ.

Haan HC, Coad A, Lardinois I. (1998). *Municipal waste management: involving micro- and small enterprises*. Guidelines for Municipal Managers. Turin, Italy: International Training Centre of the ILO, SKAT, WASTE.

Kaza S, Yao L, Bhada-Tata P, Van Woerden F. (2018). *What a waste 2.0: a global snapshot of solid waste management to 2050*. Washington, DC: The World Bank.

Lakshmi R. (2016). A burning mountain of trash in Mumbai fuels middle-class outcry. *Washington Post*.

Lerpiniere D, Wilson DC, Velis C, Evans B, Voss H, Moodley K. (2014). *A review of international development co-operation in solid waste management*. Vienna: ISWA.

Marello M, Helwege A. (2014). *Solid waste management and social inclusion of waste pickers: opportunities and challenges*. GEGI Working Paper. Boston: Boston University.

Mayhew H. (1865). *London labour and the London poor*. Harmondsworth: Penguin Books.

Medina M, Dows M. (2000). A short history of scavenging. *Comparative Civilizations Review*. 42 (42): 23–45.

Muindi K, Mberu B, Aboderin I, Haregu T, Amugsi D. (2016). *Conflict and crime in municipal solid waste management: evidence from Mombasa and Nairobi Kenya*. Working Paper #13. Nairobi: African Population and Health Research Centre (APHRC).

LWARB (2017). *Report for the London waste and recycling board (LWARB) and the greater London authority (GLA)*. Available at: www.lwarb.gov.uk/wp-content/uploads/2016/09/LWARB-International-recycling-rate-comparison.pdf (Accessed 22/10/2019).

Tearfund and IDS (2017). *Virtuous circle: how the circular economy can create jobs*. Available at: www.ids.ac.uk/publications/virtuous-circle-how-the-circular-economy-can-create-jobs-and-save-lives-in-low-and-middle-income-countries/ (Accessed 22/10/2019).

United Nations (2016). *The world's cities in 2016 data booklet*. New York: United Nations.

Velis C. (2015). Circular economy and global secondary material supply chains. *Waste Management & Research*. 33 (5): 389–391.

Velis C. (2017). Waste pickers in Global South: informal recycling sector in a circular economy era. *Waste Management & Research*. 35 (4): 329–331.

WIEGO (2010). *Organizing informal waste pickers: a case study of Bengaluru, India*. Available at: http://www.wiego.org/sites/default/files/resources/files/Chengappa-Organizing-Informal-Waste-Pickers-India.pdf (Accessed 12/9/2019).

Wilson D. (2007). Development drivers for waste management. *Waste Management & Research*. 25 (3):198–207.

Wilson D, Velis C, Cheeseman C. (2006). Role of informal sector recycling in waste management in developing countries. *Habitat International*. 30 (4): 797–808.

WRAP (2017). *London key waste and recycling data 2017*. Available at: http://laportal.wrap.org.uk/LWARB-KeyData.aspx (Accessed 22/10/2019).

25

REFUGEE CAMPS AND CIRCULAR ECONOMY IN PALESTINIAN WEST BANK

Challenges and opportunities

Marco Caniato and Valérie Thöni

Circular economy and principles of integrated sustainable waste management

The concept of the circular economy (CE) has been extensively analysed and embraced by various stakeholders. Private companies and corporations realised its potential in overcoming the limits of a linear consumption scheme and risks linked to material scarcity and started promoting an accelerated adoption, operationalisation, and mainstreaming. Offering definitions, examples, and case studies, they want to prove short-term benefits and long-term opportunities, framing the discussion and influencing policymaking (Ellen MacArthur Foundation, 2012, 2015). Despite its complexity and the need of a joint effort of several actors for a proper introduction, CE can offer some viable solutions for pressing global challenges. Some joint discussions took place in the framework of the Sustainable Development Goals in order to better understand what CE should look like and how to scale it up globally (UN ECOSOC, 2018). Several high-income-level countries reacted very positively, and the EU developed an Action Plan for its introduction in various economic compartments. The actions proposed consider each step of the value chain - including production, consumption, waste management and promotion of secondary raw material market and water reuse - and define some priority areas. Innovation, investment and other horizontal measures are promoted, and indicators suggested for monitoring (EC, 2015). Administrative and regulatory instruments, as well as economic and information instruments, promoting CE are now being developed in a certain number of countries.

However, so far, not enough attention has been dedicated to developing countries and how they should manage the innovation and policy progress required for such a transition. On the one hand, there is *a keen interest in promoting resilient (economic) growth*, moving from resource extraction to increased use and improved efficiency of secondary and primary resources, respectively; on the other hand, the actual implementation requires *targeting and tailoring the transition* to the main economic sectors of each country and their peculiarities. Indeed, it does not simply affect waste management and recycling, but it has a broader impact, affecting all the economy, promoting occupation and the creation of more added value and generating environmental benefits (e.g., reduction of pollution and greenhouse gas emissions, climate change adaptation). An enabling environment should be established, with specific initiatives moving in such a direction (Preston et al., 2019).

The topic is even more neglected regarding developing countries hosting a significant number of refugees and internally displaced persons. UNHCR (2019) reports a continuously increasing number of displaced people since 2010, reaching the all-time record of almost 70.8 million at the end of 2018. On average, such people are displaced for an increasingly long period, making them not negligible for hosting countries due to their number and their long-lasting presence. Palestinians refugees – 1.98 million in the Occupied Palestinian Territory (OPT) in 2017 – account for over 42% of the overall population (Palestinian Central Bureau of Statistics, 2019a). This high proportion has a clear footprint on the resources and environment. As small entities within bigger systems (municipalities), refugee camps are "ideal" places where CE pilot initiatives could be put in place and the impacts rapidly seen. Moreover, the United Nations Relief and Works Agency for Palestine Refugees in the Near East (UNRWA) and UN agencies are in general committed to improving environmental protection in their strategies and policies within refugee camps (e.g., UNRWA, 2017).

As an important component of CE development, the integrated sustainable waste management (ISWM) approach – initially developed in the framework of developing countries – aims at the establishment of a preventative and consultative system in order to manage all the aspects and steps of solid waste management (SWM), protecting public health and the environment and promoting the 3Rs (Wilson et al., 2013). ISWM entails a new waste hierarchy paradigm, focusing on source reduction, recycling, and composting and, to a lesser extent, on waste-to-energy technologies and landfilling. In developing countries, the application of ISWM's three dimensions (environmental effectiveness, social acceptability, and economic affordability) still faces numerous challenges (Marshall and Farahbakhsh, 2013).

In the OPT, SWM is a complex case with unique specificities, which are important to bear in mind when dealing with it. The OPT today consists in the Gaza Strip (GZ) and the West Bank (WB). Included in the WB, East Jerusalem was unilaterally annexed in 1980, with areas cut by the separation wall since 2003. In the WB, the land is divided into three zones: A (18% of territory, under the control of the Palestinian Authority), B (22% of the land, about 44 villages under the supervision of the Palestinian Authority and Israeli security control), and C (60% of territory, under the total administration of the Israeli Ministry of Defense). The Palestinian SWM system is also hindered in its development and modernisation by the Israeli control over land and resources, as well as on movement of persons (main transport axes in WB are forbidden to Palestinians) and on goods and services (e.g., financial, internet).

The adoption of policies and activities directed towards CE and ISWM could be an opportunity for the OPT to overcome its dependence on imported raw and primary material, slow or narrow the waste flows towards landfilling, and develop new innovative sectors for its young educated population. However, for Palestinians, the challenge is to build a complex and sophisticated system including adequate agreements, incentives and penalties. The engagement of stakeholders willing to contribute to the establishment of the new system, collaborate with public institutions and compete in the new market is a challenge, too. There is also some cultural resistance – waste is still considered a problem and not a resource with an economic value. Furthermore, public health security considerations are often prioritised over longer-term environmental interests.

Solid waste management in the Occupied Palestinian Territory and West Bank refugee camps

The great majority of Palestinians today (about 95%) have access to a municipal solid waste (MSW) service. At the operational level, the solid waste management chain involves 447 municipalities or local government units (LGUs), 13 joint service councils (JSCs; associations of municipalities and village councils), and the UNRWA (providing service to 8 refugee camps in GS and 19 in the WB, including Shu'fat camp in East Jerusalem). As a small part, the private sector can be contracted for waste transportation and management of transfer stations or disposal sites.

In 2018, Palestinians from the WB and GS generated about 1.59 million tonnes of solid waste (including 43,025 tonnes from refugee camps), with a per capita/per day average of about 0.9 kg. With population growth and changes of habits, waste quantities will rise in the future. The composition of the waste is typical of a middle-income country, with 50% organic waste, 14% plastic, 12% cardboard/paper, 2.5% metal, and 1.8% glass (MoLG-JICA, 2019).

MSW collection (primary and secondary) usually takes place at fixed places (bins or containers) in the streets, five to six days/week in towns or three days/week in rural areas. In the refugee camps, UNRWA sanitation workers operate five days a week, ensuring primary collection (door-to-door) and street sweeping free of charge (in other areas, fees apply to residents). The cumbersome camp structure (narrow streets, high population density, and sometimes difficult terrain) hinders SW collection, mostly done manually with primitive and overloaded pushcarts. Furthermore, UNRWA waste collection points are often open and located at the periphery of the camp area and thus also used by non-residents. Special waste (e.g., bulky, construction and demolition, tyres, and medical), is generally not collected by municipalities, and accumulates near the camps, while cardboard, tyres, or waste piles are regularly burnt (UNRWA, 2017; Cesvi, 2019).

Transportation is supervised by municipalities, 12 JSCs, and UNRWA, either through transfer stations or directly to the disposal sites. Transfer stations have gained importance over the past few years, and about 57% of the total waste collected by JSCs goes through them. Their design is in general simple, often with a weigh bridge and mechanical transfer equipment, mostly without a concrete slab to protect the soil and without any cover. They are often used as waste separation points, where some materials are collected for reuse/recycling.

About 65% of the SW is disposed of into six sanitary landfills, the rest being dumped either legally or illegally, with 1% recycled and 2% composted (Thöni and Matar, 2019). The landfills face the quick reduction of their lifespan capacities due to the growing incoming SW and the impossibility of expanding or building any other site. About 77 random dumpsites are officially used for municipal waste. In refugee camps, all waste collected is disposed of in landfills or dumpsites, and disposal fees are the highest costs in the UNRWA SW budget. There is an urgent need to find a long-term solution in order to reduce the quantity of disposed waste and limit the pollution risks due to illegal practices (dumping and burning).

Little information about generation and treatment of industrial and hazardous waste can be found. Construction and demolition waste is usually dumped in separate locations, with only a fraction of it being reused. The hazardous part of medical waste (20% of total medical waste) is segregated and treated in some places, but often ends up mixed with municipal waste during the transportation and disposal phases. Waste from electronic and electrical equipment is considered entirely hazardous waste, according to Palestinian legislation. Its collection (mostly from Israel) and treatment (dismantling and cleaning) are usually done by informal, often illegal entities in the south of the WB, without proper protective personal equipment nor awareness. The business involves 1,000–2,000 permanent workers and contributes to one-third of the local economy. Metals are sent back to Israel for further export. Other hazardous wastes (oils and lubricants, chemicals, or used tyres, for example) are not recorded (Thöni and Matar, 2019).

Although included in the National SW Strategy (PNA, 2017), recycling and recovery processes lack proper guidelines and promotion. As previously mentioned, recycling activities are mainly informal, involving individuals (waste pickers) and private factories (less than 10 people in general) that are partially registered and in compliance with the law. In all refugee camps, some residents are involved in solid waste recovery (there are 23 shops and second-hand stores in Tulkarem camp, for example – Botella Mestres, 2019). The main obstacles for the development of waste recycling/reuse in the OPT are the limited domestic market and commercial opportunities, the bureaucracy required to start a business and uncertainty regarding duration and conditions of contracts signed with municipalities and other LGUs, the complicated and expensive import of equipment, the difficulty of exporting to Israel and abroad, and the negative perception of and low interest in such activities by municipalities and the private sector. A full CE approach is clearly even more complicated, though specifically mentioned in the Sustainable Consumption and Production National Action Plan in Palestine (EQA and UNEP, 2016) and should be promoted and mainstreamed at the national level.

Selected experiences in solid waste recovery

Cardboard collection and reuse

Cardboard and paper waste has some potential recycling value in the OPT, but, like other recyclables, it depends on the local demand and prices on the Israeli market. In the West Bank, about 314 tonnes are collected daily and largely landfilled (MoLG-JICA, 2019). The informal sector (waste pickers, small second-hand traders, and junk shops) is usually involved in the recovery, with no agreements with municipalities. A few private companies formally collect cardboard (e.g., Ommar Al Ard in Ramallah) to be reused in the OPT (e.g., egg trays by Al Ostry, toilet paper by Pal Paper in Jericho) or mostly to be sent to Israel. Today, about 15–20 tonnes are recycled daily.

The Green Tulkarem project (2014–2017) is an example of a successful cardboard recycling experience. Covering Tulkarem city and ten villages around it, the project focused on the efficiency of the SW management system and the integration of a recycling process for cardboard and organic waste. The cardboard separated and treated reached 3.35 tonnes/day in January 2017 (16% of total cardboard waste produced). The project had environmental impacts (5% of total MSW collected effectively separated and high quality of the segregated fraction), economic effects (savings for the JSC and local cooperative treating the organic waste), and social consequences (reinforcement of the awareness, initiative and engagement of an important proportion of people and stores). Mapping the accumulation points, reorganising collection routes, coordinating the transportation, and encouraging cooperation between all stakeholders (municipality, local institutions, and civil society, as well as the private sector) were the core elements for success.

At the refugee camp level, waste recovery usually takes place to a lesser extent. In Shu'fat camp in East Jerusalem, for example, the collection is supervised by the Refugee Popular Committee, which effectively manages to separate a certain fraction of cardboard. Open cages were installed inside the camp, thus allowing waste pickers to easily collect it. However, the quantities segregated remain too small to have a real impact on the landfilled amount.

As indicated before, refugee camps often deal with additional waste from non-residents. Dheisheh camp (8,805 people in 2017; PCBS, 2019b) in Doha-Bethlehem faces a large quantity of commercial waste (of which about 2 tonnes/day is cardboard), that is deposited in the UNRWA waste accumulation points and exhausts their capacity. A recent project proposal to develop the separation and treatment of cardboard waste with JSC Bethlehem should allow for future effective treatment of this waste.

Organic waste management

The potential of organic waste recovery is very relevant, due to the high share – around 50% (Thöni and Matar, 2019) – in MSW. Systematic investigations have never been done in refugee camps in the OPT, but composition analyses for specific camps confirm such a figure (Botella

Mestres, 2019). However, no consistent efforts have been dedicated to recovering such a fraction in camps, as far as we know.

The only available and documented experience is the Green Tulkarem project implemented by Cesvi. At the end of 2015, a separate collection of organic waste was introduced in Tulkarem and extended to the nearby villages at the beginning of 2016, including Nur Shams and Tulkarem refugee camps. Organic waste was collected from large producers, such as markets and greengrocers, and sent to a composting facility located at the outskirts of the city and operated by Thinnabeh, a local cooperative, already producing compost out of agricultural waste and animal manure. However, after a first analysis of the quantities collected and despite positive practices (e.g., organic waste used as animal fodder), some villages, agricultural areas, and the two camps were excluded from the collection system in order to improve its logistics and overall efficiency. Unfortunately, this was not enough to make the Tulkarem organic waste management system economically sustainable (CeTamb Lab, 2017). Just after the completion of the project, Tulkarem JSC decided to discontinue it, due to the lack of support from the municipality and the line ministries.

Cesvi also launched, for the same camps, projects promoting a planned and improved SWM, hygiene, and environmental awareness, investigating the possibility of introducing community composting. Unfortunately, apart from few very interested households, the level of community commitment and camp internal organisation were not enough to make it possible. Another study was conducted by Cesvi in 2019 about the introduction of separate organic management systems in OPT, such as home composting and decentralised and centralised composting facilities. It confirmed that the main challenges were related to people's acceptance, awareness, and engagement.

Plastic recovery

Plastic waste has a relatively important share in the SW composition in the OPT (about 14–17% in 2019; Thöni and Matar, 2019). Again, it suffers from a lack of investment and has generated low revenues until now, as it finds few commercial opportunities inside the OPT and the Israeli plastic recycling market.

Shu'fat refugee camp is the biggest camp of the WB (about 23,000 people in 2016, with 123,000 persons/km^2 density). Located in East Jerusalem, it has been under Israeli control since 1967 and is included in Jerusalem Municipality. Until recently, the camp was facing huge problems in terms of SW collection and storage, added to by waste from non-camp residents (around 5,000) – due to the poor service provided by the Municipality in the surrounding areas. Several projects, in collaboration with UNRWA, were put in place for efficient SW management inside the camp, and in 2019, the situation has improved, thanks also to the distribution of appropriate household bins.

In parallel, there is a growing trend in the camp for plastic segregation, for several possible reasons. The open cages installed for cardboard are also used for plastic, allowing for easy collection. Furthermore, the fact that the Municipality has an extended producer responsibility (EPR) program could incite camp residents to recover some products (especially plastic). Empower.org has also been involved in a recent campaign, with the aim to show the economic potential of plastic recycling. Finally, the Popular Committee recently received a baler machine for plastic and cardboard, which is not yet operational due to time constraints and lack of technical and economic support. But it should allow for recycling in the future through a Palestinian intermediary dealing with Israeli companies (for polyethylene terephthalate) and with a plastic factory dealing with polyethylene and polyvinyl chloride in the Hebron Governorate.

Refugee engagement, a crucial component

The previous examples point out that refugees and camp community-based organisations (CBOs) play a significant role in the introduction and success of any SWM initiative, such as waste separation, especially if they are supposed to be active participants in it.

In 2017/18, Cesvi supplied Nur Shams camp's Popular Committee (PC) with the means to separately collect and store recyclables, also providing technical support and facilitation to establish synergies regarding SWM with third parties. Positive outcomes came out of this project: municipal cardboard collection was extended to the camp, the PC played an active role in cleaning waste accumulation and collecting construction and demolition waste and bulky waste, and some synergies and discussions started with Tulkarem municipality and the informal sector. However, the main issue was that the PC was not ready or enough committed to run a business based on recyclable collection and management. Furthermore, although some promising opportunities – for example, the possibility to collect green waste from Tulkarem municipalities (synergy camp-hosting community) and the interest of local recycling companies (CBO–private company partnership) – there was no real follow-up, and the activities never started. Finally, even the use of cardboard cages was questioned by UNRWA, analysing the real cost-benefit due to the efforts required given the limited participation of shop owners.

Based on the experience in Nur Shams, Cesvi led a long engagement process to train and identify different scenarios with the Tulkarem camp's Popular Committee. Finally, the PC decided not to take any new responsibility, preferring a blurry situation and an evident enforcement gap (Botella Mestres, 2019). However, it played a certain role in facilitating some beautification initiatives – that is, cleaning and rehabilitation of critical spots – and in general in awareness raising in the camp. Furthermore, a multistakeholder committee comprising the PC, Municipality, JSC, the Palestinian Environmental Quality Authority (EQA), and governor's office was established, and regular monthly meetings took place to deal with various SWM and environmental issues regarding the camp and the surrounding area.

In Shu'fat refugee camp, the change of the waste collection scheme was possible only due to a long participatory process, with direct engagement of neighbourhoods and community leaders, resulting in the PC taking the responsibility for domestic bin distribution and management. Such a change was the most evident result but just the tip of the iceberg, considering the large number of activities implemented during a 3.5-year period and the difficulty in building the trust and engagement of the refugee community.

Elements for future opportunities

Over the past 50 years, there were few projects directly and primarily targeting SWM among the numerous activities implemented in Palestinian refugee camps. Currently, there is a growing interest in this sector and a recognition of its importance. The volatile situation – due to political, security, and socioeconomic factors – makes "easy/straight" solutions not always available, and adaptation, innovation, and flexibility are key elements for future developments.

The inclusion of the private sector in the SWM process is an essential step. In order to allow private companies to enter the market and provide services to refugee camps, the regulatory framework should be clarified and implemented. Moreover, at the governance level, each stakeholders' role, duties, and terms should be precisely defined. In this regard, private-sector involvement could encourage the PCs in the camps and other stakeholders to take the initiative and fill the enforcement gap.

To promote investments, transparency in processes and decisions is central, as is available and reliable SW data. Trust between stakeholders is a big challenge in practically all the camps, and only inclusiveness – possibly engaging actors from both within and outside the camp – will guarantee mutual confidence.

Contractual aspects are relevant as well. Elements such as actors involved, financial resources, and contract duration should be clearly defined in order to establish an economically sustainable SWM system and attract investors. In addition, clear financial mechanisms (e.g., extended producer responsibility) and a coherent application of the "polluter-pays" principle could help reduce the risks for large and long-term investments.

The Palestinian recycling market is undersized compared to the countries of the region, especially considering the limitations on imports and exports. Reducing bureaucracy and building networks between key Palestinian players, as well as with foreign companies and organisations, are essential elements for promoting innovation and businesses in the CE sector. On a different level, building more synergies and integration between the camps and surrounding municipalities may reduce mutual tensions, for example, regarding services received without paying. Also, local public service providers may benefit from this increased collaboration, creating additional value due to the new population to serve (i.e., refugees). It would also offer opportunities for more employment and new businesses.

Finally, informal actors – which are recycling forerunners, in practice – need to be included in the development of an ISWM system and future circular economy business models. They should be furthermore encouraged to integrate the formal sector through both incentives (e.g., business support, professional education, inclusion in decision processes) and penalties, including legal consequences.

Conclusion

In developing countries, the concept of a CE is very appealing, but there is a need to clarify processes, promote best practices, and exchange information for an actual implementation. CE understanding is still strongly associated with basic recycling, while SWM is usually neither well integrated nor effective in all parts of a country.

However, various experiences in the West Bank demonstrate a solid potential not only to introduce an effective SWM system but also to overcome some limitations imposed by Israeli occupation and control (e.g., import of machinery and goods, export of secondary raw material, and limited enforcement power by the PA, especially in some areas). In practice, the benefits could go beyond SWM, resulting in new business opportunities for the local economy, provided they are clear to all stakeholders. In this regard, refugee camps could be forerunners for recycling initiatives, working in collaboration with the surrounding municipalities.

Recycling of cardboard looks to be the waste fraction most ready to be mainstreamed, while plastic still needs more incentives or at least a more structured and favourable environment. Recovery of the organic fraction is probably the most relevant area for a significant reduction of waste landfilled and can contribute to the local economy (e.g., production of locally made compost supporting the agricultural sector, generation of biogas reducing importation to meet local energy needs). Organic composting has furthermore the potential for more ambitious targets (e.g., overall food loss and waste minimisation due to better planning, post-harvesting management, and sensitisation at the consumer level). Other options – for example, private-to-private agreements for recyclable movement and reuse/new processing immediately after generation – look feasible and interesting to reduce waste generation but in specific contexts such as industrial clusters, while they should not be particularly relevant for refugee camps, where no manufacturing activities take place.

Regardless of working with one or more recyclables, the institutional framework and governance should be very clear and accepted both within and outside the camps. Refugees and their representative bodies – for example, PCs in Palestinian refugee camps – should be formally and fully engaged in the initiative, with accepted roles and responsibilities. At the national level, rules and terms for running recycling businesses should be well balanced in order to frame and support the sector. Also, the actual duration of public–private contracts supporting CE initiatives, starting from recycling, should be clearly defined and protected from any possible political change. Indeed, where rules are inconsistently applied, it is easy to increase the risk of investment, at least as perceived, making the private sector less eager to step in.

Finally, comprehensive publications on this topic and a thorough evaluation of unsuccessful projects would be very useful to learn from previous mistakes. Feasibility studies considering logistics, security factors, and restriction measures could also be helpful to analyse data, evaluate lessons learned from previous and current interventions, and put forward new recycling options. Indeed, innovation in technology and methods could disclose new and interesting opportunities, promoting sustainability and business.

Acknowledgements

The authors thank Cesvi for having kindly authorised the use of unpublished documents and reports regarding projects implemented by Cesvi with the support of various donors in Tulkarem (Green Tulkarem, funded by the Italian Ministry of Foreign Affairs, MAECI, project code AID 010147/CESVI/TOC), Nur Shams refugee camp (funded by the Italian Agency for Development Cooperation, AICS, project code AID 10910), Nur Shams and Tulkarem refugee camps (funded by AICS, project code, AID 11264), and Shu'fat refugee camp (funded by the European Union, DG DEVCO, project code ENI/2015/359–899).

References

Botella Mestres N. (2019). *Assessment on popular committee's SWM institutional role in Tulkarem refugee camp.* Brescia: Cesvi.

Cesvi (2019). *Solid waste management in West Bank refugee camp study – feasibility analysis of integrated SWM operations.* Brescia: Cesvi.

CeTAmb Lab (2017). *Evaluation of the introduction of separate collection systems for organic waste and cardboard in Tulkarem.* Brescia: Cesvi.

DeCanio S, Watkins WE. (1998). *Investment in energy efficiency: do the characteristics of firms matter? The Review of Economics and Statistics.* 80(1): 95-107.

EC (2015). *Closing the loop – an EU action plan for the circular economy.* Brussels: EC.

Ellen MacArthur Foundation (2012). *Towards the circular economy: economic and business rationale for an accelerated transition.* Cowes: Ellen MacArthur Foundation.

Ellen MacArthur Foundation (2015). *Delivering the circular economy: a toolkit for policymakers.* Cowes: Ellen MacArthur Foundation.

EQA and UNEP (2016). *Sustainable consumption and production national action plan in Palestine.* Available at: https://switchmed.eu/wp-content/uploads/2020/04/01.-SCP-NAP-assessment-Palestine.pdf

Marshall RE, Farahbakhsh K. (2013). System approaches to integrated solid waste management in developing countries. *Waste Management.* 33: 988–1003.

MoLG (Ministry of Local Government) – Japan International Cooperation Agency (2019). *Data book solid waste management of joint service councils West Bank and Gaza.* Available at: https://www.jica.go.jp/english/news/press/2018/190214_01.html

PCBS (Palestinian Central Bureau of Statistics) (2019a). *Palestinians at the end of 2019.* Available at: http://www.pcbs.gov.ps/pcbs_2012/Publications.aspx

PCBS (Palestinian Central Bureau of Statistics) (2019b). *Main indicators by type of locality – population, housing and establishments census 2017.* Available at: http://www.pcbs.gov.ps/pcbs_2012/Publications.aspx

Palestinian National Authority (2017). *National strategy for solid waste management in the Palestinian territory 2017–2022.* Available at: http://www.cesvi.eu/wp-content/uploads//2019/12/SWM-in-Palestine-report-Thoni-and-Matar-2019_compressed-1.pdf

Preston F et al. (2019). *An inclusive circular economy: priorities for developing countries.* Available at: https://www.chathamhouse.org/sites/default/files/publications/research/2019-05-22-Circular%20Economy.pdf

Thöni V, Matar S. (2019). *Solid waste management in the occupied Palestinian territory.* Brescia: Cesvi.

UN ECOSOC (2018). *Circular economy for the SDGs: from concept to practice.* Concept Note. Available at: https://www.un.org/en/ga/second/73/jm_conceptnote.pdf

UNHCR (2019). *Global trends: forced displacement in 2018.* Available at: https://www.unhcr.org/5ee200e37.pdf

UNRWA (2017). *UNRWA West Bank field operational solid waste management strategy.* Available at: https://www.cesvi.org/wp-content/uploads/2018/02/UNRWA-Solid-Waste-Management-Strategy.pdf

VDI (2018). *Wettbewerbsvorteil ressourceneffizienz, definitionen, grundlagen, fakten und beispiele.* VDI-ZRE (4. Auflage). Available at: https://www.ressource-deutschland.de/fileadmin/user_upload/downloads/aktuelles/VDI_ZRE_Wettbewerbsv

Wilson DC, Velis CA, Rodic L. (2013). Integrated sustainable waste management in developing countries. *Proceedings of the Institution of Civil Engineers: Waste and Resource Management.* 166 (2): 52–68.

PART V

Recycling

26

EXPLORING HOUSEHOLD DYNAMICS FOR RECYCLING IN THE UNITED KINGDOM

A case study of recycling habits in greater London

Annabelle Boulay, Guy M. Robinson, Stewart Barr, Gareth Shaw, and Alan Metcalfe

Introduction

Household waste recycling

In the United Kingdom, the political landscape of household waste management is dominated by recognition of the many negatives associated with the so-called throwaway society (Castellani et al., 2015; Cooper, 2016) and the role played by domestic consumers in creating waste. Policy focuses on the 'waste hierarchy' (which emphasises waste prevention and reuse of materials ahead of recycling), but the recycling of domestic waste has received the most attention and political effort, as both national and local governments encourage individuals and households to divert materials into recycling streams (Bravi et al., 2020; Gandy, 2014; Waite, 2013). Research has revealed much about the attitudes and behaviours of individuals who participate in recycling schemes (e.g., Babaei et al., 2015; Botetzagias et al., 2015; Scott et al., 2015) but has also highlighted the importance of appreciating the dynamic nature of the household in determining recycling rates (Czajkowski et al., 2017; Gregson et al., 2007b). It is increasingly recognised that research should examine ways in which materials, such as food and drink packaging, move through the household via a set of negotiations between household members, technologies, infrastructures, and aesthetics. In this way, understanding household dynamics may offer an opportunity to explore households' waste-related practices and interactions between social, technical, and regulatory regimes governing such practices (Gregson et al., 2013).

We argue that to advance understanding of how recycling of waste materials by households can be increased, recycling should be treated as part of social practices within the household (Bell et al., 2015; Gibson et al., 2015; Shove, 2010). We identify these practices and address them in the wider context of households as dynamic social organisms. This includes considering ways in which items purchased regularly, such as food and drink, which often have recyclable packaging materials, move through the household from the point of purchase to final disposal (Metcalfe et al., 2013). We focus on a small sample of households who had expressed a prior commitment to recycling, which we describe as "green casual" households (Harris, 2011). We also review how participants interpret waste systems, technology, and information about recycling and analyse waste trajectories through the household.

Household waste management research: psychology and practice

Considerable research has focused on households' willingness to participate in recycling schemes and on identifying key motivations and barriers to participation, including various situational and psychological factors (Graham-Rowe et al., 2014; Varotto and Spagnolli, 2017). This work stresses the importance of sociodemographic factors, such as age, gender, and political affiliations (López-Mosquera et al., 2015; Mansvelt, 2012), as well as knowledge and experience (Saphores et al., 2012), rights and responsibilities (Lane and Watson, 2012), moral obligations to act (Culiberg, 2014), barriers (Hadjimanolis, 2013), self-efficacy and logistical factors (Monnot et al., 2014), environmental attitudes (López-Mosquera et al., 2015), and infrastructure (Keramitsoglou and Tsagarakis, 2013). These studies indicate that participation in recycling varies according to social group and across different generations (Tudor et al., 2011). Research has also focused on motivations for and barriers to participating in different types of recycling activity within the home, examining these at the level of the individual householder. This has highlighted public education, promotion of recycling schemes, economic incentives, the amount of effort required to recycle, the supply of kerbside containers, and the frequency of waste collection (Mueller, 2013). Individuals tend to act routinely and in response to social norms and expectations, which leads to path dependency, in which decisions depend on the past knowledge trajectory (Stern, 2006: 381). Decisions made are limited by the individual's current competence base, therefore restricting their responsiveness to policies intended to increase proficiency (such as increasing recycling) (White and Hyde, 2011). Habit or routine acts as "a distinct type of causal variable and consequently, behaviour change often entails breaking old habits and becomes established by creating new ones" (Stern, 2000: 417).

Most of the studies cited previously examine factors influencing action rather than studying the household as a dynamic unit. In general, their findings emphasise a need to educate the public to help enable them make rational decisions in relation to policy (Gandy, 2014). This has involved numerous measures to facilitate greater engagement by the public in reuse and recycling schemes (Garnett and Cooper, 2014), reducing waste generation (Kalmykova et al., 2016), deploying the 'best practicable environmental option' as a basis for waste management decisions (Davies, 2016), increasing awareness of end-of-life impacts of products (Finkbeiner, 2014), and building markets for secondary materials (Ackerman, 2013).

Much of this policy aims to persuade individuals and households to act in a desirable fashion, in what Shove (2010) terms the "attitude, behaviour, choice" (ABC) model of behavioural change. This has resulted in a self-sustaining research paradigm attempting to identify factors influencing specific and policy-driven actions (Spurling et al., 2013) but largely assuming that behaviour is the result of a linear, rational mode of decision-making related to specific motivators and barriers. Alternatively, Shove argues that what many view as 'behaviours' are representations of wider social practices shared between households and communities, deeply embedded in a specific historical and economic context (Hui et al., 2016; Shove et al., 2012). By focusing on individuals and by specifying the choices which need to be made, there is neglect of how household practices relate to wider society and opportunities for change arising from a less instrumental discussion of pro-environmental behaviours (Barr, 2014; Steg et al., 2014).

The dominant approach to tackling the policy challenge of waste management has been to frame the 'problem' as one dominated by the so-called throwaway society (Cooper, 2013; Gregson et al., 2007a). In this way, complex dynamics of how materials enter domestic settings, how they are used, and what they become are overshadowed by a simplistic relationship between over-consumption and discarding, without considering associations between 'things' and people and the value and meanings of material culture (Evans, 2018). Accordingly, households and the individuals therein have been uncritically conceptualised as dysfunctional links in the waste management chain, in which some recycle, but others do not (Kirakozian, 2016).

Greater attention therefore needs to be paid to themes of routine and performance within the household (Collins, 2015; Pettifor, 2012), with further work on the materiality of waste – that is, the networked, political associations that various products possess. For example, the work of Hawkins (2011) on plastic water bottles illustrates how material waste is mobilised within different discourses – showing how bottled water simultaneously plays a role in both narratives of healthy living and unsustainable consumption.

Materials and methods

Investigating the 'green casuals'

Waste practices vary significantly across households. Segmentation is one common means of distinguishing between households on this basis (Gordon, 2013). This distinguishes between groups (individuals, households) by identifying clusters based on specified criteria. For recycling research, clusters may be identified using behavioural, attitudinal, and/or demographic characteristics, using cluster analysis to identify similarities (Hennig et al., 2015). For example, Barr et al. (2013) classified over 1,600 households in the outer London borough of Kingston-upon-Thames, recognising six groups based on household behaviour and attitudes, focusing on management of household waste and 'environmentally friendly' actions. These groups ranged from 'eco-angels,' who demonstrated all-round environmentally conscious behaviours, to 'refusenics,' who had low levels of participation across all waste practices. In the middle of the spectrum were 'normative wasters,' recycling moderately large amounts of waste and making conscious decisions about waste reduction and reuse. They were keenly aware of the recycling behaviour of other households and were positive about most aspects of waste practices (p. 73). This group corresponds with one recognised by Harris Research (2011), who termed them 'green casuals': householders who had good intentions yet did not 'go out of their way' to protect the environment but were concerned about climate change and the environment, participating in domestic waste recycling schemes but unwilling to pay more for eco-friendly products.

'Green casuals' are prevented from recycling more by time, convenience, and a perception of poor facilities for effective waste management (Harder et al., 2006). They typify consumers engaged in a 'value action gap' process, whereby they espouse green values but for various reasons do not always act in accordance (Barr, 2016: 191–228). Hence, our research explicitly targeted 'green casuals,' selecting a small sample for closer scrutiny to provide a guide to what happens to waste within the 'black box' of the household.

Household recycling rates vary considerably across England and Wales (but have been static around 45% since 2011/12): from 17% to 65% for individual local authorities in 2018–19. Rates of >50% were recorded for many rural authorities, whilst rates in inner cities were generally <30%. For outer London, the rates were largely between these two extremes: >35% for western and southern outer boroughs and <35% in the east and northeast. This range provided suitable background for a London-based study, selecting households for detailed scrutiny by drawing upon households in the outer London boroughs of Kingston-upon-Thames, Merton, Redbridge, and Sutton, with respective recycling rates in 2018–19 of 49.4, 38.5, 24.9, and 49.1 (% of household waste sent for reuse, recycling, or composting) (DEFRA, 2019).

Methods

Commissioned by Coca-Cola Enterprises Ltd, the research examined the characteristics and waste practices of 'green casuals' and dynamics of household recycling. It aimed to demonstrate the value and role of focusing on practices within the household, appreciating intricate and contextual factors underpinning how waste is managed in household settings. Accordingly, the research methods focused on a qualitative approach based on diaries, in-depth interviews, photography, and direct observation of households.

Individual waste diaries act as a form of triangulation, ensuring greater validity for the data generated. Coupled with diaries and photographs, interviews are not simply recollections of things past, a specific problem with things that are disposed of and soon forgotten. Researchers and participants could reflect on specifics, such as the history behind the use of a given item, enhancing validity of the narrative in the form of detail and nuance. Photography also transforms the way in which waste is seen by householders: from being valueless to something signifying or representing an activity specific to them. Through diaries, participants could continue their involvement in the research while the researchers were absent. We provided each person in the selected households aged seven and over with a weekly diary to complete over a seven-day period. This generated multiple cross-cutting understandings and narratives, helping build a picture of household dynamics. Diaries enabled the research to follow things through the household from acquisition to disposal (Jörissen et al., 2015; Richter and Bokelmann, 2017). They permit reflection and specificity at different moments in the process. By examining the arrival, storage, use, and disposal of things, we can understand how household negotiation, habits, spatial and material layouts, and broader infrastructures – such as the materiality of packaging and the waste and recycling collection service – affect waste and recycling practices. The data collected from the diaries acted as a starting point for interviews (Figure 26.1). This repeat interview and multiple method approach achieved in-depth understanding of each household and its waste practices.

Interviews were transcribed verbatim and transcripts were coded until no new themes emerged. As a validity check, 40% of the coding was cross-checked with the participants, with differences discussed and resolved.

The researchers used a postal questionnaire and snowball technique to recruit participants. This meant that those recruited by postal survey or word of mouth introduced the researcher to new potential recruits. Once participants were identified as suitable for the research, fitted one of the demographic categories, lived either in a house or a flat, and were willing to take part, they were included in the project. All adult participants (parents or guardians for children) signed a consent form to confirm that they understood the aims, purpose, and objectives of the research. At the end of the research, each participating household received a gratuity of £300 for taking part in the study.

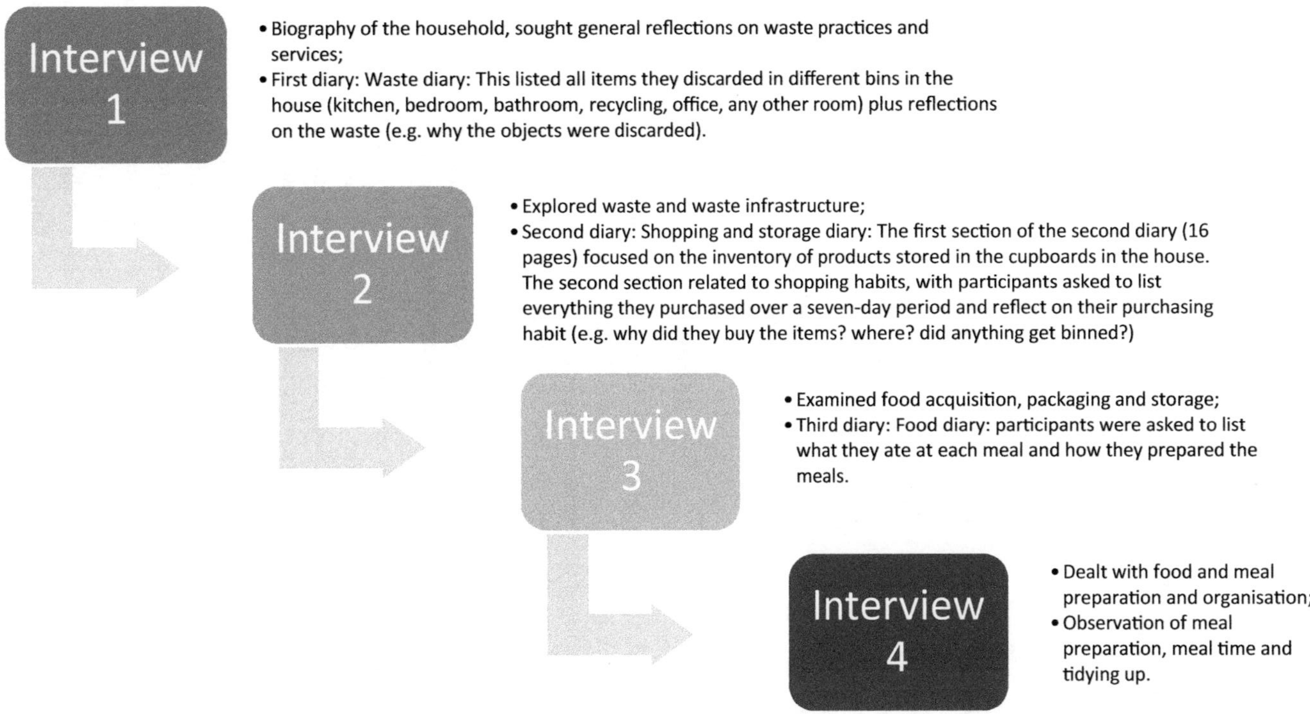

Figure 26.1 Interviews and diaries employed in investigating the 'green casuals'

Table 26.1 The sample households

Household Type	*Household Code No.*	*Names and Household Situation*
1 Young couple	B8	F, 20s, and M, 30s, married, no children, employed and student, flat
2 Retired couple	B5	F and M, 70s, married, retired, house
3 Single-person household	B6	M, 30s, single, employed, flat
4 Nuclear family with young children (<7 years old)	B1	F and M, 40s, married, 3 children (F, 6; M, 4; and M, 2), self-employed, house
5 Nuclear family with older children (aged 7–15)	B9	F, 30s, and M, 40s, married, 2 children (M, 8, and F, 7), employed and self-employed, house
	B10	F and M, 50s, married, 1 child (F, 15), employed, house
6 'Empty nesters'	B7	F and M, 60s, married, retired and semi-retired, empty nesters, house
7 House in multiple occupation	B2	F, 70s; M, 30s; and F, 20s, shared accommodation, retired and employed, house
8 35+/middle-aged no children (yet)	B4	F, 30s, and M, 50s, married, no children, employed, house
9 Single-parent household	B3	F, single mother, 2 children (M, 6, and M, 3), unemployed, house

Results

The various methods allowed insights to the habitual nature of waste and recycling practices in the home and the underpinning contexts which govern waste practices, all of which emphasised the need to focus on the household as both a social organism and as a unit where infrastructures, technologies, and aesthetics set the context for waste practices.

Household dynamics, 'decision-making,' and contested spaces of recycling practices

Recycling and waste practices more generally produce tension within certain household contexts, leading to acts of both resistance and compromise as members negotiate the necessities of daily living. These negotiations and their outcomes are important because they can affect the amount and type of materials recycled and the success or failure of recycling systems.

The interviews revealed that waste and recycling 'decision-making' are not a pre-formed and conscious part of household discourse. Rather, waste and recycling practices are rapidly assimilated into household habits that can quickly become unquestioned and 'unconscious' elements of everyday life. Habits refer to learned forms of practice undertaken on a regular basis and which rapidly become unquestioned and, most critically, difficult to change once they have become part of daily household routines. A familiar example is the 'habit' of driving children to school: something which might become easily learned (for example, consecutive days of bad weather leading to use of the car), a seemingly more convenient and simple act than using public transport or walking (as the car is regarded as a private space, with the freedom to move without the constriction of fixed timetables), and therefore it becomes clearly rationalised as the most effective choice and thus difficult to change.

Similarly, respondents did not actively make decisions about recycling, and it was only when they were asked to reflect on how they were dealing with household waste that a couple acknowledged they could easily change their waste practices and recycle more. Respondents 'dismissed' some of the questions about what happens to waste because the answer was 'obvious': "it goes there"; while others felt extremely uncomfortable discussing waste issues in their house, as they were perceived to be a private matter. It was therefore difficult to develop discussion about a given decision when 'no decision' had been made or a practice was deemed 'private.' Respondents could become unsettled when asked why they disposed of items, as the answer was so obvious to them: items are discarded when they are empty, dirty, or useless.

In all households, it was difficult to establish a clear pattern regarding 'decisions' because discarding something was a learned habit and the action often momentary in nature. As such, the nature of recycling varied in time and across individual members of the household, though recycling could also be a team effort, and the physical act of recycling could depend on who is available, who had the time, or who walked past the recycling box/container at a particular time. Decisions to recycle are embedded in habit, which is tested at a point in time.

Waste systems, technologies, and information

A second key outcome was to emphasise the challenges of complex recycling systems and the ways in which highly structured and technical waste systems can discourage practices of reuse. Such systems can be difficult to use, resulting in the rapid adoption of 'wasteful' habits. Respondents highlighted the complexity of recycling systems and their variation across local authority areas. Those with lower stated levels of environmental concern tended to discard items if they were unsure whether they could be recycled, whereas more concerned respondents recycled them.

Some respondents felt consideration should also be given to the type of containers used for recycling (see Lane and Wagner, 2013). Can recycling boxes kept indoors be carried outside easily? Is there space for a wheelie bin on a driveway? For large families, is the recycling box large enough? Convenience and the presence of infrastructure necessary for source-segregation of waste are important factors for household waste recycling (Jørgensen, 2013), but (our italics) this also "*highlights the need of addressing these aspects where waste is generated, i.e., already inside the household*" (Bernstad et al., 2013: 1317).

Information about recycling and the identification of the recyclability of materials was often perceived as being too superficial and unclear. Respondents argued that information for recycling or the recyclability of materials must be easily identifiable and visually appealing. The lack of information on recycling led one household to identify what could be recycled by looking into their neighbour's recycling box.

Un-packing the packaging

Another theme was the importance of packaging, which linked closely to the nature of the household's food shopping and eating habits. Understanding how shopping practices are undertaken and their relationship with packaging and eventual disposal is critical when exploring recycling, because households constitute an important point in the journey between product manufacture and discard.

Shopping was regarded by participants as an important but fundamentally transactional and utilitarian practice; that is, shopping largely served the purpose of simply and rapidly acquiring food and materials for daily living. Respondents claimed that what they purchased was determined primarily by product price, quality, taste preference, or local food. One household with small children purchased pre-prepared meals on a regular basis not only for convenience but also because it enabled 'a quick shop' with hardly any social interaction with the shop assistants. Similarly, for many households, shopping had to be quick, as it was not perceived as a social or 'valued' activity, and respondents claimed that they often bought the same products repeatedly to save time.

Working households were more likely to purchase and eat pre-prepared meals. However, the observation of household meals revealed that ready meals were often supplemented by some 'home cooking.' Respondents did not generally consider the packaging or its recyclability. They dealt with packaging in different ways: some would throw away the outer packaging on their return from shopping; others would keep this outer packaging to store their purchase. However, there was no clear logic in the decision to keep or remove packaging, and other factors such as time (to put shopping away) may have influenced this. Because respondents contended that generally they purchased the same items regularly, the disposal of packaging itself was something of a habit.

There was some ambivalence about the amount of packaging. Although there was consensus that there is 'too much packaging' of products, packaging was also seen as a necessity for protecting foodstuffs (therefore being associated with hygiene and safety) and preventing non-food items from being damaged. Respondents also argued that packaging and the recyclability of packaging material is the manufacturer's responsibility, claiming they had no control over the amount of packaging used by manufacturers and therefore their choices were very limited if they wanted to reduce the 'waste' from this source.

Waste trajectories

Many people within the waste and recycling industries emphasise the cyclicality of materials, but in contrast, the respondents often retained a linear conception of the movement of materials from manufacture to disposal. This was particularly evident with how children viewed acquisition and discarding of material objects, which they often illustrated with pictures in their research diaries. These linear perceptions of recycling did not vary with the household composition; whether the respondents were single, a couple, or a family, all had a similar understanding of materials consigned to waste in the household.

The concept of waste trajectories highlights one particular 'recycling imagination' within households. In most cases, there is a belief that once materials are set aside and deposited for recycling, that is the end of the process. There was often a declared absence of knowledge or concern about what happened to materials that were placed in recycling bins after those bins had been uplifted.

A sense that waste and materials sent for recycling were 'out of sight and out of mind' also existed alongside stated concerns and scepticism about recycling and its value, most likely derived from negative media coverage of the recycling industry and respondents' personal experiences of recycling. This reveals a further set of 'recycling imaginations,' the most important of which was reference to popular media portrayals of recycling materials being 'wasted' ('recycled' materials being sent to landfills) or exported for treatment. Respondents argued that these examples undermined their motives for recycling. In addition, the high cost associated with purchasing recycled products was noted. The value of recycling was therefore often questioned.

In terms of the recycling process, some respondents had knowledge of how specific recycled materials were used, especially through being transformed into the same products (e.g., a plastic bottle being melted and reformed as another plastic bottle), while others were aware of more elaborate ways to recycle materials (e.g., plastic bottles being transformed into fleeces). Financial incentives for recycling were also discussed, and most people remembered the returnable bottle scheme on selected glass bottles that used to be in operation in the United Kingdom. However, younger children, and especially teenagers, stated they would only consider returning recyclable materials if the financial incentive was 'worth it.'

Discussion and conclusions

Most households do not explicitly 'manage waste'; instead, they are simply managing their daily lives. Households largely recycle out of habits they have acquired. Recycling within the household has become an automatic response, a habit, to meet the demands of municipal waste management as proscribed by their local authority. The research reveals habits are both conscious and subconscious 'learned acts' that become automatic responses or routine behaviours in specific situations. More complex behaviours that have become 'routinised' may be described as "habitual behavioural patterns" (McGregor et al., 2019) or "semi-automatic response patterns" (Ajzen, 2002). Most of the individuals interviewed had the intention to recycle, but this is a complex issue: for some, the intention to recycle certain items such as glass may be broken down into an intention to keep bottles and jars for reuse, an intention to store the bottle in a container, or an intention to take the bottle to a bottle bank.

It is apparent that waste practices and recycling are dominated by the set of choices households have available to them. This applies to shopping habits (the packaging used for different products) and practices of disposal and recycling (what can and cannot be recycled). Behavioural change will only occur if the architectures surrounding household practices (such as waste systems and technologies) are adjusted to enable and promote changes in social practices. Indeed, the issue of space within the household is of high importance. Space for storing waste material can be minimised when people have the habit of frequently disposing of waste, and the location of the recycling box also affects the quantity of items to be recycled. Most respondents used an intermediary recycling point within the house, and when the bag/box was full, they would empty it in the recycling box/bin. Recycling is more effective if it requires little effort. Participants commented that if their recycling box/bin is adequate for their needs, then recycling is less problematic. There appear to be significant differences in the amount recycled according to the design of the waste storage container. In fact, those using a well-designed container that could be kept outside recycled more than those given containers that had to be kept indoors and taken out on collection days (Lane and Wagner, 2013). Recycling has become ever more commonplace, and its easiness and effortless aspects encourage recycling within the household. Our research agrees with Gonzalez-Torre et al. (2003) that frequent collection and bin/box capacity encourage more recycling amongst participants. It is also important to take into account the 'cost' of recycling for the individuals, as 'extra' effort and the unpleasant nature of the task may interfere with the motivation or desire to recycle (Ho, 2018).

The research emphasises the importance of connecting material worlds. Different 'recycling imaginations' exist about what the interviewees often regarded as an unknown and complex set of processes which constitute recycling after materials have been collected from the household. There is potential for changing practices when people have a connection to materials they consume (e.g., the increasing importance placed on the provenance of food products). Connecting the material worlds of 'recycling' between the home and recycled products could offer more opportunities for reducing the 'mysteries' surrounding recycling.

Overall, the research highlights four key findings to consider when identifying strategies to increase recycling rates amongst households. Household waste recycling is a dynamic and deeply contextual process that varies between households. Waste practices are learned habits, and the discarding of items is a momentary act that largely does not involve a conscious decision, so the notion of 'decision-making' is highly problematic. Waste practices are spatially and temporally variable, with sites of recycling and discarding appearing throughout the household, not just in the kitchen or back yard/garden. Pre-teenage children clearly act as protagonists for promoting recycling in households, while teenagers, in contrast, can often 'throw off' 'good' habits learned and practiced at school.

The flow of materials through households, and the discarding of materials, is partly driven by dynamic household politics. The compliance with or resistance by household members to practices of recycling determines what is recycled. Habit and learned practices become embedded into daily life, and so households do not necessarily 'decide' what to do with 'waste' each time they need to dispose of something; rather, they rapidly adopt learned habits within different household spaces. The way people deal with waste is governed by the systems and technologies that surround them, be they the type, number, and aesthetic of the recycling containers with which they are provided or the places in which they live, offering opportunities to deal with waste through practices that have a long history (e.g., yard burning or bonfires). Critically, although some elements of existing recycling systems are perceived as simple, differences in the specific materials collected support a perception that recycling is something imposed by a higher authority.

The waste systems in place as well as the technologies used for recycling are crucial to changing behaviour. Participants often cited their use of space and the size and number of recycling containers as a challenge, and hence there are major implications for the ways in which household spaces are planned and used for promoting behavioural change. For example, key spaces like kitchens or yards often have room for waste disposal and/or recycling 'designed out' rather than 'designed in.' Households highlighted the challenges of complex recycling systems and the ways in which highly structured and technical waste systems can both discourage practices of reuse and be difficult to use, resulting in the rapid adoption of wasteful habits.

Shopping habits were based on convenience, price, and perceived quality. Nonetheless, consumers felt they were powerless to influence the amount and types of packaging that are available and were critical of inconsistencies in, and the quality of, labelling. Respondents argued that packaging producers need to provide clearer labelling that is consistent and unambiguous. They argued that being more efficient in the use of packaging and reducing the variety of materials used was important in promoting a more holistic approach to waste management. The complex materialities of packaging influence perceptions of recycling as a problematic activity and the ways in which the type and amount of packaging frame attitudes towards recycling. Moreover, recycled products were regarded as being overpriced. Respondents often questioned and criticised current modes of recycling (as determined by local authorities), not only from a pragmatic perspective (i.e., what materials can be recycled and when and where) but also as a process. They often saw little value in recycling, questioning the economic value of recovering materials that were then refashioned into more expensive consumer goods. There was mistrust about 'what things become,' given the high media profile of news stories supposedly highlighting the futility of recycling. Recycling was still regarded as an 'end point' rather than the beginning of a new production process.

The knowledge of what things become could help some individuals to be more 'connected' to their waste and therefore more willing to recycle. There was often an absence of knowledge about what happened to materials placed in recycling bins. The value (both economic and environmental) of recycling was questioned, and we identified a number of 'recycling imaginations' that illustrated different beliefs about how recycling operated after materials were collected from the household.

Households are the key social unit in understanding recycling, which itself results from a set of negotiations and compromises between household members. Our research highlights the multilayered characteristics of the decision-making that occurs during recycling. The decision to recycle an item depends on factors such as age, time, nature of the item, and where the items have been binned in the first instance. Our research also highlights that lifespan impacts the decision to recycle. Younger children were keen to engage in recycling, while older teenagers disengaged from the process. Therefore, it is important for researchers to focus on the perception of the image of recycling and waste among teenagers. Education should be able to build a better understanding and good habits by using new technologies and creative arts to make the management of solid waste a more attractive activity in which to engage. Indeed, policymakers should also focus on the image of solid waste management and recycling among teenagers in order to maintain or increase recycling rates in the future.

Finally, the contemporary architectures and technologies of waste can limit recycling in the household. Further work is needed on the 'space of discard' and especially the aesthetic of the recycling containers, as well as creating homogeneous rules on discarding waste and recycling. We believe that behaviour will change only when the architecture that surrounds household practices is adjusted and simplified. Policymakers should engage with designers and housing developers to improve the architectures and aesthetics of recycling.

Acknowledgements

The research was approved by the Research Ethics Committee of the University of Exeter. The authors would like to thank all the participants who took part and Coca-Cola Enterprises for funding this project.

References

Ackerman F. (2013). *Why do we recycle? Markets, values, and public policy*. Washington, DC: Island Press.

Ajzen I. (2002). Residual effects of past on later behaviour: habituation and reasoned action perspectives. *Personality and Social Psychology Review*. 6: 107–122. doi:10.1207/S15327957PSPR0602_02.

Babaei AA, Alavi N, Goudarzi G, Teymouri P, Ahmadi K, Rafiee M. (2015). Household recycling knowledge, attitudes, and practices towards solid waste management. *Resources, Conservation and Recycling*. 102: 94–100. doi:10.1016/j.resconrec.2015.06.014.

Barr SW. (2014). Practicing the cultural green economy: where now for environmental social science? *Geografiska Annaler: Series B*. 96 (3): 231–243. doi:10.1111/geob.12048.

Barr SW. (2016). *Environment and society: sustainability, policy, and the citizen*. Aldershot and Burlington, VT: Ashgate.

Barr SW, Guilbert S, Metcalfe A, Riley M, Robinson GM, Tudor TL. (2013). Beyond recycling: an integrated approach for understanding municipal waste management. *Applied Geography*. 39: 67–77. doi:10.1016/j.apgeog.2012.11.006.

Bell S, Judson E, Bulkeley H, Powells G, Capova KA, Lynch D. (2015). Sociality and electricity in the United Kingdom: the influence of household dynamics on everyday consumption. *Energy Research & Social Science*. 9: 98–106. doi:10.1016/j.erss.2015.08.027.

Bernstad A, Malmqvist L, Truedsson C, la Cour Jansen J. (2013). Need for improvements in physical pretreatment of source-separated household food waste. *Waste Management*. 33: 746–754.

Botetzagias I, Dima AF, Malesios A. (2015). Extending the theory of planned behavior in the context of recycling: the role of moral norms and of demographic predictors. *Resources, Conservation and Recycling*. 95: 58–67. doi:10.1016/j.resconrec.2014.12.004.

Bravi L, Francioni B, Murmura F, Savelli E. (2020). Factors affecting household food waste among young consumers and actions to prevent it: a comparison among UK, Spain and Italy. *Resources, Conservation and Recycling*. 153. doi:10.1016/j.resconrec.2019.104586.

Castellani V, Sala S, Mirabella N. (2015). Beyond the throwaway society: a life cycle-based assessment of the environmental benefit of reuse. *Integrated Environmental Assessment and Management*. 11 (3): 373–382. doi:10.1002/ieam.1614.

Collins R. (2015). Keeping it in the family? Re-focusing household sustainability. *Geoforum*. 60: 22–32. doi:10.1016/j.geoforum.2015.01.006.

Cooper T. (2013). Sustainability, consumption, and the throwaway culture. In: Walker S, Girard J. (Eds.), *The handbook of design for sustainability*. London and New York: Bloomsbury Publishing. 137–155.

Cooper T. (2016). *Longer lasting products: alternatives to the throwaway society*. Boca Raton, FL: CRC Press.

Culiberg B. (2014). Towards an understanding of consumer recycling from an ethical perspective. *The International Journal of Consumer Studies*. 38 (1): 90–97. doi:10.1111/ijcs.12068.

Czajkowski M, Hanley N, Nyborg K. (2017). Social norms, morals, and self-interest as determinants of pro-environment behaviours: the case of household recycling. *Environmental and Resource Economics*. 66 (4): 647–670. doi:10.1007/s10640-015-9964-3.

Davies AR. (2016). *The geographies of garbage governance: interventions, interactions and outcomes*. London: Routledge.

Department for the Environment, Food and Rural Affairs (DEFRA). (2019). *Local authority collected waste generation from April 2000 to March 2019 (England and regions) and local authority data April 2018 to March 2019*. Available at: www.gov.uk/government/statistical-data-sets/env18-local-authority-collected-waste-annual-results-tables (Accessed 22/3/2020).

Evans DM. (2018). Rethinking material cultures of sustainability: commodity consumption, cultural biographies and following the thing. *Transactions of the Institute of British Geographers*. 43 (1): 110–121. doi:10.1111/tran.12206.

Finkbeiner M. (2014). Product environmental footprint – breakthrough or breakdown for policy implementation of life cycle assessment? *International Journal of Life Cycle Assessment*. 19 (2): 266–271. doi:10.1007/s11367-013-0678-x.

Gandy M. (2014). *Recycling and the politics of urban waste*. London: Routledge.

Garnett K, Cooper T. (2014). Effective dialogue: enhanced public engagement as a legitimising tool for municipal waste management decision-making. *Waste Manage*. 34 (12): 2709–2726. doi:10.1016/j.wasman.2014.08.011.

Gibson C, Head L, Carr C. (2015). From incremental change to radical disjuncture: rethinking everyday household sustainability practices as survival skills. *Annals of the Association of American Geographers*. 105 (2): 416–424. doi:10.1080/00045608.2014.973008.

Gonzalez-Torre PL, Adenso-Diaz B, Ruiz-Torres A. (2003). Some comparative factors regarding recycling collection systems in regions of the USA. *Journal of Environmental Management*. 69 (2): 129–138. doi:10.1016/S0301-4797(03)00109-9.

Gordon R. (2013). Unlocking the potential of upstream social marketing. *European Journal of Marketing*. 47 (9): 1525–1547. doi:10.1108/EJM-09-2011-0523.

Graham-Rowe E, Jessop DC, Sparks P. (2014). Identifying motivations and barriers to minimising household food waste. *Resources, Conservation and Recycling*. 84: 15–23. doi:10.1016/j.resconrec.2013.12.005.

Gregson N, Metcalfe A, Crewe L. (2007a). Identity, mobility and the throwaway society. *Environment and Planning D*. 25: 682–700. doi:10.1068/d418t.

Gregson N, Metcalfe A, Crewe L. (2007b). Moving things along: the conduits and practices of divestment in consumption. *Transactions of the Institute of British Geographers*. 32: 187–200. doi:10.1111/j.1475-5661.2007.00253.x.

Gregson N, Watkins H, Calestani M. (2013). Political markets: recycling, economization and marketization. *Economy and Society*. 42 (1): 1–25. doi:10.1080/03085147.2012.661625.

Hadjimanolis A. (2013). Household recycling in Cyprus: attitudes, barriers, and behaviour. *World Review of Entrepreneurship Management and Sustainable Development*. 9 (2): 291–306. doi:10.1504/WREMSD.2013.052364.

Harder M, Woodward R, Bench M. (2006). Two measured parameters correlated to participation rates in curbside recycling schemes in the UK. *Environmental Management*. 37 (4): 487–495. doi:10.1007/s00267-004-0124-8.

Harris Research (2011). *In-home recycling decision making. The opportunity to change recycling at home: quantitative and qualitative research overview*. PowerPoint presentation of summary results to Coca-Cola Enterprises.

Hawkins G. (2011). Packaging water: plastic bottles as market and public devices. *Economy and Society*. 40 (4): 534–552. doi:10.1080/03085147.2011.602295.

Hennig C, Meila M, Murtagh F, Rocci R. (2015). *Handbook of cluster analysis*. Washington, DC: CRC Press.

Ho E. (2018). Exploring the social dimensions of multi-residential recycling. *Resources, Conservation and Recycling*. 132: 77–78. doi:10.1016/j.resconrec.2018.01.027.

Hui A, Schatzki T, Shove E. (2016). *The nexus of practices: connections, constellations, practitioners*. London: Taylor and Francis.

Jörissen J, Priefer C, Bräutigam KR. (2015). Food waste generation at household level: results of a survey among employees of two European research centers in Italy and Germany. *Sustainability*. 7 (3): 2695–2715. doi:10.3390/su7032695.

Jørgensen F. (2013). Green citizenship at the recycling junction: consumers and infrastructures for the recycling of packaging in twentieth-century Norway. *Contemporary European History*. 22 (3): 499–516.

Kalmykova Y, Rosado L, Patrício J. (2016). Resource consumption drivers and pathways to reduction: economy, policy and lifestyle impact on material flows at the national and urban scale. *Journal of Cleaner Production*. 132: 70–80. doi:10.1016/j.jclepro.2015.02.027.

Keramitsoglou KM, Tsagarakis KP. (2013). Public participation in designing a recycling scheme towards maximum public acceptance. *Resources, Conservation and Recycling*. 70: 55–67. doi:10.1016/j.resconrec.2012.09.015.

Kirakozian A. (2016). The determinants of household recycling: social influence, public policies and environmental preferences. *Applied Economics*. 48 (16): 1481–1503. doi:10.1080/00036846.2015.1102843.

Lane GW, Wagner TP. (2013). Examining recycling container attributes and household recycling practices. *Resources, Conservation and Recycling*. 75: 32–40. doi:10.1016/j.resconrec.2013.03.005.

Lane R, Watson M. (2012). Stewardship of things: the radical potential of product stewardship for re-framing responsibilities and relationships to products and materials. *Geoforum*. 43 (6): 1254–1265. doi:10.1016/j.geoforum.2012.03.012.

López-Mosquera N, Lera-López F, Sánchez M. (2015). Key factors to explain recycling, car use and environmentally responsible purchase behaviors: a comparative perspective. *Resources, Conservation and Recycling*. 99: 29–39. doi:10.1016/j.resconrec.2015.03.007.

MacGregor S, Walker C, Katz-Gerro T. (2019). 'It's what I've always done': continuity and change in the household sustainability practices of Somali immigrants in the UK. *Geoforum*. 107: 143–153. doi:10.1016/j.geoforum.2019.09.013.

Mansvelt J. (2012). Consumption, ageing and identity: New Zealander's narratives of gifting, ridding and passing on. *New Zealand Geographer*. 68 (3): 187–200. doi:10.1111/j.1745-7939.2012.01233.x.

Metcalfe A, Riley M, Barr SW, Tudor T, Robinson GM, Guilbert S. (2013). Food waste bins: bridging waste infrastructures and practices. In: Evans D, Campbell H, Murcott A. (Eds.), *Waste matters: new perspectives on food and society*. Oxford: Wiley. Session 17.

Monnot E, Reniou F, Rouquet A. (2014). Recycling household waste: a classification of the logistics used by consumers. *Research and Applications in Marketing*. 29 (3): 75–97. doi:10.1177/2051570714540553.

Mueller W. (2013). The effectiveness of recycling policy options: waste diversion or just diversions? *Waste Management*. 33 (3): 508–518. doi:10.1016/j.wasman.2012.12.007.

Pettifor H. (2012). *Patterns of household practice: an examination into the relationship between housework and waste separation for households in the United Kingdom*. ISER Working Paper Series. No. 2012-14. Available at: https://EconPapers.repec.org/RePEc:ese:iserwp:2012-14.

Richter B, Bokelmann W. (2017). Explorative study about the analysis of storing, purchasing, and wasting food by using household diaries. *Resources, Conservation, Recycling*. 125: 181–187. doi:10.1016/j.resconrec.2017.06.006.

Saphores JDM, Ogunseitan OA, Shapiro AA. (2012). Willingness to engage in a pro-environmental behavior: an analysis of e-waste recycling based on a national survey of US households. *Resources, Conservation and Recycling*. 60: 49–63. doi:10.1016/j.resconrec.2011.12.003.

Scott A, Oates C, Young W. (2015). A conceptual framework of the adoption and practice of environmental actions in households. *Sustainability*. 7 (5): 5793–5818. doi:10.3390/su7055793.

Shove EA. (2010). Beyond the ABC: climate change policy and theories of social change. *Environment and Planning A*. 42: 1273–1285. doi:10.1068/a42282.

Shove EA, Pantzar M, Watson M. (2012). *The dynamics of social practice: everyday life and how it changes*. London: Sage.

Spurling NJ, McMeekin A, Southerton D, Shove EA, Welch D. (2013). *Interventions in practice: reframing policy approaches to consumer behaviour*. Manchester: Sustainable Practices Research Group, University of Manchester.

Steg L, Bolderdijk JW, Keizer K, Perlaviciute G. (2014). An integrated framework for encouraging pro-environmental behaviour: the role of values, situational factors, and goals. *Journal of Environmental Psychology*. 38: 104–115. doi:10.1016/j.jenvp.2014.01.002.

Stern NH. (2006). *Stern review: the economics of climate change*. London: HM Treasury. 30.

Stern P. (2000). Toward a coherent theory of environmentally significant behaviour. *Journal of Social Issues*. 56 (3): 407–424.

Tudor TL, Robinson GM, Riley M, Guilbert S, Barr SW. (2011). Challenges facing the sustainable consumption and waste management agendas: perspectives on UK households. *Local Environment*. 16 (1): 51–66. doi:10.1080/13549839.2010.548372.

Varotto A, Spagnolli A. (2017). Psychological strategies to promote household recycling: a systematic review with meta-analysis of validated field interventions. *Journal of Environmental Psychology*. 51: 168–188. doi:10.1016/j.jenvp.2017.03.011.

Waite R. (2013). *Household waste recycling*. London: Routledge.

White KM, Hyde MK. (2011). The role of self-perceptions in the prediction of household recycling behavior in Australia. *Environment and Behavior*. 44 (6): 785–799. doi:10.1177/0013916511408069.

27

CIRCULAR START-UPS

Five business model archetypes as frontrunners of circular disruption

Marvin Henry and Julian Kirchherr

Introduction

This chapter aims to shed further light on the circular business model (CBM) strategies and circular innovations pursued by circular start-ups (CSUs) to contribute to the theoretical body of sustainable innovations and to existing frameworks operationalising CBMs. Thus far, research on CBMs has mainly focused on approaches to CE by large incumbents, such as Google, Carlsberg, Renault, or Unilever, possibly because of their higher visibility and the influence they have on their respective markets (e.g., Bocken et al., 2017; Ma et al., 2014; Niero et al., 2017; Stewart and Niero, 2018; Urbinati et al., 2017; Williams, 2007). In contrast, to the best of the authors' knowledge, no empirical scholarly work thus far has been carried out on CSUs directly. In this research circular start-ups are defined as new, independent, and active companies pursuing a CBM. Circular entrepreneurship is still a quite novel concept in scientific literature (Zucchella and Urban, 2019).

Therefore, the objective of this research was to build on existing literature on circularity strategies and circular (sustainable) business model innovation to develop a business model typology of the potentially most innovative players in private sector CE. The questions addressed by this chapter are as follows:

RQ1: Which CBM strategies are adopted by CSUs?
RQ2: What types of CBM innovations do CSUs pursue?

Methods

The authors address the questions by analysing the CBM strategies and innovations of 128 CSUs located in three major CSU ecosystems in Europe: the Randstad region in the Netherlands,[1] London, and Berlin. The dataset was newly compiled and evaluated for the purpose of this study and has not been used for any other peer-reviewed publications before, and to the best of the authors' knowledge, there has been no research conducted on CSUs' actual business models. Therefore, the data can be proclaimed as novel.

Regional and organisational scope

The three hubs under scrutiny were chosen because they host a vibrant movement towards circularity driven by public and private organisations. Table 27.1 provides examples of initiatives aiming at strengthening the development of CE and/or circular start-ups in these locations. In addition, at the time of the study, they were key hubs for start-up activity, often positioned in the top five European start-up hubs (e.g., European Startup Initiative, 2017). These start-up hotspots are thus likely to host the circular innovations that are essential to consider if one seeks to understand what the CE will look like in the future. In addition, many CSUs were expected to be found in these locations. Only firms corresponding to the definition of CSUs were included in the analysis (see Chapter 11).

A list of 128 CSUs was created to present a comprehensive overview of CSUs in the three geographical areas under scrutiny. The search for CSUs was performed until no additional organisation could be found, with the view of developing a list that was as exhaustive as possible. A total of 68 (54%) companies were located in the Randstad region, 28 (22%) from Berlin, and 31 (24%) from London. The companies were identified through the existing resources of their respective municipal authorities and via CSU awards and circular hubs. Data were collected using three main sources, namely:

- Publicly available information from the websites and social media of start-ups as well as relevant press articles
- Existing case studies in the grey literature describing these start-ups
- For part of our sample, semi-structured interviews with founders. These interviews were intended to obtain a more in-depth understanding of the business models of CSUs and the CBM strategies they implement.

Overall, interviews were conducted with 30 of the identified CSUs: 10 in the Netherlands, 14 in Berlin, and 6 in London. They were selected to cover the diversity of sectors of activity and of

Table 27.1 Circular economy initiatives in the scrutinised hubs

Region/city	*Rationale*	*Source*
Randstad (Amsterdam, Rotterdam, The Hague, Utrecht)	• Setup of acceleration program in Amsterdam to facilitate knowledge exchange, funding, and networking for anyone pursuing innovative CE concepts • Rotterdam launched a CE initiative and hosts a circular innovation hub (BlueCity010), and one of the municipalities communicated measures to embed CE in the region through attracting CSUs	Amsterdam Smart City, 2017; Gemeente Rotterdam, 2016; Hofnaegels, 2016
Berlin	• Berlin is the base for more than 400 companies identifying themselves as circular and has more than 8,000 employees in this sector • Recently, CRCLR[2] was launched in Berlin, a think- and do-tank dedicated to CE on more than 2,500 sqm	Berlin Business Location Centre, 2017
London	• The London Waste and Recycling Board (LWARB) launched the 'Circular London' and 'Advance London' initiatives in 2018, aiming at strengthening collaborations and stakeholder engagement around CE and enabling small- and medium-sized companies to leverage the benefits of CE	London Waste and Recycling Board, 2017

2 CRCLR is a hub and think- and do tank for CE in Berlin.

business models that were found in the entire dataset. Interviews, which lasted between 45 and 75 minutes, were conducted face to face or via telephone or Skype. All interviews were audio-recorded and transcribed. Data about companies' background were also collected, including the sector of activity, date of foundation, number of employees, client focus (business-to-business [B2B], business-to-customer [B2C], or both), and, when available, annual result or revenue.[3]

Data analysis

The data were analysed using content analysis (Krippendorff, 2013: Figure 27.1). First, all extracts which referred to the business models of CSUs were collected and stored in the coding software NVivo. Next, a coding framework was developed to enable the authors to convert qualitative information into numeric data (Bourque and Dudek, 2004). To do so, both deductive categorisation, which uses higher-level existing categories from the literature (e.g., Urbinati et al.'s [2017] distinction between upstream and downstream companies or Khircherr et al.'s

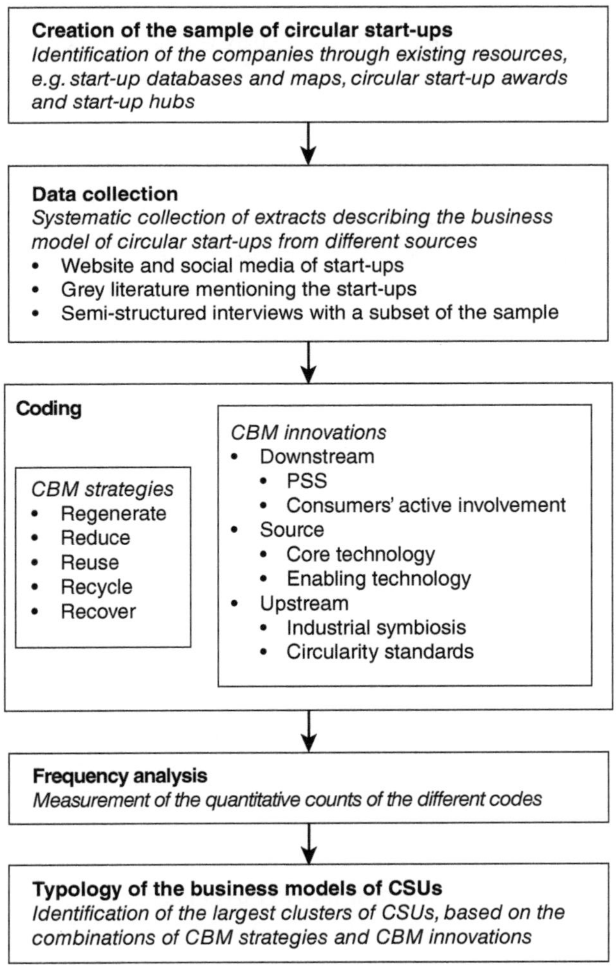

Figure 27.1 The procedure followed

[2017] 4R framework), and inductive categorisation, which uses lower-level codes (e.g., specific CBM innovations and strategies developed by CSUs) grouped into higher-level codes, were employed to develop the coding categories (Hsieh and Shannon, 2005). Coding categories had to be mutually exclusive and collectively exhaustive.

The information collected about the business model of CSUs was then carefully examined and manually coded according to this coding framework, following an initial set of coding rules. For example, a start-up substituting product components with more sustainable materials was coded as the CBM strategy 'Reduce,' while a company using external residual resource streams was coded into the innovation category 'Industrial symbiosis.' To ensure the validity and reliability of the research design, the coding exercise was first completed by the two authors separately, who each coded half of the sample and then discussed results with each other. In the case of diverging coding results, the coding rules were reconsidered, possibly leading to their modification – a good practice to enhance reliability in manual coding (e.g., Hruschka et al., 2004; Neuendorf, 2016).

In practice, the R-strategies often cannot be clearly separated. A product made of multiple components frequently requires the combination of several Rs (Reike et al., 2018), or a business model may simultaneously embrace various CBM strategies. Therefore, the coding procedure allowed for a dominant R-strategy and multiple secondary R-strategies. For instance, a start-up can adopt 'Reduce' as a dominant strategy by developing less resource-intensive material and, at the same time, encourage the reuse of its products. The decision regarding the dominant R-strategy was made on a qualitative basis (i.e., what is considered the key value driver of the business model) and was complemented with the inputs from the founders of the respective CSUs. Similarly, the coding procedure allowed for a multiplicity of circular innovations to depict the practical (hybrid) realities of the business models analysed. For instance, a start-up could adopt a product design that facilitates reparability (innovation in the core technology) while organising part of its activities as a product service system (PSS). There was no dominant innovation type per start-up determined due to lack of insight and missing scientific context. Since some of the innovation types fall in the same innovation categories as the framework presented in Chapter 11, the absolute number of different innovation types is bigger than the number of innovation categories developed by the CSUs.

As a next step, a frequency analysis was performed to measure the quantitative counts of the different codes. This enabled the authors to assess the relative significance of the CBM strategies and innovations developed by CSUs. In addition, CBM strategies of CSUs could be compared to those adopted by corporate actors, based on the data from a previous study using a similar methodology for 46 corporates (Stewart and Niero, 2018). Since there are no studies categorising corporates' CBM innovations in a similar manner based on the framework used in this research, there was no comparison made between corporates and CSUs in this respect. Finally, the relationship between CBM strategies and innovations was analysed by looking at the type of CBM innovations per CBM strategy category. By doing so, it was possible to identify the five most relevant combinations of CBM innovations and strategies in the dataset. These five combinations provided the basis for developing a typology of business models of CSUs, which enabled the authors to unambiguously classify more than 90% of CSUs in the dataset.

Results and discussion

Sector and market overview

The most dominant sector within the dataset is agriculture/food (27%). Other sectors strongly represented in the hubs are manufacturing/materials engineering and built environment/design (Figure 27.2). The emphasis on food is not surprising from a market perspective, since

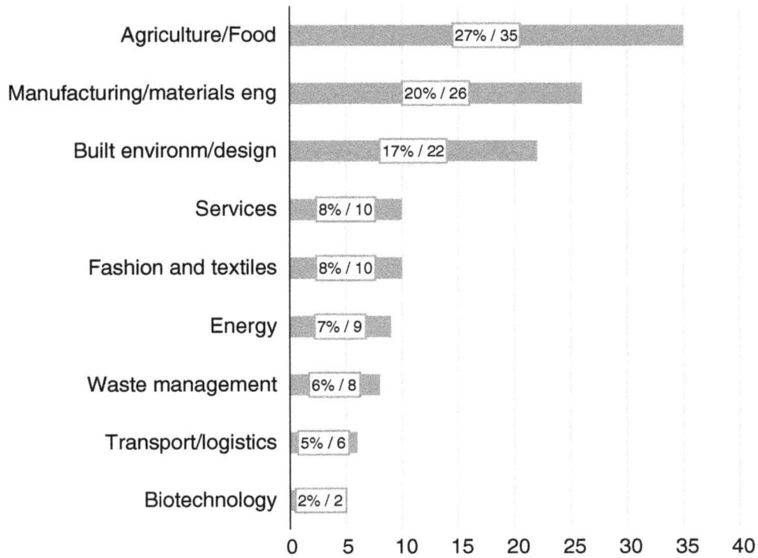

Figure 27.2 Circular start-up sector overview. Relative share in %/absolute share in #

inefficiency in food-supply chains is pressing and evident, especially in developed countries (Buzby and Hyman, 2012; Ribeiro et al., 2018; Thi et al., 2014), and corporates active in the food sector do not tackle waste management appropriately, mostly due to food-grading systems and lack of actor coordination (European Commission, 2016; Martin-Rios et al., 2018; Stewart and Niero, 2018). In total, 46% of CSUs examined deploy business models in the biological cycle, while only 7% of corporates that are active in only one of the cycles do (Stewart and Niero, 2018; 75% of CSUs are in the biological cycle if sectors are matched with the reference study, which focuses on corporates in fast-moving consumer goods). Indeed, multiple value models, in which sequential cash flows can be generated from waste in biological cycles, are predominantly adopted by start-ups. Big companies seem to fail to integrate such novel ways of value creation (Bocken, 2017).

Circular business model strategies and innovations

The findings on CBM strategies show that CSUs tend to adopt higher-ranked CBM strategies than incumbents (Regenerate, Reduce; Figure 27.3). The companies (mostly incumbents) scrutinised in the reference study were only included when listed in the Ellen MacArthur Foundation's CE100 directory (Stewart and Niero, 2018) and can, therefore, be considered frontrunners implementing CE. Previous research suggests that, although CE is increasingly integrated into the corporate sustainability agenda, the focus is on end-of-life management, while the adoption of business models incorporating higher levels of circularity is less prevalent (Bocken, 2017; Stewart and Niero, 2018). Out of the higher-ranked CBM strategies, incumbents are more strongly represented only in the 'Reuse' concept. This can be explained by two facts: first, reverse logistics, which would allow for a second use cycle of a technical product, are complex and costly (Neely, 2008; Ramanathan, 2011; Ravi et al., 2005); larger companies tend to have better resources to set up an adequate take-back management (Veleva and Bodkin, 2018).

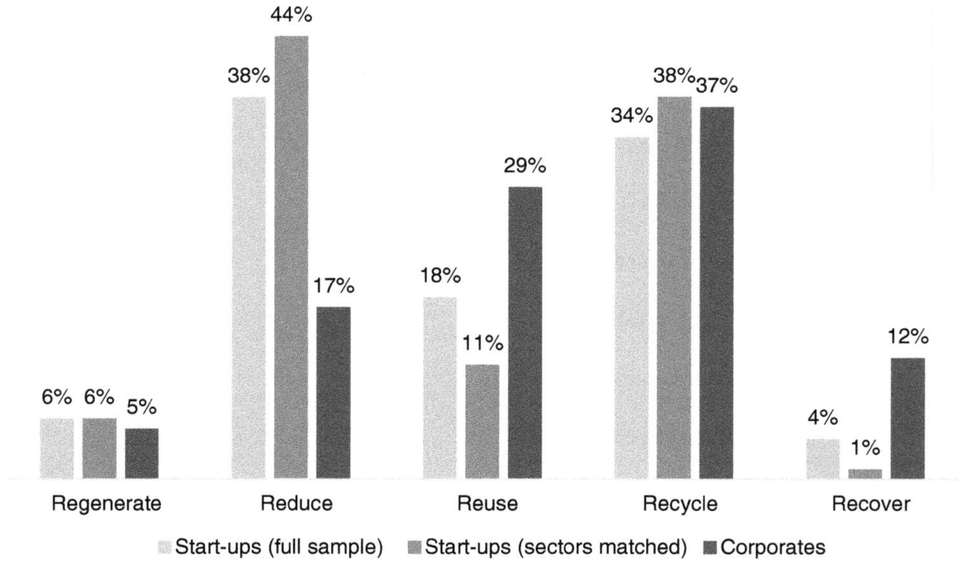

Figure 27.3 Circular business model strategies: circular start-ups vs incumbents

Second, the technological cycle is dominant among incumbents innovating in CE (Stewart and Niero, 2018), and 'Reuse' is the CBM strategy that mostly covers tech-related business models; goods in the biological cycle are rather of single-use nature, at least for the same or similar purpose of use.[4] In terms of incumbents' activity in 'Regenerate,' Stewart and Niero (2018) give indications on corporates' approaches taken through including the strategy 'Design for Regenerative Systems' (Moreno et al., 2016) in their analysis (base for calculation in Figure 27.3; see Stewart and Niero (2018) for the calculation approach).

The innovations realised by the start-ups are mostly around socioinstitutional change (see Figure 27.4); two-thirds of the start-ups innovating in core technology combine this with socioinstitutional innovation. Thus, CSUs act in alignment with the theoretical view that new technology must build up the required socioinstitutional embedding and has to be diffused in society to be successful (Geels, 2005; Boons and Lüdeke-Freund, 2013), despite the higher complexity of driving factors of socioinstitutional change compared to technological change (Fuenfschilling and Truffer, 2013). CSUs address this primarily by actively involving customers (~40% of CSUs) and working with interorganisational waste streams (~50% of CSUs). The high share of customer involvement activities partly contradicts Boons and Lüdeke-Freund's (2013) findings on the successful marketing of sustainable innovations, since it can be argued that Collaborative Consumption or Return and Repair includes a shift – or at least the sharing – of the focal organisation's socioecological burdens with its customers. However, it sheds light on the central question posed by Hall et al. (2010) on how start-ups and entrepreneurs will influence a societal transformation towards sustainability. Combined with the close ties of the CSUs with societal actors, (core) technological innovation is driving more than half of the start-ups. Core technology innovations, the most represented innovation category, range from nickel-iron batteries to fungus-based construction materials or bubble barriers to filter plastic from rivers. Last, it can be observed that CSUs do not necessarily formally engage with suppliers in circular supply chain management (circularity standards), which is considered pivotal for the marketing of sustainable innovations (Boons and Lüdeke-Freund, 2013; Seuring and Müller, 2008).

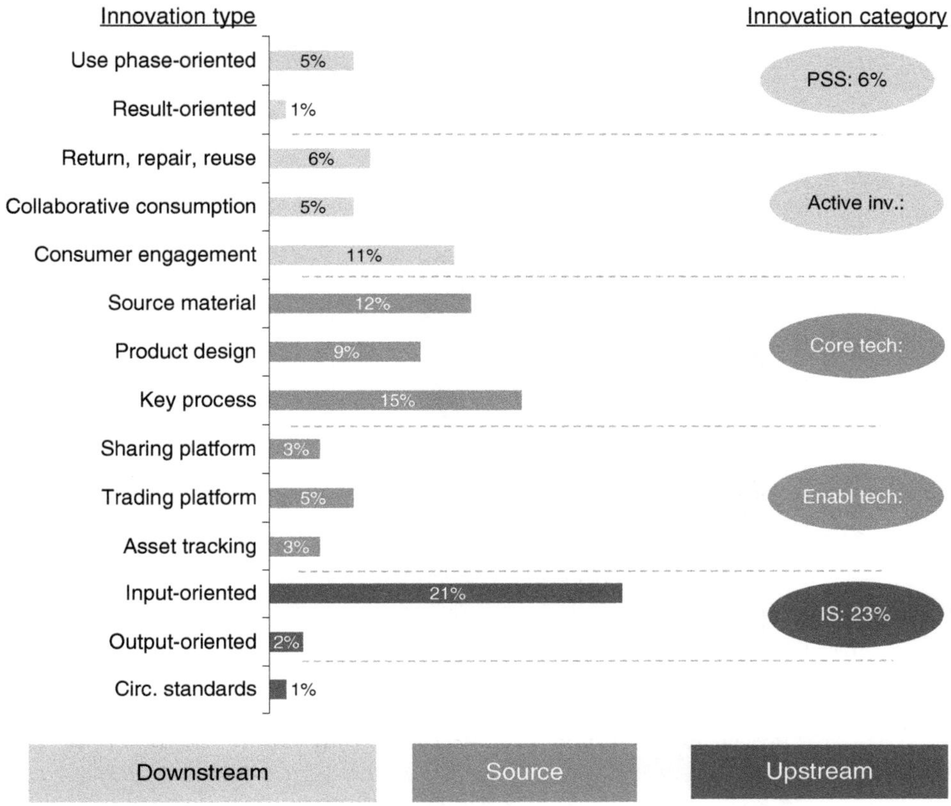

Figure 27.4 Circular business model innovation types. Percent of all innovation types

Towards a typology of circular start-ups

The typology of CSUs is based on the various innovation categories pursued and on the dominant R-strategies per venture (Figure 27.5) identifying the biggest clusters in the dataset and thereby creating additional structural insight into the approaches taken to develop CBMs.

Based on this analysis and derived from the data gathered, a typology of CSUs contrasting five main archetypes was established. The following presentation and discussion of results is structured along these five typologies (Table 27.2). In case one CSU fell into several innovation categories, the R-strategy was used as the decisive criterion. This is due to the fact that the authors assigned a dominant CBM strategy to each CSU, and this was taken into account for the development of the typology.

Based on the typology outlined (see Table 11.2 in Chapter 11), there is a marginal number of seven cases of CSUs that would fit into more than one cluster. For these, the authors applied a qualitative assessment based on the concrete type of innovation pursued and the key elements of the business model. For instance, there are two PSS-operating CSUs also innovating in core technology, substituting one component of their product with reusable source material or designing for reuse. These are still in the cluster of service-based start-ups, since the core technology innovation is a means to an end for them. Only four start-ups did not fit the

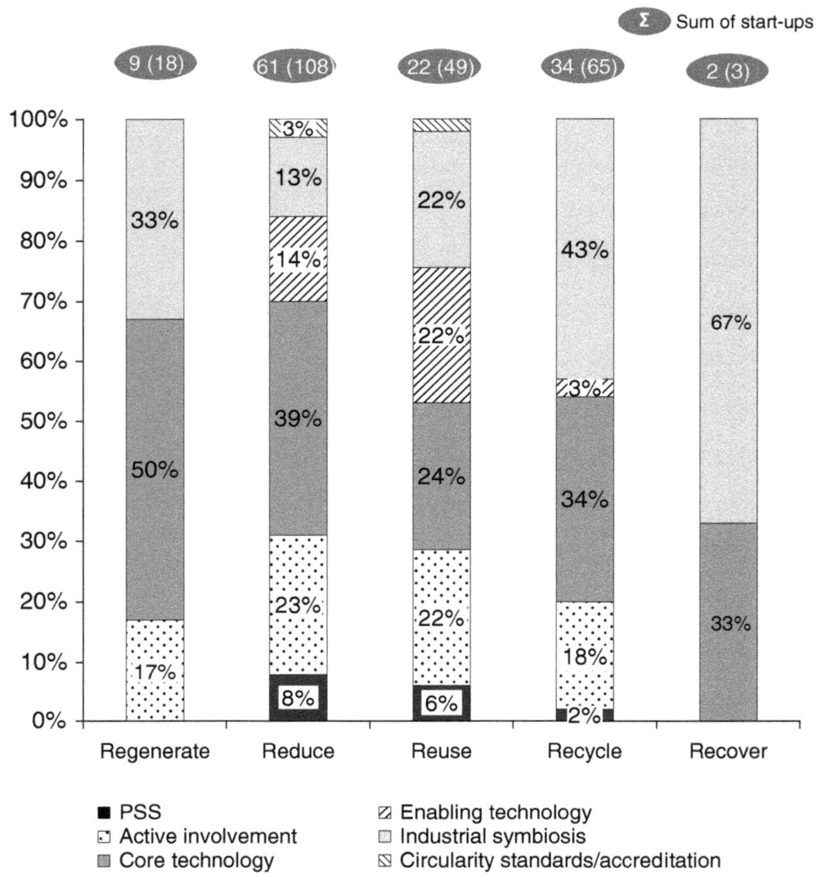

Figure 27.5 Innovation categories of circular start-ups per R-cluster. Percent of total innovation categories in respective R-strategy cluster (n = 243)

Table 27.2 Categorisation criteria and definition for circular start-up typologies

CSU archetype	*R-strategy*	*Innovation category*	*Definition*	*Share of data sample (n = 128)*
Design-based	*Reduce*	*Core technology*	Aiming to increase usage efficiency or avoid residual resource streams, design-based CSUs adopt circular innovations mostly in the pre-market phase of their product or service through source material minimisation, product design, or production process efficiency.	**36%**
Waste-based	*Recycle, Recover*	Industrial symbiosis	Waste-based CSUs seek to extract value from unexploited external waste streams (e.g., recycled plastic, CO_2, surplus food), mostly based on innovative process solutions.	**27%**

CSU archetype	*R-strategy*	*Innovation category*	*Definition*	*Share of data sample (n = 128)*
Platform-based	Various	*Enabling technology*	Platform-based CSUs pursue business models built around B2B, B2C, or customer-to-customer (C2C) marketplaces for (excess) resources. Thus, they facilitate trading or sharing of products, knowledge, infrastructure, or services.	**19%**
Service-based	Various	*PSS*	Service-based CSUs embed products in a service-system without customer ownership of the physical good, aiming for higher and better controllable usage efficiency.	**9%**
Nature-based	*Regenerate*	Various	Nature-based CSUs operate nature-based systemic solutions to deliver services (or products) with the objective of lowering input of non-renewable natural capital and increasing investment in renewable natural processes.	**6%**
Other	*Reduce, Reuse, Recycle*	Various	CSUs operating natural urban air filters or circular city tours or applying several CBM strategies and innovation types.	**3%**

Note: Mutually exclusive archetypes; dominant criterion for archetype stated in italics

typology presented in Table 27.1 (e.g., circular city tours, bio-tech filters for urban air) and are not included in any of the archetype clusters.

Design-based start-ups

This cluster of start-ups includes business models combining the R-strategy 'Reduce' with innovation in core technology and is the biggest group within the sample (37%). The high share of design-based start-ups can be explained by the high circular ambition of CSU founders, thus the dominance of the 'Reduce' strategy, as well as the correlation of 'Reduce' and the most dominant innovation types (core technology). More than two-thirds of start-ups (42 of 61) with a dominant 'Reduce' strategy innovate in core technology. This echoes the dominant scientific view connecting 'Reduce' predominantly with innovation in source material and product design and, therefore, mostly with activities during the pre-market phases of product life cycles (Lieder and Rashid, 2016; Reike et al., 2018; Sihvonen and Ritola, 2015; Worrell and Reuter, 2014). This CSU cluster includes companies striving for source material minimisation (e.g., fungus as construction material), process efficiency (e.g., low-cost infrastructure biomass growing), or innovators in product design (e.g., electricity-independent medical refrigerators).

Despite the common focus on the pre-market phase, the findings also suggest that design-based start-ups could be relevant micro-level innovators regarding integrating upstream-related – and, especially, downstream-related – activities in CBMs. More than half of the start-ups in this cluster complement their core technology innovations with innovations in active customer involvement or industrial symbiosis (IS). Amsterdam-based Kartent is one example of such a start-up. Kartent designed a festival tent made entirely from cardboard. This product design innovation is embedded in a business model that requires customers to change consumption patterns (ordering the tent for a festival, returning after usage) and utilises existing supply chains to create value from the waste streams (used tents processed into boxes for logistics industry or festival gadgets). Literature suggests that it is a key challenge for corporates to especially include customers in their circular endeavours and create awareness among them (Kirchherr et al., 2017; Young et al., 2018). This, however, is considered critical for the development of CBMs and an eventual transition to a CE (Hazen et al., 2017; Linder and Williander, 2017; Veleva and Bodkin, 2018), whereby CSUs could be relevant and necessary pioneers in successfully applying such business models.

Waste-based start-ups

Waste-based start-ups are the second-largest group (27%) of organisations in the dataset. They pursue the lower-ranked R-strategies 'Recycle' and 'Recover' as their dominant circular strategies and commonly innovate in IS (>95% of all Recycle and Recover start-ups). IS is a commonly known concept (Chertow, 2000) and is increasingly considered a tool for the implementation of CE (Baldassarre et al., 2019; Bocken et al., 2014; Domenech et al., 2019; Mortensen and Kørnøv, 2019; Prosman et al., 2017). Recycled materials mostly stem from post-consumer products or post-producer waste streams and are called secondary materials, since they often lose their fit to the original use case and can be re-applied anywhere (Worrell and Reuter, 2014; Graedel et al., 2011; King et al., 2006). Start-ups in this category include organisations turning recycled plastic into marble tiles or community-based organisations, such as recyclers of surplus food.

Only one-fifth of the start-ups using the 'Recycle' and 'Recover' strategies that innovate via IS are primarily in the waste-management sector. The sectoral focus of the start-ups is on agriculture/food (43%) or manufacturing/materials engineering (32%), whereby recycling/recovering seems to be rather a means to an end than an end in itself. More than two-thirds of companies in this cluster combine IS with core technology innovations in source material or key processes. While key process innovations in 'Recycling' can involve growing mushrooms from coffee waste, there are also more breakthrough process innovations included, such as turning CO_2 and sunlight into organic compounds or the process of creating a leather-like material from leftover fruits. An example for a source-material innovation is Rotterdam-based Refil, which develops fully recycled 3D-printing filaments made from car dashboards and old PET bottles. The CSUs active in 3D printing produce a waste stream–based filament or offer it through local recycling systems (e.g., of waste plastic). These types of offerings correspond to Despeisse et al.'s (2017) findings and represent two of the three focus categories of activities combining 3D printing and CE (the third using 3D printing for repair or remanufacturing).

The development of supply chain networks is often initiated by the players themselves due to lack of legislative clarity about by-products and waste streams. Across Europe,[5] there are a significant number of self-organising industry networks striving for circularity through IS (Domenech et al., 2019). In the expert interviews, founders stated that they tend to face regulatory problems, as in some markets, like Germany, resources that have been declared as waste cannot be reused but must be disposed of or require licenses allowing for further processing ('Waste-material sourcing requires specific licenses, and the tax on it is too high, in my opinion,' Interviewee A). Thus the companies using waste as resources tend to establish informal networks to disrupt the established supply chain, for instance, as half of the Berlin participants in the expert interviews stated when asked for major external challenges ('[the source material] is officially a waste product, so there are regulatory barriers regarding working with this material. Here, we rely on our entrepreneurial spirit and just go ahead without having regulatory clarity,' Interviewee B; '[We are facing a] lack of regulation in terms of declaration of waste material and how to handle it,' Interviewee C).

Platform-based start-ups

This archetype of start-up has, at its core, business models built around marketplaces for excess resources, mainly from the construction, fashion, or home-appliances sector. The dominant R-strategies represented here are 'Reduce' and 'Reuse.' The fact that platform-based business models are the third-largest category (20%) in the overall sample of CSUs can be explained by the general low labour intensity and thus low entry barriers, as well as the diffusion of peer-to-peer platforms as key enablers of the growing sharing and circular economy (Ellen MacArthur Foundation, 2015; Ritter and Schanz, 2019). Sharing platforms are the dominant innovation type among start-ups pursuing a 'Reduce' strategy within this cluster (7 out of 11 'Reduce' start-ups). Taking a closer look at the forms of value creation on the platforms, van Dijck et al. (2018), Potting et al. (2017) and Konietzko et al. (2019) seem to be confirmed in their views that (online) platform business models are rather facilitators of economic transactions (trading, sharing) than bases for co-creation – that is, sharing knowledge or debating/learning together. Only 3 of the 26 start-ups in this cluster focus on co-creation (e.g., Materiom, an open-source material database), while more than 90% of platform-based business models enable sharing or trading.

In the context of platforms and their roles in a CE, Konietzko et al. (2019)[6] mention the value of usage data – for example, to inform maintenance or repair needs of products. Additionally, the Ellen Macarthur Foundation et al. (2019) discuss the example of a trading platform for secondary resources with a focus on pricing algorithms and forecasting. A total of four of the 26 platform-based start-ups have a distinct focus on this kind of value creation via asset tracking. While one of the CSUs in this cluster develops a platform solution helping organisations and their networks transitioning towards circularity through business and artificial intelligence (AI),[7] the others use technology to track resources and product components. Berlin-based Design for Circularity developed a radio-frequency identification (RFID) label on clothes that grants consumers and sorting companies access to a member-based platform containing information on the inherent materials as well as on the recycler the product should go to at end of life. Another example of advanced technology used for asset tracking is Excess Materials Exchange, which uses blockchain technology to facilitate the exchange of excess material among companies. All four CSUs make the data accessible via a multiuser platform, and three of them define their offerings as platforms. Due to this evidence, and since the literature connects asset tracking with platforms in CE (Konietzko et al., 2019), this type of innovation is clustered as platform based.

Service-based start-ups

Service-based CSUs operate PSSs in order to better control and increase usage efficiency of their continuously owned products. Interestingly, only one PSS-oriented company in the entire dataset runs a CBM where ownership is transferred to the customer (Fairphone). This indicates that CSUs tend to focus on need-fulfilment systems rather than selling the maximum possible number of products (Tukker, 2015). PSSs financially incentivise the companies to invest in the durability and reuse of products and materials (Baines et al., 2007; Mont and

Tukker, 2006; Williams, 2007). Following this focus on more intensive product use, the main R-strategy pursued by this type of start-ups is 'Reduce.' The driving innovation of developing a PSS is mostly combined with active involvement of consumers through returning or repairing products (10 out of 12 CSUs, e.g., a returnable packaging service for drinks) or collaborative consumption (4 out of 12 CSUs, e.g., shared construction spaces and machinery). Involving consumers in reverse supply-chain activities (return, repair, reuse) is considered a promising avenue towards a wider diffusion of CE through PSS business models, since it can address the barrier of increased capital requirements of a reverse logistics network (Wastling et al., 2018). Product recovery activities are currently underdeveloped in the CE/PSS domain, and early customer involvement is considered critical in PSS solution design (Kühl et al., 2018; Luiten et al., 2001; Manzini et al., 2001). One-third of start-ups in this category design products for PSS, since they also innovate in product design (e.g., modular phones, modular and biodegradable diapers, reusable batteries).

Despite many authors considering PSSs one key driver of circularity in business models and their belonging to the most-cited practical application methods of the latter (e.g., Antikainen et al., 2016; Bocken, 2017; Kühl et al., 2018; Lieder and Rashid, 2016; Stahel, 2010; Tukker, 2015; Yang and Evans, 2019), only 9% of start-ups pursue this innovation category of moving from customer ownership to performance- and service-based revenue models. The most significant reason for this is that the application of typically asset-heavy PSS business models is often more problematic for small- and medium-sized businesses (SMEs) than large corporates due to lack of (financial) resources and lack of formalised planning (Augusto de Jesus Pacheco et al., 2019; Besch, 2004; Neely, 2008). In general, especially performance-based PSS are considered very difficult to implement due to the difficulty of establishing indicators and the requirements of continuous feedback loops with customers (Pereira Pesoa and Jauregui Becker, 2017) Furthermore, the biological cycle is dominant among CSUs, and PSSs are strongly associated with manufacturing companies in the technical cycle (Augusto de Jesus Pacheco et al., 2019; Copani and Behnam, 2018; Neely, 2008; Yang and Evans, 2019) – 11 out of 12 service-based start-ups operate in the technical cycle, which supports the literature perspective.

Nature-based start-ups

The distinctive features of companies in this archetype is that they pursue the highest-ranked category, 'Regenerate,' and thus not only try to avoid any environmental harm but also seek to maintain and increase the delivery of (products and) services based on natural ecosystems and nature-based solutions (Maes and Jacobs, 2017). Six percent of all start-ups fall into this category of imitating biological cycles or systems. The only sectors represented are agriculture/food and built environment/design. Examples of such start-ups include Berlin-based 'aquaponic' start-ups, synergising fish farming and plant growing in a symbiotic, self-supporting environment without the usage of soil[8] (Campanhola and Pandey, 2019; Kyaw and Ng, 2017). Further representatives of this category are start-ups working on building-integrated plant cultivation and green roofs or London-based Entocycle, which produces protein by feeding food waste to protein-rich, vertically farmed insects to nurture animals or fertilise industrial crop production.

Maes and Jacobs (2017: 123) argue that the usage of nature-based ecosystem services always leads to 'decreased input of non-renewable natural capital and increased investment in renewable natural processes.' This combination of decreased resource usage and adaptation of natural processes corresponds to the results of the data gathering, since all nature-based start-ups combine the dominant 'Regenerate' strategy with 'Reduce,' and innovation in core processes is the dominant innovation type. All start-ups focusing on building-integrated solutions work with residual streams from the buildings they plan to redesign, which explains the high share of IS. It is noteworthy that 80% of the start-ups in this cluster focus on urban systems, which are considered to have the potential to generate major socioeconomic benefits addressing the challenges of urbanisation. This is a key application area of nature-based solutions in Europe (European Commission, 2018; Frantzeskaki, 2019; Lafortezza and Sanesi, 2019; Maes and Jacobs, 2017; Vujcic et al., 2017). The low uptake of nature-based solutions thus far is due to missing definition and impact assessment of the concept itself (Canitez, 2019; Kabisch et al., 2016; Krauze and Wagner, 2019) and to lack of awareness among and support from city authorities and their respective urban-development plans (Faivre et al., 2017; Kronenberg, 2015).

Conclusions

This research aimed to conceptualise and create a typology for circular start-ups to explore innovative organisations working on the circular economy. Based on the data gathered and the frameworks applied, a typology for CSU business models could be introduced that encompasses more than 90% of all 128 CSUs scrutinised in the data sample. The major clusters in this typology are:

- design-based CSUs, adopting circular innovations mostly in the pre-market phase through source material minimisation, product design, or production process efficiency
- waste-based CSUs, seeking to extract value from unexploited external waste streams;
- platform-based CSUs, pursuing sharing/trading business models built around B2B, B2C, or C2C marketplaces
- service-based CSUs, embedding products in service-systems to increase usage efficiency
- nature-based CSUs, increasing the delivery of (products and) services based on nature-based systemic solutions.

The developed typology contributes to the literature on entrepreneurship, sustainability, and sustainable (business model) innovation by better positioning circular start-ups against related concepts, such as sustainable and environmental start-ups, and by examining the actual business models deployed by circular entrepreneurs. Thus, it strengthens the link between the previous bodies of research. The analysis shows that the theoretically dominant R-frameworks in the context of circular business models need to be expanded by the additional category 'Regenerate.' The introduction of the concept of nature-based solutions in the conceptual realm of CE strategies is one of the key theoretical contribution of this chapter. Additionally, despite being the most common conceptualisation, the R-frameworks are still a relatively abstract and rigid concept when it comes to the transformation of business models for CE. The common scientific perspective that the combination of multiple strategies, methods, and approaches is necessary for a shift towards a CE (Bocken et al., 2016) is further driven, elucidated, and nuanced in this research. The authors developed a CBM framework that goes beyond existing concepts in terms of the 'how-to' of CE through the simultaneous mapping of CBM strategies and innovation types, thus allowing for – and introducing – CBM archetypes pursuing more than one CBM strategy and innovation type. This research adds to existing frameworks in the field of CBM innovation and enables a more detailed perspective on the key actors for circular innovation along the value chain by adding the 'Source' category to existing CBM innovation frameworks (Urbinati et al., 2017).

The novel empirical contributions of this chapter include the observation that CE strategies pursued by start-ups seem to be of higher hierarchical levels than the ones pursued by corporates. Cumulatively, CSUs compare higher in the strategies 'Regenerate' (6%[9] vs 5%) and 'Reduce' (44%vs 17%). Additionally, CSUs operate more dominantly in the biological cycle than CE-oriented corporates (75% vs 7%). PSS business models, even though highly valued by many scholars as CE enablers, are only represented to a small extent among CSUs. This is caused by financial constraints faced by CSUs applying this asset-heavy innovation type, as well as the mismatch of the rather technically related PSSs and the primarily bio-based CSU business models.

This conceptualisation of CSUs will help to understand possible pathways and necessary conditions for the further spreading of the concept of CE and thus of sustainable development in practice (and theory). Corporates and start-ups striving to become more circular can derive structured transition paths towards CBMs (e.g., guidance on dominant innovation types applied when pursuing a 'Reduce' strategy). The results shown can serve as inspiration for circular entrepreneurs in their incubation phase when looking for promising, low-investment business model adaptations for circularity. For instance, the lack of coordinated action for the management of residual resource streams is an indicator for circular founders to invest in their network and social capital early on to create a competitive advantage (see Paunescu et al., 2014; Peng et al., 2018). Policymakers benefit from a further definition and differentiation of practical concepts in circularity/sustainability. For instance, the current high-level perspective on the implementation of nature-based solutions – taken, for example, by the EU (European Commission, 2015) – could be substantiated with a better understanding of the underlying practical application methods. Additionally, the empirical and conceptual character of this research allows for a better channelling of efforts on policy intervention, since the presented typology of CSUs enables a targeted addressing of relevant stakeholders and their respective needs.

This research is one of the first of its kind in elucidating business models developed by circular entrepreneurs; however, the lack of understanding of the actual sustainability performance of these very business models is a limitation related to it. Thus, hypotheses made on the hierarchy and effectivity of circularity strategies and business models are not substantially backed by representative and comparable metrics. Furthermore, since approximately two-thirds of CSUs were not analysed based on direct interviews but outside in, there is a risk of not every circularity-related activity being captured in the data gathering. This would require a deeper analysis of each start-up or direct contact with the respective organisations. However, the authors consider that the publicly available information about the scrutinised companies contains all the major activities on CE, since this is a distinct business focus of most of them.

Further research on this topic could consider a comparative analysis of the different geographical areas covered in this study. There were indications (e.g., dominant sectors, CBM strategies pursued) that the different municipal/governmental policies and business ecosystems in the areas analysed led to distinct characteristics of the CSUs active there. A deeper qualitative analysis of the underlying causes and respective market/legislative setups is necessary to gain structured insight into these regional differences. This allows for more effective intervention management. The prevalent outside-in analysis and selected expert interviews used for this research would, therefore, need to be enriched, for example, by interviews with policymakers or CE experts. Additionally, an extension of the data sample (i.e., including business models and innovation types of CE-oriented corporates) would help to better understand the start-ups' roles in transitions towards the CE. Furthermore, it is not clearly defined and structurally assessed regarding which barriers and success factors impact the diffusion of CSU business models. Further research on this could inform policymakers or investors on where to focus their efforts to further drive CE.

The authors are excited about the opportunity to analyse CSUs and their respective CBMS and hope to trigger further scientific work on this relevant topic with this research.

Notes

1 The Randstad region comprises Amsterdam, Rotterdam, Utrecht, and The Hague.
2 CRCLR is a hub and think- and do-tank for CE in Berlin.
3 The final dataset was established in February/March 2019; founders' interviews were conducted between September 2017 and October 2018.
4 An exclusion to this might be catalytic goods like cooling water, that is, "materials that are necessary to speed up chemical reactions without being used up in the process" (Stahel, 2010: 21). They are mostly relevant in the chemical industry, in which none of the CSUs in the data sample operate.
5 Italy, Germany, Scandinavia.
6 Apart from trading/sharing and co-creation, Konietzko et al. (2019) define a third role platforms can play in the CE transition, which is to operate PSSs. PSS-based business models are a distinct typology in this analysis. These types of start-ups are clustered and explained in Chapter 11.
7 UK-based Winnow uses AI to monitor food waste in food service companies.
8 'Aquaponics is one of the food production systems that combined recirculating aquaculture system and hydroponic system by the utilization of waste material excreted from fish as a liquid fertilizer for valued hydroponic vegetables and seaweeds' (Anpo et al., 2019).
9 Only those CSUs of the dataset are considered that operate in the same sectors as in Stewart and Niero's (2018) reference study, that is, food/beverages, packaging, fashion, and consumer/household goods (56 CSUs).

References

Amsterdam Smart City (2017). *Circular city*. Available at: https://amsterdamsmartcity.com/themes/circular-city.

Anpo M, Fukuda H, Endo M. (2019). *Plant factory using artificial light*. Amsterdam, The Netherlands: Elsevier.

Antikainen M, Valkokari K. (2016). A framework for sustainable business model innovation. *Technology Innovation Management Review*. 6: 5–12.

Augusto de Jesus Pacheco D et al. (2019). Overcoming barriers towards sustainable PSSs in small and medium-sized enterprises: state of the art and a novel decision matrix. *Journal of Cleaner Production*. 222: 903–921.

Baines TS et al. (2007). State-of-the-art in PSSs. *Journal of Engineering Manufacture*. 221: 1543–1552.

Baldassarre B, Schepers M, Bocken N, Cuppen E, Korevaar G, Calabretta G. (2019). Industrial symbiosis: towards a design process for eco-industrial clusters by integrating circular economy and industrial ecology perspectives. *Journal of Cleaner Production*. 216: 446–460.

Berlin Business Location Centre (2017). *The circular economy in Berlin*. Available at: https://www.businesslocationcenter.de/en/energytechnologies/circular-economy/

Besch K. (2004). PSSs for office furniture: barriers and opportunities on the European market. *Journal of Cleaner Production*. 13: 1083–1094.

Bocken N et al. (2014). A literature and practice review to develop sustainable business model archetypes. *Journal of Cleaner Production*. 65: 42–56.

Bocken N et al. (2016). Product design and business model strategies for a circular economy. *Journal of Industrial and Production Engineering*. 33: 308–320.

Bocken N et al. (2017). The circular economy – exploring the introduction of the concept among S&P 500 firms. *Journal of Industrial Ecology*. 21 (3): 487–490.

Boons F, Lüdeke-Freund F. (2013). Business models for sustainable innovation: state-of-the-art and steps towards a research agenda. *Journal of Cleaner Production*. 45: 9–19.

Bourque E, Dudek G. (2004). Procedural texture matching and transformation. *Computer Graphics Forum: Eurographics*. 23 (3): 461–468.

Buzby JC, Hyman J. (2012). Total and per capita value of food loss in the United States. *Food Policy*. 37 (5): 561–570.

Campanhola C, Pandey S. (2019). *Sustainable food and agriculture – an integrated approach*. Boston: Academic Press. ISBN:978-0-12-812134-4.

Canitez F. (2019). Pathways to sustainable urban mobility in developing megacities: a socio-technical transition perspective. *Technological Forecasting and Social Change*. 141: 319–329.

Chertow MR. (2000). Industrial symbiosis: literature and taxonomy. *Annual Review of Energy and the Environment*. 25 (1): 313–337.

Copani G, Behnam S. (2018). Remanufacturing with upgrade PSS for new sustainable business models. *CIRP Journal of Manufacturing Science and Technology*. doi:10.1016/j.cirpj.2018.10.005.

Despeisse M et al. (2017). Unlocking value for a circular economy through 3D printing: a research agenda. *Technological Forecasting and Social Change*. 115: 75–84.

Domenech T et al. (2019). Mapping IS development in Europe – typologies of networks, characteristics, performance and contribution to the circular economy. *Resources, Conservation and Recycling*. 141: 76–98.

Ellen MacArthur Foundation (2015). *Growth within: a circular economy vision for a competitive Europe*. Available at: www.ellenmacarthurfoundation.org/assets/downloads/publications/EllenMacArthurFoundation_Growth-Within_July15.pdf.

Ellen MacArthur Foundation et al. (2019). *Artificial intelligence and the circular economy – AI as a tool to accelerate the transition*. Available at: www.ellenmacarthurfoundation.org/assets/downloads/Artificial-intelligence-and-the-circular-economy.pdf.

European Commission (2015). *Towards an EU research and innovation policy agenda for nature-based solutions and re-naturing cities in final report of the horizon 2020 expert group on 'nature-based solutions and re-naturing cities'*. Brussels: EC.

European Commission (2016). *Closing the loop – an EU action plan for the CE*. Brussels. Available at: http://eur-lex.europa.eu/legal-content/EN/TXT/?uri=CELEX:52015DC0614.

European Commission (2018). *Horizon 2020 – work programme 2018–2020–12: climate action, environment, resource efficiency and raw materials*. Brussels: EC.

European Startup Initiative (2017). *Startup heatmap Europe*. Available at: www.startupheatmap.eu/analytics.

Faivre N, Fritz M, Freitas T, de Boissezon B, Vandewoestijne S. (2017). Nature-based solutions in the EU: innovating with nature to address social, economic and environmental challenges. *Environmental Research*. 159: 509–518.

Frantzeskaki N. (2019). Seven lessons for planning nature-based solutions in cities. *Environmental Science and Policy*. 93: 101–111.

Fuenfschilling L, Truffer B. (2013). The structuration of socio-technical regimes – conceptual foundations from institutional theory. *Research Policy*. 43 (4): 772–791.

Geels FW. (2005). *Technological transitions and system innovations: a co-evolutionary and socio-technical analysis*. Cheltenham: Edward Elgar.

Gemeente Rotterdam (2016). *Roadmap circular economy Rotterdam*. Available at: www.rotterdamclimateinitiative.nl/documents/2016/roadmap-circular-economy.pdf.

Graedel TE et al. (2011). What do we know about metal recycling rates? *Journal of Industrial Ecology*. 15 (3): 355–366.

Hall JK, Daneke GA, Lenox MJ. (2010). Sustainable development and entrepreneurship: past contributions and future directions. *Journal of Business Venturing*. 25 (5): 439–448.

Hazen BT, Mollenkopf DA, Wang Y. (2017). Remanufacturing for the circular economy: an examination of consumer switching behaviour. *Business Strategy and the Environment*. 26 (4): 451–464.

Hofnaegels K. (2016). Das tropicana-kollektiv. *Smart Magazine*. Available at: https://de.smart-magazine.com/bluecity010-rotterdam/.

Hruschka DJ, Schwartz D, St. John DC, Picone-Decaro E, Jenkins RA, Carey JW. (2004). Reliability in coding open-ended data: lessons learned from HIV behavioral research. *Field Methods*. 16: 307–331.

Hsieh HF, Shannon SE. (2005). Three approaches to qualitative content analysis. *Qualitative Health Research*. 15: 1277–1288.

Kabisch N et al. (2016). Nature-based solutions to climate change mitigation and adaptation in urban areas: perspectives on indicators, knowledge gaps, barriers, and opportunities for action. *Ecology and Society*. 21 (2): 39 f.

King AM, Burgess SC, Ijomah W, Mcmahon CA. (2006). Reducing waste: repair, recondition, remanufacture or recycle? *Sustainable Development*. 14: 257–267.

Kirchherr R et al. (2017). Conceptualizing the CE: an analysis of 114 definitions. *Resources, Conservation and Recycling*. 127: 221–232.

Konietzko J, Bocken N, Hultink EJ. (2019). *Online platforms and the circular economy. innovation for sustainability: business transformations towards a better world*. Cham: Palgrave Macmillan. 435–450.

Krauze K, Wagner I. (2019). From classical water-ecosystem theories to nature-based solutions – contextualizing nature-based solutions for sustainable city. *Science of the Total Environment*. 655: 697–706.

Krippendorff K. (2013). *Content analysis: an introduction to its methodology*. Thousand Oaks, CA: Sage. 3rd Ed.

Kronenberg J. (2015). Why not to green a city? Institutional barriers to preserving urban ecosystem services. *Ecosystem Services*. 12: 218–227.

Kühl C, Tjahjono B, Bourlakis M, Aktas E. (2018). Implementation of circular economy principles in PSS operations. *Procedia CIRP*. 71: 124–129.

Kyaw TY, Ng AK. (2017). Smart aquaponics system for urban farming. *Energy Procedia*. 143: 342–347.

Lafortezza R, Sanesi G. (2019). Nature-based solutions: settling the issue of sustainable urbanization. *Environmental Research*. 172: 394–398.

Lieder M, Rashid A. (2016). Towards circular economy implementation: a comprehensive review in context of manufacturing industry. *Journal of Cleaner Production*. 115: 36–51.

Linder M, Williander M. (2017). CBM innovation: inherent uncertainties. *Business Strategy and Environment*. 26 (2): 182–196.

London Waste and Recycling Board (2017). *London's circular economy route map*. Available at: www.lwarb.gov.uk/wp-content/uploads/2015/04/LWARB-London%E2%80%99s-CE-route-map_16.6.17a_singlepages_sml.pdf.

Luiten H, Knot M, van der Horst T. (2001). Sustainable PSSs: the Kathalys method. *Proceedings Second International Symposium on Environmentally Conscious Design and Inverse Manufacturing*. 190–197.

Ma S, Wen Zong-guo, Chen J, Wen Zhi-chao (2014). Mode of circular economy in China's iron and steel industry: a case study in Wu'an city. *Journal of Cleaner Production*. 64: 505–512.

Maes J, Jacobs S. (2017). Nature-based solutions for Europe's sustainable development. *Conservation Letters*. 10: 121–124.

Manzini E, Vezzoli C, Clark G. (2001). PSSs: using an existing concept as a new approach to sustainability. *Journal of Design Research*. 1 (1): 31–38.

Martin-Rios C et al. (2018). Food waste management innovations in the foodservice industry. *Waste Management*. 79: 196–206.

Mont O, Tukker A. (2006). Product-service systems: reviewing achievements and refining the research agenda. *Journal of Cleaner Production*. 14 (17): 1452–1454.

Moreno M, De los Rios C, Rowe Z, Charnley F. (2016). A conceptual framework for circular design. *Sustainability*. 8 (9): 1–15.

Mortensen L, Kørnøv L. (2019). Critical factors for IS emergence process. *Journal of Cleaner Production*. 212: 56–69.

Neely A. (2008). Exploring the financial consequences of the servitization of manufacturing. *Operations Management Research*. 1 (2): 103–118.

Neuendorf KA. (2016). *The content analysis guidebook*. Los Angeles: Sage.

Niero M et al. (2017). Combining eco-efficiency and eco-effectiveness for continuous loop beverage packaging systems. *Journal of Industrial Ecology*. 21 (3): 742–753.

Paunescu C, Raluca Badea M. (2014). Examining the social capital content and structure in the pre-start-up planning. *Procedia Economics and Finance*. 15: 560–568.

Peng B, Tu Y, Wei G. (2018). Governance of electronic waste recycling based on social capital embeddedness theory. *Journal of Cleaner Production*. 187: 29–36.

Pereira Pesoa MV, Jauregui Becker JM. (2017). Overcoming the product-service model adoption obstacles. *Procedia CIRP*. 64: 163–168.

Potting J, Hekkert M, Worrell E, Hanemaaijer A. (2017) *Circular economy: measuring innovation in the product chain*. Policy Report for PBL Netherlands Environmental Assessment Agency. The Netherlands.

Prosman EJB, Wæhrens V, Liotta G. (2017). Closing global material loops: initial insights into firm-level challenges. *Journal of Industrial Ecology*. 21 (3): 641–650.

Ramanathan R. (2011). An empirical analysis on the influence of risk on relationships between handling of product returns and customer loyalty in e-commerce. *International Journal of Production Economics*. 130 (2): 255–261.

Ravi V, Shankar R, Tiwari MK. (2005). Analyzing alternatives in reverse logistics for end-of-life computers: ANP and balanced scorecard approach. *Computers and Industrial Engineering*. 48 (2): 327–356.

Reike D, Vermeulen WJV, Witjes S. (2018). The circular economy: new or refurbished as CE 3.0? – exploring controversies in the conceptualization of the circular economy through a focus on history and resource value retention options. *Resources, Conservation and Recycling*. 135: 246–264.

Ribeiro I, Sobral P, Pecas P, Henriques E. (2018). A sustainable business model to fight food waste. *Journal of Cleaner Production*. 177: 262–275.

Ritter M, Schanz H. (2019). The sharing economy: a comprehensive business model framework. *Journal of Cleaner Production*. 213: 320–331.

Seuring S, Müller M. (2008). From a literature review to a conceptual framework for sustainable supply chain management. *Journal of Cleaner Production*. 16 (15): 1699–1710.

Sihvonen S, Ritola T. (2015). Conceptualizing ReX for aggregating end-of-life strategies in product development. *Procedia CIRP*. 29: 639–644.

Stahel WR. (2010). *The performance economy*. Hampshire: Palgrave Macmillan.

Stewart R, Niero M. (2018). Circular economy in corporate sustainability strategies: a review of corporate sustainability reports in the fast-moving consumer goods sector. *Business Strategy and the Environment*. 27 (7): 1005–1022.

Thi N, Biswarup S, Chen CC, Gopalakrishnan K, Lin CY. (2014). Food waste to bioenergy via anaerobic processes. *Energy Procedia*. 61: 307–312.

Tukker A. (2015). Product services for a resource-efficient and circular economy – a review. *Journal of Cleaner Production*. 97: 76–91.

Urbinati A, Chiaroni D, Chiesa V. (2017). Towards a new taxonomy of circular economy business models. *Journal of Cleaner Production*. 168: 487–498.

van Dijck J, Poell T, de Waal M. (2018). *The platform society – public values in a connective world*. Oxford: Oxford University Press. ISBN:0190889772.

Veleva V, Bodkin G. (2018). Corporate-entrepreneur collaborations to advance a circular economy. *Journal of Cleaner Production*. 188: 20–37.

Vujcic M et al. (2017). Nature based solution for improving mental health and well-being in urban areas. *Environmental Research*. 158: 385–392.

Wastling T, Charnley F, Moreno M. (2018). Design for circular behaviour: considering users in a circular economy. *Sustainability*. 10 (6): 1743 ff.

Williams A. (2007). Product service systems in the automobile industry: contribution to system innovation? *Journal of Cleaner Production*. 15: 1093–1103.

Worrell E, Reuter M. (2014). *Handbook of recycling: state-of-the-art for practitioners, analysts, and scientists*. Boston: Newnes.

Yang M, Evans S. (2019). PSS business model archetypes and sustainability. *Journal of Cleaner Production*. 220: 1156–1166.

Young CW, Russell SV, Robinson CA, Chintakayala PK. (2018). Sustainable retailing – influencing consumer behaviour on food waste. *Business Strategy and the Environment*. 27 (1): 1–15.

Zucchella A, Urban S. (2019). *Circular entrepreneurship*. Cham: Palgrave Macmillan.

28

ENABLERS AND BARRIERS FOR INDUSTRIAL SYMBIOSIS

Lessons learnt from twenty-five case studies

Doroteya Vladimirova and Karen Miller

Introduction

The results provided in this chapter support the definition of pathways to move from anecdotal one-off resource synergies to a systematic application of resource synergies as a means to increase resource productivity and competitiveness. To understand industrial symbiosis (IS) more specifically, Yap and Devlin (2017) proposed a multiple-level framework as an analytical lens for IS, which studied enablers and barriers at the levels of society, network, and enterprise. We have adapted this framework to structure the chapter.

Enablers

The enablers are organised in a different format to the triggers, as they follow different dimensions. The reader will, however, see how these two sections relate to one another.

The role of government/regional authorities

One of the most powerful forces evident in the case studies was that of direct state or regional intervention in the form of financial contributions or investment in IS. Public investment in Case Study (CS) 2, particularly in infrastructure, utilities, and services, created a favourable context for IS development. Further details are provided in CS 20 and 22 on the positive impact of government intervention.

Policy is another tool that provides an appropriate framework for IS development (CS 24). Interactive processes identified in CS 2 between the government, other institutions, and the firms themselves helped towards their alignment and collaborative efforts. One notable, specific element of support for IS has been national and regional government policies derived from regulation. Examples are the Danish government's regulations on pollution control, emissions reduction, and resource and energy efficiency (CS 3) and the German government's ban on direct disposal to landfill (CS 5).

Government investment in supportive innovation processes (CS 20) and in coordinating bodies featured prominently in many of the studies. Of note is the UK government's decision to invest heavily in the establishment of the National Industrial Symbiosis Programme (NISP), which operated (and continues to do so) as the connector and knowledge broker between firms and research bodies (CS 7, 9, 10, and 13). Regional, city, and even local governments help in another way by acting as a form of bridge between national government and businesses. This feature has proved vital, as they are aware of local conditions and contexts and can help shape national policy accordingly. Additionally, this awareness has allowed them to develop more tailored policies and support emergent synergies (CS 2).

Favourable conditions have also been created through the use of green marketing. Specific marketing has proved a powerful tool and helped raise environmental awareness in municipalities (as in CS 19 of Lidköping, Sweden), serving to attract new businesses to the region.

Overall, from all the case studies, it is evident that consistent investment and policymaking and regulation have been the most effective form of support from government at any level.

Intermediaries/knowledge agents

What emerged from many of the case studies was the vital role played by intermediaries. For example, the NISP and the Business Council for Sustainable Development-United Kingdom (BCSD-UK) were fundamental to catalysing the development of new IS systems in the United Kingdom and ensuring effective ongoing function of these systems (CS 4). Their primary role was to identify, connect, and coordinate firms and stakeholders within the Humber region Industrial Symbiosis Programme (HISP), with the scientific support of researchers from the University of Lund, Sweden. Other examples of intermediary activity are provided in CS 2, 6, 7, 9, 10, 11, 13, 15, and 16.

In some cases, intermediaries have worked in great detail, providing informational support to identify synergy feasibilities (CS 11) and negotiating the terms of engagement to enable the setup of systems. In CS 4, the intermediaries brokered the contracts and timings, which were critical, as there were multiple firms in the network. Their role helped to build confidence and trust, both of which are essential factors in propagating IS. In many cases, intermediaries have brokered relationships and built shared knowledge and understanding, in addition to securing investment and funding (CS 10).

Another aspect of many intermediaries' facilitative role was assessing what other specialist knowledge was required and where could it be sourced from, for example, in CS 2 with the Technical University of Lisbon. See also CS 6, where the Italian National Agency for New Technologies, Energy and the Environment (ENEA) established a programme between various leather and agro-producers and research institutes. Universities, research organisations, and specialist consultancies themselves have played a vital part in enabling synergies to be established through the provision of highly specialised knowledge (CS 12) and the involvement of researchers (CS 11). In CS 11, the researcher actively worked alongside the NISP to take the learnings from previous IS projects, including those in other geographies, to assist in the establishment of new synergistic networks.

Intermediaries have assisted further by providing a negotiation pathway to authorities such as the Scottish Environment Protection Agency (SEPA) in CS 7, which was contacted so as to reclassify hessian that was previously listed as waste on the pollution, prevention and control license to a by-product so that it could be reused. Frequently the agents that work as intermediaries, or within agencies that provide this function, have extensive knowledge due to their industry backgrounds. This knowledge can relate to resource stream capabilities, chemistry, extraction, and reprocessing techniques and technologies (see CS 13, for instance). Background knowledge of legislation also proved valuable in bridging the gap between an existing state and one involving symbiosis. Knowledge sharing across networks and geographies (CS 10) added to the importance of intermediaries.

Within CS 9 and CS 10 were examples of project coordinators based in local companies with backgrounds as high-ranking managers, who provided both credibility and visibility, encouraging IS activities. In China (CS 20), the involvement of managers embedded within firms appeared to be pivotal – they usually had a role within local government. This finding underlines the importance of having 'conduits' between the various levels (firm, local, and national government). In some cases, managers had formal roles of coordination, as in the cases just mentioned, while in other cases, managers of different firms formed informal relationships that enabled initial synergies to be coordinated and thence larger IS networks to flourish. The Kalundborg system (CS 3) offered evidence of this phenomenon.

Social/relationships

Fundamental to many of the other enablers were social enablers that allowed IS to develop. Trust, either between individuals or firms or with the intermediaries/coordinating bodies, was a strong factor, for example, in CS 9. In this case, the Midlands Environmental Business Communications (MEBC) had an established presence in the area with several existing connections, encouraging other companies to trust the process and thus allowing IS to kick-start in a meaningful way. CS 18 revealed how trust between various firms and the city government of Helsingborg created a strong system of interdependencies. In both CS 9 and CS 18, trust was built on the basis of embeddedness in particular vicinities and long-termism.

Trust building between stakeholders in a specific place extended further in CS 8, which focused on Bussi sul Tirino, in the Abruzzo region of Italy. Local community involvement and trust-building activities were evident here, and this case revealed how important long-term participation and social relationship building were in identifying and seeding the development of symbiotic opportunities. Specifically in this case, the local population had deep-seated negative views based on previous experiences.

Close working relationships appeared critical in many case studies and extended to other stakeholders, for example, intermediaries, such as the NISP and regional development agencies (RDAs) in the United Kingdom (CS 9), and academic institutions. In CS 9, a further notable example of the power of trust in generating engagement in IS was evident in that a confidentiality agreement between those involved was deemed unnecessary due to the credibility of North West Chemical Initiative (NWCI). In cases such as Kalundborg (CS 3), self-organised and brokered relationships between individuals, firms, and authorities provided ideal conditions for synergies to develop.

Confidence in collaboration was noted in CS 4 as an enabler, and it was based on cognitive rather than social proximity. Certainty in CS 4 was centred on the economic benefits, resource security, and reliability of technology made transparent by BCSD-UK, who acted as the programme coordinator. From this point, we see how intermediaries perform vital roles in establishing IS, in this case acting in a 'hands-on' leadership capacity.

Leadership

Leadership has acted at various levels to encourage the development of IS systems. At an individual level in CS 10, a highly credible managing director of a large-scale refinery with a collaborative mindset assumed the project leadership role. In the same study, the project coordinators had backgrounds as senior managers in the companies involved. Leadership of IS therefore was not only evident at an executive level but also disseminated within organisations.

From an organisational-level perspective, a leadership role was occupied by a major company in CS 9 at the start of an IS programme with a strong commitment. This generated a gravitational pull centred on a collaborative culture that encouraged other stakeholders to participate.

Strong sustainability leadership at a more macro level was found in CS 18 as local/regional/city strategies in Sweden helped overcome IS barriers related to geographic location. This was through active city planning in addition to facilitating strong relations with neighbouring municipalities, resulting in synergies that generated the district heating corridor between Helsingborg and Landskrona.

Geographical/spatial

Geographic proximity is a critical factor in the development of synergies (CS 23), for instance, in cases where firms have been able to take advantage of waste materials that converge in the vicinity (CS 2) or where they have been close to disposal structures and/or resource partners. Many cases underpinned these findings (CS 1, 3, 4, 18).

Relative isolation and the small size of a region/place helped, in particular cases, to create favourable conditions, as social interactions both between workers and managers occurred regularly, as demonstrated in the Kalundborg industrial cluster example (CS 3). The nature of the industries in a place/region played a part in the development of IS; in some, the homogeneity was a positive attribute; in others, heterogeneity was (CS 8). All cases depended on the specific needs of the industries, authorities, and broader communities present.

The role of a lead firm in a location was noted in CS 9 and 25, where they established initial synergies and then provided a 'gravitational pull' effect that enabled the network to extend. In a similar way, a specific place-focused investment (CS 22) or provision of a special status (CS 1, 24) by national governments or regional/city authorities – often due to economic, environmental, or social needs (or all of these) – generated a powerful incentive for firms to aggregate and develop synergistically.

Process

What emerged from some of the case studies was that various processes were important factors in driving IS. In some cases, there was evidence of a spontaneous process of IS activity; in Ulsan City (CS 24), companies made one-to-one negotiations for economically beneficial exchanges, driven by the need to respond to stringent South Korean government environmental legislation.

Many other synergistic cases were the result of systematic, planned activity, typically coordinated by an intermediary or organising committee who used a process of information collection to gather the necessary technical, legal, and economic information. This process then led to the identification of the ideal firms, and/or other stakeholders such as municipalities, for involvement in synergies – for examples, see CS 1, 2, 4, 7, 10, 11, 13, 15, and 16.

The typical overall process adopted by intermediaries emerged as an important enabler. Of note were the following elements: the dissemination of information in a clear/open format to appropriate firms/stakeholders; the setting up and running of awareness-raising programmes, such as formal launch/awareness-building workshops; the negotiation of formal agreements to realise synergies (and develop confidence); the agreement of timings between stakeholders; and the careful monitoring and feeding back of information to guide initiatives and/or develop further synergies (CS 2).

Evident in CS 4 was the process of experience transfer to other geographies. In this case, the lead firm's previous experience of IS was used to disseminate knowledge and seed wider

IS practice in other geographies. The company was a global oil and gas producer with experience in Tampico (Mexico) that was transferred to Humberside and further areas of the United Kingdom.

Technology

Technology has driven understanding of the most effective area to intervene in and to monitor through the establishment of initial benchmarks or baselines (CS 2) and the live feeding of data over time. This was fundamental to shifting towards widespread synergistic practice (CS 2, 3). The Internet of Things (IoT) allowed intelligent decision-making processes through, for example, the streaming of data. Remote data from embedded IT components reduces the need for human interactions and interventions and has enabled IS networks – see CS 2 in particular for details. Industry 4.0, which has been enabled by the IoT, has extended the ability to link singular smart factories to modular networks by the monitoring of physical materials and processes. Optimal future scenarios have been produced which allow for decentralised decision-making in these contexts. Digital communication technologies (ICT) have allowed information to flow dynamically around networks (CS 9). In so doing, technology has become a powerful synergies enabler.

Barriers

From the case studies, a wide variety of obstacles were identified. As already alluded to, the reader will identify linkages between these and the triggers and enablers.

Materials

The most basic barrier was uncertainty surrounding the procurement of sufficient inputs of the right quality at a predictable flow rate (CS 14 and 18), in the first case (CS 14) with a regular supply of organic materials from Italian olive oil farms; in the second (CS 18), there were issues centred on the composition of waste materials, particularly those that had been produced using a complex combination of materials. As another example, the composition of tyres in CS 15 required highly specialist knowledge and technologies in order to extract materials that could be reused. Technical challenges were also identified in CS 6, as poultry dejections (PoDe), while suitable for use in preparing hides in the leather tanning process, were unsatisfactory due to the high olfactory impact.

The classification of materials as waste rather than useable by-product arose in several of the studies, most explicitly in CS 7 with hessian sacks and the difficulty in reclassifying them with regulators. Some other raw material requirements presented barriers, as they were challenging to fulfil. This was the reason that in CS 19, not all of the potential waste material feedstocks could be used to produce biogas marked with the Nordic Ecolabel.

Capacity and capability

From a firm-level perspective, there were various constraints that came to the fore. The first was the data collection necessary for an IS programme in relation to the time available, due to competing activities – predominantly the running of the core business (CS 10). A similar issue surfaced in CS 8, in the form of insufficient human resources being available to seek out and understand the avenues for waste streams as feedstocks for others, which acted as an information barrier. Linked to this point, skills shortages more broadly have been identified (CS 10) relating to understanding how IS could be made to work. This occurred in CS 17, where shortfalls arose in innovation skills, and in CS 7, where the need emerged for knowledge transfer capabilities in coordinating bodies.

Potential mismatches in the capacity/capabilities of partners in an IS network created another problem evident in CS 16. Here, a larger lead company potentially had greater volumes of polyvinyl chloride (PVC) tarpaulins than could be absorbed by a smaller manufacturer of bespoke products for the boating, banner, and trailer market.

From CS 18 and 19, it was found that firms that are part of larger international groups had limited autonomy, because major (strategic) decisions needed head office approval. Typically, the head office would be geographically distant. This led to situations where centralised top management was not interested in the possibilities of synergies at local levels.

Complexity and continuity

The transition to an IS system requires a multitude of changes across all dimensions in an organisation. These changes have been met by strong resistance forces that operate within existing industrial regimes (CS 3). Inertia emerged in CS 17 in the form of binding to dominant practices or technologies. The strategic approaches of most businesses more broadly typically direct the highest levels of attention towards the primary product (the core business) – see CS 1, 3, and 25. Alternative uses of waste streams have therefore been viewed as a diversion of capital and human resources from core business, making IS seem unattractive. During periods of economic uncertainty, intransigence becomes more pronounced. Uncertainty surfaced in several cases as a major barrier; in CS 9, funding/investment issues led to a loss of momentum of IS implementation, which was compounded when the lead player ceased trading in the Humberside (UK) region. In CS 18, the insecure nature of continued government support was a barrier in Helsingborg (Sweden), exacerbated by the deregulation of the district heating market, forcing companies to be driven by short-term strategies.

Communication and coordination

Given the importance ascribed to the role of intermediaries and coordinators in the triggers and enablers sections previously, it is evident that any shortcomings in their function would raise barriers. In CS 13, the lack of an intermediary/coordinator to undertake vital functions was described. These functions were identified as information gathering, identification of synergy feasibilities, and the support given to set up a viable system, in addition to helping to secure investment. Poor communications led to incorrect assumptions in CS 9, as the HISP was believed only to be relevant for large energy provider/users in a region rather than being of value to other sectoral firms. These assumptions slowed down the exploratory process and caused firms not to engage. The paucity of effective and continuous communication emerged from another study (CS 19), where there was a poor understanding due to technical language barriers.

Leadership and mindset

Loss of leadership was problematic in CS 9, when a major company in an IS network ceased trading in the region. Social inertia was also found when incumbent management failed to direct the efforts of engineers and technologists, who are required for transformation to IS

(CS 3). A similar effect related to policymakers' attention (CS 3), which may have been caused by what is identified, in CS 1 and 8, with self-interest and a lack of motivation due to an absence of clear economic/environmental/social benefits. Social inertia was compounded by stable platforms (business models) that reinforced the leadership/management mindset that the current mode of operation worked effectively in CS 17.

Fierce competition in certain industries created another mindset barrier that has blocked collaborative efforts. This occurred most notably in non-manufacturing industries (CS 10), whereas manufacturers typically needed to cooperate in order to remain economically viable. Mindsets resulting from previous negative experiences of collaboration are cited in CS 8 as a further impediment to symbiosis. One particularly vivid example of attitudinal negativity is provided in the study of Lidköping (Sweden), where CHP plant employees had negative opinions towards the use of sewage, even though it provided synergistic opportunities (CS 19).

Geographical/spatial

Within specific locations, the nature of the existing businesses created barriers; in CS 19, the area consisted of heterogeneous businesses rather than a homogeneous industry, which blocked cooperation between companies. This was primarily due to a lack of knowledge of other companies' activities. Large distances between companies acted as a barrier by decreasing the economic viability of collaboration and causing a negative environmental impact through the transportation of materials (CS 25). From CS 3, the difficulties of confined geographic locations were identified in that unless there was co-existence, severe difficulty would be realised in attempting to (re)locate heavy/power industries to a common region. The difficulties are a consequence of the complexity and sheer physical scale inherent in firms from these industries.

Shortages of basic materials such as water have precluded any further industrial development in certain places, which is raised in CS 3. Similarly, a lack of infrastructure, utilities, and services (and maintenance) in a zone/area/region in CS 3, and lack of suitable facilities in CS 5, emerged as barriers to IS. Opposition from local communities to additional industrialisation was evident in CS 8, and a lack of active community engagement was evident in CS 25. Communities can thus act as powerful impediments.

Technology

Three specific technological elements emerged from the case study analysis: first, the obstacle generated by what was termed format heterogeneity in the data fusion process – or, more simply, the use of different identifiers for an identical concept. Second, the availability of reliable material recovery technologies. This barrier related to the third, which was the lack of initial and ongoing investment in appropriate technologies and skilled technologists, engineers, and IT specialists (CS 3).

Economic and environmental

Multiple barriers emerged in relation to economic/environmental factors. The first was raised in CS 8 and 20 and related to weak/unstable demand factors. This was partly due to relatively low costs for waste disposal/utility resources (CS 25), along with difficulties experienced by firms in identifying what value was embedded in the waste streams of other firms. Another barrier was firms not having sufficient resources to be able to find out what synergistic linkage opportunities existed and what would be required to implement them

(CS 1, 13, and 18). A further hurdle was the need to partition the economic/environmental benefits and negative consequences for all system parties in order for them to engage. The sheer complexity of partitioning and calculating benefits (the overall system impacts must be calculated before apportioning to each element) was revealed in CS 12. Funding issues were of note in CS 9, as these led to a loss of motivation, which underpins the point earlier about how negative previous experiences become deeply problematic when trying to re-engage stakeholders in IS.

Policy/regulation (firm/network level) and contracts

In CS 2, where Chamusca is controlled by the Lisbon and Tagus Valley Regional Development Agency (CCDR-LVT), permits for industrial development proved difficult to acquire. Existing environmental regulation, for example, where intensive approval procedures for by-product reuse are stipulated (CS 9 and 25), and water and energy utility regulations (CS 25) therefore have the potential to create substantial blockages.

Contractual issues were noted, such as in CS 9, with the requirement to sign a contract at an early stage specifying a given percentage of any economic gains to go to a coordinating body. Confidentiality issues added to the disincentives and had the capacity to derail synergistic propositions (CS 25). All of these contractual elements had the potential to create anxiety and lead to stakeholders continuing to be locked into a business-as-usual paradigm.

Policy, regulation, and investment at a government/regional/local/city authority level

A significant set of barriers related to governmental/authority regulations and policy directives limited the ability of firms to alter their business processes, for example, the deregulation of the district heating market forcing companies into short-sighted strategies in CS 18. Another example was the complicated synergies for the CHP plant in CS 19 due to the prohibition of the use of fly ash in Sweden. Underlying these is a lack of knowledge of new sectors/materials and processes (CS 4). Another deeply embedded barrier was the influence of an ingrained market model where the benefits of IS were not recognised. This is found in CS 18 regarding public procurement laws.

A lack of specific IS policy was evident in CS 23 – in other cases, uncertainty in relation to the future direction of national politics towards the environment/waste/IS compounded many of the issues identified as barriers. The change of direction, for instance, with the removal of government funding for the NISP, a key intermediary in the United Kingdom, emerged as a source of uncertainty (CS 13).

Conclusions

We outlined an analytical framework for identifying triggers, enablers, and barriers for industrial symbiosis, scrutinising 25 industrial case studies. In Chapter 8, we defined triggers at national governmental, regional, and firm levels. In this chapter, we presented a set of enablers and barriers for industrial symbiosis based on the analysis of 25 diverse case studies.

This research is part of the wider EU H2020 SCALER research project (Scaling European Resources with Industrial Symbiosis), which sought to scale up industrial symbiosis throughout Europe.

Reference

Yap NT, Devlin JF. (2017). Explaining industrial symbiosis emergence, development, and disruption: a multilevel analytical framework. *Journal of Industrial Ecology*. 21 (1): 6–15. doi:10.1111/jiec.12398.

Case study references

CS 1: Notarnicola B, Tassielli G, Renzulli PA. (2016). Industrial symbiosis in the Taranto industrial district: current level, constraints and potential new synergies. *Journal of Cleaner Production*. 122: 133–143. doi:10.1016/j.jclepro.2016.02.056.

CS 2: Costa I, Ferrão P. (2010). A case study of industrial symbiosis development using a middle-out approach. *Journal of Cleaner Production*. 18: 984–992. doi:10.1016/j.jclepro.2010.03.007.

CS 3: Adamides ED, Mouzakitis Y. (2009). Industrial ecosystems as technological niches. *Journal of Cleaner Production*. 17 (2): 172–180. doi:10.1016/j.jclepro.2008.04.003.

CS 4: Velenturf APM. (2016). Promoting industrial symbiosis: empirical observations of low-carbon innovations in the Humber region. *Journal of Cleaner Production*. 128: 116–130. doi:10.1016/j.jclepro.2015.06.027.

CS 5: Vollmeier T, Castorini A. (2015). *Case study Spremberg: RDF-fueled CHP plant for a paper mill.* Available at: www.enea.it/it/seguici/pubblicazioni/pdf-volumi/ExperiencesofIndustrialSymbiosisinItaly_Proceedings.pdf.

CS 6: Dall'Ara A, Billi L, La Peruta MT, Cepparrone A, Roig M. (2015). *PODEBA: an industrial symbiosis case*. Available at: www.enea.it/it/seguici/pubblicazioni/pdfvolumi/ExperiencesofIndustrialSymbiosisinItaly_Proceedings.pdf.

CS 7: Organisation for Responsible Business (n.d.). *NISP*. Available at: www.orbuk.org.uk/article/the-national-industrial-symbiosis-programme-nisp.

CS 8: Taddeo R, Simboli A, Morgante A. (2012). Implementing eco-industrial parks in existing clusters: findings from a historical Italian chemical site. *Journal of Cleaner Production*. 33: 22–29. doi:10.1016/j.jclepro.2012.05.011.

CS 9: Mirata M. (2004). Experiences from early stages of a national industrial symbiosis programme in the UK: determinants and coordination challenges. *Journal of Cleaner Production*. 12: 967–983. doi:10.1016/j.jclepro.2004.02.031.

CS 10: Mirata M. (2004). Experiences from early stages of a national industrial symbiosis programme in the UK: Determinants and coordination challenges. *Journal of Cleaner Production*. 12: 967–983. doi:10.1016/j.jclepro.2004.02.031.

CS 11: Mirata M. (2004). Experiences from early stages of a national industrial symbiosis programme in the UK: determinants and coordination challenges. *Journal of Cleaner Production*. 12: 967–983. doi:10.1016/j.jclepro.2004.02.031.

CS 12: Martin M. (2015). Quantifying the environmental performance of an industrial symbiosis network of biofuel producers. *Journal of Cleaner Production*. 102: 202–212. doi:10.1016/j.jclepro.2015.04.063.

CS 13: Organisation for Responsible Business (n.d.). *NISP*. Available at: www.orbuk.org.uk/article/the-national-industrial-symbiosis-programme-nisp.

CS 14: Zabaniotou A, Rovas D, Libutti A, Monteleone M. (2015). Boosting circular economy and closing the loop in agriculture: case study of a small-scale pyrolysis – biochar based system integrated in an olive farm in symbiosis with an olive mill. *Environmental Development*. 14: 22–36. doi:10.1016/j.envdev.2014.12.002.

CS 15: Laybourn P. (2013). *Opportunities through industrial symbiosis: UK NISP and global experience*. Available at: www.endustriyelsimbiyoz.org/wp-content/uploads/2013/02/industrial-symbiosis_uk-nisp-and-global-experience_31.01.2013.pdf.

CS 16: Invest Northern Ireland (n.d.). *Industrial symbiosis: improving productivity through efficient resource management*. Available at: https://secure.investni.com/static/library/invest-ni/documents/industrial-symbiosis-guide-for-businesses-in-northern-ireland.pdf.

CS 17: Short SW, Bocken NMP, Barlow CY, Chertow MR. (2014). From refining sugar to growing tomatoes: industrial ecology and business model evolution. *Journal of Industrial Ecology*. 18 (5): 603–618. doi:10.1111/jiec.12171.

CS 18: Celander F et al. (2013). *Industrial symbiosis in Helsingborg*. Available at: www.industriellekologi.se/documents/Helsingborg.pdf.

CS 19: Angren J et al. (2012). *Exploring industrial symbiosis in Lidköping, Sweden.* Available at: www.industriel lekologi.se/documents/Lidkoping.pdf.

CS 20: Yuan Z, Shi L. (2009). Improving enterprise competitive advantage with industrial symbiosis: case study of a smeltery in China. *Journal of Cleaner Production.* 17: 1295–1302. doi:10.1016/j.jclepro.2009.03.016.

CS 21: Yang S, Feng N. (2008). A case study of industrial symbiosis: Nanning Sugar Co, Ltd. in China. *Resources, Conservation and Recycling.* 52: 813–820. doi:10.1016/j.resconrec.2007.11.008.

CS 22: Hashimoto S, Fujita T, Geng Y, Nagasawa E. (2010). Realizing CO_2 emission reduction through industrial symbiosis: a cement production case study for Kawasaki. *Resources, Conservation and Recycling.* 54: 704–710. doi:10.1016/j.envpol.2010.10.022.

CS 23: Dong L, Fujita T, Zhang H, Dai M, Fujii M, Ohnishi S, Geng Y, Liu Z. (2013). Promoting low-carbon city through industrial symbiosis: a case in China by applying HPIMO model. *Energy Policy.* 61: 864–873. doi:10.1016/j.enpol.2013.06.084.

CS 24: Mathews JA, Tan H. (2011). Progress toward a circular economy in China: the drivers (and inhibitors) of eco-industrial initiative. *Journal of Industrial Ecology.* 15 (3): 435–457. doi:10.1111/j.1530-9290.2011.00332.

CS 25: van Beers D, Corder G, Bossilkov A, van Berkel R. (2007). Industrial symbiosis in the Australian minerals industry: the cases of Kwinana and Gladstone. *Journal of Industrial Ecology.* 11 (1): 55–72.

29

A PROPOSED APPROACH FOR A SOLID WASTE COLLECTION SYSTEM IN AN AFRICAN RURAL TOWN

A case study from Kenya

Mentore Vaccari

Introduction

For many cities, solid waste management is their single largest budget item and largest employer. For example, municipalities spend 20–50% of the available budgets on solid waste management, but the service often covers less than 50% of the population (Chifari et al., 2017). In developed countries, the operating costs for integrated waste management are generally more than €90 per tonne, while in developing countries, it is around €30 per tonne. In developing countries, some 80% of the spending is on collection, while it is around 10% in developed countries (World Bank, 2017). The costs of transportation alone are often around €18–45 per tonne. With cost recovery differing significantly across income levels, full or nearly full cost recovery is primarily limited to developed countries (World Bank, 2020). While monies are usually provided for capital spend (CAPEX), operational expenses (OPEX) are often limited (World Bank, 2019). Sustainable financing is a major obstacle for municipalities trying to improve their waste management services, as they are unable to generate the necessary fee collection rate until they improve the service and, at the same time, households are unwilling to pay for poor service (World Bank, 2017).

According to the World Bank (2017), cities generate approximately 1.3 billion tonnes of solid waste per year. However, with current urbanisation trends, this is predicted to rise to 2.2 billion tonnes per year by 2025. There is a global shortfall of around €37 billion annually for municipal solid waste (MSW). However, managing waste will become more expensive. It is expected that expenditures will grow from around €185 billion to around €345 billion. These financial pressures will be primarily severely felt in low-income countries already struggling to meet basic social and infrastructure needs, particularly for their poorest residents.

Within developing countries, issues around public health and the recovery of resource value tend to be key driving factors for waste management systems (Peltola et al., 2016; Vaccari et al., 2019). Indeed, the urban poor face a number of institutional, legal, financial, and technical barriers to accessing adequate services (Jaligot et al., 2016). However, according to the World Bank (2017), the management of MSW is often inefficient and underperforming in developing countries. Even if it is only provided to a percentage of the public, it benefits the overall public welfare. Thus, the higher the coverage of the service provision, the higher the public welfare. Indeed, the World Bank (2019) notes that more effective management of waste represents a significant opportunity for cities to reduce greenhouse gas emissions and improve health and well-being in communities (e.g., reduced flooding after a heavy storm). Open dumping and burning have significant adverse effects on water bodies, air, and land resources, and attract disease vectors, which contributes to increased disease burdens (World Bank, 2019). Improved management of solid waste can also provide other benefits such as job creation and income generation opportunities for local citizens (Vaccari and Perteghella, 2016). This is particularly important in the case of developing countries, which often face a significant number of challenges in effectively managing their waste.

Using a case study of Sololo in Kenya, this chapter reports on the results of a study that sought to evaluate selected approaches to enhance the collection and treatment of solid waste management within developing countries. The study was especially important, as it focused on the rural areas of the case study, which is an aspect that often receives limited attention. In addition, unlike previous studies, it also focused on waste from households as well as commercial organisations in the town. Finally, it proposes an option to overcome one of the key challenges often faced, that of limitations in funding.

Waste collection and management in developing countries

Understanding how best to improve the delivery of waste management services is crucial to ensure cost efficiencies and meet legislative compliance (Pérez-López et al., 2016; World Bank, 2017). The development of strategies to effectively manage municipal solid waste should begin with an understanding of the characteristics and properties (e.g., quantity, composition, density, moisture levels, etc.) of the waste produced (Zurbrügg et al., 2014; UNEP, 2015; Aleluia and Ferrão, 2016). Generally, the organic content of MSW in developing countries tends to be high (>65%) (World Bank, 2020). This high organic content should be taken into consideration in the design of collection, transport, and treatment systems. There are a number of other factors that impact the efficiency of solid waste collection and management systems in developing countries. For example, many developing countries do not possess the necessary legislation, policies, procedures, and regulations (World Bank, 2019). There can also be limitations in route planning and public access to adequate information about collection schedules, inadequate waste management infrastructure (Jaligot et al., 2016; World Bank, 2017), and inadequate financial resources (World Bank, 2020). The level of existing segregation of the waste is also an important consideration, as this impacts the quantity and potential value of the materials recovered (Aleluia and Ferrão, 2016). At a strategic level, there is often a lack of adequate governance systems for effective waste management in developing countries (World Bank, 2019; Filho et al., 2016). Waste collection rates in developing countries vary widely, from, for example, 20% in countries such as Bangladesh to 90% in others like Azerbaijan (Aleluia and Ferrão, 2016). Collection rates will also vary between cities (and affluent neighbourhoods) and rural areas within the countries.

Various technological approaches have been suggested to improve collection systems in developing countries, including the use of vehicle routing (VR), the use of geographic information system (GIS)–based technologies (Buenrostro-Delgado et al., 2015), and value chain optimisation (Jaligot et al., 2016). However, vehicular access is often problematic or indeed impossible in rural areas, due to either there being no roads (at some times of the year) or very narrow roads (World Bank, 2017). Thus, high-tech recommendations are often not the ideal option. With respect to the management of the waste, while the high organic content lends itself to the use of anaerobic and aerobic treatment such as anaerobic digestion and composting, respectively, landfilling is the primary method employed (World Bank, 2017). However, the landfills are often poorly designed, sited, and managed and cannot be considered sanitary (Vaccari et al., 2019).

Thus, given the number and range of factors, the development and implementation of effective waste collection systems have, therefore, to be holistic and integrated in their approach, incorporating environmental, socioeconomic, and technical factors (Filho et al., 2016). The adoption of more efficient services, and the implementation of systems to recover value from any waste produced, can lead to significant financial and environmental benefits (UNEP, 2015; World Bank, 2019).

The case study area

The town of Sololo is located in the northeast of Kenya, next to the Ethiopian border. It is one of the main towns of the Moyale district and one of its main administrative centres. At the time of the study, it was made up of 19 settlements (villages) and had about 7,500 inhabitants. Each settlement was administrated by a chief and his collaborators in agreement with the Council of Elders of the town. Moyale district is located in an arid or semi-arid, flat area, crossed by the road connecting Nairobi and Addis Ababa. During the rainy season, the area is often flooded, and Sololo becomes isolated. The local economy is based on livestock, agriculture, and small commercial activities. All these activities are conducted mainly to provide subsistence for the family, and the population of the district often needs external aid for its survival. Effective waste management was completely absent in Sololo at the beginning of the study, with all solid waste either deposited along the roads or in private courtyards or burnt in the open air.

Methods

The project was divided into two main activities, namely an analysis of the production and composition of the waste and a survey on the management of domestic waste by means of semi-structured interviews and questionnaires. The findings from these were then employed in the design of a more effective system for solid waste management in Sololo.

Evaluating waste generation patterns

Based in part on previous similar studies (e.g., Vaccari and Perteghella, 2016), an analysis of the waste production and composition was conducted to examine the effectiveness and efficiency of the system. Based on consultation with the collaborators of the chief and the Council of Elders, five compounds and three points of informal collection along the streets were selected at random in order to collect the waste to be analysed. Empty containers were distributed to householders, and they were asked to keep their waste for three days, after which the containers were collected by the two researchers. Commercial and sweeping wastes were collected from the streets in buckets. Rough wet waste (about 100 kg on the whole) was sieved using a 2-cm mesh net. The waste was then divided into two fractions (smaller and larger than 2 cm), which were weighed separately. Waste belonging to the second fraction (larger than 2 cm) was further separated into six categories (organic matter, plastic, paper and cardboard, glass and metals, textiles, and inert materials, like stones and debris) and each category weighed.

Understanding the views of the community

Questionnaire surveys were conducted with 100 families in order to assess their behaviours and attitudes towards solid waste management. The households were chosen at random and the questionnaires hand delivered and collected back by the two researchers over a period of two weeks. In total, 715 people (i.e., about 9.5% of inhabitants) were part of the surveyed families: 103 men, 114 women, and 498 children. As an average, the family included one man, one woman, and five children. Table 29.1 lists the questions asked.

The 16 owners of commercial activities (nine bazaars, three grocery shops, two clothes shops, one butcher shop, and one bar) located along the main road were also interviewed. These five activities represented all of the major commercial activities in the village at the time of the study. In addition, the directors of the four schools (two primary and two secondary schools) and the director of the hospital were interviewed in order to get information about composition of waste produced on those premises and how it was disposed of. Some 890 children attended the two primary schools and 290 the secondary school. There was only one hospital that covered the most common medical, surgical, obstetric, and paediatric emergencies and diseases. The services provided by this hospital, which had 168 beds, were the only ones available to the resident population.

Developing the collection system

First, the Sololo territory was mapped by using a GPS-based survey. Two GPS receivers, Trimble Geo XT, were used, and the data collected were elaborated using Pathfinder Office software. This gave an indication of the spatial dimensions for the collection routes, including the area of the territory and distances between the houses and the disposal site. In consultation with the chief, collaborators, and Council of Elders, two different alternatives were explored and evaluated, considering one or two collections of waste per week. Based on previous similar studies (e.g., ISWA, 2013) the disposal site was assumed to be 1 km from the town centre.

Two different methods of collection were designed, one for areas that were accessible by donkey carts and another for those that were not. In the first area, there were about 2,900

Table 29.1 Questions used in the questionnaire surveys with families

#	*Question*
1	Which is the name of the village where you live?
2	How many men, women and children are in the family?
3	What are the job, the gender, and the age of the head of household?
4	Where do you dispose of your waste?
5	How do you dispose of your waste?
6	Which are the main materials contained in your waste?
7	Which animals do you keep at home?
8	How do you dispose of your organic waste?
9	How do you dispose of your plastic waste?
10	Do you reuse/recover material from waste?
11	What kind of fuel do you use to cook and heat?
12	Where do you put the small miscellaneous personal waste items (e.g., cigarette butts and chewing gum)?
13	Which are the risks associated with plastic ingestion by animals?
14	Which are the risks associated with plastic combustion?

Table 29.2 Unitary costs used to estimate capital and operational costs

Item	*Unit*	*Cost*
Donkey	€	100
Cart	€	235
Donkey-bag	€	17
Rake	€	2.50
Street container	€	50
Jerry can	€	2.50
Donkey shed	€	1,000
Guard house and latrines	€	1,000
Cart maintenance shop	€	5,000
Road to landfill	€	1,000
Landfill	€	5,000
Donkey farmer	€/y	660
Waste collector	€/y	800
Maintenance man	€/y	800
Landfill worker	€/y	800
Director	€/y	1,320
Donkey feeding	€/y	180

inhabitants, whereas in the second, there were about 5,500 people. The average speed at which the donkey carts moved was calculated by timing the distance travelled by two carts over a period of five days. Jerry cans with a capacity of 20 L were assumed, as these were readily available and common at the time of the study. The average number of jerry cans that could be transported by each cart was determined at the same time as the speed.

Capital and operational costs were estimated in consultation with the collaborators of the chief. Capital costs were calculated considering the construction of a road to the landfill, a shed for the donkeys, a guard house, and a maintenance shop for the carts. Workers' wages and food for the donkeys, as well as depreciation of the facilities, were considered to evaluate the operational costs.

Table 29.2 lists the unit costs used to estimate capital and operational costs of the scheme.

Depreciation costs were calculated considering an estimated lifetime of 15 years for donkeys; 3 years for carts and donkey-bags; and 2 years for rakes, street containers, and jerry cans.

Results

Waste production and composition

Household waste production was estimated to be 0.31 kg/(inh d). The waste contained a high percentage of dust (the courtyards were unpaved) and goat faeces, as represented by the sieved fractions for each of the three categories of waste (Table 29.3).

For the household waste, the organic waste fraction was highest (27.8%) and included animal feed discarded and animal faeces from the cows, donkeys, and camels. Overall, about 50% of the waste consisted of animal faeces. Plastics were the second largest fraction in the household waste (4.4%). Sweeping waste was rich in organic waste (primarily goat faeces) and potentially recyclable materials (e.g., plastics [18.3%], paper [8.3%], and glass and metals [6.7%]). The commercial waste was composed primarily of paper and cardboard (21.4%).

Table 29.3 Composition of household, sweeping, and commercial solid waste

Waste fraction	*Household waste (wt %)*	*Sweeping waste (wt %)*	*Commercial waste (wt %)*
Organic matter	27.8	33.3	9.5
Plastic	4.4	18.3	9.5
Paper and cardboard	1.1	8.3	21.4
Textiles	2.2	0.0	2.4
Glass and metals	1.1	6.7	2.4
Inert materials	0.0	3.4	11.9
Sieved fraction (<2cm)	63.3	30.0	42.9
TOTAL	100	100	100

Inert materials (11.9%) (e.g., stones), as well as plastics and organic waste (both 9.5%), were the next-highest fractions.

The views of the community

The 100 surveyed families were located in 14 out of 19 villages present in Sololo. The head of the household was mainly male (84%), and the average age was 48 years old. Some 29 heads of household were jobless, 11 were retired, and 10 had occasional jobs. The remaining participants worked mainly as security agents in Nairobi (13%), entrepreneurs (10%), teachers in the local schools (4), and nurses in the local hospital (3).

The vast majority of the families (88) accumulated waste within their premises, often in a dedicated pit, and burnt it periodically. The other families discarded their waste outside in the street. Some 37 of those surveyed believed that burning the waste saved the lives of animals, while 44 were of the view that in this way, waste is reduced. However, just over a third of the respondents (35) thought that the fumes generated by the burning of the waste plastic were harmful to human beings.

None of the families had ever weighed their waste and the materials contained in it. However, all noted that the main fractions were organic matter (food waste, prunings) and plastic. Only eight households mentioned the presence of metals and two the presence of paper.

The food waste was used to feed the animals, where present. Plastic jars were reused to collect, transport, and store water. All the other waste was dumped. Some 62 families had an animal on their premises, mainly goats (43), donkeys (28), cows (26), and chickens (21). Four families had sheep, and one had camels. Eighty-seven families reported that they used animal faeces as fertilizer in their own compound. Around 50% of those who had donkeys stated that the animal's faeces were not considered suitable as fertilizer and were therefore burnt. Finally, three respondents claimed to use animal excreta to curb and limit the flooding of the compound during the rainy season.

Table 29.4 lists the types of waste materials produced by the 16 commercial organisations surveyed. It can be seen that packaging waste (mainly paper and plastics) were the most widespread items.

Fourteen of the 16 interviewees said that they burnt their waste. Most (13) burnt it directly in the road, while 1 did so at the back of the store. Two merchants reported that they paid a person (5 Kenyan shillings, around €0.05, per day) to take away their waste. However, it was still burnt in a space at the beginning of the market road.

Table 29.4 Types of waste produced by the commercial activities

Activity	*Waste material*		
Bazaar	vegetables	plastic	paper
Bazaar	paper	plastic film	plastic bottles
Bazaar	textiles	paper	plastic
Bazaar	paper	plastic	
Bazaar	paper	plastic	
Bazaar	textiles	paper	
Bazaar	paper	plastic	
Bazaar	vegetables	plastic	
Bazaar	paper	plastic	
Grocery shops	paper	plastic	
Grocery shops	vegetables	plastic	
Grocery shops	vegetables	paper	
Clothes shops	textiles	paper	
Clothes shops	paper	plastic	
Butcher shop	meat	bones	plastic
Bar	paper	metals	

The majority of the waste from the schools was reported to be composed of food waste, paper, and plastic. In all the schools, waste was accumulated in dedicated pits and periodically burnt. Only one of the primary schools gave some of their food waste to local breeders to feed the animals.

Waste generated by the hospital was made up of syringes, latex gloves, medicines, glass and plastic containers, packaging cartons, food waste, prunings, and wood ashes. It was thrown into a 3-meter-deep pit, fenced to prevent access to animals, and burnt daily.

Design of a system for waste collection and disposal

Given the existing waste generation patterns and waste-burning practices, a system of door-to-door collection was developed in conjunction with local representatives. It was proposed to use plastic jerry cans as containers for primary collection and teams made up of two pickers and two donkeys hauling a cart for secondary collection. In addition, an area about 1 km from the town was identified for the construction of a sanitary landfill. The average estimated speed of the donkey carts was 2.5 km/h. Each cart was able to transport 20 jerry cans, considering the maximum carrying capacity of the animals. Based on the quantity of waste, the collection would be carried out by teams made up of two workers and two donkeys hauling a cart each. Given the high organic content of the waste, jerry cans would be distributed to householders and collected once or twice per week, and collection would be carried out six hours per day, six days per week. Table 29.5 illustrates the number of teams necessary for the different alternatives.

Other teams should be employed to collect waste from the market, schools, and streets. About two teams would be needed to collect commercial and institutional waste and two for sweeping waste. The estimated costs for waste collection and disposal are outlined in Table 29.6.

Table 29.5 Number of teams necessary to carry out solid waste collection in the different waste collection alternatives

Frequency of collection	*Area*	*Number of teams*
Once per week	Accessible	3
	Not accessible	4
	Total	**7**
Twice per week	Accessible	6
	Not accessible	7
	Total	**13**

Table 29.6 Estimated capital and operational costs for waste collection and disposal in Sololo

Frequency of collection	*Capital costs*		*Operational costs*	
	Total [€]	*Per capita [€/y]*	*Total [€]*	*Per capita [€/inh*y]*
Once per week	23,000	2.70	55,650	6.55
Twice per week	25,700	3.01	71,650	8.42

Discussion

Sololo produced approximately 0.31 kg/(inh d) of solid waste. The waste generation rate is around half of the average production rate of low-income countries of around 0.6 kg/(inh d) (World Bank, 2012). This lower rate may potentially be due to the level of poverty that existed in the village, with most heads of households unemployed, retired, or only in temporary employment. Faecal matter represented about 50% of the overall waste. The remaining fraction consisted mainly of dust, organic matter, and plastic. Similarly to other studies in developing countries (Vaccari et al., 2019), the percentage of organic waste was high across the main categories of waste, particularly for households and street sweepings. Plastics were the second-largest fraction in the household waste (4.4%), because plastic bags were used to transport food bought at the market. However, plastics were even higher in the sweepings, primarily due to discarded plastic bags used for drinking water. The non-household waste was composed primarily of paper and cardboard (21.4%), due primarily to the use of cardboard boxes as a form of packaging to transport materials. In the case of the schools, the waste generated was due primarily to teaching activities and canteens for the housed students.

Some of the waste items from the households were reused. For example, in households, the food waste was given to animals and the plastic bottles used to collect water. Given that these two streams were the largest percentage of waste from households, this served to reduce the overall quantity of waste to be disposed of. Some of the animal excreta was also used as a fertilizer. In the case of the commercial entities and schools, none of the waste was reused or recycled. Nevertheless, despite reuse efforts amongst households, almost all of the families either accumulated or dumped waste near their homes. This was most likely because, at the time of the study, similarly to other similar towns, there was no organised service for waste collection and disposal (World Bank, 2017; Aleluia and Ferrão, 2016).

A key challenge was that wastes from households, markets, hospitals, and schools were primarily burnt in the street (World Bank, 2019). Given the size of the town, this meant that this burning took place in close proximity to people, including children. Interestingly, approximately a third of those surveyed were of the view that this practice of burning was harmful to health. Evidently, therefore, levels of awareness of the risks of the potential links between the burning of waste and the health impacts existed. However, the lack of a collection scheme meant that burning was viewed as the only approach to get the rid of the waste, as burying it would mean that it still existed.

A proposed collection scheme which was developed in consultation with the community and based on the local conditions aimed to target both the organic fraction and plastics. The low-tech approach also aimed to overcome the logistical issues in the town and address the inherent logistical issues faced with the terrain. In this way, the approach would not only reduce the waste that was discarded and burnt but also reduce the impacts on the health of the inhabitants in the town (Peltola et al., 2016). The use of donkeys as a means of transport meant that, in addition, employment would also be created for some of the villagers and value recovered from the waste. However, financially, similarly to other studies (e.g., World Bank, 2019), both the CAPEX and OPEX were too high for the local community. These 'high' costs were primarily due to wages of the waste collectors. Given the living standards and the availability of monies in Sololo, adopting a fee-based approach would not be viable. The suggested approach to be employed to overcome this challenge is that of results-based financing. This is a financial mechanism through which the payment for solid waste services is conditioned on the achievement and verification of pre-agreed targets, for example, between the funder and the contractor or community (World Bank, 2017). It focuses attention on measurable outputs in service delivery. It has been employed successfully by the World Bank in a number of developing countries to strengthen waste collection and transport, including in Jamaica, Mali, and Tanzania (World Bank, 2019). In this way therefore, it might be possible for the collection service in Sololo not only to be efficient but also to overcome the financial challenges and achieve a degree of sustainability over the long term (World Bank, 2020). This approach should be combined with awareness building and technical support for capacity building within the community.

Conclusions

Effective waste management in rural areas in developing countries poses significant logistical, environmental, and socioeconomic challenges. These challenges are expected to increase (World Bank, 2017). However, at the same time, the development of effective solid waste management services in rural areas is crucial, as not only can they improve the management of waste, but they also have much wider socioeconomic and public health benefits (Aleluia and Ferrão, 2016; World Bank, 2017).

In consultation with the community in Sololo, this project developed and proposed a simple approach for implementation. However, even though the system met the logistical and environmental needs of the community, the costs were deemed too high. A suggestion for overcoming this challenge has been outlined using a results-based approach. The implementation of an improved approach would lead not only to improved practice in Sololo but also opportunities for improved livelihoods through, for example, the development of improved public health and the production of by-products from the waste. Given the similarities with other rural communities, it is an approach that could potentially also be employed in other such communities.

Acknowledgements

The author wishes to thank Pino Bollini, Veronica Di Bella, Cinzia Scaini, and Sara Tononi for their support during the research.

References

Aleluia J, Ferrão P. (2016). Characterization of urban waste management practices in developing Asian countries: a new analytical framework based on waste characteristics and urban dimension. *Waste Management*. 58: 415–429.

Buensrotro-Delgado D, Ortega-Rodriguez JM, Clemistshaw KC, González-Razo C, Hernández-Paniagua IY. (2015). Use of genetic algorithms to improve the solid waste collection service in an urban area. *Waste Management*. 41: 20–27.

Chifari R, Lo Piano S, Matsumoto S, Tasaki T. (2017). Does recyclable separation reduce the cost of municipal waste management in Japan? *Waste Management*. 60: 32–41.

Filho WL, Brandli L, Moora H, Kruopienė J, Stenmarck Å. (2016). Benchmarking approaches and methods in the field of urban waste management. *Journal of Cleaner Production*. 112–115: 4377–4386.

ISWA (2013). *Guidelines for design and operation of municipal solid waste landfills in tropical countries*. Available at: www.iswa.org/home/news/news-detail/browse/16/article/new-publication-guidelines-for-the-design-and-operation-of-municipal-solid-waste-landfills-in-tropi/109/terms-and-conditions/.

Jaligot R, Wilson DC, Cheeseman CR, Shaker B, Stretz J. (2016). Applying value chain analysis to informal sector recycling: a case study of the Zabaleen. *Resources, Conservation and Recycling*. 114: 80–91.

Peltola T, Aarikka-Stenroos L, Viana E, Mäkinen S. (2016). Value capture in business ecosystems for municipal solid waste management: comparison between two local environments. *Journal of Cleaner Production*. 137: 1270–1279.

Pérez-López G, Diego P, Zafa-Gómez JL, Plata-Diaz AM. (2016). Cost efficiency in municipal solid waste service delivery: alternative management forms in relation to local population size. *European Journal of Operational Research*. 255: 583–592.

United Nations Environment Programme (UNEP) (2015). *Global waste management outlook*. Available at: www.unep.org/ietc/OurWork/WasteManagement/GWMO/tabid/106373/Default.aspx.

Vaccari M, Perteghella A. (2016). Resource recovery from waste by Roma in the Balkans: a case study from Zavidovici (BiH). *Waste Management & Research*. 34 (9): 866–874.

Vaccari M, Tudor TL, Vinti G. (2019). Characteristics of leachate from landfills and dumpsites in Asia, Africa, and Latin America: an overview. *Waste Management*. 95: 416–431.

World Bank (2012). *What a waste: a global review of solid waste management*. Available at: http://documents1.worldbank.org/curated/en/302341468126264791/pdf/68135-REVISED-What-a-Waste-2012-Final-updated.pdf

World Bank (2017). *Solid waste management*. Available at: www.worldbank.org/en/topic/urbandevelopment/brief/solid-waste-management.

World Bank (2019). *Solid waste management*. Available at: www.worldbank.org/en/topic/urbandevelopment/brief/solid-waste-management.

World Bank (2020). *What a waste 2.0*. Available at: https://datatopics.worldbank.org/what-a-waste/trends_in_solid_waste_management.html.

Zurbrügg C, Caniato M, Vaccari M. (2014). How assessment methods can support solid waste management in developing countries – a critical review. *Sustainability*. 6 (2): 545–570.

30

CIRCULAR ECONOMY OPPORTUNITIES IN AFRICA – EMERGING SECTORS AND MISSING NARRATIVES

Sally-Anne Käsner and Sarah O'Carroll

Circular economy development in Africa

Three key sectors have emerged as circular economy (CE) priorities in Africa: plastics, agriculture, and electronics. Plastics is likely to be the first material stream to truly experience a CE transition because of the volume of work being championed at the EU level. Following plastics, CE opportunities are already being realised in the electronics and agriculture sectors.

It is hoped that support from institutions such as the UNEP, UNIDO, the European Commission, the World Economic Forum, the Waste and Resources Action Programme (WRAP) in the UK, the Ellen MacArthur Foundation, and the African Circular Economy Network (ACEN) will be great enablers for CE transitions on the continent, providing technical expertise for policy development including roadmaps, strategies, regulation, and economic incentives.

Plastics

Like many other regions around the world, plastics have been the lead sector in Africa for a CE transition. Rwanda, Ghana, and South Africa are leading the circular transition in the plastics sector using a variety of interventions, including bans, extended producer responsibility, and Plastic Pacts to do so. Going forward, the EU's Plastics Strategy (part of the EU's Circular Economy Package and the European Green Deal), associated funding, and technical expertise, coupled with global businesses responding to the strategy, is expected to continue to be a key driver for the development of a plastics CE action in Africa.

The case of South Africa: leveraging extended producer responsibility activities and introducing a Plastic Pact

South Africa probably has the longest history associated with circular economy activities in the plastics sector, with some of the highest local recycling rates in the world. This has largely been driven by voluntary extended producer responsibility (EPR) organisations like Plastics SA, the Packaging Council of South Africa (Packaging SA), and Polymer Recycling Groups (PETCO, POLYCO, etc.) that have been in operation since the early 2000s (Packaging South Africa, 2018).

Plastic waste collection and recycling has become a key issue in South Africa over the last decade, following the publication of a Government report in 2012 that indicated that 90% of waste generated in South Africa was disposed of at landfill, and of that, 49% constituted packaging waste (South African Department of Environmental Affairs, 2012).

In 2018, the World Wildlife Fund for Nature and Plastics Recycling South Africa started working towards the development of a South African Plastics Pact similar to those in the United Kingdom, France, and Chile. Launched on 30 January, it is the first Plastics Pact to be launched in Africa as part of the Ellen MacArthur Foundation's New Plastics Economy programme. A local not-for-profit, GreenCape, is delivering the pact and 23 organisations across South Africa's plastics value chain have already joined the initiative and are collaborating together to:

- Eliminate unnecessary and problematic plastic packaging through redesign and innovation
- Move from single-use to reuse where relevant
- Ensure all plastic packaging is reusable, recyclable, or compostable
- Increase the reuse, collection, and recycling or composting of plastic packaging
- Increase recycled content in plastic packaging

(South African Plastics Pact, 2020)

Many of the members of the South African Plastics Pact are also working with the broader packaging industry in South Africa to develop an industry-led waste management plan for the packaging industry.

The case of Ghana: leveraging global action to develop a local policy framework

Ghana joined the World Economic Forum's (WEF) Global Plastic Action Partnership (GPAP) in October 2019 (World Economic Forum, 2019). The country's National Plastic Action Partnership will work in collaboration with the GPAP to transform the plastics economy in Ghana. Similar to the Ellen MacArthur Foundation's Plastic Pacts (currently operating in the United Kingdom, France, and Chile), the GPAP brings together public, private, and civil society and academic sector stakeholders to stop plastic pollution from source to sea by 2025 and therefore focuses on coastal economies. Ghana is the first African country to join the initiative. The Ministry of Environment, Science, Technology and Innovation (MESTI) has been tasked to develop a National Plastic Management Policy to ensure the sustainable management and reduction of plastic waste in the country in a manner that supports economic growth and boosts job creation and progress towards meeting the Sustainable Development Goals (IISD SDG Knowledge Hub, 2019). This will be the first policy of its kind in Ghana, and it tackles key circular economy topics of redesign, recovery and recycling, behaviour change, and the mobilisation of finances for investment and innovation to transform the industry (Ghana Web, 2019).

The case of Rwanda: utilising regulation for behaviour and market change

Rwanda was the first country in Africa to implement a single-use item ban, banning the use of plastic bags in 2004 and enacted in 2008 as a response to the disposal methods in use that have significant human health and environmental impacts: burning and open dumping (Plastic Oceans, 2018). Businesses in the country have converted to paper bags, and luggage is inspected at border control to ensure that plastic bags do not enter the country.

In 2019, the law relating to the prohibition of manufacturing, importation, use and sale of plastic carry bags and single-use plastic items, Article 3, was updated and includes inspections and the threat of a hefty fine at the airport, and the borders of neighbouring countries DRC, Tanzania, Kenya, and Ethiopia are all working to ensure that plastic bags do not enter their countries, but the black market continues to be a challenge. As long as plastic bags are produced elsewhere in the world, the black market will continue to be a problem for any country.

Agriculture

The agriculture sector is a significant contributor to GDP, employment, and exports for African countries. The adoption of chemical fertilizer-based farming has depleted soils of small-holder farmers and, with irregular rainfall and the increased frequency of droughts, resulted in large post-harvest losses (World Bank, 2015). As a result, regenerative farming methods and value-adding activities to organic waste have dominated the circular economy narrative in this sector. In fact, this could be seen as the re-adoption of traditional farming practises that rebuild soils to reduce the need for pesticides and fertilizers. Research, development, and innovation (RDI) is critical to the success of circular economy transitions around the world. There is no one-size-fits-all approach, and different technologies and approaches will be needed to accelerate the circular economy transition in Africa in comparison to those used in Europe. It is encouraging to know that RDI is being undertaken for the African context, both in Europe and within Africa itself; there are also now several circular food value chain enterprises operating in Africa that offer the product-as-a-service business model or use waste-to-nutrient recycling approaches.

Research, development, and innovation in AgTech based on the local context

(B)Energy is a social enterprise that has innovated traditional anaerobic digestor technologies for the African context. (B)Energy currently operates in Benin, Mali, Rwanda, Malawi, Kenya, and Ethiopia and provides biogas technologies to communities that are affordable and mobile to provide access to energy to local communities. The novelty relies in the (B)Pack, a biogas backpack that can be carried by an individual for sale of biogas to households. Importantly, the enterprise also provides training and micro-finance to entrepreneurs to support the growth of sustainable biogas businesses ((B)Energy, 2019a, 2019b, 2019c).

Agriprotein is probably one of Africa's greatest success stories. The firm has developed a large-scale fly farming technology that processes organic waste into animal feed (MagMeal), an extracted fat (MagOil), and a rich residual soil conditioner (MagSoil) (Agriprotein, 2019). The company's MagMeal product earned the title of BBC Food Chain Global Champion and the CleanEquity award at the Global Cleantech Top 100 in 2017 and the Australian government's Blue Economy Challenge in 2016 (Agriprotein, 2017).

Songhai Farm in Benin is historically the concept that has influenced permaculture in Africa and beyond thanks to its early approach to regenerative farming since 1985. Songhai aims at restoring nobility to farm work and to help young people choose not to suffer and to provide service in creating wealth for their families, their countries, and their continent, through functional training based on knowledge, skills, and a value system based on three thematics: bio-production, bio-processing, and bio-consumption (Songhai Farm, 2019).

Hello Tractor, based in Nigeria, is an AgTech social enterprise that works to make small-scale farms more productive by providing an on-demand tractor service to farmers through a smartphone app. The enterprise has grown quickly over the last four years, with 500 tractors currently available on the app, capturing 75% of the private tractor market in Nigeria, and it is now expanding across five markets in Africa (Hello Tractor, 2019).

Electronics

Despite the Basel Convention being ratified in 1991, Africa is one of the world's leading destinations for WEEE. Ghana and Nigeria, in West Africa, receive the largest amounts of WEEE, but other countries like the Ivory Coast and the Democratic Republic of Congo are also burdened by this illegal dumping from the United States and Europe.

Much of the WEEE received in African countries is often dumped or burned by the unregulated informal sector, releasing heavy metals and toxic chemicals that have significant environmental and human health impacts. The circular economy is increasingly seen as a framework to tackle the problem of WEEE because of its focus on designing out waste and pollution and keeping materials in use for as long as possible.

The case of Nigeria: developing a circular economy for electronics policy framework

Nigeria generates some 1.1 million tonnes of WEEE every year from imports and local WEEE generation, supporting 100,000 jobs in the informal economy in recycling and reuse activities. A portion of this waste stream is recycled by registered companies like Hinckley Recycling or repaired for Nigeria's vibrant second-hand market, but the vast majority is still dumped (TRT World, 2019).

In June 2019, the Nigerian government, the Global Environment Facility, and UN Environment announced a US$15-million initiative to kickstart a circular electronics economy in the country. The project, part of the World Economic Forum's Platform for Accelerating the Circular Economy (PACE), is led by the National Environmental Standards and Regulations Enforcement Agency (NESREA) and will support the E-waste Producers Responsibility Organization (EPRON) to accelerate the development of a circular economy for electronics in Nigeria. The project will seek to expand reuse and recycling activities and improve the livelihoods of those informal workers in the sector, promoting regional and international dialogue and engaging global manufacturers (UN Environment, 2019).

The case of South Africa: the need for a roadmap

It is not currently known how many mobile phones are sent to landfill or incineration each year in South Africa. Even though mobile phones form a small part of the WEEE stream in South Africa, they possess greater amounts of valuable materials in comparison to other types of WEEE, such as computers, photocopiers, and printers (Desmond, 2015). The constraints to adopting more circular approaches are wide ranging, and some relate to mobile phone design, manufacture, use, and disposal processes. A coordinated effort is required by government, regulators, manufacturers, network operators, retailers, and consumers to ensure that constraints to recycling are identified and strategies developed to overcome them.

There is also a need to recognise the complexity of designing, manufacturing, and recycling mobile phones and the extent of the global value chains that result. One route to overcoming the constraints to efficiently recycling mobile phones in South Africa is the creation of a broader CE manifesto or roadmap on a country, regional, or city basis, as has been created for a few countries and cities in the Global North, for example, Finland and London in the United

Kingdom. Such a roadmap for the electronics sector would need to form part of a broader strategy for the adoption of circular principles across all sectors of business and within government policy. As in other African countries, the role of the informal sector will become an additional dimension for a South Africa circular economy framework for electronic waste management and resource utilisation.

The missing narratives in Africa's circular economy

As the circular economy develops in Africa, there are some narratives that are currently missing from the discourse. These include the role of the informal sector and the design of products and services, the need to promote and develop African as part of a CE connected service provider economy, and opportunities in the urban building and water systems for a CE transition.

Leveraging the role of the informal waste management sector

The informal sector plays a critical role in waste management in Africa. Informal waste collectors, often referred to as waste pickers or scavengers, rely on the value of waste materials for their livelihoods. As countries in Africa develop formal waste management systems, they should also consider the significant role of the informal waste sector. Not only do informal waste collectors (or waste pickers or scavengers) rely on valuable waste materials such as glass, polyolefins, and metals for their livelihoods, in many cases, they are also highly efficient. Key challenges of the integration of the informal sector into the formal waste management sector include:

- The policy landscape, often resulting in the municipality owning all household waste, thus rendering waste picking an illegal activity
- The poor social acceptance of waste pickers
- Lack of evidence to support the efficiency of informal waste collection methods
- The unsafe and unhygienic waste collection methods used in the informal waste sector
- The poor business case for the collection and disposal of municipal waste management if high-value recyclables are removed from the waste stream.

The case of Accra, Ghana: an informal waste management system

The city government in Accra, Ghana, is the first city in Africa to incorporate its informal waste collection system into its formal waste collection system. In 2016, the Accra Metropolitan Authority (AMA) initiated a programme to integrate the informal waste collectors into the city's official waste management system to increase collection of waste, close the illegal open waste sites, and ensure fair and inclusive employment. The city provides waiting areas at three new transfer stations for the waste collectors to sort the waste they collect and store their recyclables, which they can sell to buy-back centres (Bloomberg Philanthropies, 2019). This locally responsive intervention has had significant benefits for the city:

- Informal collectors have increased waste collection coverage from 75% to 90% and recycling rates from 5% to 18%
- Collection rates in the informal sector have increased from 300 to 650 tonnes daily
- The municipality has saved US$5.46 million in operational costs by leveraging this low-cost solution (Nordic Sustainability, 2019).

Packa-Ching and WeCyclers: technology and incentives to access recyclables in informal settlements

An initiative of POLYCO, Packa-Ching uses a mobile, incentive-based, cashless payment system to collect recyclables from informal settlements in South Africa. A truck with a trailer attached parks in a community one day each week to collect separated recyclable plastic, paper, metal cans, and glass. Individuals bring their materials to the truck, where they are checked and weighed and the monetary value is paid into an online account, which can be redeemed at participating retailers or transferred to others using a mobile phone.

Packa-Ching seeks to directly address the challenge that waste management infrastructure is often not available in informal settlements, individuals don't know the value of waste, and there are no incentives for individuals within these communities to recycle. After the pilot project in Langa, Cape Town, proved a success, Packa-Ching quickly expanded to two other locations: Ivory Park in Johannesburg and Buffalo City in East London (Packa-Ching, 2019).

Similar to South Africa, Lagos also lacks waste management infrastructure. In 2012, only 40% of the city had access to waste management infrastructure and only 13% of collected waste was recycled. WeCyclers aims to address this challenge using a fleet of low-cost cargo bicycles to collect household recyclables in Lagos, Nigeria. Low-income households are incentivised to collect for the recyclables, which will be sold to the local recycling market. Collected materials are taken to mini material recovery facilities around the city for processing before being sold to the local recycling industry. Households are paid in points using a mobile technology platform for their recyclables, which they can redeem for food and other household items.

The enterprise has seen great success since its initiation in 2018, diverting more than 400 tonnes of waste from landfill to productive uses (WeCyclers, 2019). In October 2019, Wecyclers signed a memorandum of understanding with Nestle, one of the largest food and beverage companies in the world, to expand WeCyclers operations to more communities in Lagos (WeCyclers Nigeria, 2019).

The role of design in circular economy transitions

As evidenced by the case studies presented in the earlier section, the current circular economy discourse in urban Africa is very downstream focused, with the narrative largely focused on end-of-pipe activities such as EPR and recycling. It is disappointing that upstream activities such as circular design and eco-design, reuse, repair, and remanufacturing are not prevalent in the current discourse. Historically, design has been the domain of the developed North, where many of the world's largest manufacturing companies are headquartered. Lack of political will within African countries to tackle the issue of waste management for decades has possibly reinforced this narrative.

Circular Design Nairobi, in Kenya's capital, works to support the entrepreneurial potential of local designers and train the design community in circular economy practises to find solutions to some of the city's biggest challenges: mobility, construction, and waste management (Circular Conversations, 2019).

In order for Africa's CE to flourish, it will be important to foster circular design activities, not only products but also business models that prioritise access via sharing and reuse, repair and refurbishment, and remanufacturing.

Circular economy opportunities in building systems

Africa will overtake Asia as the world's most rapidly urbanising region by 2025 (UN, 2014). In addition, the majority of the housing stock needed for the 2050 population will need to be built; currently 47% of the population live in slums.

The role of the circular economy in building systems has hardly been explored in Africa, and where it is part of the narrative, it is in pocketed silos. Africa is also known for its incremental housing developments, something to be leveraged in a CE transition where buildings should be modular, their primary use changed over time, and material banked for the future.

Circular economy opportunities in water systems

As Africa is one of the most water-scarce continents, where droughts are increasing in frequency and there is still much infrastructure to be developed, water systems also present a key opportunity for CE integration.

Large centralised water systems are the current focus, where decentralised, modular systems for rainwater collection, stormwater capture, and surface water runoff would offer diversity and greater resilience. Solutions that mimic nature (also called biomimicry) have great potential for urban water systems. Plants can be particularly effective to filter and treat wastewater, as evidenced in Langrug at the Genius of Space project and Kraaifontein Waste Management Facility, both in South Africa. The Kraaifontein Waste Management Facility uses a bioswale (a shallow trench filled with vegetation) that has been designed to treat polluted stormwater from the facility in South Africa (Civil Engineering, 2012).

Questions that remain unanswered

There is a great need to document Africa's existing circular economy as well its future circular economy. As the world's most circular continent, there is an opportunity to better understand its current activities and context to ensure that this circularity is the foundation of its future enhanced circularity.

The current information asymmetry that exists about CE in Africa is a significant barrier to the development of a transition that supports existing best-practice activities. This evidence base could serve as supporting information for the expansion of existing reuse, repair, remanufacturing, refurbishment, and recycling economies on the continent rather than the development of new large-scale waste-to-energy projects. It could also be used for best practice sharing with other regions with similar socioeconomic contexts such as Latin America and South-East Asia.

For Africa's future CE, research is needed to understand the impacts of a transition to a circular economy. Similar to the Ellen MacArthur Foundation's 'Growth Within' study for Europe (Ellen MacArthur Foundation, 2015), a study is needed to estimate the potential for CE in Africa, ideally at the country level.

Some of this research will be addressed by the European Commission's study entitled 'How Can EU-Africa Cooperation Programmes and Partnerships Be Fostered to Promote the Transition to a Circular Economy?' However, as much about Africa's circular economy is unpublished and based on Indigenous principles and culture, the level of on-the-ground activities will likely be unknown via this majority desktop study.

Further to this, additional research is also needed to understand the following:

- To what extent can the CE assist African countries to meet the Sustainable Development Goals?

- To what extent can the CE contribute towards mitigating negative impacts of climate change in Africa?
- How will global value chains change into the future? Which global value chains will transition to more local, decentralised value chains? How will these transitions impact African economies?
- What skills are needed to support the CE, and how can these skills be developed in Africa?
- What would the impact be on the informal sector? Will different opportunities for the informal sector and a 'just transition' be facilitated through training and other capacity-building and awareness-raising activities?
- What is the impact on African producers from an increase in the level of circularity employed by global corporations based in Europe?

Until CE in Africa is better understood by decision-makers in the private and public sectors and the potential and positive and negative impacts of a continental transition are better understood, Africa's CE will continue to be at great risk, and the African Union's Agenda 2063 cannot be realised.

References

Agriprotein (2017). Sustainable feed manufacturer. In: *Fast track 2017: revolutionary insect protein wins acclaim for AgriProtein as top business disruptor*. Available at: https://agriprotein.com/press-articles/sustainable-feed-manufacturer-in-fast-track-2017/ (Accessed 3/10/2019).

AgriProtein (2019). *About*. Available at: https://agriprotein.com/ (Accessed 3/10/2019).

(B)Energy (2019a). *(B)Energy*. Available at: www.be-nrg.com/b-energy/ (Accessed 5/10/2019).

(B)Energy (2019b). *(B)Products*. Available at: www.be-nrg.com/b-products/ (Accessed 5/10/2019).

(B)Energy (2019c). *(B)Press*. Available at: www.be-nrg.com/b-press/ (Accessed 5/10/2019).

Bloomberg Philanthropies (2019). *World's seven best climate projects announced as 2019 C40 cities Bloomberg Philanthropies awards winners*. Available at: www.bloomberg.org/press/releases/worlds-seven-best-climate-projects-announced-as-2019-c40-cities-bloomberg-philanthropies-awards-winners/ (Accessed 12/10/2019).

Circular Conversations (2019). *Circular Kenya: when the magic happens. A discovery journey into community-led circular innovations with Wekesa Zablon Wanyama*. Available at: www.circularconversations.com/blog/2019/5/20/circular-kenya-when-the-magic-happens (Accessed 8/10/2019).

Civil Engineering (2012). *The Kraaifontein waste management facility*. Available at: www.civildesigner.com/news/saice/kraaifontein.pdf (Accessed 8/10/2019).

Desmond P. (2015). *Towards a circular economy in South Africa – what are the constraints to recycling mobile phones?* MA dissertation. Institute of Development Studies, West Sussex.

Ellen MacArthur Foundation (2015). *Growth within – economic and business rationale for an accelerated transition*. Isle of Wight: Ellen MacArthur Foundation.

Ghana Web (2019). *Akufo-Addo launches national plastic management policy*. Available at: www.ghanaweb.com/GhanaHomePage/business/Akufo-Addo-launches-National-Plastic-Management-Policy-October-1-782009 (Accessed 6/10/2019).

Hello Tractor (2019). *About*. Available at: www.hellotractor.com/home (Accessed 5/10/2019).

IISD (2019). *Ghana, G20, EU take action on plastics*. Available at: https://sdg.iisd.org/news/ghana-g20-eu-take-action-on-plastics/ (Accessed 8/10/2019).

Nordic Sustainability (2019). *Cities100*. Available at: https://issuu.com/nordicsustainability/docs/cities100_2019_report (Accessed 8/10/2019).

Packa-Ching (2019). *About*. Available at: www.packaching.co.za/how-we-do-it/ (Accessed 8/10/2019).

Packaging South Africa (2018). *SA PET industry hits new high with 2.15 billion bottles recycled*. Available at: https://www.packagingsa.co.za/2018/05/17/sa-pet-industry-hits-new-high-with-2-15-billion-bottles-recycled/

Plastic Oceans (2018). *Rwanda plastic bag ban*. Available at: https://plasticoceans.org/rwanda-plastic-bag-ban/ (Accessed 4/10/2019).

The SA Plastics Pact (2020). *About*. Available at: www.saplasticspact.org.za/who/ (Accessed 20/2/2020).
Songhai (2019). *Songhai in action*. Available at: www.songhai.org/index.php/en/ (Accessed 7/10/2019).
South African Department of Environmental Affairs (2012). *National waste information baseline report*. Available at: http://sawic.environment.gov.za/documents/1880.pdf (Accessed 6/11/2019).
TRT World (2019). *Nigeria has become an e-waste dumpsite for Europe, US and Asia*. Available at: www.trtworld.com/magazine/nigeria-has-become-an-e-waste-dumpsite-for-europe-us-and-asia-24197 (Accessed 8/10/2019).
UN (2014). *Population facts*. Available at: https://www.un.org/en/development/desa/population/publications/pdf/popfacts/PopFacts_2014-3.pdf
UN Environment (2019). *Nigeria turns the tide on electronic waste*. Available at: www.unenvironment.org/news-and-stories/press-release/nigeria-turns-tide-electronic-waste (Accessed 8/10/2019).
WeCyclers (2019). *About us*. Available at: http://wecyclers.com/about-us/ (Accessed 8/10/2019).
WeCyclers Nigeria (2019). *Nestle Nigeria and WeCyclers to tackle plastic waste pollution in Lagos*. Available at: http://blog.wecyclers.com/2019/09/24/nestle-nigeria-and-wecylers-to-tackle-plastic-waste-pollution-in-lagos/ (Accessed 8/10/2019).
The World Bank (2015). *Agriculture in Africa: telling facts from myths: overview*. Available at: www.worldbank.org/en/programs/africa-myths-and-facts (Accessed 4/10/2019).
World Economic Forum (2019). *Ghana becomes first African nation to join ambitious partnership to end plastic pollution*. Available at: www.weforum.org/press/2019/10/ghana-becomes-first-african-nation-to-join-ambitious-partnership-to-end-plastic-pollution/ (Accessed 8/10/2019).

PART VI

Reuse

31

MODULAR SMARTPHONES AND CIRCULAR DESIGN STRATEGIES

The shape of things to come?

Sabine Hielscher, Melanie Jaeger-Erben, and Erik Poppe

Modular smartphones and circular design strategies

It is widely acknowledged that transitions towards a circular economy (CE) are based on the complementary development of circular business models and design strategies (Stahel, 2016; Agrawal and Bellos, 2016; Bocken et al., 2017; Ellen McArthur Foundation, 2018). Among the most prominent design strategies are modular product architectures that break products down into components that can be independently and easily replaced or reused (Agrawal and Ulku, 2013; Proske et al., 2016). Modular product designs have been around for over 50 years (Starr, 1965). In the beginning, it was mainly seen as an opportunity to decrease lead times in the production process, and later, it became a means to offer customised products to the masses. It was only with the rising of the CE debate in the last decade that modular product designs were linked to more sustainable consumption patterns. From the perspective of producers and manufacturers, modularity allows an extension and simplification of services, since modules can be replaced more easily in case of malfunctions and/or upgraded in case of changing customer needs or technological advancements. In addition, the product's storage and redistribution can become more straightforward. Some authors also see modularity as an enabler of innovations (Ulrich, 1995; Baldwin and Clark, 1997).

Much research on modular product designs focuses on smartphones. Smartphones are considered a suitable product to make use of the advantages derived from modular designs. They offer a wide functionality spectrum that can be translated into independent modules or components. Moreover, due to relatively short replacement cycles, the lifetimes of smartphones could potentially be lengthened if only modules are replaced. Surveys have shown that smartphones are used on average no longer than two to two and a half years (Jaeger-Erben and Hipp, 2018; Wieser and Tröger, 2018). Most smartphones still work when being replaced (ibid.). The reasons for their replacement often relate to functional aspects of the phone (e.g., too-small storage space) or personal aspects (e.g., a general appreciation of 'newness' by the owner).

Modular concept ideas for smartphones became popular with Phonebloks around 2013. The design was based on a fully modular and upgradeable smartphone. Although the design idea received much attention and inspired other attempts to develop modular smartphone concepts, it was never commercially implemented. Other attempts have focused on upgradeability and personalisation, such as project ARA, a prototype of a modular phone by Motorola and later Google. It was developed to a high degree but never entered the market. The first devices with a modular design that have entered the market are the Fairphone 2 and SHIFT6m. These designs mainly focus on easy reparability and thereby the restoration of the device's functions and possible upgrades for the phone. In contrast, a more dynamic modular design strategy would allow for a flexible use of a diverse set of parts and modules (as intended by the Phonebloks concept). Within these technological developments, questions about whether smartphone users would make use of smartphones that are easy to repair and upgrade are never really reflected upon. It seems reasonable to assume that the appropriate type of modular design differs depending on people's use patterns. So far, much of the literature concentrates on the producers and retailers of modular products. Although a variety of forms of modularity have been investigated, there are only a few empirical examinations of how people domesticate (i.e., take up and use) modular smartphones in their daily life.

From our point of view, understanding the historical developments of smartphone designs and existing user practices with smartphones is key to developing notions of modularity as a sustainable design strategy and business model options for a CE. What actually is a 'modular' user? This question goes beyond a mere examination of the public acceptance of modular devices. Sustainable consumption potentials only come into play if the devices widely diffuse and are used in accordance with the design strategies in mind (e.g., repairing the phone). Modular devices not only require users to accept the phone but also specific competences and skills (e.g., knowledge of repair and upgrade options), values (e.g., desire to keep the phone for long periods), and willingness for more frequent interactions with product service providers (e.g., upgrade services).

Proske and Jaeger-Erben (2019) have discussed the relations between different types of modularity, and people's use patterns and potentials for prolonging the use time of products. They have described four different types of modularity focused on functionality aspects:

- maintenance that allows easy repair
- adaptation, which allows for customisation and a better fit to people's needs
- upgradability that enables a replacement of components
- expandability that makes it possible to add new components.

All four forms of modularity have the potential to increase the longevity of a device, since components can be replaced or added instead of buying a completely new product. Repair services could work more efficiently if components can be replaced more easily. People might feel more attached to their products if the functions fit their needs better. The possibility to 'downsize' a product could enable sufficient use of smartphones. Proske and Jaeger-Erben (2019) have also discussed possible risks and rebound effects ranging from decreased technological reliabilities of the modular device to overburdening the user because of the need to spend more time 'to get to know' the phone. While their considerations are conceptual and focus on the ecological efficiency of modular smartphones, in this chapter, we investigate the role of modularity in existing smartphone use patterns and delve into some empirical examples of how people live with their phones in everyday life. We investigate modularity as a sociotechnical arrangement embedded into people's everyday lives.

The second section considers the historical socio-technical evolution of mobile phones to show how the co-shaping of technological and consumer trends and patterns of product (service) designs do not provide an easy starting point for the modularity of smartphones in a CE. The third section delves more deeply into the media studies literature and into three empirical cases to be able to better understand potentials for the different types of modularity. The fourth section discusses how the three contrasting empirical cases of smartphone use reveal that different forms of modular designs and accompanying services would need to be implemented to move towards a more sustainable consumption of smartphones. The final section offers preliminary conclusions, reflecting on how an efficient and sufficient use of modularity as a part of transitions towards a CE cannot be enabled by modular product designs and services alone.

A brief history of the sociotechnical developments of mobile phones

In this section, we present the results derived from an analysis of the sociotechnical evolution of mobile phones. Not surprisingly, there is much to be said about the historical developments of mobile communication technologies (Huurdeman, 2003; Agar, 2013; Martin and Garcia-Swartz, 2015). This section gives only a brief outline of these developments, particularly concentrating on the design developments and imagined use patterns of mobile phones. We aim to show that the developments of mobile phone designs do not provide an easy starting point for the modularity of smartphones in a CE. History shows that the convergence of different mobile communication technologies has always been accompanied by a stronger integration of the devices and not by modularisation.

The evolution of a new technology (1861–1990)

The first telephone invented by Alexander Graham Bell and, much later, its mobile phone version were not instant successes. Innovators often did not know what type of innovation they had developed.

> Alexander Graham Bell was so convinced his pioneering telephone would be such an unwanted intrusion that in the 1880s, he initially promoted his invention at expositions and fairs as an entertainment system that conveyed music and theatrical performances over headphones to those who couldn't afford to buy tickets to the real thing.
> (McNish and Silcoff, 2015: 516)

The example shows an inherent characteristic observable in all emerging technological fields that products do not appear out of the blue and always evolve in the interplay between innovators (producer) and lead users. While the 'knowing-how' creates the basis for certain technologies, the 'knowing-what' or 'what for' is about the successful identification of its use and function. Technological innovations are best to be understood as an evolutionary process (Eger and Ehlhardt, 2018), where user expectations, use patterns, and acceptance are crucial and a limiting factor in the design of products (Norman, 2013).

In 1910, an interdisciplinary group of leading German scientists and thinkers published their vision of this future in the book *The World in a Hundred Years* (translated from German). In his essay, 'The Wireless Century,' the author Robert Stoss (1910/2017: 35) anticipated some features of a modern smartphone: "the citizens of the wireless age will walk around everywhere with their 'receiver' placed somewhere in the hat or elsewhere." He even goes beyond simple mobile communication, as he predicted the spoken newspaper, multimedia streaming, and mobile online shopping.

From the middle of the 19th century onwards, the first mobile car telephones were developed. These first-generation devices had to connect to the local network, thus making it impossible to make phone calls whilst driving the car at the same time. Throughout the Second World War, smaller backpack-mounted devices that were nicknamed 'Walkie Talkies' were developed (Magnuski, 2005). Over time, radio technology further advanced but could not overcome its main limitation for a wider commercial use (Agar, 2013). In 1973, Motorola revealed the 'game changing' portable telephone, the shoe-sized and 800-gram-heavy 'Dyna-Tac.' It was nicknamed 'the brick.' However, it was not until 1983 that a commercial version came onto the market.

Commercialisation, segmentation, and product differentiation (1991–2007)

The actual innovations took place in the background. They consisted of, first, a network allowing several devices to communicate without interfering with each other and, second, a fixed infrastructure of wires, transmitters, and receivers to enable mobile communication. The first full-digital network (2G), in particular, the GSM-Network (Global System for Mobile Communications), was not set up until early 1990s in Europe (Klemens, 2010). Although GSM was a great commercial success, network vendors initially registered a subscription of less than 5% of the total population in Europe (Klemens, 2010). It was not until 2002 that the number of mobile subscribers overtook the number of fixed-line subscribers on a global scale (Srivastava, 2005).

Finally, standardisation efforts (Weber, 2008) linked to a shared network led to market growth, as manufacturers could produce mobile phones on a global scale rather than for niche markets. In 1991, Orbitel produced the first GSM-approved phone. It weighed 2 kg and was more like a 'bulky' transportable phone. Motorola and Nokia followed a little later with much smaller, cost-efficient consumer handsets. The new lighter handsets (the so-called candy bar) replaced the predominant design concept of the heavy 'brick' (Eger and Ehlhardt, 2018: 104). In the first half of the mid-1990s, the design and manufacturing of phones was characterised by several technical improvements and constraints (e.g., bigger batteries with longer run-time).

In the early market phase, manufacturers focused on vertical innovations, for example, by integrating components along the supply chain and up-scaling production to reduce costs (Koski and Kretschmer, 2007). Throughout the 1990s, there was a clear convergence towards light and compact mobile phones with improved technical performance. The pioneering advantage of early innovators diminished over time. The increasing competition between actors forced manufacturers to differentiate their products to be able to attract more consumers. The product differentiation phase created a variety of new designs (e.g., flip phones, sliders, and clamshells) and dozens of new features that extended the core function of a simple telephone with which people could only call each other (Eger and Ehlhardt, 2018: 106). In parallel with the developments of feature phones, 'personal digital assistants' (PDAs) were introduced into the market at the beginning of the 1990s. PDAs or handhelds were portable devices and downsized pocket personal computers (PCs) (Wheeler, 2004). They combined several features for personal information management, such as calendars, notes, and wireless data transmission (email, Internet). In contrast to mobile phones, the focus was on applications for business users that could support mobile and remote office work. Most handhelds used monochrome touch screens or small keyboards in combination with smaller screens for their user interface. Considering the parallel developments of mobile phones and PDAs, it was no surprise that both products began to converge. In 1997, Ericsson came up with a prototype for a new device that was named the 'smartphone' (i.e., Ericsson GS 88).

Dominance of the multipurpose smartphone (2007–2019)

At the beginning of the 21st century, it became hard to keep up with all new developments. Product designs began to proliferate in all directions, colours, sizes, features, different user interfaces, and navigation concepts. Experimenting with different designs is a necessary part of any product evolution and often ends in a dominant design (Eger and Ehlhardt, 2018).

The introduction of the iPhone by Apple marked a turning point for the smartphone industry. The iPhone had an integrated operating system (OS), a web browser, and the iTunes store for downloading audio and video. It had a touch screen (instead of a keyboard) with a software-based keyboard. One year later, the iPhone 3G was launched, along with a virtual marketplace for downloading additional software applications. The iPhone was mainly based on technologies that had already been on the market. According to Giachetty (2018: 9), Apple was able to redefine the market boundaries in two ways: 1) it combined different industry branches (e.g., smartphone, Internet, and portable music), and 2) it combined a product with service offerings (e.g., app stores). Instead of inventing completely new technologies, Apple has consistently converged what had already been there (Agar, 2013). After initial hesitations, major manufacturers followed Apple's principles and adapted their smartphone designs. Physical keyboards gradually disappeared, form factors became uniform, and all manufacturers relied on flexible operating systems that could be extended with third-party apps.

The historical development of mobile phones shows that performance densities and integration of parts and features have intensified enormously within the market. Moreover, novel use patterns have been 'integrated' into mobile phones (e.g., clocks and diaries). Over time, phases of dominant designs seem to be followed by product differentiations to gain market advantage. For most devices, not even the battery can easily be replaced. Displays, which are often damaged when dropped, are also not easy to repair. The architectural innovation of the smartphone (as a converged system of different technologies) was accompanied by an increasing integration of single components that are glued to the phone. Unfortunately, all attempts to mainstream hardware modularisation to be able to lengthen the useful time of mobile phone have failed so far (Hankammer et al., 2017; Proske and Jaeger-Erben, 2019).

The next section introduces the notion of media convergence derived from media studies (Peil and Sparviero, 2017; Mueller and Roeser, 2017) that conceptualises the 'coming together' of previously distinct media technologies (i.e., smartphone as alarm clock, diary, and/or health tracker). It also describes three empirical case studies of people and their mobile phones.

Investigating the convergence and deconvergence of mobile phones

The idea of the 'one personal all-round medium' or 'supermedium' has been used as a buzzword for many years to promote technological innovations linked to notions of efficiency, synergy, and simplification. As argued by Peil and Sparviero (2017: 9), "highly personalised and portable online media like the smartphone integrate telecommunications, content and information technology, and thus represent in an almost paradigmatic manner one of the original notions of technological convergence."

The 'one person all-round medium' and associated notion of media convergence have been highly debated with some scholars, arguing it is not the integration of media technologies but "their disintegration, multiplication and increased complexity that are believed to be central features of recent developments" (Peil and Sparviero, 2017: 6). Research has started to emerge that has built an understanding of the interplay of media convergence (i.e., merger) and deconvergence (i.e., disparity). The research has outlined how not only producers (and technological design) but also consumers and their use patterns shape the convergence of technologies.

People make use of "extremely sophisticated media environment and a multiplicity of media technologies, each of which comes with a specific focus or core competence and a broad range of functionalities" (Peil and Sparviero, 2017: 9). These developments are signified through a growing disparity rather than a process of merging in varying forms of media access, personalisation of media content depending on individual interests, and disparate media biographies (ibid.). Moreover, the interplays between old and new technologies and online and offline practices are key to understanding convergence and deconvergence.

Smartphones as multifunctional tools enable a great variety of everyday practices, for example, communicating, taking photographs, listening to music, and keeping diaries. At first glance, it seems that smartphones allow for a reduction of household devices due to media convergence. People no longer need cameras, alarm clocks, radios, and so on. This might not always be the case. Miller (2020) has developed a case for media deconvergence when describing the smartphone to be 'convivial,' considering that, for instance, people can choose how they want to communicate with others through either SMS, forums, or social media.

Little research has gone into investigating the 'conviviality' of smartphones and deconvergence processes in everyday life in relation to modular smartphone design developments. In the next section, three types of users are introduced – the frugal, overburdened, and seeking-optimisation users – grounded in in-depth interviews to investigate processes of convergence and deconvergence in relation to design strategies like modularity for maintenance, adaptation, customisation, and upgradeability (Proske and Jaeger-Erben, 2019). For this analysis, we draw on ongoing research[1] that investigates pathways to a sustainable CE to better understand the dynamics between conviviality and modularity in everyday life. As part of this project, problem-centred interviews (Witzel, 2000) have been conducted about people and their smartphones in everyday life. The selection of interviewees was based on theoretical sampling; that is, categories about possible contrasting use patterns of smartphones were continuously tentatively developed, discovered, and refined. So far, a total of five interviews have been conducted, transcribed, and encoded using the ATLAS.ti qualitative analysis software. A coding scheme was created and validated based on 15 additional semi-structured interviews that were carried out as part of a previous project on the use of smartphones in daily life. Thus, a total of 20 interviews were analysed based on open coding and constant comparison as defined in the grounded theory methodology (Strauss and Corbin, 1998).

Living with a smartphone in everyday life: possibilities for modular smartphones

Our initial qualitative analyses have produced three tentative types of how people live with their smartphones, which differ in their everyday life practices and use patterns of functions and applications associated with the smartphone (other dimensions might be relevant for the analysis but are not discussed in this chapter). The resulting types are to be understood from the perspective of Weber's 'ideal types' as analytical categories and thus as abstractions that are not one-to-one descriptions of actual empirical examples. Rather than claiming that all empirical examples within one category are equal, what is claimed is that certain empirical examples fit more neatly with one analytical category (thus one type) than with another (Collier et al., 2008).

The frugal user

Barbara is an academic in her late 30s. She lives with her husband and two children. She uses her smartphone for her work and private life, as she mainly works from home. Barbara does not feel that she uses her smartphone a lot, often leaving it at home. She does not tend to use the smartphone's full capacity because she tries to limit the amount of functions that are actually in use. Barbara attempts to use her smartphone as long as possible. Her current use patterns are representative of a frugal user.

As part of Barbara's usual daily routines, she does not pick up her phone until she sits down and starts working in her home office. Her work phone redirects calls to her smartphone. Overnight, Barbara prefers to not have her smartphone on her bedside table. She does not regularly know where she has left it the day before and has to look for it in the morning. During a usual workday, she needs to spend about an hour on her smartphone. Her usage increases if she has to leave her home office and, for instance, meet someone. This is when Barbara makes use of some of the other functions on the phone, such as looking at maps or listening to some music. During her spare time, Barbara regularly misses calls and messages or her battery runs out without her noticing. This is also the time where she often decides not to take her smartphone with her. Although she does not feel reliant on the phone, she still has several applications on her phone so that she can communicate with friends and family, more easily travel via public transport, take pictures and record videos, and keep up-to-date with the news, time, and weather.

Barbara mainly makes use of some of the basic functions of her smartphone such as the address book, messaging service, communication tools, camera, and World Wide Web. She added one or two other applications that are relevant for her daily travel, that is, public transport and map applications. Barbara occasionally uses the pre-installed newsfeed and weather forecast. More recently, she discovered an application that allows her to stream music onto her smartphone, which, since then, she has appreciated very much. She uses Bluetooth to connect to her external loudspeaker. A few of the applications that she downloaded Barbara no longer uses, such as a mediation app. She wants to delete these at some stage. Barbara keeps a paper-based diary because she is too worried about losing the entries of a digital version. She only checks her email on her laptop. This is really the device that she uses for her work. To keep her smartphone as long as possible, Barbara regularly deletes and organises applications, transfers pictures and other data to her laptop, conducts updates, and keeps the protective case around the smartphone.

Barbara purposefully does not make her smartphone a 'personal all-in-one medium.' Although Barbara enjoys some of the conveniences the phone provides, including making it easier to get around, she does not want to become dependent on the phone for her everyday life. A modularity that would allow her to be able to 'adapt' her phone (to downsize it to a few functions) could potentially work for this frugal user. This would be a smartphone that can be customised to people's needs and does not use up much storage so that the frugal user can keep it for longer. A modular phone that can be easily 'maintained' would also fit the frugal user. Such a maintenance and adaptability form of modularity would be grounded in people using only a limited amount of functions and a low dependency on the phone for everyday life activities, including the willingness of the person to look after it to be able to prolong the phone's useful life.

The overburdened user

Nicole is a senior consultant who is in her 40s and lives with her partner and young child in a two-bedroom house. She owns her own private smartphone and received an additional one from her workplace. Nicole often feels overburdened by the number of smartphones on offer and their settings, functions, and applications. Nicole could develop the competences to understand her smartphone better, but for her, it is sheer 'laziness' that stops her from engaging more deeply with the phone.

When Nicole talks about her daily smartphone use, it becomes apparent that it plays an integral part of her daily life. The smartphone is her alarm clock and weatherman several times a day. It also helps her to organise her private life through sharing diaries and note-taking applications with her partner and using communication apps to make ad-hoc childcare arrangements. She also keeps in touch with friends and family who live abroad. Her smartphone is connected to her laptop. This way, she can use the communication applications on her smartphone and laptop. In the office and at home in the evening, the smartphone becomes an entertainment device by playing music, podcasts, and the evening news. The music comes out of a shared stereo that is connected to an application on her smartphone that streams the music. At the weekend, she likes to take photos with her smartphone, films her guitar teacher so that she can copy the songs at home, and sometimes uses it to look up recipes to cook dinner. Nicole finds it not straightforward to use the smartphone for these activities, seeing that the screen often goes black whilst she cooks. Moreover, Nicole still likes to use her digital camera to take some pictures rather than using the one on her smartphone.

Nicole's daily life with her smartphone makes apparent how the smartphone has become a device that has converged different technologies, such as alarm clocks, diaries, television, camera, and cookbook. This does not necessarily mean that the smartphone has replaced these devices. Nicole still owns a watch, physical diaries and cookbooks, a large screen on which she and her partner watch television, and digital cameras. The smartphone has not necessarily reduced the number of devices at home but rather changed some of her daily practices. For instance, she can watch the news on several devices, depending on where and how she wants to see it. The way she takes pictures has also changed. Photos that are important to her (such as holiday pictures) are taken with her digital camera, whereas things that just need to be quickly captured (e.g., a work diagram) can be taken with the smartphone. Ways of taking pictures have multiplied in addition to meanings and competences surrounding what it means to 'capture' something with a camera. The photographs that Nicole takes on her digital camera can be transferred via Bluetooth to her smartphone. She often does this so that she can share them via a communication application on her phone with her family and friends. The applications on Nicole's smartphone are also linked to the music stereo and her laptop, creating an invisible network between digital devices, Nicole, and her partner.

The smartphone is not only a device in itself, for instance, allowing us to phone a friend, but also an intermediary, connecting different digital devices (e.g., playing music from the smartphone through the stereo) and facilitator, allowing for varying practices to emerge (e.g., keeping a digital diary). This connectivity and facilitation often only become visible if something no longer works on the phone. Nicole has replaced her smartphone whenever the storage capacity was full and some software updates were no longer possible to be able to keep the 'functionality' of the phone going. It might be possible to suggest the more connections and facilitations exist, the more vulnerable the smartphone becomes to losing its functionality. Although Nicole seems to rely on her smartphone to carry out her daily life, she is not willing to spend a lot of time maintaining her phone. Any form of modularity might overburden this type of user. Maintenance and upgradability that are delivered through a service rather than purely through the design of the smartphone could potentially attract this type of user to modular smartphones.

The seeking-optimisation user

Natalie is a head of educational counselling and in her late 40s. She lives with her husband and two children in a flat in a small town. She owns several digital devices and currently uses a laptop, tablet, and two smartphones; that is, she sometimes switches between the two.

Natalie regularly reflects on her media ensemble (i.e., different types of electrical devices) at home and how she could reduce it and/or optimise its use. The tablet probably enables most of her digital activities: watching sports, listening to the radio and audio books, looking up the news, and planning family projects. Although her smartphone can provide the same functions, she prefers to use the tablet at home, as it has a larger screen. Still, every now and again, one of her children takes the tablet, and she has to revert back to using the smartphone. The size and mobility of phones mean that whenever they go on holiday, Natalie is glad to have it. She can upload applications that allow her to organise public transport, theatre, cinema, exhibitions, and travel on her smartphone. This is straightforward because she has already set them up in a cloud system and is regularly able to load them on her smartphone and take them off again if she no longer needs them for a while. During holiday trips, her smartphone becomes a 'multi-functional tool' for her and the family. She does not really need this type of tool during her everyday life. If she could have one rather than four digital devices, this would really help her to 'structure' her life.

Examining more closely Natalie's ways of organising her applications on the smartphone and connectivity between devices, the 'all-in-one device' seems to be hidden within her cloud system. She has tried to change her operating system several times because she does not believe in the companies' values of her current system. Even though she spent several hours on it, Natalie was unable to create the same connectivity between devices that allowed her to sync all of her data across them. With her current operating system, the cloud somehow becomes a spatial device that holds all of the information that can be uploaded and taken off several physical devices. These physical devices are like 'empty shells' that are being filled with data by cloud systems. They become exchangeable because any device can hold this information if it has the needed storage capacity. Finding this 'all-in-one device' seems to be a project, where Natalie constantly changes her media ensemble over time, re-introducing devices that have no longer been in use (i.e., the television), swapping devices (i.e., two smartphones), and buying new ones (i.e., new smartphone).

Although she has an overall wish to bring more structure into her life through an 'all-in-one device,' her media ensemble does not really seem to stabilise over time, partly demonstrating that this project is not straightforward to realise in daily life. What also becomes apparent is the changeability of people's life situations that require different types of media ensembles. Natalie used to have a job where she had to do a lot more work on the go. This required her to regularly check email and carry several documents on her smartphone. Nowadays, this is no longer the case, but whenever the existing media ensemble in the home 'gets stretched' by one of her children taking one of the devices or them going on holiday, the current number of devices provides a flexibility that allows her to fulfil her caring role as a mother.

Natalie's search for optimisation seems to be grounded in sometimes reducing the amount of devices, making one device the main one, and creating a cloud system that smoothly links and syncs all of her devices whilst at the same time protecting the amount and types of data that are being shared with the outside world within the family home. Although Natalie has repaired and upgraded the battery on one of the smartphones (i.e., applying the maintenance and upgradability modularity), this has not really prolonged the useful life of the smartphone. This is her second smartphone that she gets out every now and again. The phone has a different operating system to the other phone, and the cloud system does not work as well on it. For Natalie, the non-connectivity to her existing cloud system makes the smartphone not particularly user friendly. She therefore only uses it occasionally. A mixture between adaptation modularity, which allows for customisation and a better fit to the consumer needs at the time of purchase, and expandability modularity, which makes it possible to add new components, could be a technological development that might help to prolong Natalie's smartphone's useful life. This would also require her to finally find her 'all-in-one device' so that the functions on the other devices could be reduced, keeping functions and application on the devices relatively stable.

People living with smartphones in everyday life and reflections on modular smartphones

Our analyses outline the 'ideal types' of smartphone user in relation to different forms of modularity, focusing on several dimensions of use. The comparison of the three contrasting forms of living with a smartphone reveals a diversity of meanings, competences, and material arrangements that are associated with smartphones in everyday life. Growing old with your smartphone does not seem to be a straightforward process in everyday life, seeing that the technological design of modular smartphones does not seem to be able to keep up with the 'conviviality' of the smartphone. Our findings show that the frugal user type would benefit from maintenance and adaptability modularity, whereas the seeking-optimisation user type would value expandability modularity. The overburdened user type would potentially feel overwhelmed by all types of modularity, but a service model that provides maintenance and adaptability modularity could be an attractive way forward. Still, it is unclear whether these types of users and linked forms of modularity would actually prolong the useful life of smartphones. Natalie's example shows that although one of her smartphones is easily repairable and upgradable, its life is not necessarily prolonged. She does not consider it 'user friendly' and uses her second smartphone most of the time because it allows for an easy connectivity, whereas the easy-to-repair-and-upgrade phone only comes out of the drawer every now and again.

Even though a frugal user type might use the smartphone less frequently and limit the amount of functions and applications on the phone (see Barbara's description), the utilisation of adaptability and maintenance modularity to promote sufficient use might not be straightforward. Similar to the other user types, the frugal user still discovers new functions and applications (such as streaming music) on the phone over time, whereas other ones (such as the meditation application) become irrelevant. A customised phone (linked to adaptability modularity) requires the user to be extremely reflective about their existing and potential future practices that are enabled by the smartphone to prolong the customisation and associated use as long as possible. Although digital technologies such as smartphones are not neutral, notions of longevity and sustainability are not inherent in them (even if these are modular smartphones). These technologies get interpreted and adapted into existing patterns of people' everyday life. How people use and live with their smartphones is deeply embroiled with, for instance, existing digital infrastructures, media ensembles, family relations, and changes to work situations and lifestyles. The developments towards an increasing 'conviviality' of use patterns and integrated designs of smartphones do not necessarily make it straightforward to introduce different types of modularity to prolong mobile phones' useful life.

Thus, a smooth transition from a presently rather unsustainable use of the static modularity of a smartphones towards a sustainable consumption and CE-enabling use of a dynamic modularity seems very unlikely. We discuss the implications of our results so far in the last section.

Discussion and conclusions

Even though our findings are part of an ongoing research project that needs to be further refined and validated, we can draw some initial conclusions that are relevant to future research on transitions towards a CE.

Our results show that the sociotechnical developments of mobile phones over time have been signified by integrated technological designs (rather than interchangeable, i.e., modular, ones). Mobile phone parts, such as batteries, which could be easily replaced a few years ago, are now integrated into the mobile phone design where only experts can change the parts when they no longer function. At the same time, smartphones have taken on more and more functions of other everyday objects (e.g., alarm clocks, newspapers, and stereos), that is, media convergence, allowing for a performance of multiple daily practices. The convergence of functions and applications is not the whole story when examining smartphones in everyday life. Miller (2020) has argued to consider the smartphone 'convivial,' allowing people to choose from several possibilities of how they want to communicate with others, compose thoughts, plan the day, and entertain oneself through the device, that is, media deconvergence. Although this conviviality goes along with notions of modular smartphones that adapt to people's lives, our findings have shown that the idea of smartphones growing old with their owners does not seem to be easily transferred into everyday life. Types of modularity might by no means be clear-cut strategies for a CE.

There are currently niche markets that cater to smartphone users who like the idea of a more sustainable phone. Within these niche markets, a lot of thinking seems to go into the technological developments of modular smartphones and related business models to increase the potential for more sustainable production and consumption patterns of smartphones. Considerations of how modular smartphones are being domesticated into people's daily life, that is, what types of meanings are created and how the phone is placed into existing patterns of daily life, often seem neglected within these developments. The setup of groups and networking events (e.g., Meetups) between modular smartphone users might present opportunities to experiment with diverse sets of sociotechnical configurations of modular smartphones and their use (and even more sustainable ones). Such settings might open up several technological pathways and use patterns for modular smartphones, including the potential considerations of social, political, and environmental goals and alternatives to current consumption systems.

Still, setting up these meetings and networks might not be enough to encourage more sustainable production and consumption patterns, in particular if social, economic, and political issues (e.g., who attends these meetings and how participants domesticate their phones into their everyday lives) are being considered alongside technological advancements (e.g., how to make the phone more repairable). An additional issue that arises from our investigation is about roles and responsibilities taken by different actors within the CE. Examining the sociotechnical development of smartphones over time, the increasingly integrated technological design features become visible, raising questions about smartphones' reparability, compatibility of parts, availability of software updates, and so on. Regulations that counter some of these market developments in terms of eco-design requirements or standardisation of components could be considered when discussing extensions of smartphones' useful life. Easy and affordable replacements of screens and batteries can prevent some early product replacements. Recent activities by the 'right-to-repair' movement advocating for people being able to fix their electronic products (including smartphones) draw attention to producers' responsibility of creating more sustainable smartphones. These developments might be steps towards normalising the prolonged use of smartphones in everyday life, moving towards a CE.

Note

1 The study presented here is part of two connected research projects: The three-year research project MoDeSt (www.modest-projekt.de) and the five-year research group "Obsolescence as a challenge for sustainability" (www.challenge-obsolescence.info). Both are funded by the German Federal Ministry of Education and Research (research streams ReZiProK and Social-Ecological Research).

References

Agar J. (2013). *Constant touch: a global history of the mobile phone.* London: Icon Books.

Agrawal V, Bellos I. (2016). Servicizing in supply chains and environmental implications. *Environmentally Responsible Supply Chains.* Springer. 109–124.

Agrawal V, Ulku S. (2013). The role of modular upgradability as a green design strategy. *Manufacturing Service Operations Management.* 15 (4): 640–648.

Baldwin C, Clark B. (1997). Managing in the age of modularity. *Harvard Business Review.* 75 (5): 84–93.

Bocken N, Olivetti E, Cullen J, Potting J, Lifset R. (2017). Taking the circularity to the next level: a special issue on the CE. *Journal of Industrial Ecology.* 21 (3): 476–482.

Collier D, Laporte J, Seawright J. (2008). Typologies: forming concepts and creating categorical variables. In: Box-Steffensmeier JM, Brady HE, Collier D. (Eds.), *Oxford handbook of political methodology.* Oxford: Oxford University Press. 152–173.

Eger AO, Ehlhardt H. (2018). *On the origin of products: the evolution of product innovation and design.* Cambridge: Cambridge University Press.

Ellen McArthur Foundation (2018). *Modular design and new business model create a more circular music experience.* Available at: www.ellenmacarthurfoundation.org/case-studies/modular-design-and-a-new-business-model-create-a-circular-music-experience (Accessed 5/5/2020).

Giachetty C. (2018). Explaining Apple's iPhone success in the mobile phone industry: the creation of a new market space. In: Giachetty C. (Ed.), *Smartphone start-up's.* Cham: Palgrave Macmillan. 9–48. doi:10.1007/978-3-319-67973-0_2.

Hankammer S, Jiang R, Kleer R, Schymanietz M. (2017). Are modular and customizable smartphones the future, or doomed to fail? A case study on the introduction of sustainable consumer electronics. *CIRP Journal of Manufacturing Science and Technology.* doi:10.1016/j.cirpj.2017.11.001.

Huurdeman AA. (2003). *The worldwide history of telecommunications.* Hoboken, NJ: John Wiley and Sons Inc.

Jaeger-Erben M, Hipp T. (2018). All the rage or take it easy – expectations and experiences in the context of longevity in electronic devices: descriptive analysis of a representative online survey in Germany. In: Obsolescence Research Group (Ed.), *OHA texts 1/2018.* Available at: https://challengeobsolescence.info/wp-content/uploads/2018/03/Research-Group-OHA_Description-Online-Survey_2018.pdf (Accessed 5/5/2020).

Klemens G. (2010). *The cellphone: the history and technology of the gadget that changed the world.* Jefferson, NC and London: McFarland and Company.

Koski H, Kretschmer T. (2007). Innovation and dominant design in mobile telephony. *Industry and Innovation.* 14 (3): 305–324, doi:10.1080/13662710701369262.

Magnuski HS. (2005). *Radio set SCR-300 – a war department technical manual.* TM 11-242. Available at: www.scr300.org/.

Martin C, Garcia-Swartz D. (2015). *From mainframes to smartphones: a history of the international computer industry.* Cambridge, MA and London: Harvard University Press.

McNish J, Silcoff J. (2015). *Losing signal: the spectacular rise and fall of BlackBerry.* Toronto: Harper Collins. eBook edition.

Miller D. (2020). *The Lancet commission on the value of dying.* UCL blog. Available at: https://blogs.ucl.ac.uk/assa/2020/02/07/the-lancet-commission-on-the-value-of-dying/.

Mueller KF, Roeser J. (2017). Convergence in domestic use? The interplay of old and new media at home. In: Sparviero S, Peil C, Balbi G. (Eds.), *Media convergence and deconvergence.* Cham: Palgrave Macmillan.

Norman D. (2013). *The design of everyday things.* New York: Basic Books.

Peil C, Sparviero S. (2017). Media convergence meets deconvergence. In: Sparviero S, Peil C, Balbi G. (Eds.), *Media convergence and deconvergence.* Cham: Palgrave Macmillan.

Proske M, Jaeger-Erben M. (2019). Decreasing obsolescence with modular smartphones? An interdisciplinary perspective on lifecycles. *Journal of Cleaner Production.* 23: 57–66.

Proske M, Winzer J, Marwede M, Nissen N, Lang KD. (2016). *Obsolescence of electronics – the example of smartphones, electronics goes green 2016+.* Berlin: IEEE.

Srivastava L. (2005). Mobile phones and the evolution of social behaviour. *Behaviour and Information Technology.* 24 (2): 111–129.

Stahel WR. (2016). The circular economy. *Nature.* 531 (7595): 435.

Starr MK. (1965). Modular production: a new concept. *Harvard Business Review.* 43 (6): 131–142.

Stoss R. (1910/2017). Das drahtlose Jahrhundert. In: Brehmer A. (Ed.), *Die Welt in hundert Jahren.* Hildesheim: Georg Olms Verlag.

Strauss A, Corbin J. (1998). *Basic of qualitative research techniques and procedures for developing grounded theory*. London: Sage.
Ulrich K. (1995). The role of product architecture in the manufacturing firm. *Research Policy*. 24 (3): 419–440.
Weber H. (2008). *Das Versprechen mobiler Freiheit. Zur Kultur- und Technikgeschichte von Kofferradio, walkman und handy*. Bielefeld: Transcript Verlag.
Wheeler WR. (2004). *Integrating wireless technology in the enterprise: PDAs, Blackberries and mobile devices*. Oxford: Elsevier.
Wieser H, Tröger N. (2018). Exploring the inner loops of the CE: replacement, repair, and reuse of mobile phones in Austria. *Journal of Cleaner Production*. 172: 3042–3055.
Witzel A. (2000). The problem-centered interview. *Forum: Qualitative Social Research*. 1 (1).

32

THE USE OF BY-PRODUCTS IN NEW MATERIALS

Rory Doherty, Elizabeth Gilligan, Charlie Farrell, Sreejith Nanukuttan, and Ruth Morrow

Introduction

The use of by-products or secondary raw materials in the production of everyday products is a key principle of circular design and the circular economy. The concept of industrial ecology where the waste or by-products from one industry could be reintroduced as components or 'building blocks' of new products from the same or other industries is centred around by-products. In the past, by-products were often identified as waste rather than potentially valuable products. Article 3 (1) of the EU Waste Framework Directive (2008/98/EC) defines waste as 'any substance or object which the holder discards or intends or is required to discard.' This has implications for the potential use of by-products. Generally, it is easy to define waste as any material/goods that the owner does not want anymore; however, in practice it is difficult to formulate a concise legal definition that suits all purposes. The main issues are: When is a by-product waste? And when does a material cease to be waste (post-treatment)? Further EU guidance on the matter specifies when certain waste ceases to be waste and obtains the status of a product or by-product.

The Waste Framework Directive 2008/98/EC identifies certain specified waste shall cease to be waste when it has undergone a recovery (including recycling) operation and complies with specific criteria to be developed in line with certain legal conditions, in particular:

- the substance or object is commonly used for specific purposes
- there is an existing market or demand for the substance or object
- the use is lawful (substance or object fulfils the technical requirements for the specific purposes and meets the existing legislation and standards applicable to products)
- the use will not lead to overall adverse environmental or human health impacts

Only once end-of-waste criteria have been achieved can a substance previously considered waste become a 'secondary raw material' 'by-product,' or product.

This legislation and associated policies attempt to create a level playing field that promotes both economic and environmental benefits that encourage the uptake and use of by-products.

Sustainable Development Goals and the use of by-products

The circular economy and circular design are complementary to the UN Sustainable Development Goals (SDGs). When considering by-products and SDGs, a hierarchy similar to a waste hierarchy can be developed (Farrell et al., 2020) with Goal 12, 'Responsible Consumption and Production,' as the most relevant, topping the pyramid. Unnecessary production of harmful or damaging by-products creates problems for circular design use of by-products. Responsible production of by-products that are not harmful to human health or damaging to the environment creates new raw materials for society. These materials can be further designed to have cascades of use, which relates to responsible consumption. In turn, Goal 12 promotes and facilitates the next level of Sustainable Development Goals of 'Industry Innovation and Infrastructure,' which design and cascade the by-products into new products; 'Sustainable Cities and Communities,' which use, utilise, reuse, and recycle the by-products which in turn promotes 'Decent Work and Economic Growth. The upper levels of the hierarchy in Figure 32.1 ultimately produce the outputs of 'Good Health and Well Being,' 'Life on Land,' and 'Life below Water' through the production and use of non-hazardous and non-toxic by-products and their cascades and 'Climate Action' through carbon and energy savings associated with reuse and non-reliance on primary raw materials.

To develop this SDG hierarchy within a truly circular 3R framework (reduce, reuse, and recycle) requires that circular design or specifically, in this case, by-product design, be considered and implemented by the industries that create the by-products in the first place. Ultimately, good circular design will remove the majority of waste streams, leaving only nontoxic and non-hazardous by-products that can be easily remanufactured, reused, and recycled. However, we are not at that stage yet. Economic models have been developed that promote recovery of by-products through landfill mining in order to incentivise by-product uptake (Schreck and Wagner, 2017). Other authors have argued that primary production of materials will still be required even if considerable recycling efforts are achieved (Fellner et al., 2017), with current policies based around exporting of waste acting as a considerable barrier to recovery and use of by-products. The quality of by-products is also a factor in

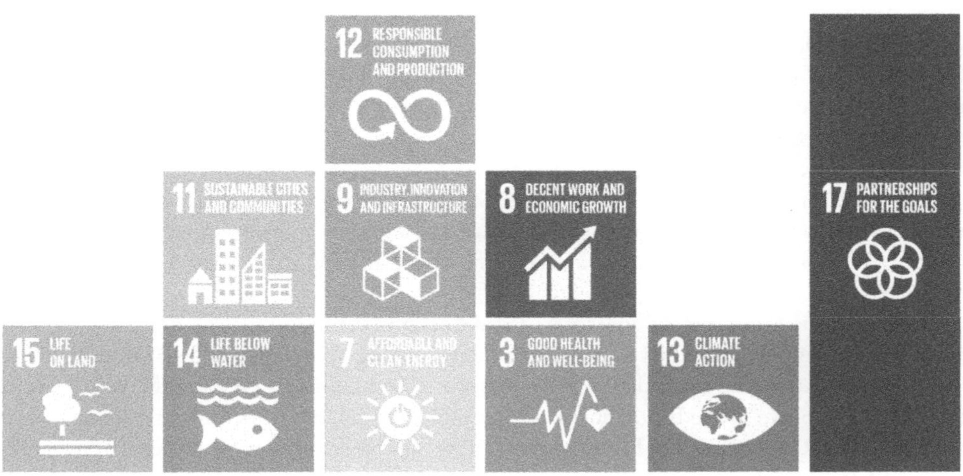

Figure 32.1 A hierarchy of Sustainable Development Goals associated with by-products and the circular economy

determining the 'circularity' of new products made from these materials (Steinmann et al., 2019), with social, economic, environmental, and technical parameters that need to be considered as part of the design and performance of the cascading product (Iacovidou et al., 2017). It is often the case that safety or environmental parameters will dictate the design specification rather than the need for the by-product to be used. This is discussed further in the next section.

Registers of by-products

Developing worldwide registers and databases of by-products and their properties can allow for better eco-design of new products. Registers of critical raw materials for waste electrical and electronic equipment, batteries, end-of-life vehicles, and mining waste are available but are often scattered across institutions, companies, and regulators and thereby difficult to access and transpose. Recent research projects (such as PROSum) are attempting to mine and map these datasets across Europe, promoting the concepts of urban mines, but these databases are still in the early stages of development (Huisman et al., 2016). Other research projects such as By-BM (Sas et al., 2017) and Norm4Building (Schroeyers et al., 2018) have developed worldwide databases of by-products that take into account environmental characteristics and state-of-the-art regulation to aid the design of building and construction materials that incorporate by-products.

By-products in construction

The construction industry presents excellent opportunities for the circular recovery of materials from other industries that can be used to develop 'green' building products. The cement industry and portland cement production are energy and carbon intensive and polluting (Lei et al., 2011). These activities range from the extractive impacts of limestone and similar products to the high-energy input in cement kilns required to convert limestone $CaCO_3$ to calcium oxide, driving off carbon dioxide (CO_2) in the process. CO_2 emissions from global cement and concrete production have increased by 2.4% per year in the last decade (Friedlingstein et al., 2014), with cement production making up 5.5% of the total CO_2 emissions in 2013 of 0.54 gigatons (544,500,000 tonnes of CO_2) (Le Quéré et al., 2014). This rate of increase tracks with the high end of emission scenarios and is set to continue unless stringent measures are applied or alternative carbon capture technologies are developed (Washbourne et al., 2012; Hartmann et al., 2013; Fuss et al., 2014).

There has been research highlighting the use of waste such as sewage sludge and refuse-derived fuels as an alternative to fossil fuels in the clinker burning process (Aranda Usón et al., 2013), with the incorporation of a gasification step to improve burn efficiencies and produce additional secondary raw materials such as ash that are incorporated into the cement design.

A concrete is a mixture of aggregates (sand and stone), often with reinforcement (steel/fibres) that is physiochemically bound by a cement that hardens and strengthens when hydrated into a structure used for construction. Cements are often modified or substituted with reactive materials to improve strength, and chemicals are added to improve workability, flow, and setting characteristics. Concrete is used in practically every kind of construction project, from buildings through to all types of infrastructure for societal needs. Concrete design has had several breakthroughs that have changed construction methods and have brought societal impacts on how we build and live together.

Reinforced concrete

Reinforced concrete was invented by Joseph Monier in 1849. By combining concrete with iron and later steel, this added tensile strength to the concrete and changed the way concrete could be used (Palley, 2010). Reinforced concrete, created by embedding metal rods/mesh (iron or steel) into the form work before pouring the concrete, was developed to change the characteristics of concrete and improve its tensile strength. This made concrete a versatile material that could be used for a wide variety of building elements. Concrete became a material of choice for architects, and this created a business case that allowed for research developing concrete technology.

By-products in reinforced concrete

Production of steel for reinforced concrete in the United Kingdom is through the electric arc furnace process, where 90% of United Kingdom–produced reinforcement is from recycled steel (Mineral Products Association, 2018). The emissions associated with the electric arc furnace steelmaking process are 0.6 tCO_2 per tonne of steel, the majority of which are attributed to energy input, which is less than the basic oxygen furnace (BOF) steelmaking process, at 2.8 tCO_2 per tonne of original steel from iron ore (Columbia Climate Center, 2012).

Fibre-reinforced concrete

Fibres are used to reinforce building materials that are weaker under tension than compression to prevent cracking. This technique has been used for over 10,000 years, for example, the addition of straw into sun-dried mud bricks. Fibre reinforcement in concrete was first considered in the 1960s (Romualdi and Batson, 1963). Since then, research has looked into glass, steel, melted basalt fibres, and polypropylene in concrete. In the 1970s, fibre-reinforced concrete (FCR) came out of the lab (Thomas, 1972) and into the realms of architecture and construction with projects like the Sydney Opera House in 1973 (Murray, 2003) that used glass fibre-reinforced concrete. By the 1980s–90s, the use of synthetic fibres such as polypropylene (PP) (Hibbert and Hannant, 1982), polyethylene (PE), poly vinyl-alcohol (PVA), and polyester (PEST) had been developed (Zheng and Feldman, 1995), and these are increasing in popularity to the present day.

By-products in fibre-reinforced concrete

There have been several approaches to using by-products are a replacement for raw material fibres in FRC; these can be divided into approaches that use waste plastics to replace synthetic fibres such as PE and PVA (Pacheco-Torgal et al., 2012). Care must be taken with the type of plastic fibre used, with polypropylene fibres found to have an adverse effect on the air content of the concrete, reducing workability; however, impact resistance increased (Siddique et al., 2008). A slightly different mode of thought reverts back to the original use of biofibres (e.g., straw in mud bricks); here, fibrous biomass from several sources has been considered, from hemp to oil palm broom (Momoh and Osofero, 2019), as alternatives to synthetic fibres. This requires additional processing of the biomass by pulping, often with chemical additions to prepare and homogenise the fibres (Pacheco-Torgal and Jalali, 2011). Pyrolysis has also been suggested as a biomass pre-treatment to improve performance prior to incorporation into the concrete matrix (Arsne et al., 2007). One of the main factors that will affect biomass fibres in concrete is degradation of the fibres over time in a highly alkaline environment, an area that requires additional research (Pacheco-Torgal and Jalali, 2011)

Superplasticisers

Superplasticisers are admixtures to concrete that allow for a reduction in the water content, slow the setting time, and create better freeze/thaw characteristics whilst retaining the desired flow (Aïtcin, 2016). Most superplasticisers (mainly accelerators and retarders) were developed and patented between the 1980s and 1990s (Ramachandran, 1996). At that time, water reduction was obtained by using the plasticisers based on the chemicals polynaphthalene sulfonate (PNS) and polymelamine sulfonate (PMS) (Flatt and Schober, 2012; Mantellato et al., 2016).

By-products as superplasticisers

Waste paint has been found to be a suitable replacement for superplasticisers, with replacement of water with paint of up to 12% to achieve desired workability characteristics when making concrete masonry blocks (Almesfer et al., 2012). When waste paints have been used as superplasticisers in pavement block design, they improve resistance to chloride ingression and attack from winter de-icing activities alongside the economic benefits of not requiring expensive and bespoke chemical admixtures (Nehdi and Sumner, 2003).

By-products as cement and aggregates

Early attempts at recycling materials in concrete viewed the concrete as a method of encapsulating or stabilisation, often replacing the aggregates, with the recyclate playing little or no part in the design and performance of the concrete. This used the low permeability of cementitious binders, stabilising materials that would otherwise pose a risk to health or the environment. The use of such stabilisation techniques is often applied to difficult waste streams such as contaminated land or hazardous waste (Voglar and Leštan, 2011). However, it fails to consider the waste a useful by-product that could improve design characteristics or future use of the concrete but rather is a long-term disposal method. Ideally, the substitution of the primary raw materials with by-products or secondary raw materials should be based on the concept that the materials that are used as replacements outperform the original materials technically, environmentally, economically, and socially. The replacement and/or substitution of portland cement, however, is a prime target for the application of industrial by-products. Many types of cement have some form of mineral or addition that could be supplemented by industrial by-products; standard examples accepted by industry include coal fly ash, pulverised fly ash (PFA), and ground granulated blast-furnace slag (GGBS) that have pozzolanic properties that react with calcium hydroxide (hydrated lime) to improve the performance characteristics of concrete (Ganesh and Sree, 2000).

Coal fly ash has increased in popularity within the cement industry, with over 9 million tonnes used in construction annually in the EU. Coal fly ash can be used as part of the clinkerisation process to produce cement or as a pozzolanic addition, partially replacing portland cement, to improve strength, lower water demand, and reduce cracking. In some cases, coal fly ash can also replace the fine sand fraction of the aggregate (Siddique, 2003). As the need for high-grade concrete for structures that take higher loads increases, for example, design for high-rise buildings, then the need for coal fly ashes that optimise raw material costs will increase. Coal bottom ash is commonly used as an aggregate replacement in road bases, but when it is ground, it has the potential to be used as a partial replacement for portland cement (Singh et al.,

2020). Research for new cement replacements or constituents includes ashes from a variety of industries, such as municipal waste ash (Pan et al., 2008), paper mill ash (Vegas et al., 2006), and biomass ash (Teixeira et al., 2019).

Designing by-products in 'green' cements

Due to the high carbon impact of portland cement production, there is growing research around alternative low-carbon impact or green cements. As mentioned previously, the production of clinker (the active ingredient of cement) results in the global production of approximately 1.4 billion tonnes of CO_2 per year. Another 1 gigaton of CO_2 results from the large amounts of energy required to produce cement (Andrew, 2018). This is equal to ~7% the total global CO_2 emissions from energy and industry. If circular principles are adopted, then designing out the polluting material and replacing it with waste or by-products is the first step in addressing current issues with cement production (The Ellen MacArthur Foundation, 2019). Alkali-activated materials (AAMs) are a type of 'green cement' that replaces the need for portland cement and uses by-products from other industries. There are two active components in the process – a highly alkaline material, usually caustic or alkaline salts, such as sodium hydroxide or sodium silicate that acts as an activator, and a cementitious by-product. These by-products can be from a variety of industrial sources, such as fly ashes, blast furnace slag, phosphorous slag, coal fly ash, silica fume, and non-ferrous slag (Shi et al., 2011), as well as aluminosilicate waste such as clays and lateritic quarry waste (Shubbar et al., 2019). The production of activators such as sodium silicate from recycled glass cullet has also been demonstrated (Vinai and Soutsos, 2019). The reaction between the cementitious by-product and the activator produces an amorphous to semi-crystalline material that, when designed correctly, can quickly produce high-strength concrete (Palomo et al., 1999). The formation of the amorphous material is an additional environmental benefit, as spare metal ions will be irreversibly bound into the cement matrix, rendering them non-bioavailable, with low leachability (Abdel-Gawwad et al., 2019). However, there are other environmental aspects that need to be considered when using industrial by-products that are ashes or slag. Often, industrial processing of materials can concentrate metal fractions in the ashes and slag. This can also concentrate naturally occurring radioactive materials (NORMs), such as ^{238}U,^{235}U, ^{232}Th, and ^{40}K, to levels that could pose a risk to human health. The radioactive progeny of NORMs such as ^{222}Rn or radon, which is a radioactive gas that can cause lung cancer, must be known if the by-products are to be considered for use. Often, the levels of NORMs are determined by either the ore or the coal product that was used to refine the original product. Where in the world the by-product originated or was processed could have an effect on the level of NORMs present. This means that worldwide databases of by-products to be used in cements or AAMs are required in order to screen materials to identify any risks that are posed and to allow safe use and good design of materials with by-products (Sas et al., 2017). From such databases, it can be seen that primary raw materials (some types of granite) can also have highly elevated concentrations of NORMs. By-products that are generated due to high temperature combustion processes (coal combustion, cement generation, and waste incineration) are generally found to have lower radon emanation than by-products of lower-temperature processes (e.g., red mud from alumina production). When considering how the end user will interact with the product (e.g., use of by-products in a domestic environment), more detailed testing and monitoring rather than just screening of the by-product are required (Sas et al., 2019b). Environmental monitoring of the by-product itself may not be enough to mitigate potential risks. There is the possibility that the by-product may react or change as it is processed into a new material. This was found to be case when some types of fly ash, which appeared to have low risk, were transformed into alkali-activated materials. Fly ashes contain cenospheres, which are glassy particles; these are damaged by the aggressive alkali activation process, which releases trapped radon gas. This is only identified when the new material is also monitored (Sas et al., 2019a).

Conclusion

There is a need to teach engineers and product developers/designers the concepts of eco-design, circular design, and user-centred or 'human-centred' design (Giacomin, 2014). In particular, by-product producers and future product designers/engineers need to:

- Engage by-product producers and future users
- Understand future users' needs
- Design the by-product to suit the future user needs

This may seem like an onerous task on the part of the by-product producer, but ultimately it changes waste with significant handling and disposal costs into a valuable product. Designing the by-product may also seem a strange concept, but by altering upstream processes and taking a more holistic overview of the production process, there is the opportunity not only to create a valuable original product but also a valuable by-product. The concept of eco-design for products (Knight and Jenkins, 2009) can be modified to take account of by-product design and identify significant environmental aspects in relation to future phases of the life cycle of any potential product:

- secondary raw material selection and use
- manufacturing a new product from the by-product
- new product use
- future cascades of reuse/recycling/recovery

For each of these phases, environmental aspects can further be assessed:

- use or presence of substances classified as hazardous to health and/or the environment
- predicted consumption of materials, energy, and other resources
- anticipated emissions to air, water, or soil
- consumption of energy, water, and other resources throughout the life cycle

In order to promote future cascades of reuse, recycling, and recovery information relating to how to return the product at end of use, information on the possibility of remanufacturing and reuse and information for facilities concerning disassembly, recycling, and recovery should be made readily available. Ultimately, the designer will want to produce products and by-products that have a "true product sustainability"; this a concept where the product or by-product can be identified as also having a contribution to solving societal problems (Dyllick and Rost, 2017).

References

Abdel-Gawwad HA, Mohamed SA, Mohammed MS. (2019). Recycling of slag and lead-bearing sludge in the cleaner production of alkali activated cement with high performance and microbial resistivity. *Journal of Cleaner Production*. 220: 568–580. doi:10.1016/J.JCLEPRO.2019.02.144.

Aïtcin P-C. (2016). The importance of the water–cement and water–binder ratios. In: Aïtcin PC, Flatt RJ. (Eds.), *Science and technology of concrete admixtures*. Cambridge: Woodhead Publishing. 3–13.

Almesfer N, Haigh C, Ingham J. (2012). Waste paint as an admixture in concrete. *Cement and Concrete Composites*. 34: 627–633. doi:10.1016/J.CEMCONCOMP.2012.02.001.

Andrew RM. (2018). Global CO_2 emissions from cement production. *Earth System Science Data*. 10: 195–217. doi:10.5194/essd-10-195-2018.

Aranda Usón A, López-Sabirón AM, Ferreira G, Llera Sastresa E. (2013). Uses of alternative fuels and raw materials in the cement industry as sustainable waste management options. *Renewable and Sustainable Energy Reviews*. 23: 242–260. doi:10.1016/J.RSER.2013.02.024.

Arsne MA et al. (2007). Chemically and thermally treated vegetable fibers for reinforcement of cement-based composites. *Materials and Manufacturing Processes*. 22: 214–227. doi:10.1080/10426910601063386.

Columbia Climate Center (2012). *Mitigating iron and steel emissions*. Columbia University. Available at: http://climate.columbia.edu/files/2012/04/GNCS-Iron-Steel.pdf.

Dyllick T, Rost Z. (2017). Towards true product sustainability. *Journal of Cleaner Production*. 162: 346–360. doi:10.1016/J.JCLEPRO.2017.05.189.

Ellen MacArthur Foundation (2019). *Completing the picture how the circular economy tackles climate change*. Cowes: Ellen MacArthur Foundation.

Farrell CC, Osman AI, Doherty R, Saad M, Zhang X, Murphy A, Harrison J, Vennard ASM, Kumaravel V, Al-Muhtaseb AH, Rooney DW. (2020). Technical challenges and opportunities in realising a circular economy for waste photovoltaic modules. *Renewable and Sustainable Energy Review*. In Press.

Fellner J, Lederer J, Scharff C, Laner D. (2017). Present potentials and limitations of a circular economy with respect to primary raw material demand. *Journal of Industrial Ecology*. 21: 494–496. doi:10.1111/jiec.12582.

Flatt R, Schober I. (2012). Superplasticizers and the rheology of concrete. In: *Understanding the rheology of concrete*. Cambridge: Woodhead Publishing. 144–208.

Friedlingstein P et al. (2014). Persistent growth of CO_2 emissions and implications for reaching climate targets. *Nature Geoscience*. 7: 709–715. doi:10.1038/ngeo2248.

Fuss S et al. (2014). Betting on negative emissions. *Nature Climate Change*. 4: 850–853. doi:10.1038/nclimate2392.

Ganesh Babu K, Sree Rama Kumar V. (2000). Efficiency of GGBS in concrete. *Cement and Concrete Research*. 30: 1031–1036. doi:10.1016/S0008-8846(00)00271-4.

Giacomin J. (2014). What is human centred design? *The Design Journal*. 17: 606–623. doi:10.2752/175630614X14056185480186.

Hartmann J et al. (2013). Enhanced chemical weathering as a geoengineering strategy to reduce atmospheric carbon dioxide, supply nutrients, and mitigate ocean acidification. *Reviews of Geophysics*. 51: 113–149. doi:10.1002/rog.20004.

Hibbert AP, Hannant DJ. (1982). Toughness of cement composites containing polypropylene films compared with other fibre cements. *Composites*. 13: 393–399. doi:10.1016/0010-4361(82)90149-5.

Huisman J et al. (2016). ProSUM: prospecting secondary raw materials in the urban mine and mining wastes. In: *2016 electronics goes green 2016+ (EGG)*. Berlin: IEEE. 1–8.

Iacovidou E et al. (2017). Metrics for optimising the multi-dimensional value of resources recovered from waste in a circular economy: a critical review. *Journal of Cleaner Production*. 166: 910–938. doi:10.1016/J.JCLEPRO.2017.07.100.

Knight P, Jenkins JO. (2009). Adopting and applying eco-design techniques: a practitioners perspective. *Journal of Cleaner Production*. 17: 549–558. doi:10.1016/J.JCLEPRO.2008.10.002.

Lei Y, Zhang Q, Nielsen C, He K. (2011). An inventory of primary air pollutants and CO_2 emissions from cement production in China, 1990–2020. *Atmospheric Environment*. 45: 147–154. doi:10.1016/J.ATMOSENV.2010.09.034.

Le Quéré C et al. (2014). Global carbon budget 2014. *Earth System Science Data*. 7: 521–610. doi:10.5194/essdd-7-521-2014.

Mantellato S, Eberhardt AB, Flatt RJ. (2016). Formulation of commercial products. In: Aïtcin P-C, Flatt R. (Eds.), *Science and technology of concrete admixtures*. Cambridge: Woodhead Publishing. 343–349.

Mineral Products Association (2018). *Concrete credentials: concrete, concrete products and constituent materials*. London: Mineral Products Association.

Momoh EO, Osofero AI. (2019). Behaviour of oil palm broom fibres (OPBF) reinforced concrete. *Construction Building Materials*. 221: 745–761. doi:10.1016/J.CONBUILDMAT.2019.06.118.

Murray P. (2003). *The saga of Sydney Opera House*. London: Routledge.

Nehdi M, Sumner J. (2003). Recycling waste latex paint in concrete. *Cement and Concrete Research*. 33: 857–863. doi:10.1016/S0008-8846(02)01084-0.

Pacheco-Torgal F, Ding Y, Jalali S. (2012). Properties and durability of concrete containing polymeric wastes (tyre rubber and polyethylene terephthalate bottles): an overview. *Construction and Building Materials*. 30: 714–724.

Pacheco-Torgal F, Jalali S. (2011). Cementitious building materials reinforced with vegetable fibres: a review. *Construction and Building Materials*. 25: 575–581. doi:10.1016/J.CONBUILDMAT.2010.07.024.
Palley R. (2010). *Concrete: a seven-thousand-year history*. New York: Quantuck Lane Press.
Palomo A, Grutzeck MW, Blanco MT. (1999). Alkali-activated fly ashes: a cement for the future. *Cement and Concrete Research*. 29: 1323–1329. doi:10.1016/S0008-8846(98)00243-9.
Pan JR, Huang C, Kuo JJ, Lin SH. (2008). Recycling MSWI bottom and fly ash as raw materials for Portland cement. *Waste Management*. 28: 1113–1118. doi:10.1016/J.WASMAN.2007.04.009.
Ramachandran VS. (1996). Chemical admixtures—recent developments. In: *Concrete admixtures handbook*. Park Ridge, NJ: Noyes Publications. 137–184.
Romualdi J, Batson G. (1963). Mechanics of crack arrest in concrete. *Journal of Engineeering Mechanics Division*. 89: 147.
Sas Z et al. (2017). Radiological evaluation of by-products used in construction and alternative applications: part I. Preparation of a natural radioactivity database. *Construction and Building Materials*. 150: 227–237. doi:10.1016/j.conbuildmat.2017.05.167.
Sas Z et al. (2019a). Radiological characterisation of alkali-activated construction materials containing red mud, fly ash and ground granulated blast-furnace slag. *Science of the Total Environment*. 659: 1496–1504. doi:10.1016/j.scitotenv.2019.01.006.
Sas Z et al. (2019b). Radiological evaluation of industrial residues for construction purposes correlated with their chemical properties. *Science of the Total Environment*. 658: 141–151. doi:10.1016/j.scitotenv.2018.12.043.
Schreck M, Wagner J. (2017). Incentivizing secondary raw material markets for sustainable waste management. *Waste Management*. 67: 354–359. doi:10.1016/J.WASMAN.2017.05.036.
Schroeyers W et al. (2018). The NORM4Building database, a tool for radiological assessment when using by-products in building materials. *Construction and Building Materials*. 159: 755–767. doi:10.1016/J.CONBUILDMAT.2017.11.037.
Shi C, Jiménez AF, Palomo A. (2011). New cements for the 21st century: the pursuit of an alternative to portland cement. *Cement Concrete Research*. 41: 750–763. doi:10.1016/j.cemconres.2011.03.016.
Shubbar AA, Sadique M, Kot P, Atherton W. (2019). Future of clay-based construction materials – a review. *Construction and Building Materials*. 210: 172–187. doi:10.1016/J.CONBUILDMAT.2019.03.206.
Siddique R. (2003). Effect of fine aggregate replacement with class F fly ash on the mechanical properties of concrete. *Cement and Concrete Research*. 33: 539–547. doi:10.1016/S0008-8846(02)01000-1.
Siddique R, Khatib J, Kaur I. (2008). Use of recycled plastic in concrete: a review. *Waste Management*. 28: 1835–1852. doi:10.1016/j.wasman.2007.09.011.
Singh N, Shehnazdeep, Bhardwaj A. (2020). Reviewing the role of coal bottom ash as an alternative of cement. *Construction and Building Materials*. 233. doi:10.1016/j.conbuildmat.2019.117276.
Steinmann ZJN, Huijbregts MAJ, Reijnders L. (2019). How to define the quality of materials in a circular economy? *Resources, Conservation and Recycling*. 141: 362–363. doi:10.1016/J.RESCONREC.2018.10.040.
Teixeira ER et al. (2019). Recycling of biomass and coal fly ash as cement replacement material and its effect on hydration and carbonation of concrete. *Waste Management*. 94:39–48. doi:10.1016/J.WASMAN.2019.05.044.
Thomas JAG. (1972). Fibre composites as construction materials. *Composites*. 3: 62–64. doi:10.1016/0010-4361(72)90376-X.
Vegas I et al. (2006). Obtaining a pozzolanic addition from the controlled calcination of paper mill sludge: performance in cement matrices. *Materiales de Construcción*. 56: 49–60. doi:10.3989/mc.2006.v56.i283.8.
Vinai R, Soutsos M. (2019). Production of sodium silicate powder from waste glass cullet for alkali activation of alternative binders. *Cement and Concrete Research*. 116: 45–56. doi:10.1016/J.CEMCONRES.2018.11.008.
Voglar GE, Leštan D. (2011). Efficiency modeling of solidification/stabilization of multi-metal contaminated industrial soil using cement and additives. *Journal of Hazardous Materials*. 192: 753–762. doi:10.1016/J.JHAZMAT.2011.05.089.
Washbourne CL, Renforth P, Manning DAC. (2012). Investigating carbonate formation in urban soils as a method for capture and storage of atmospheric carbon. *Science of the Total Environment*. 431: 166–175. doi:10.1016/j.scitotenv.2012.05.037.
Zheng Z, Feldman D. (1995). Synthetic fibre-reinforced concrete. *Progress in Polymer Science*. 20: 185–210.

33

USING CIRCULAR SUPPLY CHAINS TO CREATE COMMUNITY BIOGAS

Ananya Mukherjee

Introduction

> the traditional model of industrial activity in which individual manufacturing processes take in raw materials and generate products to be sold plus waste to be disposed of should be transformed into a more integrated model: an industrial eco-system. In such a system the consumption of energy and materials is optimized, waste generation is minimized and the effluents of one process . . . serve as the raw material for another process.
>
> *Frosch and Gallopoulos (1989: 44)*

Sustainability is the underlying principle of circular supply chain models. Traditional, linear economic models of 'use and throw' have been in existence in many industrial operations and are not only wasteful but create negative externalities by increasing global greenhouse gas emissions. An estimated waste of 400 billion tonnes of solid waste can be generated by 2030, a third of which is food waste, causing natural resource depletion in the long run (Frosch and Gallopoulos, 1989; Ellen MacArthur Foundation, 2017). The circular economy, on the other hand, has high economic potential through the process of recycling by-products produced from raw materials such as biomass and converting them to products that can be used as primary products within the supply-chain system as part of industrial symbiosis instead of being discarded as waste. Biomass, for instance, moves through different stages of the supply-chain, that is, from agriculture, plantations, and forestry to farm products, textiles, and the manufacturing industry to a wide variety of retail and commercial products (Andersen, 2007; Ellen MacArthur Foundation, 2013, 2017; Farooque et al., 2019).

The advantage of such an ecosystemic approach is not only a cleaner environment but also the use of innovative techniques which enable industrial systems to adapt, restore, and regenerate themselves, leading to substantial economic benefits and sustainability in the long run. However, the success of such a circular supply chain model is largely dependent on technological innovation within which industrial outputs and residues are broken down into products that can be reused within the system using catalysis. This has the potential to generate low-carbon energy from renewables. The reuse and recycling of industrial outputs within all the stages of the supply chain underpins the work of ISABEL (Triggering Sustainable Biogas Energy

Communities through Social Innovation) on community biogas. As many tonnes of waste are produced in Europe, reusing and recycling them has the potential to reduce Europe's carbon footprint alongside producing cheap, clean renewable energy. The underlying concept is to integrate recycling of food waste, produce biogas, and create energy communities through social innovation and circular supply chain systems in order to support its uptake and enlargement.

The impetus to this approach is provided by the EU's Renewable Energy Directive (RED), which has set a binding target of 20% of its total energy consumption from renewable energy sources by 2020. In addition to the targets, the RED regulates the use of renewable electricity, heating, cooling, and transport. To achieve this, EU countries have committed to reaching their own national renewable targets, ranging from 10% in Malta to 49% in Sweden. Although the majority of European community Renewable Energy Sources (RES) projects are linked to solar and wind energy, biogas has also been a part of this development, with a slow acceptance in many regions of Europe. As transition to such a system is complex and not without a range of different challenges, this makes its ready adoption and use as a cheap, green, affordable, renewable energy product difficult. In addition, the concept of a circular economy model where waste is converted to heat, which not only reduces the amount of waste but also creates affordable energy, is a relatively novel concept in Europe and necessitates several systems to be in place prior to operationalising the model. It requires factoring in certain socio-economic and technological factors alongside the active participation and engagement of the human component or citizen participation, which makes it complex, interdisciplinary, and innovative at the same time.

This chapter discusses ISABEL's circular supply chain model using industrial symbiosis and the circular economy model to reuse and reduce waste, the result being cheap, clean, and green energy, leading to reduced emissions and a smaller carbon footprint. The underlying goal is the recovery of green waste such as biomass, food, and horticultural waste, which can create new revenue sources and renewable energy that can be used for individual and industrial consumption using innovative technology such as anaerobic digesters. The added benefit is the nutrient-rich by-product or the digestate from the digester, which has been used to grow vegetables and horticultural products on a small scale for local community consumption. The next section unpacks the meaning of the ecosystems approach prior to discussing how the ISABEL programme used this approach and the circular economy model, to create a circular supply chain system by drawing on a few case studies from Germany and the United Kingdom, where the project operated.

Ecosystems approach to creating community biogas

Energy communities have become a part of the local energy rhetoric, as they can significantly reduce carbon emissions and contribute to the transition to a low-carbon economy in Europe. Although the majority of the renewable energy projects are linked to wind and solar energy in Europe, biogas is part of the RED and has become the framework for all the European Union member countries. This is mainly because of European Directive 98/2008/EC, which clearly states reuse and recovery of resources from waste materials for use as primary products within industries is part of its waste management and pollution prevention policy (EU, 2008). In addition, RED Article 12 states biogas is part of the greenhouse gas emission reduction strategy, and, aside from the environmental benefits, it is also part of regional infrastructural development in rural areas, particularly based on the principle of decentralised production of heat and power (EU, 2008). This makes biogas a renewable resource and its installation a more acceptable proposition, available for adoption as the technology is available and ready with a new prospect of citizens becoming the future suppliers of raw materials. However, the operation of the supply-chain relationship, necessary to set up a community biogas initiative, is based on the premise that there is a functioning industrial symbiosis operating using the ecosystems approach. This necessitates the understanding and adoption of ways which would not only help industries overcome the challenges surrounding the treatment and processing of waste by-products as primary products but also in creating ecosystemic communities. Some of these challenges will be discussed in the next section and will be preceded by a discussion on the ecosystems approach.

The ecosystems approach can be defined as the interdependent relationship between different parts or players within the context of community energy, leading to a holistic and integrated functioning of the system as a whole. It is regenerative and departs from the traditional profit-motivated economy which is centred around reducing competitive advantage of competitors within the market, causing external negativities and natural resource depletion in the process. The paradigm of an ecosystem of interacting players, within a community energy project, targets various parts of the system, such as financial, human, sociopolitical, and intellectual knowledge-based resources, functioning as a cooperative which offers a conducive environment for the development of renewable energy initiatives (Alvarez and Puente, 2017). The available literature on circular economy discourse shows the ecosystemic approach, which essentially means the greening of the supply chain system, developed in parallel with the circular economy model (Lowe, 1993; Ying and Li-Jun, 2012; Ellen MacArthur Foundation, 2013; Genovese et al., 2017) and using the concept of industrial ecosystems.

The concept of industrial ecosystems developed to counter the negative impact of linear economic models, where 'take-make and dispose' has been the widely prevalent approach. Instead, reduce, reuse, and recycle has been the 'mantra' for industrial ecosystem models in which economic production and activities are viewed as part of a systemic loop which allows wealth to be generated in a much more sustainable way (Ellen MacArthur Foundation, 2019). However, this is largely dependent on setting up community ecosystems, which essentially means setting up a network of stakeholders who are committed to co-creating a community-owned initiative which is essentially decentralised and local. As co-production is key, its successful operation is dependent on the establishment of a symbiotic relationship between parts of the system. This creates an advantage, that is, the need to create a product which is solely centred on the localised problem of the consumer.

To design such an ecosystem would mean excluding free-riders and requiring full participation of producers and consumers at the same time (Ostrom, 1990; 2012). For instance, in order to keep the community safe, it is equally important for the locals to be aware of any suspicious activity within the neighbourhood and report to the authorities, as it is for law enforcement officials to be vigilant about any ongoing anti-social activity within the vicinity. Systems fail to function because the interdependence between producers and consumers is not always recognised. Taking the example of protected fisheries, these resources are violated and poached despite strict regulations in place because citizens or locals fail to report any suspicious activity they may have witnessed. Once the interconnection between systems or the ecosystemic interdependence between law enforcement and citizen participation is realised, then the possibility of implementing laws and curbing poachers becomes more realistic. This factor is also dependent on conditional cooperation between individuals who are willing to initiate cooperative communal action (Ostrom, 2000) based on an estimated understanding that others are willing to reciprocate in a similar way. The next section discusses a case study from Sweden where a group of individuals sharing a common vision of a community eco-village created a community biogas initiative using an ecosystems approach.

The Suderbyn project: a case study

The Suderbyn project illustrates how a permaculture and community biogas initiative began at the Suderbyn Eco-village with a small group of people who came together by design as partners and stakeholders to co-create and co-design a renewable energy initiative. Produced from local biomass and food waste on the island of Gotland, Sweden, the initiative was successful because the group as a whole had a shared goal of developing a zero-carbon and sustainable lifestyle, and the vision was to create a prosperous living environment whilst minimising the environmental footprint. The factors that contributed to the successful creation of the eco-village are collaboration between partners and various stakeholders, a clear set of boundaries, and creating a culture of co-development and co-creation whereby actors learnt by doing through trial and error. Co-production in this context created an advantage, that is, the need to create a product which was solely centred on the localised problem of the consumer, and the solution designed was also a localised one, that is, to produce affordable energy which was based on a circular loop framework of resource-product-waste-renewable resource (Ying and Li-Jun, 2012; Ellen Macarthur Foundation, 2013). The energy produced is completely off grid, localised, and from local food and biomass waste, which are the basic raw materials which enable the eco-village to become self-reliant and also switch to a low-carbon economy. Such decentralised systems gave makers an opportunity to create a solution where the consumer is also a participant in indirectly creating a demand for the solution. The maker movement originated from such a systemic need for localised solutions where there were high incentives to create and co-produce (Aligica and Tarko, 2012) products using local knowledge and often through trial-and-error learning methods (Ostrom, 2012).

Community biogas and the circular supply chain system

The principle of community biogas is centred around a system of creating value from waste using a closed-loop system which not only reduces waste but also prevents the depletion of natural resources by producing secondary products from the primary product, which is then recycled as part of the decentralised supply-chain system as part of a closed-loop system. This process requires going through stages where production and consumption of resources are no longer considered an end in itself, as in linear economies, but a means to an end. Creating products is part of the circular economy where symbiotic relationships between parts lead to creation of new products which can be reused. However, there are a few key steps that are needed prior to reducing, recycling, and reusing.

As biogas communities are part of the new-age maker movement, their success is centred around cooperation from various community individuals and stakeholders. These include business partners, energy companies, industrial partners to farmers, and even community households. Collaboration between these actors is essential in creating a community self-help group to address a local problem, in this case a decentralised energy system generated from recycling of food and biomass that would make the communities self reliant. The underlying principle is industrial symbiosis, which optimises material flows, creating energy as well as managing waste through a symbiotic relationship based on trust, mutual cooperation, and social innovation. The supply-chain system needs to be transparent and based on a symbiotic relationship in which the raw (waste) materials are collected from local households and fed into an anaerobic digester. Processing this waste from food and biomass creates a digestate which is not only rich in nutrients such as nitrates but also produces methane gas, which is another by-product of the waste recycling process. This can be stored in battery containers for later usage either as heat and gas and can be used not only for cooking but also as heat.

The production of biogas is dependent on an established industrial symbiotic relationship not only with different community stakeholders but also with organised industrial systems in order to process the waste raw materials into different by-products. This process is dependent on an organised and symbiotic relationship between different stakeholders, which are essential prerequisites to designing an ecosystem of collaborators.

The success of such an industrial symbiosis is, however, dependent on the adoption and diffusion of socially innovative ideas, which not only has the potential to create competitive new business models but also new income opportunities alongside social, environmental, and economic benefits (Dees et al., 2004; Dutta et al., 2018). The emphasis is mainly on collaboration, partnerships, and interdependence of parts, which is instrumental in the flow of resources as a cyclical process. It operates as a functional unit based on the principles of sustainability, inclusivity, and reduction of negative externalities, which complements the role of different stakeholders and players, interacting with rather than eliminating them (Taddeo et al., 2017). The success of these endeavours is dependent on institutions and their willingness to embrace new ideas (Andersen, 2007) and on the adaptive capacity of individuals and groups to embrace this change. Arguably, such transformative behaviour takes considerable time and often challenges traditional institutional practices and beliefs. Added to that is each community's geographical and social characteristics, which add to the complexity, as they vary widely. While innovation and more specifically social innovation can help design approaches which can address these complexities and differences by adopting bespoke solutions in each region, one of the bigger challenges in designing such projects is "mutual trust" among interacting individuals (Ramaswamy, 2010) and developing collaborations. In other words, overcoming challenging and competitive rivalries is also part of the social innovation strategy. The first step, thus, is to map the operational plan of community biogas projects locally in order to overcome existing risks and explore opportunities. The second step is to develop bottom-up collaborations and encourage a democratic and decentralised form of participation and interaction in society which is fluid, providing room for adaptation and improvisation, which are common when designing bespoke solutions for complex situations.

The next section discusses two different case studies of biogas communities and highlights the various ways in which industrial symbiosis has been used to trigger such initiatives using socially innovative methods. Prior to that, it is pertinent to mention that information for this work was gathered using participatory evaluation workshops with stakeholders in each region to ascertain how industrial symbiosis was being used within the circular economy model to set up a community biogas initiative. Participants in the workshop were asked to design their model in a vision-setting exercise using the circular economy model, and based on their design, the following information was extrapolated.

Using industrial symbiosis to trigger community biogas initiatives in Germany and United Kingdom

Germany

Germany has been the frontrunner of renewable energy initiatives because of the widely prevalent renewable energy incentives such as the Renewable Source Act, which regulated feed-in-tariffs for electricity. ISABEL's special focus was on the Baden-Wurttemberg region, characteristically known for its biogas landscape. The study showed pre-existing biogas villages with large infrastructural features within the German biogas landscape. The theme is centred around large biogas plants, energy from which is connected to both the local electricity grid and the district heating network to meet local industry needs and those of the local communities. Energy is derived from multiple sources, such as animal manure, energy crops, and wood pellets, recycled and converted into gas, which is then plugged into the district heating grid and the industrial waste heat source.

Stakeholders designed a circular loop model where a diverse range of products can be produced and reused besides biogas, that is, vegetable and horticultural produce from the nutrient-rich residue from the anaerobic digester. It reflected the fact that in Germany, the emphasis is more on the wider environmental and social benefits of using anaerobic digesters and that there is a shift in focus from energy crops to livestock waste and other forms of biomass waste. Another important point that comes out is the fact that biogas not only addresses the energy needs of the community but also offers more financial assurance, as the price and costs are predictable. In addition, a digester makes the life of a farmer much easier by integrating it with the existing farm system. For instance, every farm has its own regime, layout, and operating system, and this should be integrated with the biogas system and any other supplementary energy source such as wood chips and so on. Once integrated, this will become an enabler in supporting the farmers by reducing their workload and increasing general health and well-being.

The previous section discussed the relevance of mutual trust, cooperation, and strong social networks as necessary prerequisites to a successful social innovation project where production conditions are reversed from 'production for the masses' to 'production by the masses' (Murray et al., 2008). However, such a transition is not without its fair share of challenges. Starting from low motivation from the wider society and community to land-use competition for the growth of energy crops such as maize and lack of incentives and supportive legislation from the government, all of these factors have hindered the progress and uptake of community biogas initiatives in the region. Stakeholders also identified the diverse social background of the community as an obstacle to the progress of community biogas work, because not all individuals may share the same motivation to create a biogas community. The second problem they identified is reduced efficiency as a result of the mixed background of the community, which could go against the building of a strong and cohesive biogas community. Solutions and opportunities that will make the community biogas initiative a success are the presence of a community motivator, participation of energy suppliers, and best-practice examples. Broad participation by citizens and individuals in the biogas recycling programme was also identified as another opportunity in order to make it successful.

The case study from Germany illustrates the functioning of an industrial ecosystem which is reliant on optimising use of raw materials, leading to energy creation from waste products. When raw waste materials are processed and transformed into liquid digestate, it generates energy or biogas which has its own utility value. The residue from the processing of waste matter becomes another product which is used for growing vegetables and various horticultural purposes. The symbiotic relationship between various stakeholders, industry, and the locals based on mutual trust, cooperation, and shared knowledge forms part of the social innovation which can add value and create benefits, including new business models, aside from creating long-term sustainability and environmental and social benefits. Even though these are the fundamental principles behind setting up a community biogas initiative, our study in the United Kingdom showed a slightly different biogas landscape scenario.

United Kingdom

The United Kingdom is second in biogas production within Europe and first when it comes to landfill gas. The national regulatory environment is supportive of bioenergy, as the United Kingdom has recently introduced significant promotional measures for biomethane, giving the domestic biogas market a significant push. ISABEL's community biogas work was specifically focused on the Yorkshire and Humber region. Even though the feed-in tariff and the Renewable Heat Incentive initially supported the anaerobic digestive industry in England, Scotland, and Wales, there were other policy and legislative hurdles which stakeholders identified when developing a community biogas initiative.

The stakeholder engagement activity showed diverse views among participants who viewed community biogas initiative as a circular loop model. Community biogas initiatives in the United Kingdom are mostly small-scale, either within an urban or rural area. Stakeholders described a rich picture of a small anaerobic digester adjacent to a composting unit which is dependent upon availability of waste from the local residents, who would need to be plugged in to the community initiative. Stakeholders suggested that it is important to understand the diversity within the community in order to develop a decentralised system where they could decide with whom to work. Technical support is important and could be part of a decentralised training programme, as is building capacity. It is critical to setting up such a community-based initiative. The digestate processing was envisaged as the centre point for processing surplus digestate with the schools, social housing, community centre, business, retail markets, restaurants, prisons, parks, and universities situated in the adjoining areas. The process is reflective of a mature systems-thinking approach with a centralised small-scale biogas programme within an urban space in the UK biogas landscape, in contrast to the large-scale biogas model in Germany.

The second model also described a small-scale biogas initiative within a vegetable allotment setting within an urban setting. Vegetable allotment centres are owned by the community and are given out on lease by the local authorities to the general public to grow their own food and vegetables in an eco-friendly way. Stakeholders wanted to set up the biogas plant within the vegetable allotment centre in an urban environment. They would begin by collecting food waste from the local housing sites and composting the food waste to produce biogas. The vegetable allotment centre would help diversify production, as vegetables and plants can be grown for various purposes and also to meet seasonal needs. The biogas produced could be used to meet the needs of the allotment centre and to create a polytunnel where the surplus heat can be used. A nutrient recovery unit would also be useful within the allotment centre where the heat can be used to treat the nutrients, recover the water, and use it for some other purpose. This opportunity will help save the local council, valuable resources and reduce the impact of waste management for them, as they will be using garden waste and food waste as feedstock for the anaerobic digester.

The two regional case studies, however, highlight a huge potential in creating energy communities using industrial ecosystems and circular supply chains. The closed-loop system is characteristic of industrial ecosystems with particular emphasis on reducing waste through reuse and recycling processes. This requires systematic acquisition of raw materials and products, including innovative design plans which will ensure the construction of industrial ecosystems, including the building of social capital enabling these ecosystems to thrive. However, both case studies highlighted that the success of such a collaborative approach is reliant upon the successful engagement and participation of multiple stakeholders with a shared vision, which can become a challenge. A participatory approach to such a process stems from the investment of the ecosystem of actors willing to work in a cooperative and interconnected way. Attracting individual players and engaging them in the system is not only central but can also be a major challenge. Aside from this, there are several other problems that hindered the progress and development of these community biogas initiatives, that is, legislative, administrative, and financial. Lack of supportive legislation, administrative, and financial support from local government, incentives necessary to provide the impetus for a community biogas initiative, has been instrumental in the slow uptake of the circular supply chain model. Other challenges include a disconnect between relevant policy that could support businesses, government, and people through incentives such as tax breaks, which is hindering communities from developing decentralised supply chains to initiate community biogas initiatives. These issues include reduction in renewable heat incentives and feed-in tariffs and necessitates new policies to promote such a system.

The other major hurdle which is often underestimated is people's tendency to accept systemic change (Aguinaga, 2016). A shift from previous linear economic models to an industrial ecosystem is gradual and requires making gradual systemic changes which will pave the way for behavioural changes among stakeholders. This has been a crucial finding from the ISABEL project. Communities have been sceptical and resistant to change and on occasion have even displayed negative attitude towards 'community digesters,' reiterating the fact that developing supply chains using industrial symbiosis is subject to time and considerable investment in human and social capital. (Ellen MacArthur Foundation, 2013; Aguinaga, 2016). This also necessitates a targeted advocacy strategy in the region which could expedite the progress by raising awareness on how decentralised energy systems can create affordable energy and reduce carbon emissions at the same time, including the potential to create new jobs.

The case studies from Germany and United Kingdom are exemplars of new business models where industrial ecosystem approaches have been used to co-create and co-design decentralised energy networks. These are not only inclusive, based on collaborative approaches, mutual trust, cooperation, and partnership among multiple stakeholders, but also exhibit organic leadership features which are essential to initiate, innovate, lead, and co-design such complex ecosystems (Ramaswamy, 2010).

Conclusion

Industrial ecosystems and circular supply chains have vast potential to reduce wastage and create new products through the circular loop model. Some of the key essentials necessary to develop a successful community biogas initiative are community participation; a shared vision and the commitment to create decentralised energy systems from recycling green waste; available funding, including strong business cases; technical and local authority support; clear, consistent, and supportive policies; regulatory frameworks; and an appropriate location which will have ready community acceptance. Not only does it save resources and lower costs through connected supply chain systems but also creates new income sources through social and technical innovation. More importantly, as ecosystemic approaches are complex processes which require the intersection of many different disciplines in a holistic way, it is important to identify and build these relationships at the outset to ensure that the different supply chains are connected through collaboration and partnership. Such new paradigms creating systems to address local problems through local solutions by community stakeholders has the potential of creating new business models where the production system is run "by the people" rather than "for the people" (Murray et al, 2008). Such new models of industrial ecosystems have the potential to co-create decentralised and democratic energy systems where complex systems are designed, produced, and operated by the people using innovation which could be a combination of socioeconomic and technological endeavours based on key factors such as trust, respect, and mutual cooperation (Lowe and Evans, 1995; INSEAD, 2009; Dutta et al., 2018).

Systemic change is an ever-evolving process and can take time, especially where the concept is new. Such transformations involve changing perceptions through concerted efforts by consciously taking measures to bring about change and also by creating a favourable environment.

A favourable environment is generated from supportive policies and legislation which promote change, followed by various incentives which motivate people's behaviour favourably towards the desired outcome. Added to that are resources which are conducive to the development of an environmental condition, for instance, the paradigm of an ecosystem of interacting players, in the community energy project sense, thus targeting financial, human, sociopolitical, and intellectual knowledge-based resources, which can offer important elements for creation of the appropriate conditions for the success or failure of a renewable energy project.

Finally, the ISABEL programme illustrated how gradual transformation in complex systems can be brought about by understanding what people want and what they need so that this can align with the overarching goal of the initiatives. As systemic change is an ever-evolving process over considerable time, it also involves organic leadership and organisational skills which will enable such new concepts to develop. It would be erroneous to assume that such complex supply chain systems can be created by a one-size-fits-all formula when adaptive, bespoke solutions will create a much better outcome.

References

Aguinaga E. (2016). *Developing a regional circular value ecosystem based on residues and wastes: the case of Higueras village, Mexico*. PhD thesis. EGADE Business School, Technologico de Monterrey, Mexico.

Aligica Paul D, Tarko V. (2012). Polycentricity: from Polanyi to Ostrom, and beyond. *Governance: An International Journal of Policy, Administration, and Institutions*. 25 (2): 237–262.

Alvarez R, Puente-Ruiz C. (2017). Development of the tool symbiosis to support the transition towards a circular economy based on industrial symbiosis strategies. *Waste Biomass Valour*. 8: 1521–1530.

Andersen MS. (2007). An introductory note on the environmental economics of the circular economy. *Integrated Research System for Sustainability Science*. 2: 133–140.

Dees JG, Anderson BB, Wei-Skillern J. (2004). Scaling social impact: strategies for spreading social innovations. *Stanford Social Innovation Review*. Spring. 24–32.

Dutta S, Lanvin B, Wunsch-Vincent S. (2018). *Global innovation index: energising the world with innovation*. Available at: www.wipo.int/edocs/pubdocs/en/wipo_pub_gii_2018.pdf.

Ellen MacArthur Foundation (2013). *Towards the circular economy*. Available at: www.ellenmacarthurfoundation.org/publications/towards-the-circular-economy-vol-1-an-economic-and-business-rationale-for-an-accelerated-transition.

Ellen MacArthur Foundation (2017). *Urban bio-cycles*. Available at: www.ellenmacarthurfoundation.org/publications/urban-biocyles.

Ellen MacArthur Foundation (2019). *Completing the picture: how the circular economy tackles climate change*. Available at: www.ellenmacarthurfoundation.org/publications/completing-the-picture-climate-change.

European Union (EU) (2008). Directive 2008/98/EC on waste and repealing certain directives. *The Official Journal of the European Community*. 31: 3e30.

Farooque M, Zhang A, Thurer M, Qu T, Huisingh D. (2019). Circular supply chain management: a definition and structured literature review. *Journal of Cleaner Production*. 228: 882–900.

Frosch RA, Gallopoulos NE. (1989). Strategies for manufacturing. *Scientific American*. 261: 144–152.

Genovese A, Acquaye AA, Figueroa A, Koh SCL. (2017). Sustainable supply chain management and the transition towards a circular economy: evidence and some application. *Omega*. 66: 344–357.

INSEAD (2009). *Are you innovation ready?* Available at: https://centres.insead.edu/eu-competitiveness-initiative//research/documents/INSEAD-Logica-Are-You-Innovation-Ready.pdf.

Lowe E. (1993). Industrial ecology – an organising framework for environmental management. *Environmental Quality Management*. Fall Issue. 3 (1): 73–85.

Lowe E, Evans LK. (1995). Industrial ecology and industrial ecosystems. *Journal of Cleaner Production*. 3 (1–2): 47–53.

Murray R, Mulgan G, Caulier-Grice J. (2008). *How to innovate: the tools on social innovation*. The Young Foundation. Available at: https://youngfoundation.org/wp-content/uploads/2012/10/How-to-innovate-the-tools-for-social-innovation.pdf.

Ostrom E. (1990). *Governing the commons: the evolution of institutions for collective action*. Cambridge: Cambridge University Press.

Ostrom E. (2000). Collective action and the evolution of social norms. *Journal of Economic Perspectives*. 14 (3): 137–158.

Ostrom E. (2012). *The future of the commons: beyond market failure and government regulation*. New York: Institute of Economic Affairs, American Economic Association.

Ramaswamy V. (2010). Competing through co-creation: innovation at two companies. *Strategy and Leadership*. 38 (2): 22–29.

Taddeo R, Simboll A, Ioppolo G, Morgante A. (2017). Industrial symbiosis, networking and innovation: the potential role of innovation poles. *Sustainability*. 9 (2): 169.

Ying J, Li-Jun Z. (2012). Study on green supply chain management based on circular economy. *Physics Procedia*. 25: 1682–1688.

34

CIRCULAR ECONOMY INITIATIVES IN INDIA

A case study approach

V. Madha Suresh

Introduction

In the last decade, India has experienced economic growth averaging 7.4% annually (Statista, 2020). During 1985–2015, its real gross domestic product (GDP) witnessed a growth rate of approximately 6%, while its resource consumption during the same period increased at a compound annual growth rate (CAGR) of around 4% (FICCI, 2019). If this trend continues, it could become one of the four biggest economies worldwide in about two decades. However, it faces significant challenges, including rapid urbanisation, resource scarcity, and a large portion of the population living below the poverty threshold. For an emerging economy like India, the strong link between economic growth and resource consumption suggests that this growth will in the coming years most likely be accompanied by a rise in the demand for resources as well as in the quantity of waste produced. For example, it will nearly triple its demand for primary raw materials to 15 billion tonnes by 2030 if it follows the current linear economy model (Mohanraj, 2019). However, it is the third-largest carbon emission producer globally. Behind China, India is by far the largest economy contributing to the world's ocean plastic (Kaplan, 2019).

Key challenges faced

India faces a number of challenges in moving towards a circular economy. The country produces around 62 million tonnes of waste each year, of which approximately 22% is processed (Prachi, 2019). By 2045, it is expected to be producing around 450 million tonnes of waste. Around 1.7 million individuals in the country work in the (formal and informal) waste sector. With respect to municipal solid waste (MSW), the generation rate is estimated to be 143,000 tonnes per day, with around 35,602 tonnes per day (24.8%) being treated (NPC, 2019). The number of wards with 100% door to door collection is approximately 61,846 (73% of wards), while the number with 100% source segregation is around 30,749 (36% of wards) (NPC, 2019).

Goyal et al. (2018) list four key challenges: (1) India has inefficient recycling processes, as well as limited waste recovery infrastructure and industry-level capabilities. These issues are restricting awareness, adoption, and growth of the circular economy concept; (2) the development and implementation of legislation and policies is generally disjointed, as the processes are managed by different entities, including the Ministry of Environment, Forests and Climate Change in association with state pollution control boards, state governments, and municipalities; (3) there is an inferior social perception about people engaged in the waste management industry; and (4) there is a lack of government focus toward formal recognition of the waste management industry. Meanwhile, Mohanraj (2019) is of the view that another key challenge is a lack of comprehensive understanding on what the circular economy really is.

In India, significant resource challenges are faced with the loss of food and plastics waste along the value chain (FICCI, 2018a). Almost 40% of the farm produce in the country is wasted annually. Unlike developed countries where food wastage is primarily an issue at the consumer and post-consumer stages, in India, it is spread across the value chain. While per-capita consumption of plastic is less than half the global average, demand in India is expected to rise at a CAGR of 9% and increase from 17.8 million tonnes to around 20 million tonnes by 2021, driven primarily by demand for packaging. This growth in consumption is likely to be accompanied by a rise in post-consumer plastic waste production. According to the Central Pollution Control Board of India, 5.6 million tonnes of plastic are disposed of annually, 43% of which is attributable to packaging. Moreover, only approximately 60% of the waste is recycled, with the rest going to landfill.

Due in large part to increasing buying power and the availability of cheap phones, there is a new mobile subscriber base of 1,003.49 million in India, with around 60 million new subscribers added every year (Trak, 2017). This has led to an increase in the quantity of waste electrical and electronic equipment (WEEE) produced. However, only 5% of India's total WEEE gets recycled due to poor infrastructure and lack of legislative framework (Chaudhary and Vrat, 2018). This inefficiency results in significant damage to the environment and the health of stakeholders in the sector. A key segment of the population who are at risk are children, with around 400–500,000 in the age group of 10–15 years thought to be engaged in various WEEE activities in waste yards and recycling workshops without adequate protection and safeguards (Economic Times, 2016).

Adopting a circular economy

EMF (2016) identified three key areas that should be focused upon in the transition to a circular economy, namely cities and construction, food and agriculture, and mobility and vehicle manufacturing. Adopting a circular economy pathway could yield annual benefits of around US$624 billion by 2030–2050 (EMF, 2016; FICCI, 2019). This is equivalent to 30% of India's GDP. Cash-out cost in the three focus areas would be US$218 billion (11% of India's GDP) lower in 2030 and US$624 billion (30% of India's GDP) lower in 2050, compared with the current scenario (NPC, 2019). In addition to creating cost savings for businesses and households, a circular transition would reduce negative externalities. For example, greenhouse gas (GHG) emissions would be 44% lower by 2050 compared to the current scenario, and congestion and pollution would fall significantly, leading to health and economic benefits to citizens.

A number of legislative measures have been put in place in the country. Comprehensive amendments to various waste management rules, including those on municipal solid waste, plastic waste, hazardous waste, bio-medical waste, and electronic waste, have been developed (NPC, 2019). India's National Manufacturing Policy focuses on promotion and adoption of green technologies and manufacturing, especially amongst its medium- and small and medium-sized enterprises (MSMEs). It has embarked upon an initiative of creating 100 smart cities across the country, and waste management and resource conservation are a significant part of this initiative. The Zero Defect, Zero Effect scheme aims to improve the quality of Indian manufacturing while minimising any negative externalities. The Ministry of Environment, Forest and Climate Change established the Indian Resource Panel to partner with other governmental ministries and private/public organisations to facilitate the use of recycled materials, act as a hub for resource efficiency, and provide the right policy and technological support. A national action plan on climate change (e.g., comprising National Solar Mission, National Mission for Enhanced Energy Efficiency, and National Water Mission) aims to reduce the environmental footprint of the economy. The government also introduced Plastic Waste Management Rules 2018 to reduce and manage plastic waste. One key feature of this was the introduction of extended producer responsibility (EPR). Under EPR, producers are made responsible for collecting and processing their manufactured products at the end of their lifetime. The national and several state governments also have a firm commitment to the Swachh Bharat Mission (Clean India Mission), with, for example, severe penalties for non-compliance with the EPR (e.g., losing one's license to operate). In addition to these macro-initiatives, the government has also introduced several sector-specific rules to help catalyse circular business models. For example, Mission Zero Waste aims to convert waste into valuable products utilising scientific approaches in 400 targeted cities in the country. However, confusion and inconsistency of implementation of these policies exist.

In 2019, the National Institution for Transforming India (NITI Aayog) released the Status Paper and Way Forward on Resource Efficiency and Circular Economy (MoEFCC, 2019). The document outlines 30 recommendations for transformation to resource efficiency and a circular economy in India. These recommendations include the formulation of a national policy on resource efficiency and circular economy, establishment of a Bureau of Resource Efficiency (BRE), and R&D for scalable technologies. Four sector-specific papers on resource efficiency in steel and aluminium, construction and demolition waste, and the electrical and electronics sector have also been released. The government has developed a viability gap fund/grant to encourage value creation from activities such as composting, waste to energy, utilisation of plastics waste in road construction, and more sustainable management of construction and demolition waste. A number of initiatives have been put in place, including the use of recycling for plastics, bricks made from fly ash, furniture made from tetrapak, the introduction of eco-friendly cutlery, biodegradable shopping bags, and encouraging the reuse of treated wastewater (NPC, 2019).

Circular economy case studies

There are several examples related to food and plastics (FICCI, 2018a). For example, the company Olio is working towards creating a digital app-based aggregation and sharing platform for leftover food. The service connects people with each other and with local establishments so that surplus food and other items can be shared instead of being thrown away. Mimica Lab has designed a biologically accurate expiry label for food products. The label decays at the same rate as the food, thus providing an accurate indicator of freshness and preventing wastage. Tesco has introduced resealable packaging with separate compartments to store food for later to increase the shelf life of packaged food items. Rallis India has developed a bio-fertilizer made from recycled waste from sugar mills and distilleries that is enriched with nutrient additives and microbes and can reportedly improve yields per acre by 15–20%. Between 2017–2022, the Fruit Circular Economy (FCE) initiative undertaken by Coca-Cola India will encourage sustainable and more localised procurement of agricultural raw materials. The initiative will entail an investment of around US$1.7 billion to facilitate the expansion of the local procurement of agricultural inputs (e.g., oranges and mangoes).

The Bangalore-based NGO Hasiru Dala has partnered with waste pickers from the informal sector to provide services to more than 22,000 households and create more than 800 jobs.

In 2019, the Indian fashion project (I was a Sari) won the first Circular Design Challenge Award at the Lakmé Fashion Week Summer-Resort 2019 in Mumbai, India (Mintel, 2019). As part of this project, an underprivileged community of women in Mumbai have made a range of products using upcycled, pre-loved saris. The winning collection was composed of recycled plastic tarpaulin used to make a raincoat, bag, umbrella, 3D appliquéd garment, and a pair of shoes, among others.

Organisations like INORA, Kalpavruksha Farm, Organic India, and Govardhan Ecovillage are utilising precision farming techniques that have been delivering increased productivity and reduced waste (EMF, 2016).

Haathi Chaap developed a unique business model for manufacturing paper and related products. The company created a market niche and differentiated itself from other hand paper and paper manufacturing companies by incorporating the use of elephant manure as a raw material. The shift resulted in significant savings in operational costs and provided scalable opportunities for employment in the surrounding communities. The residual water from washing the elephant manure also served as a by-product in the form of a rich source of organic fertilizer for farms.

Goonj developed reuse opportunities involving 'waste' cloth to provide dignity, hygiene, skills development, and income generation opportunities for villagers, especially women, where, for example, around 68% of women lack access to affordable sanitary napkins. It also created a system for channelling excess resources from urban households to impoverished, rural, and disaster-stricken areas, as well as local infrastructure development opportunities for underserved villages. The company undertook mass awareness campaigns, facilitated skills development training, and mobilised the village and slum communities for local development activities like building schools, roads, bridges, and toilet facilities in exchange for clothes, furniture, household goods, and medical supplies. It manages more than 1,000 tonnes of excess urban material every year, ranging from clothes to school-related materials to old doors, windows, and computers. It has produced more than 3 million sanitary pads from waste cloth and delivered the same to women in villages and slums across India. It has also created income-generating activities across villages by imparting skills to the women for creating mattresses from waste clothing, thereby converting more than 0.5 million kg of waste cloth into mattresses.

Attero established an end-to-end electronic asset management company and clean-tech pioneer, which focused on eco-friendly reuse, refurbishing, recycling, and urban mining of electronic waste and customised solutions for electronic asset management. It did so by developing a nationwide collection system, eco-friendly recovery of materials from WEEE using clean tech approaches, and an online platform for sales. The company created awareness among individuals, corporations, and electronic product manufacturers regarding the safe disposal and efficient recycling of WEEE. It collaborated with institutions like the International Financial Corporation, corporations, electronic manufacturers, and academic institutions to build collaborations such as Clean E-India with informal waste collectors into an organised and efficient national network. Attero undertakes recycling, refurbishing, and reconditioning of WEEE, thereby facilitating access to precious metals locally, as well as providing online availability of refurbished and surplus electronic equipment that is in good condition.

In terms of financing, FICCI (2018c) states that Coca-Cola India (along with its fruit suppliers and processors) will contribute more than US$1.7 billion to foster circularity in India's agri-ecosystem over the next five years.

There are also a number of examples in the mining and metals sectors (FICCI, 2018b). For example, Arcelor Mittal introduced a rental business model for its steel sheet piles. These piles can be used for construction applications such as dams, underground car parks, tunnels, and bridges. Customers can rent and reuse the sheet piles, thereby reducing their manufacturing costs. The business model allows Arcelor Mittal to extend its ownership of the assets across the value chain through to the end of the useful product life, thereby facilitating easier collection and recovery. This product-as-a-service model has allowed the organisation to implement a closed-loop business model with a recovery rate of approximately 99%. Ambuja Cement utilises alternative fuels such as municipal solid waste, biomass, mixed industrial waste, refuse-derived fuel, and waste oil to power their cement kilns and have achieved a carbon reduction of approximately 10.3 million tCO and additional gross value of approximately US$4.1 million, while Tata's Jaguar and Land Rover sell the waste aluminium from vehicles back to the supplier (Novelis), which has enabled the recovery of over 50,000 tonnes of aluminium in one year and has also helped Novelis reduce its GHG emissions by 13%.

Towards a circular economy

In recent times in India, there has been a shift towards more sustainable products. For example, studies by Assocham and Techsci and GT Nexus indicate that some 60–62% of consumers in the high-income category prefer organic products (especially fresh produce like fruits and vegetables). There is also a demand for ethically and sustainably sourced products. FICCI (2018) suggests the need therefore to leverage this change to create greater awareness about the clean and resource-efficient options offered by circular business models. In addition, there should be use of biomaterials such as bio-plastics and digital technologies (e.g., Internet of Things – IoT – tags) to facilitate tracking of resources and assets, thereby enabling the development of shared-platform business models. Greater collaboration along value chains, especially between upstream suppliers and downstream customers, should also be facilitated to design new products, services, and processes which reduce wastage and close material flow loops.

Goyal et al. (2018) state that companies need to focus on the design of products and services, which can address the multiple social or environmental needs of Indian society. They should also design an integrated reverse logistics setup for the collection of raw materials, followed by the sorting and transformation of these materials into product or service offerings. Distribution channels should be designed depending upon the target segment. The choice of distribution channels should vary, from setting up online e-commerce platforms and retail stores to building a network of local institutions and individuals for last-mile reach.

Chaudhary and Vrat (2018) note that most components for phones are made outside India. Therefore, to create a circular economy for WEEE, it is important to create an integrated value chain, with a focus on manufacturing indigenous products and components. This would create local availability of raw materials and components and higher local value and thus sustainable end-to-end markets. The authors also estimated that an WEEE recycling unit of 6,000 million tonnes per annum in India could potentially realise a profit margin of approximately US$6.3 million annually.

Both FICCI (2018) and Kaplan (2019) contend that there is a need for more financing in the system. For example, Kaplan (2019) states that there is a need for catalytic capital to finance waste management projects such as recycling technologies in order to grow the marketplace and present investors with attractive opportunities, while FICCI (2018) is of the opinion that private equity funds and venture capitalists are well positioned to channel funding into circular economy initiatives, therefore increasing their attractiveness to other funders such as retail banks. Kaplan (2019) contended that, due to its size, India has a huge potential for investment in recycling and waste recovery. Unlike Malaysia and other countries in the region, India's recycling industry does not need to rely on imports or exports. It has its own recycling system and can manage all of the recyclates domestically. For example, plastic waste imports into the country are comparatively small, at around 48,000 tonnes in 2018. Improvements in waste recycling and recovery would lead to benefits for all. For example, waste collectors from the informal sector would have a better quality of life, health, and working conditions.

Conclusions

In India, while there are many challenges involved in moving to a more circular economy, the government has put in place policies and pieces of legislation to effect the change. In addition, there are a number of organisations in a range of sectors that are already developing and successfully implementing initiatives. With the correct strategies, the country is well placed to take advantage of further opportunities.

References

Chaudhary K, Vrat P. (2018). Circular economy model of gold recovery from cell phones using system dynamics approach: a case study of India. *Environment, Development and Sustainability*. 1–28.

Economic Times (2016). *Delhi-NCR may generate 1 lakh metric tonnes of e-waste per annum: Assocham*. Available at: http://economictimes.indiatimes.com/tech/hardware/delhi-ncr-may-generate-1-lakh-metric-tonnes-of-e-waste-perannum-assocham/articleshow/51212872.cms.

Ellen MacArthur Foundation (2016). *Circular economy in India: rethinking growth for long-term prosperity*. Available at: www.ellenmacarthurfoundation.org/assets/downloads/publications/Circular-economy-in-India_2-Dec_2016.pdf (Accessed 21/11/2019).

FICCI (2019). *Accelerating India's circular economy shift: circular economy opportunity in India*. New Delhi: Athena Information Solutions Pvt. Ltd.

FICCI (2018a). *Accelerating India's circular economy shift: circular economy in agriculture, food and beverages sector*. New York: Athena Information Solutions Pvt. Ltd.

FICCI (2018b). *Accelerating India's circular economy shift: circular economy in metals and mining sector*. New York: Athena Information Solutions Pvt. Ltd.

FICCI (2018c). *Accelerating India's circular economy shift: enablers and success factors for adoption of circular economy in India*. New York: Athena Information Solutions Pvt. Ltd.

Goyal S, Esposito M, Kapoor A. (2018). Circular economy business models in developing economies: lessons from India on reduce, recycle, and reuse paradigms. *Thunderbird International Business Review*. 60: 729–740.

Kaplan (2019). Now's the time to invest in recycling and circular economy in India. *Forbes Magazine*. 2/8/2019.

MoEFCC (Ministry of Environment, Forest, and Climate Change) (2019). *NITI Aayog pitches for transition to resource efficiency and circular economy as an economic paradigm for new India*. Available at: https://eeas.europa.eu/delegations/india/57035/niti-aayog-pitches-transition-resource-efficiency-and-circular-economy-economic-paradigm-new_en (Accessed 21/11/2019).

Mintel (2019). *Circular economy makes headway in Indian fashion*. Available at: www.mintel.com/blog/new-market-trends/circular-economy-makes-headway-in-indian-fashion.

Mohanraj P. (2019). *Circular economy: an alternative development pathway for India*. Available at: https://indiaincgroup.com/circular-economy-an-alternative-development-pathway-for-india-india-global-business/ (Accessed on 21/11/2019).

NPC (National Productivity Council of India) (2019). *Circular economy for productivity and sustainability*. Available at: https://currentaffairs.gktoday.in/national-productivity-week-february-12-18-2019-02201965935.html.

Prachi (2019). *Turning to a 'circular' economic model to combat the climate crisis*. Available at: www.youthkiawaaz.com/2019/09/turning-to-a-circular-economic-model-to-combat-the-climate-crisis/ (Accessed 21/11/2019).

Statista (2020). *India: real gross domestic product (GDP) growth rate from 2009 to 2021*. Available at: https://www.statista.com/statistics/263617/gross-domestic-product-gdp-growth-rate-in-india/

Trak (2017). *Indian telecom stats: 1 billion mobile subscriber base reached, active base cross 902 M*. Available at: http://trak.in/tags/business/2016/01/05/indian-telecom-stats-1billion-mobile-subscriber-base/.

35

PRODUCT-SERVICE SYSTEM BUSINESS MODELS AND CIRCULAR ECONOMY

Miying Yang

Introduction

Most current production and consumption systems are linear, adopting a "take, make and dispose" model and thereby contributing to the depletion of natural resources (WEF, 2014). To achieve sustainable growth, firms need to shift from linear to circular economy models. A number of companies (e.g., Unilever, Toyota, P&G) and third-sector organisations (e.g., the Ellen MacArthur Foundation) are seeking solutions to accelerate scale-up and transition efforts towards more circular operations. This change requires not only product, process, and technological innovation but also a new business model that can improve the circularity of the manufacturing and supply chains through activities of reduce, reuse, remanufacturing, refurbishment, and recycling. Also, circular economy cannot be achieved by a specific firm. It requires collaboration between organisations across the supply chains and other stakeholders from similar and/or diverse sectors. A change in one organisation's business model will affect the business activities of other organisations in its supply chains. Therefore, a systemic approach to managing better utilisation of the materials, energy, and other valuable resources through higher rates of recycling, reuse, and remanufacturing is imperative for success.

Business model innovation for circular economy

Circular economy

A circular economy is defined as *an industrial system that is restorative and regenerative by design* (Ellen MacArthur Foundation, 2015). In this system, the flows of materials, energy, and information are closed loop. The word 'circular' is an antonym of 'linear' and is related to the concept of a cycle, so a circular economy is a cyclical closed-loop system in which environmental protection and economic development are balanced (UNEP, 2006). There are two types of cycles which are particularly important in the circular economy: the biological nutrients cycle, in which they are designed to re-enter the biosphere safely, and the technical nutrients cycle, in which they are designed to circulate without entering the biosphere through end-of-life activities (Ellen MacArthur Foundation, 2012). Thus, the 'Reduce, Reuse, Recycle' (3R) principles are essential to the notion of a circular economy (Murray et al., 2015).

The circular economy has been studied on three levels: the firm, interfirm, and entire industrial levels (Murray et al., 2015). The management of manufacturing and supply chain operations at all three levels plays an important role in moving towards a circular economy ideal. This may be accomplished by the activities of reduce, reuse, recovery, remanufacturing, and recycle within supply chains (Cooper, 1999). UNEP (2006) proposed the general characteristics of circular economy as: low consumption of energy, low emission of pollutants, and high efficiency. Hu et al. (2011) emphasised that the circular economy needs to be resource productive. The Ellen MacArthur Foundation's investigation with a practical emphasis summarises three principles, five key characteristics, and four sources of value creation in a circular economy (Ellen MacArthur Foundation, 2012, 2014, 2015). The four sources of value creation are:

1 *Power of the inner circle*, referring to minimising material usage and reducing cost through the inner circles such as production, reuse, and refurbishment, and then through outer circles such as recycling;
2 *Power of circling long*, aiming to maximise the number of circles as much as possible and prolong product longevity;
3 *Power of cascaded use*, applying 'waste-is-food' logic and suggesting a different use of the used products through symbiosis approaches; and
4 *Power of pure circles*, using an uncontaminated material stream so that the redistribution efficiency and material productivity can be increased.

Circular business model

Business model refers to the logic of how a companies does business (Magretta, 2002; Teece, 2010). It describes how a firm creates and delivers value for its stakeholders within the value network and how it captures value from them (Richardson, 2008; Zott et al., 2011). Osterwalder and Pigneur (2010) developed a business model canvas using nine components to describe a business model: value proposition, customer segment, customer relations, key resources, key activities, partners, channels, cost structure, and revenue streams. Designing and developing business models that are conducive to a circular economy need to identify new sources of value creation in the context of the nine components, for example, turning the waste from production facilities and end-of-life products into revenue streams.

A circular business model is defined by Linder and Williander (2015) as "a business model in which the conceptual logic for value creation is based on utilizing the economic value retained in products after use in the production of new offerings." Some other researchers, such as Loomba and Nakashima (2012), believe that circular flow not only includes products after use but also production waste and by-products. The Ellen MacArthur Foundation (2012) identified four sources of value creation within the circular economy, which implies that business models for circularity could create value from the inner circle, circling longer, cascading use, and the pure circles. To some degree, every business model is both linear and circular (Mentink, 2014; Lewandowski, 2016). Lewandowski (2016) investigated how circularity could be embedded into each of the business model components and proposed a conceptual framework for circular business models. Planing (2014) regarded business model innovation as one of the fundamental building blocks for the transition to the circular economy. Mentink (2014) described the required changes of business model components for developing circular business models and proposed a tool for business model innovation in a circular economy. Laubscher and Marinelli (2014) identified six components of embedding circular economy principles into business models, among which reversed supply-chain logistics is regarded as the most important component of circular business models. A number of methods and tools have been developed to support firm transition towards a more circular operations, such as the business model scan developed by Van Renswoude et al. (2015), 7-P model proposed by Scott (2015), and sustainable value analysis tool by Yang et al. (2017b). In general, the research on circular business models is in its infancy yet continues to attract attention from operations management, innovation, and strategy scholars.

Product-service systems for the circular economy

Product-service systems

Product-service systems (PSS) is a business model through which manufacturers sell integrated products and services rather than products alone (Goedkoop et al., 1999). According to the ratio of products/services, PSS can be classified into different types (Hockerts and Weaver, 2002; Tukker, 2004): (1) product-oriented PSS, when manufacturers sell products while providing related services, such as maintenance and consultancy; (2) use-oriented PSS, when manufacturers sell the utility or accessibility of products without transferring the ownership to customers, such as leasing, renting, and sharing; and (3) result-oriented PSS, when manufacturers retain the ownership of products and sell the results of products, such as selling printed documents rather than printers.

The transit from traditional product-based firms to service orientation is complex and contextual. Bustinza et al. (2013) proposed that this requires a reconsideration of the management of supply and demand chains. Vendrell-Herrero et al. (2015) explored how servitisation affects upstream–downstream firm interdependencies in the context of digital business models and found that digital services could empower upstream and downstream firms differently under certain circumstances.

Many researchers have considered PSS promising sustainable business models because they have the potential to reduce total production and consumption throughout the entire product life cycle (Maxwell and Van Der Vorst, 2003; Yang et al., 2013; Yang and Evans, 2019). A number of firms have witnessed the potential of PSS to bring significant revenue and meanwhile reduce negative environmental impact (Baines et al., 2007; Tukker, 2004; Yang et al., 2017a). Some researchers have investigated how sustainability could be embedded into PSS development (Manzini and Vezzoli, 2003; Geum and Park, 2011; Yang, 2015), and a number of methods and tools have been developed to assist companies in developing sustainable PSS business models or increasing sustainable value creation in PSS, such as Matzen and McAloone (2006) and Yang et al. (2014).

Product-service systems and circularity

The Ellen MacArthur Foundation (2012) put forward the idea that in order to achieve a circular economy, it is essential for manufacturers or retailers to transform to 'functional service models,' in which customers buy services or the use of products rather than the ownership of products. The manufacturers retain the ownership of products and therefore have a greater motivation to extend the cycle of products and increase the reuse and remanufacturing rates of the used products. Some other researchers also regard PSS as a promising form of circular business model. For example, Mentink (2014) implied that selling performance-based services is an approach that moves firms towards greater circularity in their business model.

The Ellen MacArthur Foundation (2015) proposed a RESOLVE framework, in which sharing is regarded as one of the important actions for businesses moving towards a circular economy. The Ellen MacArthur Foundation (2012) took washing machines as an example and calculated the circularity of the washing machines in different business models. The result shows that a leasing model could achieve more cycles than normal models. Yang et al. (2018), proposed that different types of PSS could affect the circularity of supply chains in a different way.

Case Study

From selling gas generators to gases

This chapter presents a large state-owned air separation unit manufacturers in China. The company used to be a manufacturer producing machines which separate air into different gases, such as oxygen, hydrogen. The machines are mainly used as part of engineering equipment in areas such as metallurgy, petrochemical and coal chemical industries, and aero industries. The traditional business model of the company is making and selling air separation units to its customers. Now the company has extended its business to include the sale of integrated engineering equipment as a total product-service solution (e.g., petrochemical equipment) and even further to the sale of industrial gases.

Co-existence of product-service systems business models

In addition to the traditional product-based model, the company has implemented three types of PSS business models.

(1) Product-oriented product-service system: selling products and providing technical services

The company sells the air separation product units and also provides technical services, including installation, maintenance, and repair, as extra market propositions. This is the most common PSS business model operated in the company. It is a product-oriented PSS because the ownership of products is transferred to the customer, and the technical services are included as part of the original sales package.

(2) Use-oriented product-service system: leasing products and providing technical services

The company leases air separation units, or an entire engineering system, to customers under contracts that usually last for a certain number of years. Leasing contracts are mainly tailored to customers with special financial needs (e.g., without the financial capability to buy equipment or build projects). This forms a use-oriented PSS, since the company retains ownership of the products and customers just pay for the use of products and services.

(3) Result-oriented product-service system: selling industrial gases

The company also provides 'industrial gas' rather than 'gas generators.' This is result-oriented PSS because the company owns the gas generators and customers pay for the gases consumed.

The company started the first industrial gas project in 2003, but not until 2010, when the company went public, did it start the second one and make an extensive investment into industrial gas projects. Industrial gas projects have increased rapidly in recent years. As of 2015, more than 38 gas centres have been built in industrial parks in China. The gas centres produce industrial gases such as O_2, N_2, CO_2, and H_2; rare gases (e.g., Ar, He); and special gases to customers in various industrial sectors in industrial parks. There are four ways of providing gases: bottled gas, liquid gas (e.g., cold air separation of liquid), gasification, and pipeline industrial gas supply for industrial parks.

From linear to circular: how product-service systems affect the circularity in the supply chain

In order to understand the impact of these different business models on supply chain operations, the interviewees were asked to describe the supply chains of each business model. Figure 35.1 shows the supply chains architectures of different business models.

The circularity of the supply chain operations of each business model is analysed.

1 Traditional product-based model

As shown in (a) of Figure 35.1, the supply chain of the traditional product-based model mainly includes design, procurement, production, distribution, consumption, and disposal processes. It follows a linear 'take, make and dispose' pattern. The company purchases materials and parts from suppliers, produces the products, and sells them to the customers all over China, who then discard the products when they no longer function. There are no materials and information flowing back to the company. The rate of remanufacturing and reuse is very low. Sometimes whole products are discarded only because small parts are broken.

2 Product-oriented product-service system

In this business model, the company provides technical services for the products during the usage phase. Compared to the supply chain of product-based business model, product-oriented PSS extend the product life through regular maintenance and repair, as shown in (b) of Figure 35.1. Both the company and customers have a high motivation to have technical services because this increases the service revenue for the company and also reduces the cost for customers compared to building their own service teams. The advantage of the manufacturer providing technical services is that the company understands its own products better and can provide more professional services than customers or other third-party companies.

3 Use-oriented product-service system

For use-oriented PSS, the company leases products to the customers and provides technical services. Compared to the previous two business models, the leasing projects have more cycles to reuse, recover, remanufacture, and recycle the used gas generators, as shown in (c) of Figure 35.1. The reason is that the company retains ownership of the products and is incentivised to extend the product longevity and capture value from end-of-life products as much as possible. The longer the customers use a product, the more revenue the company could get from leasing.

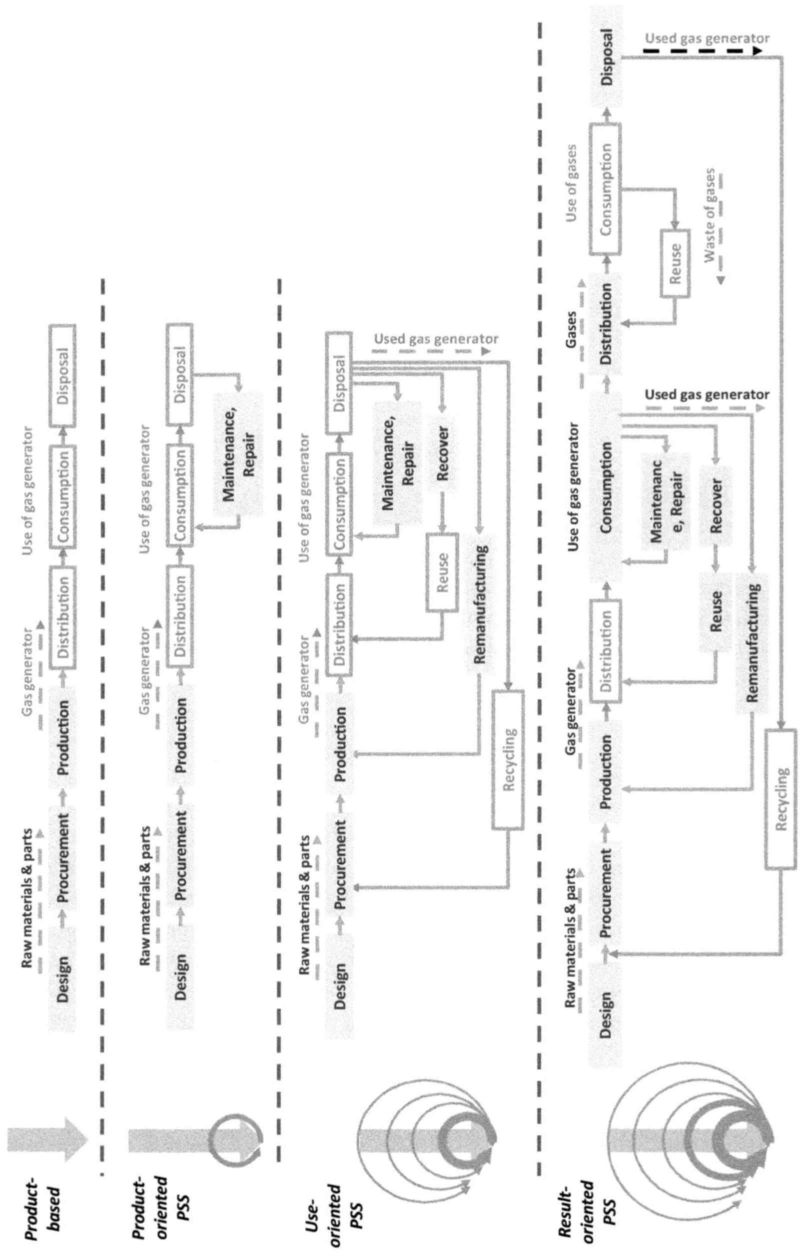

Figure 35.1 Supply chain architectures of different business models

4 *Result-oriented product-service system*

In this business model, the company sells industrial gases rather than gas generators. Instead of making and distributing gas generators to customers, the company built its own gas centres in industrial parks which are close to customers. The company became the actual consumer of gas generators, and the produced gases were distributed to various customers in the industrial parks. As shown in (d) of Figure 35.1, the power of maintenance, repair, reuse, recover, remanufacture, and recycle is much stronger than in the previous three business models, because the company is the actual user and can control the usage phase. In addition, the company created value from by- or co-products (e.g., N_2 is regarded as a by-product of production for O_2). The potential value from gases, which was previously ignored by customers (since customers did not have the expertise to fully use gases), is now captured by the manufacturer in the result-oriented PSS business model. In this way, the manufacturer internalises the potential value of the products' products and is incentivised to maximise value from them. This also applies to co-products. In this business model, the company has greater incentive to increase the use of gases. It has utilised different gases for different customers, coordinated use of gases among customers during peak and off-peak times, and therefore reduced gas emissions. Therefore, in this business model, the company not only establishes an inner circle of reuse of the products (i.e., gas generators) but also the reuse of the products' products (i.e., gases).

The case study shows that the closer to result-oriented PSS, the tighter and more efficient the cycle is, which means the repair, reuse, and remanufacturing system is faster, the rate is higher, and the product has to be changed less to come back to the cycle again. The more efficient the cycles, the higher the potential savings on the material, energy, labour, and operations and the lower the waste of emissions. The result-oriented PSS also extends the reuse of products to the reuse of products' products. The main reason for this is that the manufacturer has ownership and is the actual user of the products, so it can control the usage and end-of-life phases of the products.

Circularity as a source of value creation of product-service systems

Circularity through increasing the inner cycle, cycling longer, cascaded use, and pure circles are regarded as the four main sources of value creation (Ellen MacArthur Foundation, 2015). In order to compare the circularity of different business models, we use these four sources (explained in the subsection on product service systems) as the criteria to analyse the product-based business model and the three types of PSS business models in Figure 35.2.

Figure 35.2 shows that the product-based business model in the company adheres closely to linearity in its supply chain. Each of the PSS business models has inherent circularity to varying degrees. Both the product-oriented PSS and use-oriented PSS predominantly create value through increased use of inner circles and circling longer, whereas the use-oriented PSS has a stronger impact on the two sources. It means that use-oriented PSS has more frequent inner circles and circling longer than product-oriented PSS.

In addition to the inner circle and circling longer, the result-oriented PSS could also create value from cascaded use through a symbiosis approach of co-products and by-products in other value chains, for example, the application of co-produced gases in new fields.

This type of circularity only happens in result-oriented PSS in this case because only in this business model can the manufacturer control the co-products and by-products of its original products and create value through cascaded use of the co-products.

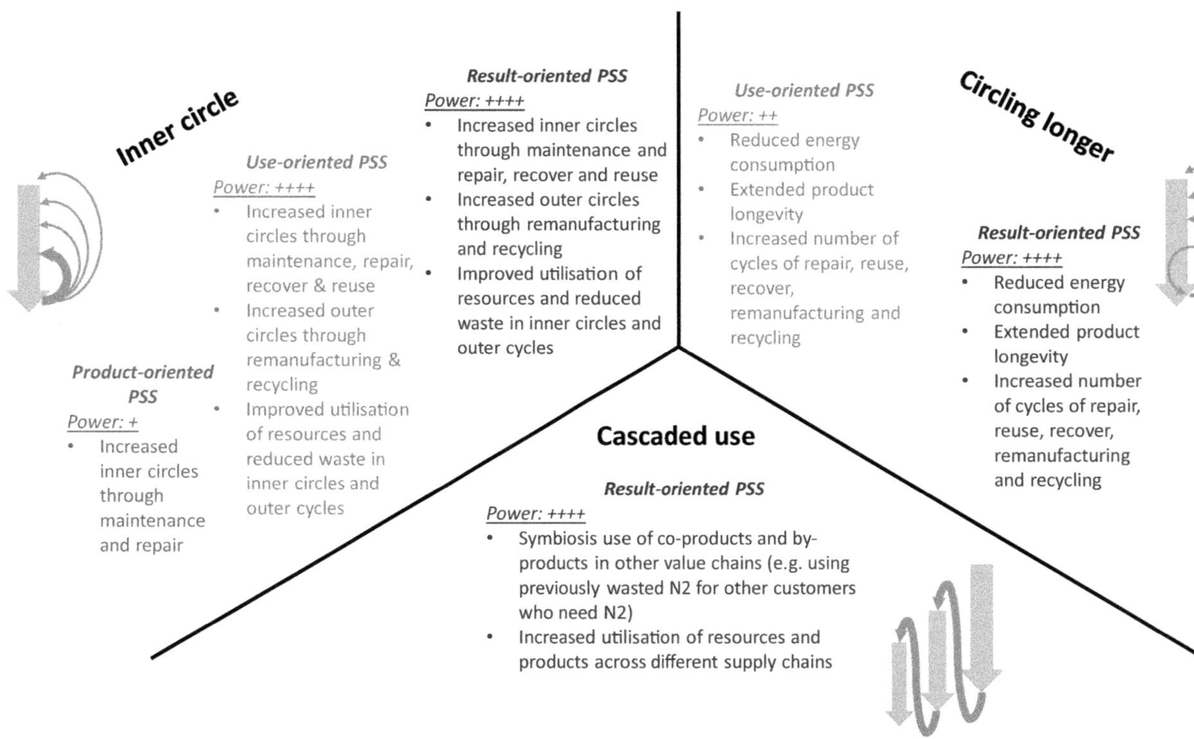

Figure 35.2 Sources of value creation for circularity of different business models in the company

Figure 35.2 also shows that pure circles seem to be irrelevant to the PSS business models in these cases, but the interviewees showed higher motivation to achieve pure reuse of materials under the result-oriented business models.

Underlying the phenomenon, the fundamental reason for the difference of circularity of the business models seems to be related to ownership of products. The company retains the ownership in use-oriented PSS and result-oriented PSS and has strong control of products over the life cycle and therefore is more incentivised to create value from the whole product life cycle (air separation units) and even the products' products (e.g., O_2, H_2, Ar, He, Xe, and Ke). The company therefore has the responsibility and incentive to reduce the environmental impact of the products in use, especially when these affect the economic value. For other PSS business models, these issues are the responsibility of customers; for example, in the product-based model, the company does not have a detailed level of information of the used products because the information is controlled by the customers. Therefore, use-oriented and result-oriented PSS business models are more appropriate for circular supply chain development because of product ownership characteristics.

Discussion and conclusions

In general, the current research in the burgeoning field of the circular economy and business models is not yet mature, as indicated by the lack of agreed-upon concepts and practices. This chapter presented the relationship between different types of PSS business models and the circular economy, in particular how each of type of PSS would affect the key sources of value creation for circularity. The four sources of value creation are highly related to end-of-life strategies, which implies their potential linkage with PSS business models, because PSS extends the opportunities of value creation from the production phase to product usage and end-of-life phases (Yang, 2015).

All three types of PSS could be enablers for circular value creation. Compared to the linear model of 'take, make and dispose,' PSS increases the rate of repair, recover, reuse, remanufacturing, and recycling, which makes the model more circular. Manufacturing companies owning the products motivates them to create economic value from the entire product life cycle, including end of life. From a customer's viewpoint, PSS means shifting from a 'buy and consume' pattern to 'rent, consume and return' or 'consume and return.' From a manufacturer's viewpoint, PSS can get closer to customers, gain continuous revenues, and prolong product lifetime. PSS increases designers' motivation to design for sustainability, which aligns with the 'design out waste' principle of the circular economy.

This chapter concludes the following aspects. First, PSS business models have the potential to trigger and enhance the circularity features of supply chains. Even though some existing literature implies that 'product as a service' and 'leasing rather than selling' could generate a positive effect on the circularity of supply chains (Ellen MacArthur Foundation, 2012, 2015), there was still a lack of solid empirical data to support this assumption, and this case study provides strong empirical evidence to fill in the gap. Second, the empirical analysis demonstrates that different types of PSS business models are associated with distinct circularity signatures within their supply chains. In particular, the use-oriented and result-oriented PSS business models are more appropriate for circular supply chain development because of product ownership matters. Third, the research shows that PSS business models could enable circular value creation through the inner circle, cycling longer, and cascaded use but not pure cycles.

This chapter therefore highlights the potential of business model innovation for the circular economy and also extends the dimensions of previous research on PSS by incorporating supply chain circularity into the analysis. Future research might further explore the mechanisms of using PSS business models for developing circular supply chains and investigate new business models that could enhance circularity through pure cycles. The research in this chapter can be used in practice to facilitate the transition to circular supply chains through developing appropriate PSS business models. The case study can inspire business model innovation for circularity in different industries. The research findings could be embedded into some business model tools, such as business model canvas (Osterwalder and Pigneur, 2010) and sustainable value analysis tool (Yang et al., 2017b) to assist practitioners in designing and analysing circular business models and supply chains.

References

Baines TS et al. (2007). State-of-the-art in product-service systems. *Proceedings of the Institution of Mechanical Engineers Part B Journal of Engineering Manufacture*. 221 (B): 1543–1552. doi:10.1243/09544054JEM858.

Bourlakis M et al. (2014). Examining sustainability performance in the supply chain: the case of the Greek dairy sector. *Industrial Marketing Management*. 43 (1): 56–66.

Bustinza OF, Parry G, Vendrell-Herrero F. (2013). Supply and demand chain management: the effect of adding services to product offerings. *Supply Chain Management*. 18 (6): 618–629.

Cooper T. (1999). Creating an economic infrastructure for sustainable product design. *The Journal of Sustainable Product Design*. 8: 7–17.

Ellen MacArthur Foundation (2012). *Towards a circular economy – economic and business rationale for an accelerated transition, greener management international*. Cowes: Ellen MacArthur Foundation.

Ellen MacArthur Foundation (2014). *Towards the circular economy: accelerating the scale-up across global supply chains*. World Economic Forum. Available at: www3.weforum.org/docs/WEF_ENV_TowardsCircularEconomy_Report_2014.pdf.

Ellen MacArthur Foundation (2015). *Delivering the circular economy: a toolkit for policymakers*. Cowes: Ellen MacArthur Foundation.

Geum Y, Park Y. (2011). Designing the sustainable product-service integration: a product-service blueprint approach. *Journal of Cleaner Production*. 19 (14): 1601–1614.

Goedkoop MJ et al. (1999). *Product service systems, ecological and economic basics, ecological and economic basics*. The Hague, The Netherlands: Ministry of Environment.

Hockerts K, Weaver N. (2002). Towards a theory of sustainable product service systems. In: *INSEAD-CMER research workshop on sustainable product service systems*. Available at: https://scholar.google.com/scholar_lookup?hl=en-US&publication_year=2002&author=K.+Hockerts&author=N.+Weaver&title=INSEAD-CMER+Research+Workshop+on+Sustainable+Product+Service+Systems.

Hu J et al. (2011). Ecological utilization of leather tannery waste with circular economy model. *Journal of Cleaner Production*. 19 (2–3): 221–228.

Laubscher M, Marinelli T. (2014). Integration of circular economy in business. *Proceedings of the Conference: Going Green – Care Innovation*. Vienna, Austria. Available at: https://www.researchgate.net/profile/Thomas_Marinelli/publication/270207909_Integration_of_Circular_Economy_in_Business/links/54a2674e0cf257a63603867e/Integration-of-Circular-Economy-in-Business.pdf.

Lewandowski M. (2016). Designing the business models for circular economy – towards the conceptual framework. *Sustainability*. 1–28.

Linder M, Williander M. (2015). Circular business model innovation: inherent uncertainties. *Business Strategy and the Environment*. 26: 182–196.

Loomba APS, Nakashima K. (2012). Enhancing value in reverse supply chains by sorting before product recovery. *Production Planning & Control*. 23 (2–3): 205–215.

Magretta J. (2002). Why business models matter. *Harvard Business Review*. 80 (5): 86–92.

Manzini E, Vezzoli C. (2003). A strategic design approach to develop sustainable product service systems: examples taken from the 'environmentally friendly innovation' Italian prize. *Journal of Cleaner Production*. 11 (8): 851–857.

Matzen D, McAloone TC. (2006). A tool for conceptualising in PSS development. In: *Design for X, Beiträge zum 17. Symposium*. Erlangen: Friedrich-Alexander-Universität Erlangen-Nürnberg, Lehrstuhl für Konstruktionstechnik. 131–140.

Maxwell D, Van Der Vorst R. (2003). Developing sustainable products and services. *Journal of Cleaner Production*. 11 (8): 883–895.

Mentink B. (2014). *Circular business model innovation: a process framework and a tool for business model innovation in a circular economy*. Delft: University of Technology.

Murray A, Skene K, Haynes K. (2015). The circular economy: an interdisciplinary exploration of the concept and application in a global context. *Journal of Business Ethics*. 140 (3): 369–380.

Osterwalder A, Pigneur Y. (2010). *Business model generation: a handbook for visionaries, game changers, and challengers*. New York: John Wiley & Sons. 1st Ed.

Planing P. (2014). Business model innovation in a circular economy reasons for non-acceptance of circular business models. *Open Journal of Business Model Innovation*: 1–11.

Richardson J. (2008). The business model: an integrative framework for strategy execution. *Strategic Change*. 17: 133–144.

Scott J. (2015). *The sustainable business a practitioner's guide to achieving long-term profitability and competitiveness*. Sheffield: Greenleaf Publishing. 2nd Ed.

Teece DJ. (2010). Business models, business strategy and innovation. *Long Range Planning*. 43 (2–3): 172–194.

Tukker A. (2004). Eight types of product – service system: eight ways to sustainability? experiences from SusProNet. *Business Strategy and the Environment*. 13 (4): 246–260.

UNEP (2006). *Circular economy: an alternative for economic development*. Paris: UNEP.

Van Renswoude K, Wolde A, Joustra D. (2015). *Circular business models. Part 1: an introduction to IMSA's circular business model scan*. Amsterdam: IMSA.

Vendrell-Herrero F et al. (2015). Servitization, digitization and supply chain interdependency. *Industrial Marketing Management*. 60.

WEF (2014). *Towards the circular economy: accelerating the scaleup across global supply chains*. World Economic forum report.

Yang M. (2015). *Sustainable value analysis for product-service systems*. PhD thesis. University of Cambridge, Cambridge.

Yang M, Evans S. (2019). Product-service system business model archetypes and sustainability. *Journal of Cleaner Production*. 220: 1156–1166.

Yang M, Evans S, Vladimirova D, Rana P. (2017a). Value uncaptured perspective for sustainable business model innovation. *Journal of Cleaner Production*. 140: 1794–1804.

Yang M, Rana P, Evans S. (2013). Product service system (PSS) life cycle value analysis for sustainability. In: *The 6th international conference on design and manufacture for sustainable development (ICDMSD)*. Available at: http://me.zju.edu.cn/meenglish/2013/0419/c15706a734459/page.htm.

Yang M, Smart P, Kumar M, Jolly M, Evans S. (2018). Product-service systems business models for circular supply chains. *Production Planning & Control*. 29 (6): 498–508. doi:10.1080/09537287.2018.1449247.

Yang M, Vladimirova D, Evans S. (2017b) Creating and capturing new value through sustainability: the sustainable value analysis tool. *Research-Technology Management*. 60 (3): 30–39.

Yang M, Vladimirova D, Rana P, Evans S. (2014). Sustainable value analysis tool for value creation. *Asian Journal of Management Science and Applications*. 1 (4): 312–332.

Zott C, Amit R, Massa L. (2011). The business model: recent developments and future research. *Journal of Management*. 37 (4): 1019–1042.

36

CIRCULAR BUSINESS MODELS IN SELECTED GEOGRAPHICAL CONTEXTS

An analysis of two cases

Gianmarco Bressanelli, Nicola Saccani, and Marco Perona

Introduction

The circular economy (CE) has been proposed as an answer to the call for decoupling economic growth from resource extraction, waste generation, and climate-change issues, since it promotes production and consumption systems in which products and materials are used multiple times and for multiple cycles (Ghisellini et al., 2016). Besides these environmental concerns, the transition from a traditional, linear economy to a CE aimed at maximising the value of products, materials, and resources constitutes an opportunity for firms to reduce costs, increase revenues, and achieve competitive advantage. In particular, the CE may increase the European gross domestic product as much as 11%, with a net benefit of about €1.8 trillion by 2030 and savings on material costs up to €0.9 trillion (Ellen MacArthur Foundation, 2012).

To gather such benefits, companies need to implement *circular business models* (CBMs), that is, business models that create, deliver, and capture value to improve resource efficiency through contributing to extending the useful life of products and components (e.g., through long-life design, repair, remanufacturing, and so forth) and closing material loops (Nußholz, 2017). However, a CE transition implies several challenges, including economic and financial viability, market and competition, product characteristics, standards and regulation, technology, and user behaviour obstacles (Bressanelli et al., 2019). In this context, and in order to introduce CBMs, companies are called to implement levers such as circular product redesign, servitisation, and supply chain management strategies (Bressanelli et al., 2019). While their role has been addressed by the literature (Bocken et al., 2018; Bressanelli et al., 2019; Pinheiro et al., 2019; Tukker, 2015), enabling factors such as government intervention or user engagement have received limited attention to date, and both academia and practitioners have limited knowledge on how to implement them. This becomes more relevant in light of the geographical dimension, since geographical contexts affect how governments and users enable CE.

This chapter aims to explore and discuss the role of geographical contexts on the enabling factors of circular business models. To this end, we investigated two case companies that have run different CBMs in two different geographical settings, namely the Netherlands and Italy.

Theoretical background: enabling factors of circular business models

Previous literature recognises government intervention and active user engagement as two main enabling factors that companies should exploit in order to implement the CE.

Governments play a relevant role in pushing the CE: for instance, China issued a circular economy law in 2008 based on the "command and control" principle. Europe, the United States, and Japan instead tried to implement actions based on market instruments (Ghisellini et al., 2016). More specifically, governments and public authorities can adopt several actions to push the CE into action (Stahel, 2016). First, legislation and mandatory regulations, as in the case of take-back agreements, polluter pays principles, or extended producer responsibilities, are some of the main enablers. Second, financial incentives such as subsidies, deposit-refund schemes for collecting and separating waste, or tax deductions should be promoted: taxes should be raised on the consumption of non-renewable resources instead of on renewable resources such as labour (Stahel, 2016). Third, green public procurement is another government enabler, which forces public bodies to source goods and services with a reduced environmental impact. Fourth, the adoption of norms and standards such as the ISO 14001 technical standard or eco-labels should be promoted. Fifth, different (environmentally focused) indicators and metrics should be adopted in order to assess the development of countries and companies, not only on the economy but also adopting a broader sustainable perspective.

An active engagement of users in a CE is another main enabling factor. In fact, the CE requires a novel approach where customers are treated as an integral part of enterprises (Lieder and Rashid, 2016). Shifting to a user-centric focus in value proposition is required for building fully circular business models (Heyes et al., 2018). Users may have an active role in pushing the transition towards a CE during the purchase, usage, and end of life (EoL) of products. During purchasing, they can be engaged by choosing greener, more efficient, refurbished, or remanufactured products (Lewandowski, 2016). During the usage phase, users may be involved by lowering resource consumption through better usage habits (Bocken et al., 2018) or by properly choosing the best time to replace energy-consuming products, counterbalancing the impact of continuing to use old products versus the impact of producing a new one that consumes less energy during usage (Ardente and Mathieux, 2014). During EoL, users play a crucial role in how products are discarded, thus facilitating the recovery processes for reuse, remanufacture, or recycling (Parajuly and Wenzel, 2017). Despite these enabling roles, it should be carefully taken into account that many users may not be attracted by servitised CBMs that offer product access instead of ownership (Rizos et al., 2016). In fact, users may fear the loss of a sense of control, availability, self-esteem, or status symbol connected with product ownership (Tukker, 2015). Moreover, since in servitised CBMs, users no longer own products, careless behaviour in product usage may also arise, increasing repair and maintenance needs and generating relational drawbacks such as legal issues (Tukker, 2015). Last, users' willingness to pay is also critical in CBM. For instance, customers usually look for substantial savings when purchasing renovated or 'second-hand' products, even though they are sold 'as good as new' (Baxter et al., 2017). Nevertheless, costs for renovating processes might prevent selling them at such a lower price.

Evaluating the cases

To explore the role of geographical contexts on the enabling factors of CBMs, two case companies in different geographical settings have been investigated. Given the exploratory nature of the research as well as the embryonic stage of development of the CE paradigm, case study

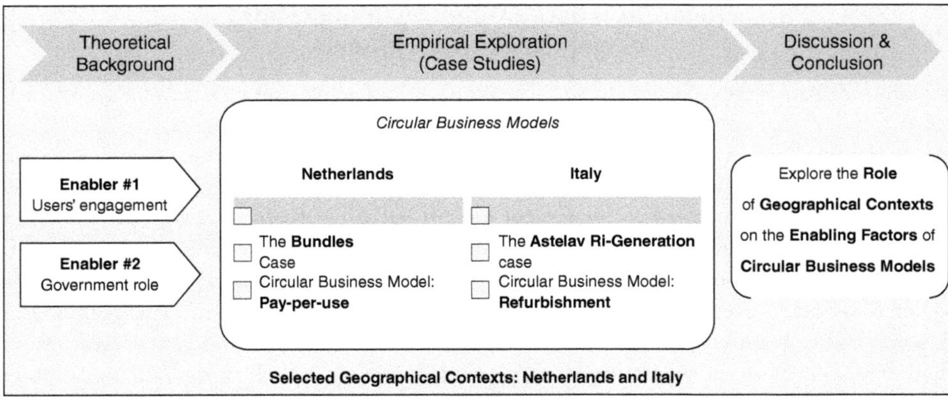

Figure 36.1 The research process

methodology was considered suitable (Yin, 2009). To ensure homogeneity of results, only cases regarding a specific industry have been selected. In particular, the household appliance industry has been chosen, since it is a promising arena for the adoption of the CE paradigm (Bressanelli et al., 2019). The investigated companies are Bundles, which offers household appliances as a service in the Netherlands, and Astelav, which started a refurbishment project of household appliances that have reached the end of life in Italy.

A checklist and a questionnaire were used to collect general data about the companies (turnover, number of employees, etc.). Then, semi-structured interviews were carried out. Each interview lasted one hour, and different company roles were consulted to enhance study reliability. All interviews were transcribed, coded, and validated with the respondents. Moreover, secondary sources such as company documentation, company websites, and publicly available information have been used for triangulation. Data analysis was carried out through the lens of the business model canvas architecture (Osterwalder and Pigneur, 2010), given the wide diffusion and consensus it has reached in academia and industry (Lewandowski, 2016). Results were then formalised under the four main CBM dimensions (Bocken et al., 2014; Lüdeke-Freund et al., 2018): (1) value proposition; (2) value delivery, encompassing channels, customer segments, and customer relationships; (3) value creation, including key partners, key activities, and key resources; and (4) value capture, encompassing revenue streams and cost structure. Figure 36.1 depicts the overall research process.

Case studies

The Netherlands: the Bundles case

Bundles is a technology service provider in the household appliance industry, which designs technologies to offer washing machines, tumble dryers, dishwashers, and coffee machines on a pay-per-use basis. The company was founded in 2014 as a start-up, and to date it employs less than 10 employees, serving about 2,500 active subscriptions in the Netherlands. A subscription model provides long-lasting, premium, and high-efficiency appliances to consumers, increasing resource efficiency as well as the lifetime of everyday equipment. More specifically, Bundles provides household appliances through a servitised CBM: instead of physically selling appliances, the company offers the opportunity to access the appliance service through the subscriptions of a contract. Each month, customers pay either a fixed cost or an amount that depends on the number of usage cycles performed. To deliver its servitised CBM, Bundles supplies appliances equipped with an Internet of Things (IoT) tool that connects products to the internet, thus transforming usually stand-alone appliances into smart and connected ones. The IoT technology enables product monitoring and data collection. These data are used to provide personal advice and hints (through a mobile app) on how to reduce consumable costs such as energy, water, detergent, and so forth. Given the servitised CBM employed, Bundles retains the ownership of appliances and collects them when users end the subscriptions. Collected appliances are checked, cleaned, and repaired before starting a new subscription cycle with a new user. Users play an active role in enabling this CBM, since they are called to choose a subscription model instead of owning their appliances.

In the following, the servitised CBM adopted by Bundles is analysed, through the lens of the business model canvas methodology. The result is presented in Figure 36.2.

Regarding value proposition, the starting point for CBM characterisation is the definition of the value for the customer. Bundles generates customer value by reducing the initial investment required to access household appliance performance: instead of immediately paying the initial household appliance price, costs are spread over the entire duration of the subscription. Moreover, the company provides high-efficiency appliances with lower energy, water, and detergent consumption, thus leading to lower costs per usage cycle. Bundles is also able to co-create services *with* customers. In fact, thanks to the IoT kit, Bundles collects data generated by the appliances during usage (such as usage parameters, energy consumed, time of usage, etc.) and uses them to provide personal advice, hints, and targeted marketing adverts. More specifically, the company constantly monitors a set of parameters such as the consumption of electricity, water, and detergent; the washing load; and the number of cycles performed. The analysis of these data allows Bundles to improve its capability to support customers in order to minimise the consumption of resources. Based on each specific user's habits, Bundles provides personal advice (e.g., suggesting the best washing cycle duration to reduce energy and water consumption, the best loading of appliances, the most appropriate usage conditions, etc.) to optimise the usage phase. Thus, users are incentivised to reduce their laundry environmental impact, since Bundles awards customers who follow this tailored advice with a reduction of the washing fee. In this regard, the collection and analysis of usage data allows Bundles to offer advanced services such as preventive maintenance, product technical and software upgrades, free repair, and technical support. Bundles also offers free appliance software upgrades. In the case of washing machines, for instance, Bundles can remotely upgrade the software that controls the appliance's usage of consumables (energy, water, detergent), incorporating the most up-to-date laundry programs and thus achieving an overall lower resource consumption. All these services are included in the subscription fee.

Regarding value delivery, Bundles principally targets customers with an "access over ownership" mindset. Bundles also pursues a social aim, since households that generally cannot afford high-quality appliances – given their high initial price – can now access the required services. To reach these customer segments, Bundles reinforces customer awareness of the concepts of sustainability and circular economy. For instance, on Bundle's website, a strong emphasis is put on the environmental savings that can be reached through the subscriptions. Last, a closer relationship (compared to traditional, sales-based business models) between Bundles and its customers is vital to enable the mutual creation of value. To this end, the IoT kit enhances continuous information sharing from customers to Bundles. Gathering and analysing information from appliances allows Bundles to gain insights about how appliances are used. Such insights are then used in order to improve the customer experience.

Regarding value creation, digital technologies are one of the key resources for the company, since: (1) the IoT kit enables the constant monitoring of the installed base usage, condition, and

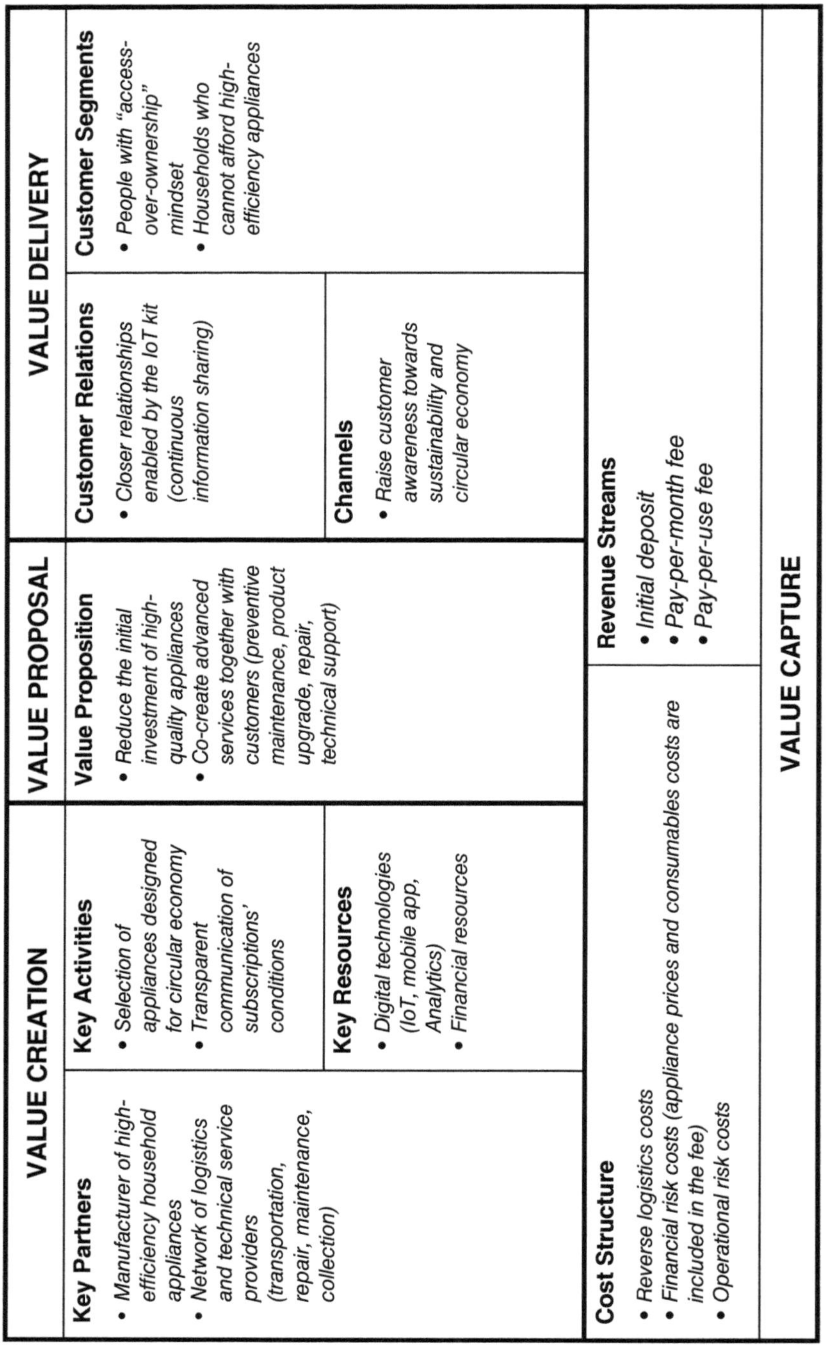

Figure 36.2 Bundles' circular business model

location; (2) the mobile app is needed to offer closer communication with customers; and (3) analytics are fundamental to the provision of advanced services such as ad-hoc advice and hints on optimal appliance usage, preventive maintenance, and so forth. Since Bundles retains ownership of the appliances, the payback period of this servitised CMB is usually longer than that of a traditional sales-based one. Thus, financial resources are another critical element of the analysed CBM. In this regard, the company counts on warranty revenue from the initial deposit that new users must pay when they start an appliance subscription. Moreover, the company constantly monitors and seeks funds from alternative financial solutions such as crowdfunding. In servitised CBMs, companies are usually responsible for all the product life-cycle costs. Thus, they generally have a powerful incentive to design products that are easy to maintain, upgrade, and reuse. Then, design activities are usually key activities in servitised CBMs. However, given its role as a service provider, Bundles does not affect the design phase of the appliances but offers household appliances manufactured by other companies. Therefore, the key activity becomes the selection of appliances that have been appropriately designed for CE, that is, that are high-efficiency, long-lasting appliances that are easy to maintain and to disassemble of at the end of life. Moreover, another key activity is related to the transparent communication of the conditions of subscriptions: through its website, Bundles provides information regarding appliance operational costs, transparency, flexibility of subscriptions, and so forth. Last, and as explained previously, the success of Bundles' CBM relies on the availability of high-efficiency appliances that have been designed for the circular economy. Since the company does not design and manufacture appliances internally, a partnership with Miele, a high-quality household appliance manufacturer, has been created. Moreover, a network to perform in-field support (including transportation, repair, etc.) is a prerequisite for a successful delivery of the company's offering. Bundles has a partnership with other companies to perform transportation, repair, maintenance, and collection when the subscription ends.

Regarding value capture, a strong emphasis is put on the cost structure: when the solution (e.g., clean laundry for washing machine and laundry appliances) is sold rather than product ownership, overheads and new costs become relevant. Moreover, consumable costs (e.g., detergent for washing machines) are included in the fee. For instance, reverse logistics costs increase. Moreover, the servitised CBM exposes Bundles to both financial and operational risks (Bressanelli et al., 2019), generating related costs. On the other hand, Bundles' revenues are generated over time and through different ways. Besides the initial deposits, cashflows are generated based on the duration of the subscriptions (pay-per-month) or on how often the appliance is used (pay-per-use). The fee for a classic Miele washing machine is €20.95 per month or €12.95 per month + €0.40 per wash.

Italy: the Astelav case

Astelav is an Italian distributor of spare parts for household appliances (washing machines, refrigerators, ovens, etc.), which supplies customers in more than 70 countries all over the world. It offers technical assistance and spare parts to more than 2,000 customers (which are retailers, distributors, and wholesalers). Thanks to its technical experience in components and spare parts of several brands, in 2017, the company started a project aimed to refurbish appliances that have become waste electrical and electronic equipment (WEEE). The goal is to reintroduce old appliances into a new usage cycle instead of disposing of them. Moreover, this project contributes to both the environment and society: through refurbishment, less waste is generated and fewer raw materials are needed to produce household appliances; on the other hand, Astelav refurbishment activities employs people experienced in repair and services activities who were out of work due to the 2007 economic crisis, as well as marginalised people. In its first year of activity, the company has renovated more than 1,000 washing machines, with a total savings for the environment of about 23 tonnes of concrete, 23 tonnes of steel, 8 tonnes of wire, 11 tonnes of bulk plastics, and 3 tonnes of aluminium. Users play an active role in enabling this CBM, since they are called to buy a refurbished appliance instead of buying a new one.

In the following, the servitised CBM adopted by Astelav is analysed through the lens of business model canvas methodology. The result is presented in Figure 36.3.

Regarding value proposition, Astelav (through the refurbishment project) generates customer value by reducing the price of household appliances through the offering of refurbished products. In fact, refurbished appliances are usually offered at a reduced price, compared to the price of brand-new products. Moreover, the refurbishment project specifically targets the environment and society: customers who buy a refurbished product also contribute to the environment (since less waste is produced) and to society (since Astelav employs both marginalised people and workers who were out of work due to the economic crisis).

Regarding value delivery, Astelav principally targets customer segments that include low-income households and temporary residents, as well as second homes, since it offers household appliances which have a very competitive price (even though they may have higher costs due to resource consumption, since these appliances are becoming old and less efficient compared to brand-new ones). To be allowed to undertake WEEE refurbishment activity in Italy, Astelav has the authorisation to process WEEE within the R2 Italian category (which includes washing machines, dishwashers, dryers, ovens, and so forth). Customers are reached via both physical and digital channels. In fact, products refurbished by Astelav can be bought online, through the company's website, or can be purchased directly in two stores in Turin (Italy): the project has started with one refurbishment facility (that is used also as a store) close to the company headquarters, and the company has recently opened a second refurbishment laboratory and shop, with the double aim to sell renovated appliances and to train people in repair and remanufacturing activities. It is important to stress that household appliance users are seen by the company as *both customers and suppliers*. In fact, some WEEE that is going to underpin refurbishment come from donations made by private consumers through the Astelav website or through collaboration with voluntary organisations. Last, customer relations are usually limited in time, since they end at the point of sale (one-shot relations). However, Astelav provides a one-year warranty on the refurbished products. Warranty repairs are carried out by Astelav-affiliated technicians.

Regarding value creation, Astelav technical experience with the components and spare parts of several brands is one of the key resources for the company. In fact, the refurbishment project is seen by the company as the opportunity to widen Astelav activity, matching the technical competences with the wide availability of spare parts. Since Astelav's core business is providing spare parts to technical assistance centres, the refurbishment project can count on a wide set of spare parts accessible at a competitive cost. WEEE sourcing and the refurbishment process are the key activities of Astelav. Regarding the WEEE sourcing process, WEEE comes from different sources: some products come from commercial agreements with logistics companies that deliver new household appliances and withdraw used products; some products are collected directly from waste collection centres, while others come from donations made by private consumers through the Astelav website or through the collaboration of voluntary organisations. Old appliances are also collected in the case of the purchase of Astelav refurbished products. The Astelav refurbishment process involves several phases. To be refurbished, an appliance should meet some basic conditions (i.e., to be under 10 years old, to not show evident rust signs, to have been in operation until recently, to have a good level of energy efficiency). First, a product test phase is run in order to check appliance operation. In this phase, spare part needs are also identified. Spare parts are sourced directly from the Astelav warehouse. Then, the actual refurbishment process is carried out: in this phase, damaged components are substituted, and only original parts are used. The last step is sanitisation, granting the utmost hygiene through cleaning cycles with sanitising products.

VALUE CREATION

Key Partners

- *WEEE collection centers*
- *Logistics and transportation service providers*
- *Voluntary organizations (that donates WEEE or provide workers)*
- *Original equipment manufacturers for spare parts*
- *Astelav-affiliated technicians*

Key Activities

- *WEEE sourcing*
- *Refurbishment process*

Key Resources

- *Company technical experience in spare-parts management*
- *Availability of spare parts*

VALUE PROPOSAL

Value Proposition

- *Reduce the price of household appliances through the offering of refurbished products*
- *Provide household appliances that generate benefits to the environment and to the society*

VALUE DELIVERY

Customer Relations

- *Limited, one-shot relations (during purchasing)*
- *One-year warranty that includes technical support*

Channels

- *Both physical and digital channels (stores and company website)*
- *Customers = Suppliers*

Customer Segments

- *Low-income households*
- *Temporary residents*
- *Second homes*

Cost Structure

- *Acquisition costs (WEEE price)*
- *Labor costs*
- *Material (spare parts) costs*

Revenue Streams

- *Price of refurbished appliances*

VALUE CAPTURE

Figure 36.3 Astelav's circular business model

Finally, the product is equipped with the user's manual and sold just like a new product. To run this CBM, Astelav counts on a large set of partners, acting in a wider ecosystem. To source WEEE, commercial agreements have been made with logistics companies that deliver new products (and collect used household appliances), as well as with waste collection centres. Astelav also established a partnership with SERMIG (Missionary Service of Young People), an association of volunteers, to develop a project focusing on refurbishing WEEE and offering job opportunities to people in difficult situations. Thus, users can donate their WEEE directly to SERMIG, and also a social purpose is pursued, since people in marginal situations are involved by offering them the chance of starting a professional course. Spare parts used during the refurbishment process are sourced directly from original equipment manufacturers, and this is possible because Astelav currently works with over 20 appliance manufacturer brands.

Regarding value capture, the main Astelav cost component is related to the acquisition costs for acquiring WEEE in good condition (which will underpin the refurbishment process). In fact, the higher the quality and the condition of the WEEE, the higher the ease (and the profitability) of the refurbishment process. On the other hand, Astelav revenues are generated directly at the point of sale by means of the price of refurbished appliances (which is lower than that of brand-new products).

Discussion and conclusions

This chapter aimed at investigating the role of geographical contexts on the enabling factors of CBMs by means of two case studies of companies that have run different CBMs in two different geographical settings: the Netherlands and Italy.

In both cases, users played a major role in enabling the CE. In the first case, users are called to choose a subscription model instead of owning their appliances, while, in the latter, they are called to buy a refurbished appliance instead of a brand-new one. Thus, cases were very different in terms of users' behaviour and type of CBM implemented. In the Netherlands, Bundles introduced a servitised CBM, specifically targeting people with an "access-over-ownership" mindset. In Italy, on the contrary, Astelav focused on a refurbished CBM targeting low-income households, temporary residents, and second homes, that is, customers who still want to buy and own a household appliance. The decision to introduce such very different CBMs can be explained by the geographical context in which the two cases are located: Northern European citizens are less reluctant to adopt servitised CBMs, especially if coupled with environmental benefits (Lieder et al., 2018). In fact, to the best of authors' knowledge, no attempts to introduce similar CBMs have been found in Italy. Consequently, the channels used to reach customers are also quite different in the two cases. Astelav employs both physical and digital channels, but the delivery of value is focused on *efficiency*, to reduce costs. On the other hand, Bundles focuses more on digital channels, especially to raise customers' awareness of the concepts of sustainability and the circular economy. Last, different effects of the geographical contexts can also be found in the different approaches to customers relations (continuous versus one-shot). The Bundles servitised CBM requires a continuous relation between the company and its customers, grounded on the constant monitoring and data sharing during usage (through the IoT). This CBM thus requires a different attitude by users, who must accept such prerequisite if they want to achieve environmental savings. On the contrary, customer relations in the Astelav CBM are principally one shot and limited in time. Again, these very different relations can be explained by the different geographical settings in which the cases are located.

Even though government and legislation are usually considered in literature as enabling factors for the CE, the Italian case showed that government intervention can unfortunately also act as a barrier. In fact, the process carried out by Astelav to obtain the required authorisations to manage WEEE was long and difficult, since Italy is still lagging regarding the adoption of the

European Directives on the end of waste. On the other side, no mentions in this regard emerged during Bundles interviews. This aspect emphasises the critical role that national government plays in enabling (or unintentionally delaying) the introduction of a CBM.

Overall, this study shed some light on the role of geographical contexts on the enabling factors of CBMs. However, this research has some limitations, too. The two case studies have been chosen for their appropriateness rather than for representativeness. Thus, the external validity of the findings cannot be ensured. Exploring how the geographical context affects the enabling factors of CBMs across other geographies and other industries will further increase knowledge on the CE, therefore leading to relevant managerial implications.

References

Ardente F, Mathieux F. (2014). Environmental assessment of the durability of energy-using products: method and application. *Journal of Cleaner Production*. 74: 62–73.

Baxter W, Aurisicchio M, Childs P. (2017). Contaminated interaction: another barrier to circular material flows. *Journal of Industrial Ecology*. 21 (3): 507–516.

Bocken NMP, Mugge R, Bom CA, Lemstra HJ. (2018). Pay-per-use business models as a driver for sustainable consumption: evidence from the case of homie. *Journal of Cleaner Production*. 198: 498–510.

Bocken NMP, Short SW, Rana P, Evans S. (2014). A literature and practice review to develop sustainable business model archetypes. *Journal of Cleaner Production*. 65: 42–56.

Bressanelli G, Perona M, Saccani N. (2019). Challenges in supply chain redesign for the circular economy: a literature review and a multiple case study. *International Journal of Production Research*. 57 (23): 7395–7422.

Ellen MacArthur Foundation (2012). *Towards a circular economy – economic and business rationale for an accelerated transition*. Available at: www.ellenmacarthurfoundation.org/publications.

Ghisellini P, Cialani C, Ulgiati S. (2016). A review on circular economy: the expected transition to a balanced interplay of environmental and economic systems. *Journal of Cleaner Production*. 114: 11–32.

Heyes G, Sharmina M, Mendoza JMF, Gallego-Schmid A, Azapagic A. (2018). Developing and implementing circular economy business models in service-oriented technology companies. *Journal of Cleaner Production*. 177: 621–632.

Lewandowski M. (2016). Designing the business models for circular economy – towards the conceptual framework. *Sustainability*. 8 (1): 43.

Lieder M, Asif FM, Rashid A, Mihelič A, Kotnik S. (2018). A conjoint analysis of circular economy value propositions for consumers: using 'washing machines in Stockholm' as a case study. *Journal of Cleaner Production*. 172: 264–273.

Lieder M, Rashid A. (2016). Towards circular economy implementation: a comprehensive review in context of manufacturing industry. *Journal of Cleaner Production*. 115: 36–51.

Lüdeke-Freund F, Gold S, Bocken NMP. (2018). A review and typology of circular economy business model patterns. *Journal of Industrial Ecology*. 23 (1): 36–61.

Nußholz J. (2017). Circular business models: defining a concept and framing an emerging research field. *Sustainability*. 9 (10): 1810.

Osterwalder A, Pigneur Y. (2010). *Business model generation: a handbook for visionaries, game changers, and challengers*. New York: Wiley. doi:10.1523/JNEUROSCI.0307-10.2010.

Parajuly K, Wenzel H. (2017). Product family approach in e-waste management: a conceptual framework for circular economy. *Sustainability*. 9 (5): 768.

Pinheiro MAP, Seles BMRP, De Camargo Fiorini P, Jugend D, Lopes de Sousa Jabbour AB, da Silva HMR, Latan H. (2019). The role of new product development in underpinning the circular economy. *Management Decision*. 57 (4): 840–862.

Rizos V et al. (2016). Implementation of circular economy business models by small and medium-sized enterprises (SMEs): barriers and enablers. *Sustainability*. 8 (11): 1212.

Stahel WR. (2016). The circular economy. *Nature*. 531 (7595): 435–438.

Tukker A. (2015). Product services for a resource-efficient and circular economy – a review. *Journal of Cleaner Production*. 97: 76–91.

Yin RK. (2009). *Case study research: design and methods*. Sage. 4th Ed. Available at: https://doi.org/10.1097/FCH.0b013e31822dda9e.

37

IMPLEMENTING LOW-CARBON STRATEGIES IN SMALL AND MEDIUM-SIZED ENTERPRISES

Auditing strategies

Johannes Fresner, Fabio Morea, Christina Krenn, and Fabio Tomasi

Introduction to energy efficiency and low-carbon energy supply

The European Energy Efficiency Directive requests the application of high-quality and cost-effective energy audits in enterprises across Europe. An existing potential of up to 20% of energy savings shall be tapped to follow the strategy of smart, sustainable, and inclusive growth and to promote the use of renewable energy. In Roadmap 2050, the European Commission presents paths to a decarbonised European industry (EC, 2019), also mentioning energy efficiency and renewable energy.

Background

The potential for energy efficiency in small and medium-sized enterprises

Energy efficiency is one of the dedicated objectives in the European strategy for smart, sustainable, and inclusive growth (EC, 2006). The industrial sector has an important role in energy efficiency. To improve industrial energy efficiency, Article 8 of the Energy Efficiency Directive (EED) requires member states to promote top quality, price-effective energy audits and energy management systems to all final customers (KfW, 2013).

Since SMEs represent a big share of energy consumption in Europe, addressing them is a priority in implementing the general objective. There are more than 23 million SMEs in the EU, which represent 99% of European undertakings. They employ more than 100 million employees. SMEs generate 60% of European GDP (EC, 2003). At the same time, 60–70% of the environmental impact relates to SMEs, as reported by the European Commission. Several sources give a potential of 10–25% potential savings of energy consumption in SMEs (Gerber, 2011; Thollander and Palm, 2012). Experts estimate that about 40% of this potential can be harvested by measures requiring no capital investment (DOE, 2015). A wide range of cost-effective energy efficiency measures is available (Thollander et al., 2013), but they remain unimplemented due to several energy efficiency barriers, as described in literature by Cagno et al. (2010).

Over two-thirds of efficiency potential is still untouched (OECD, 2017). In 2017, there was global spending of about EUR 200 billion on energy efficiency. This was invested largely on heating, cooling, and lighting improvements in buildings. Tapping the full potential of energy efficiency will require at least four times as much investment than is spent at present. The International Energy Agency warns these investments are actually slowing down, partly due to the slow implementation of energy efficiency policies.

Low-carbon strategies

When implementing measures to increase resource efficiency, the onion-skin principle (see Figure 37.1) depicts the possible measures, with the ones with the highest leverage effect regarding

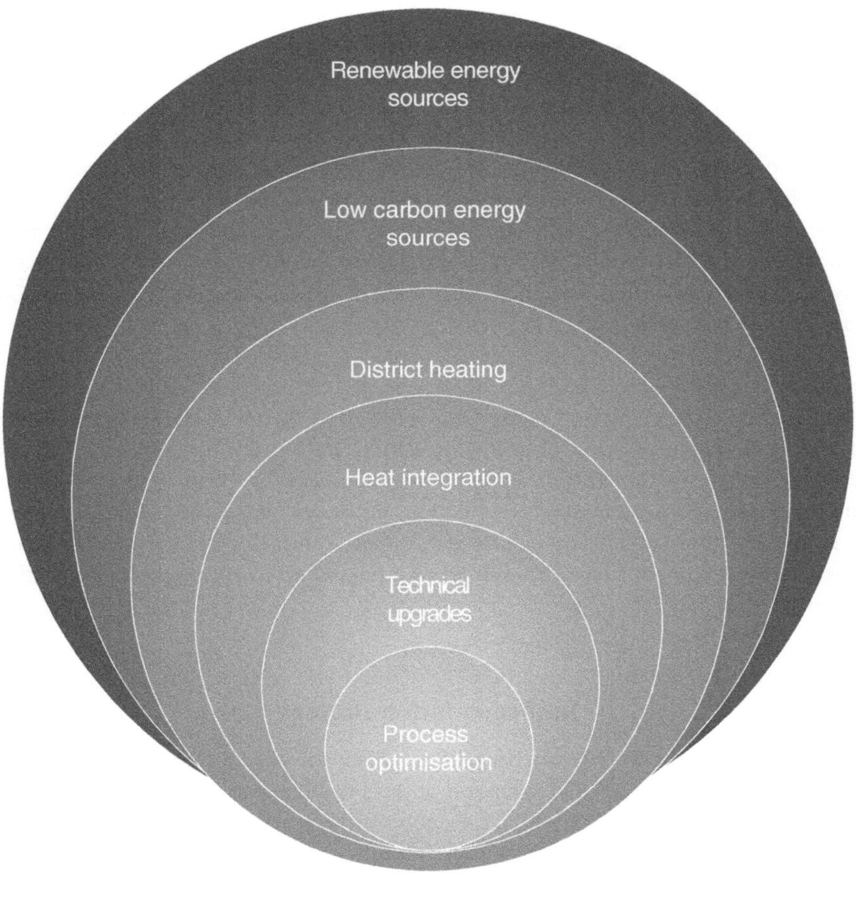

Figure 37.1 Onion-skin principle – optimisation measures

cost effectiveness in the centre. The onion-skin principle can help prioritise optimisation measures.

The onion-skin principle ranks process optimisation as the highest priority, followed by technical upgrades of process equipment, use of waste heat, and switching to low-carbon fuel and electricity and renewable energy:

- Priority 1 – Process optimisation: The first priority is to implement all measures that lead to a direct reduction of energy consumption without major investments. All processes are evaluated in terms of energy efficiency. To optimise individual process steps, one records the most important energy inputs and outputs and improves the process step efficiency;
- Priority 2 – Technical optimisation: The second priority is to evaluate whether retrofitting or replacement of technical systems economically pays due to the resource-saving potential;
- Priority 3 – Waste heat recovery: As a third priority and once all of the aforementioned potentials have been exhausted, investment in recovery and waste heat recovery will bring long-term economic benefits;
- Priority 4 – District heating: In case there are other enterprises or buildings requiring heating nearby, waste heat can be shared with them for heating;
- Priority 5 – Low-carbon energy sources: Possible fuels and electricity should be provided by low-carbon sources: Natural gas is less carbon intensive than coal and liquid fuels, and electricity from hydroelectric power plants is less carbon intensive than electricity from coal-fired power plants. Waste can be used to generate energy, either organic waste in anaerobic digesters or waste in waste incineration plants; and
- Priority 6 – Renewable energy sources: Finally, the remaining energy needs should be supplied from renewable sources, considering their potential environmental impacts. Photovoltaic panels, solar thermal panels, windmills, hydroelectric power stations, and biofuels may provide renewable energy sources.

Organisational measures to avoid unnecessary energy consumption, as well as choosing the most appropriate technology and equipment, contribute to reducing energy consumption without jeopardising production or product quality. The impact of individuals' behaviour is of great importance for the efficient use of energy. Equipment for monitoring and controlling energy flows, room temperature regulation, automatic shutdown of lights when there are no people in rooms or corridors, automatic burner control system on boilers or furnaces, and regeneration of "waste" heat from different processes will also support energy efficiency significantly.

Implementation strategy

Figure 37.2 shows a step-by-step procedure to minimise the overall carbon emissions from the energy supply to an enterprise. The adoption of this procedure will help to prepare for energy reporting, the definition of an action plan to improve energy performance and to prepare for the implications of the climate change strategy. It is fully consistent with an energy management system according to ISO 50001.

Table 37.1 describes the steps of this approach in more detail.

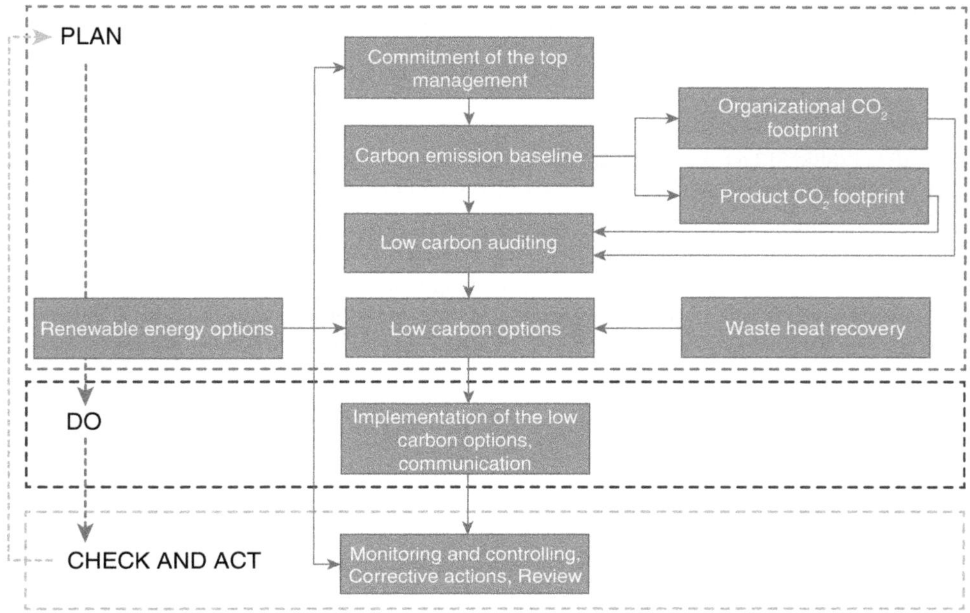

Figure 37.2 Method to implement a low-carbon production system in an industry

Table 37.1 Description of the steps to implement a low-carbon production system

Step	*Objective*	*Activity*
1	Secure management commitment.	Define a policy, define resources, establish the organisational structure for the implementation of low-carbon production.
2	Baseline: Understand your energy use.	Study your operation and identify the major areas of energy consumption. Check age, maintenance, and time of operation of equipment and monitor the power consumption over, for example, one week including a weekend to obtain a base figure against which energy efficiency improvements can be measured. Describe the main consumers, identify drivers of consumption.
2	Audit your organisation: Identify opportunities.	Compile an energy checklist. Walk round your building and complete the checklist at different times of day (including after hours) to identify where energy savings can be made.
3	Action plan: Prioritise actions.	Compile an action plan with improvements that need to be made and when, together with who will be responsible for them.
4	Make the changes and measure the savings.	Implement your energy-saving actions and measure against original consumption figures. This will assist future management decisions regarding your energy priorities.
5	Continue to manage your business for energy efficiency.	Enforce policies, systems, and procedures to ensure that your business operates efficiently and that savings are maintained in the future.

Experiences in implementing low-carbon strategies

Energy audits in 280 enterprises

The PINE project (Promoting Industrial Energy Efficiency in SMEs in the manufacturing sector, IEE/11/885/SI2.615936) was supported by the Intelligent Energy Europe program. An SME is defined by an annual turnover of less than 50 million EUR and less than 250 employees. PINE was conducted between 2012 and 2015 (Fresner et al., 2017).

The main purpose of the project was triggering energy efficiency and use of renewable energy in industrial SMEs with the aim of achieving benefits not just for the environment but also as a way to promote both technological and process innovation and improve efficiency in general. The project partnership included 15 organisations from seven European countries: Austria, Bulgaria, Cyprus, Italy, Romania, Slovakia, and Spain. In each country, the chambers of commerce promoted the project, invited and selected companies for the audits, and did (part of) the preliminary audits.

After an initial awareness-raising campaign, 40 enterprises in each European country participating in the project were selected by a public call for the preliminary auditing phase. After the scouting phase, 140 companies were selected to benefit from a full energy audit provided by the PINE technical partners. At the end of the audit, a technical report, drafted by the auditor, was delivered to each company, including various recommendations.

The auditing procedure is based on modelling the energy consumption of an enterprise. The model-based approach also provides the basis for a modular stepwise extension or detailing of the audit in interesting areas and for autonomous continuous improvement and innovation. This approach diverges from conventional approaches using manuals or guidebooks describing encyclopaedic options for energy efficiency in specific sectors or for different types of equipment. The approach helps to capture the expenses for energy, understand the structure of energy consumption, relate energy consumption to unit operations, and thus provide a basis for learning about the drivers of energy consumption, sources of losses, and opportunities for improvement.

A guidebook and auditing tools were developed to provide auditors with a practical and focused toolkit for the implementation of energy audits. These tools are available in seven languages. The method is compatible with the energy audit standard ISO 16247 for processes and buildings. Consumption was derived following the priorities of Table 37.2.

The unit process categorisation is a way to divide the energy use of an industry into smaller parts or units (Thollander and Palm, 2012). A unit process is based on the purpose of a given industrial process, for example, mixing materials, cooling or drying products, carrying goods, and so on. Unit operations are to be considered the smallest parts of an industrial production system and its related energy use. The unit processes are thus general across industrial companies, allowing comparison of specific processes between industrial companies regarding, for example, energy efficiency. Unit processes include core production units as well as utilities.

Table 37.2 Priorities to identify consumption

Priority	*Data*
1	Submeter for a unit
2	Measurement for a certain period, extrapolation
3	Calculation from nominal load, load factor, and operating hours; crosscheck with existing measurements, load curves, and eventually regression analysis

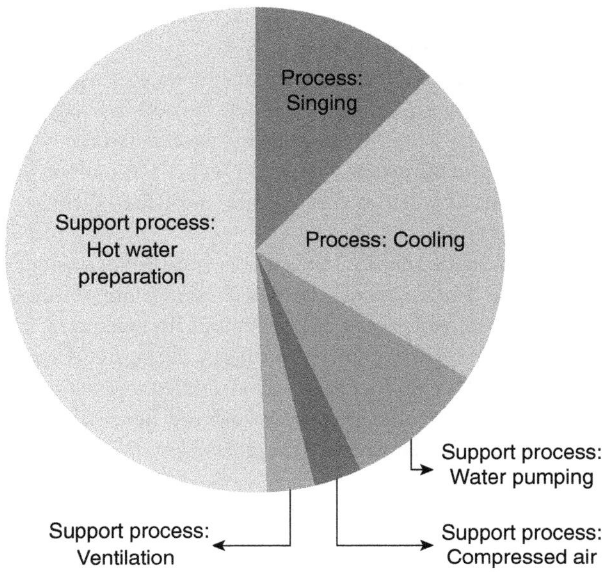

Figure 37.3 Example of energy consumption breakdown per unit operation in a slaughterhouse

According to this definition, a "unit operation" can consist of several energy-consuming pieces of equipment (e.g., the unit "spray drying" will consist of a pump, a heat exchanger, a blower, the dryer, and a product cooler).

For unit operations, basic energy balances can be derived to compare input (electricity, heat, and where applicable heat from chemical reactions) to output (energy in products, heat) to assess efficiency and calculate losses (exhaust gas, hot water, radiation, etc.). Figure 37.3 shows an example of energy breakdown.

The audit tool comprises models and assumptions for electric drives, air compressors, pumps and fans, boilers, insulation, and buildings. For these units, a set of questions is provided to guide in the identification of improvement options. From the answers, the consumption during idling, utilisation, specific consumption, and waste heat is calculated and evaluated. The respective suggestions are formulated. Using rules of thumb, savings are calculated or related to potential savings from experiences with other energy audits.

Parts of this tool were developed from modules of the Austrian initiative Klima:aktiv (www.klimaaktiv.at) and adapted with permission of the Austrian Energy Agency.

When a comprehensive energy monitoring system was not available (as is often the case in SMEs), analysing load curves showing daily or hourly consumption as a basis (e.g., during weekends, nights, or holidays) was employed to identify base loads. Linear regression on data available from general meters (e.g., monthly readings of electricity and gas meters) can provide further insights on the influence of production and external temperature on energy consumption.

Results of the energy audits

The companies came from a variety of sectors, including mining, minerals processing, metal processing, engineering, wood processing, food processing, packaging, plastic processing, tyre manufacturing, and construction. In several countries, some sectors were represented more strongly (metal processing in Austria, Bulgaria, Italy, and Slovakia; food processing and packing in Spain and Cyprus, construction in Romania).

The sample (Table 37.3) had an average primary energy consumption of 724 toe per year. Apart from Slovakia, where 100% of the audited SMEs had an energy manager, in the other countries, 25% to 35% of companies had a person specifically in charge of energy. Apart from Austria and Bulgaria, where 15% and 10% of companies, respectively, had audit experience, the audit was the first energy audit for all other firms in the other countries.

In total, these detailed audits resulted in almost 500 improvement suggestions which were selected for implementation. The calculated technical savings are about 60,000 MWh. The energy efficiency measures will involve over 10,700,000 EUR investment in sustainable energy within three years to achieve 6.500 toe primary energy savings and reduce greenhouse gas (GHG) emissions by 13,500 t CO_2. The project identified annual savings of more than EUR 3 million. Table 37.4 shows the monetary savings and the percentage of energy reductions which were reached by the implementation of the improvement suggestions. On average, the realised savings represent about 5.0% of the energy demand of the SMEs. The bigger the consumption, the more options were identified.

Table 37.3 SMEs included in the auditing program

Country	*SMEs*	*Energy consumption – fuel [MWh]*	*Energy consumption – electricity [MWh]*	*Primary energy [toe]*	*GHG emissions [tCO$_2$]*	*Primary energy per SME [toe]*	*GHG emissions per SME [tCO$_2$]*
Romania	40	665,161	83,268	78,021	155,346	1,951	3,884
Bulgaria	40	161,288	110,368	41,463	122,972	1,037	3,074
Spain	40	100,393	77,377	27,978	57,652	699	1,441
Austria	40	116,427	44,869	21,230	62,733	531	1,568
Italy	40	65,809	47,505	17,536	34,195	438	855
Slovakia	40	38,118	27,000	10,028	13,343	251	334
Cyprus	40	8,493	22,512	6,359	17,497	159	437
Total	280	1,155,689	412,898	202,614	463,739	–	–
					Average	724	1,656

Table 37.4 Monetary savings from the PINE project

Country	*Savings [EUR]*	*Actual Savings [% of energy consumption]*
Austria	1,112,865	7.3
Bulgaria	623,347	5.1
Cyprus	278,341	6.5
Italy	333,538	3.2
Romania	493,445	1.5
Slovakia	162,834	7.5
Spain	463,097	3.1
Average	495,352	5.0

According to the feedback of the persons in charge, the main reasons for not implementing several recommendations are the long payback period and the investment costs. Generally, payback periods beyond three years are not considered attractive by the involved companies. The average readiness to pay for an energy audit according to the responses from the companies is between 1,200 EUR (in Slovakia) and 3,025 EUR (in Cyprus).

More than 50% of the suggested measures were cross-cutting technologies, such as compressed air (repair of leakages, optimised pressure level, control of compressors), lighting (control, installation of LED), motors (optimised size, control), boilers (return of condensate, pressure level, air/fuel ratio, size, control, preheating of combustion air and/or feed water), cogeneration, energy recovery (from hot waste water from washing to preheat fresh water, from hot exhaust air to preheat air or product), or the installation of photovoltaic panels.

Organisational and behavioural issues were tackled as well, such as avoiding parasitic loads (e.g., equipment running idle), works planning, automatic control, assigning of responsibilities and definition of procedures with set points, and monitoring and controlling of actual consumption.

The fuel or heat savings amounted to about 70% of the total energy savings. The reasons for this are twofold. On the one hand, a large number of the analysed SMEs are from sectors which require a rather high amount of heat for their processes, and, on the other hand, electricity is responsible for higher specific costs, making savings more attractive, which has already triggered actions to save electricity in the past.

There are some improvement options which are relevant in all or most countries. On average:

- Optimised control by using frequency-controlled drives and the use of high efficiency drives have a share of about 25% of the identified saving options for electricity
- Optimisation of lighting (changing to T5 bulbs or LEDs, control of lighting) is responsible for about 20% of the potential electricity savings
- Improvements related to the reduction of compressed air consumption have a potential of about 10% electricity savings.

Other improvements, which are strongly dependent on the requirements of the individual processes in each company and their annual operating hours, deal with the use of renewable energy, the installation of monitoring and controlling systems, reactive power compensation, replacement, and exchange of single production plants. Photovoltaic electricity generation is an option where there is a supportive legal framework allowing connection to the grid and guaranteed feed-in tariffs. Wood waste and by-products from food processing can be used as fuel economically.

Regarding heat savings, heat recovery from production processes and building refurbishment have the highest potential, with about 60% of the saving potential. However, especially for these two options, investment costs are rather high, resulting in longer payback periods. Other relevant improvement actions are: isolation of the heat distribution system, improved control of the heating and boiler system, and reduction of heat losses at units requiring heat. Solar thermal systems suffer the problem of high investment cost and low hours of operation.

Tables 37.5 and 37.6 show the main results of the project. The "potential" savings correspond to the amount of energy which could be saved if all suggested improvements were implemented.

Table 37.5 Monetary results of the energy audits in the seven countries

	Austria	*Bulgaria*	*Cyprus*	*Italy*	*Romania*	*Slovakia*	*Spain*	*Total*
Number of energy audits	20	20	20	20	20	20	20	140
Number of customer satisfaction questionnaires	18	20	20	18	20	15	20	131
Average customer satisfaction	88%	89%	92%	80%	90%	90%	84%	88%
POTENTIAL sum of all actions								
POTENTIAL final energy savings [toe]	1,267	911	84	790	1,304	771	454	5,580
POTENTIAL primary energy savings [toe]	1,540	1,411	207	1,506	1,519	1,102	890	8,174
POTENTIAL GHG emission reduction [tCO_2]	3,393	2,985	672	3,140	2,510	2,490	1,721	16,912
POTENTIAL investment sustainable energy [€]	€ 699,451	€ 1,453,260	€ 1,186,000	€ 3,260,450	€ 1,187,650	€ 4,205,511	€ 4,264,911	€ 16,257,233
ACTUAL sum of actions that company implemented within project time frame								
Actual/potential	57%	17%	5%	13%	44%	5%	20%	27%
Final energy savings [toe]	771	87	4	72	589	34	110	1,668
Primary energy savings [toe]	878	233	11	201	671	54	177	2,224
GHG emission reduction [tCO_2]	2,332	455	44	369	1,303	111	336	4,949
Savings [€]	€ 713,000	€ 93,000	€ 12,650	€ 109,000	€ 262,000	€ 23,000	€ 93,000	€ 1,305,650
Investment sustainable energy [€]	€ 130,000	€ 181,000	€ 15,000	€ 134,000	€ 477,000	€ 314,000	€ 363,000	€ 1,614,000
ACTUAL sum of actions that company will implement soon or within 3 years								
Actual/potential	88%	77%	100%	70%	86%	68%	85%	80%
Final energy savings [toe]	1,106	764	84	555	1,100	514	375	4,498
Primary energy savings [toe]	1,353	1,089	207	1,051	1,308	744	759	6,511
GHG emission reduction [tCO_2]	3,075	2,342	672	2,173	2,119	1,661	1,464	13,506
Savings [€]	€ 1,004,000	€ 405,000	€ 278,341	€ 605,000	€ 493,000	€ 298,000	€ 363,000	€ 3,446,341
Investment sustainable energy [€]	€ 447,000	€ 724,000	€ 1,186,000	€ 1,598,000	€ 1,110,000	€ 3,253,000	€ 2,435,000	€ 10,753,000

Table 37.6 Technical results of the energy audits in the seven countries

Country	*Energy manager [%]*	*First energy audit [%]*	*Sector focus*	*Number of options selected for implementation*	*Units with biggest saving potential*	*Readiness to pay for audit [EUR]*
Austria	25	85	Metal processing	90	Air compressors, optimisation of lights, frequency-controlled drives, insulation, heat recovery, process control	1,625
Bulgaria	60	90	Metal processing	85	Optimisation of lights, frequency-controlled drives, improved control of heating, insulation	2,550
Cyprus	None	100	Food and packaging	40	Optimisation of lights, control of heating	3,025
Italy	30	100	Metal processing	67	Optimisation of lights, optimisation of air compressors, frequency control of drives, energy-efficient drives	1,444
Romania	20	100	Construction	73	Optimisation of lights, optimisation of air compressors, frequency control of drives, control of processes	1,450
Slovakia	100	100	Metal processing	60	Optimisation of lights, frequency control of drives	1,200
Spain	35	100	Food and packaging	84	Optimisation of lights, optimisation of air compressors, frequency control of drives, heat recovery	2,000

Feedback from the audits

After the finalisation of the audits, feedback was collected from the audited companies. The auditing approach was rated on average with 85 points out of 100 possible by the companies. Regarding the detailed audit tool, the feedback was very positive: The company representatives stated that:

- The procedure, modelling, and questions promote interactivity and discussion
- The model is developed by the company itself (or interactively with the consultant), which provides transparency of the analysis and option generation
- Priorities are transparent; resource allocation can be done considerately
- Losses are quantified
- Potential savings are quantified.

Regarding the audit procedure, the feedback was:

- It was received positively that the structure allows a modular approach, working from the bird's eye view to the details
- It is positive that a model of the energy consumption is built, providing an explanation of energy consumption depending on impact factors (such as production volume)
- The unit-oriented approach also allows synthesising a model of atypical processes
- The user is guided towards options by the checklists and the energy balances showing sources of losses
- The sensitivity of consumption in relation to future changes in production therefore can be tested
- The companies reacted positively to opportunities for funding of options being suggested during the audit.

The analysis of the results and experiences done during the full audits in the seven countries is the basis for the following interpretations by the research team:

Operational barriers to energy efficiency are a major issue in the SMEs included in our study, specifically:

- Very few SMEs in our sample have appointed an energy manager or have a specific procedure to systematically enhance energy efficiency. Many companies depend fully on their suppliers and do not have the capacity to fully understand the energy aspects of their equipment and the connection of operational practices to energy consumption in detail themselves.
- Information regarding energy-efficient technologies and economic incentives is not available to relevant decision-makers, or it is only available in a very generic form, not tailored to the company's needs.
- Personal awareness is an important factor. Raising awareness on energy efficiency in issues may boost energy efficiency, since personnel working on the production line know the details of the production process and can provide useful information on small (or large) energy waste, help identify actions to lean improvements, and acquire new – more efficient – procedures.

These barriers may be overcome only through a consistent capacity-building process within the company, as implemented by the modelling stage in the energy audit. During the audit, it was found important to build a (rough) model to explain the structure of energy consumption according to processes and unit operations to understand the dependency on driving factors and to monitor and control consumption.

A remarkable lack of data on energy consumption is common to all 280 companies in our sample. The main sources of information on energy consumption are the data available through energy bills. Information on the use of energy within the company is often available only in the form of aggregated monthly costs. Appropriate analysis of the data by the calculation of indicators (e.g., energy consumption per unit of product) may give the manager a better vision of energy consumption and potential improvement of efficiency. Monitoring and targeting techniques based on regression analysis may be very helpful in this respect.

An energy-monitoring system is certainly the best solution for a SME, and it should include submeters on key equipment, use of production-related specific consumption indicators, periodic monitoring and controlling, and evaluation of load curves. Analysis should include energy consumption of infrastructure (buildings) and utilities (water treatment, air supply, vacuum generation, air handling, lighting, etc.). The model generated during the audit can guide which priority areas to collect data in more detail or where to install submeters.

Economic factors are significant barriers to energy efficiency, including limited access to capital devoted to energy efficiency investments, expectation of excessively high profitability, and limited use of available incentives (despite financial support schemes for energy efficiency existing in all seven countries involved in the project). Cheap energy prices or low consumption means that actually achievable monetary savings are low, which, however, does not necessarily mean low interest in the audit or implementation rate of options.

SMEs with high energy consumption generally employ energy-intensive processes. In these companies, it is important to understand the process conditions which have an impact on energy consumption (idling time, internal recycling of off-spec product, automatic control to adjust temperature, media flows to actual process requirements).

Cost of investment for action solely aiming at increased energy efficiency without added value regarding quality, productivity, or increased useful life of a plant is (perceived as) a relevant barrier. The SMEs in our sample showed a strong reluctance to use energy efficiency criteria – and their profitability – as the only or main reason for an investment. The main motivation for the investment has mostly been of a technical nature (e.g., replacement of obsolete machinery), using energy efficiency criteria as an added value. Investment criteria change among companies and sectors.

There is a gold mine of options to save energy at little or no investment, mainly adopting good housekeeping measures, using the potential of works planning, and making good use of existing control systems. Employees need to be informed and trained – defining standard operating procedures and appropriate metrics and creating awareness.

Some energy efficiency measures in our sample are as straightforward as setting the right pressure level in compressed air systems, setting the right temperature for process fluids, switching off equipment when it is not used (e.g., air compressors, heating systems, lights), and separating heated areas using automatic opening doors or air barriers.

For electricity, the optimisation of lighting, efficient control of drives, and careful use of compressed air systems combined made up more than half of the saving potential in all countries. For heat, heat recovery from exhaust gas, products, air compressors and refrigeration cycles, building refurbishment, and insulation has the highest potential for savings.

An energy audit is a complex activity requiring an investment of company resources (cost of the audit as well as time of management and employees), leading to investments in energy efficiency that, finally, bring value to the company in the medium and long term. The perceived value of an energy audit is underestimated in SMEs due to the following reasons:

- Energy audits require technical skills and knowledge on energy-efficient technology that is outside the core business of the company; therefore, there is rarely qualified dedicated staff. The key decision-makers do not believe that an external consultant can identify any improvement which the staff would not have identified before.
- Furthermore, some SMEs focus only on the core production processes, overlooking the large potential improvement in energy efficiency that can be achieved in cross-cutting technologies.
- Support for implementation should be always included in the audit procedure in order to allow for further information exchange.

The SMEs showed a readiness to pay for the audits between 1,500–2,000 EUR. The consultants on average needed 30–50 hours to conduct the whole auditing procedure in a company. At daily rates between 500–800 EUR (depending on the country), this indicates that, without subsidies, full energy audits for SMEs are a difficult business area. To pay back an energy audit in four years out of measures which require no or low investment,[1] the company initially should spend 200,000 EUR annually on energy.

Conclusions

Engaging SMEs in energy efficiency programs requires a good understanding of SMEs' constraints and their needs. The support of chambers and industry associations who work directly with SMEs proved very beneficial. Working with entities that SMEs trust and consider authoritative improves program participation.

The audit methodology should be flexible enough to be adapted to different sizes, sectors, or organisations: Effective engagement of SMEs in energy efficiency relies on presenting a good business case for energy efficiency.

Care should be taken to build a model of energy consumption to explain consumption, understand the drivers, and develop conclusive and transparent improvement options. In our experience, a mixture of qualitative and quantitative reasoning can be applied. The appraisal of options, however, requires accurate data to prepare a reliable economic evaluation of the suggested options. The unit operations–oriented approach chosen in this program facilitates linking energy consumption to production units and understanding the factors with impact on energy consumption. The use of indicators, monitoring, and controlling are essential steps to enable improvements and to report on progress and results.

Acknowledgements

The PINE project (IEE/11/885/SI2.615936) was co-funded by the Intelligent Energy Europe Programme of the European Union. The Austrian component was also co-funded by SFG (Steirische Wirtschaftsförderung).

Note

1 The following assumptions are used: 5% energy cost savings out of no or low investment cost, 5% energy cost out of turnover.

References

Cagno E., Trucco P, Trianni A, Sala G. (2010). Quick-e-scan: a methodology for the energy scan of SMEs. *Energy*. 35 (5): 1916–1926.

DOE (2015). *SME guide to energy efficiency*. Available at: www.gov.uk/government/uploads/system/uploads/attachment_data/file/417410/DECC_advice_guide.pdf.

EC (European Commission) (2003). *Commission recommendation of 6 May 2003 concerning the definition of micro, small and medium-sized enterprises*. Official Journal of the European Union. Brussels: EC.

EC (European Commission) (2006). *Communication from the commission. Action plan for energy efficiency: realising the potential*. COM(2006)545. Available at: https://eur-lex.europa.eu/LexUriServ/LexUriServ.do?uri=COM:2006:0545:FIN:EN:PDF.

EC (European Commission) (2019). *Roadmap 2050*. Available at: www.roadmap2050.eu/.

Fresner J, Morea F, Krenn K, Aranda J, Tomasi F. (2017). Energy efficiency in small and medium enterprises: lessons learned from 280 energy audits across Europe. *Journal of Cleaner Production*. 142 (4, 20): 1650–1660.

Gerber E. (2011). *Energy consumption: SME lag behind (Energiekonsum: KMU hinken hinterher)*. Handelszeitung. Available at: www.handelszeitung.ch/konjunktur/energiekonsum-kmu-hinken-hinterher.

KfW (2013). *Energy costs and energy efficiency in the German SME sector*. Available at: www.kfw.de/PDF/Download-Center/Konzernthemen/Research/PDF-Dokumente-Fokus-Volkswirtschaft/Fokus-englische-Dateien/Fokus-Nr.-40-Dezember-2013-_EN.pdf.

OECD (2017). *Renewable energy for industry – from green energy to green materials and fuels*. Paris: OECD.

Thollander P, Backlund S, Trianni A, Cagno E. (2013). Beyond barriers – a case study on driving forces for improved energy efficiency in the foundry industries in Finland, France, Germany, Italy, Poland, Spain, and Sweden. *Applied Energy*. 111: 636–643.

Thollander P, Palm J. (2012). *Improving energy efficiency in industrial energy systems – an interdisciplinary perspective on barriers, energy audits, energy management, policies & programs*. Cham: Springer. ISBN:978-1-4471-4161-7.

38

CIRCULAR ECONOMY PRINCIPLES IN AFRICA

The case of off-grid solar in Kenya

Federico Magalini, Joe Segal and Marco Meloni

Introduction

In recent years, the term "circular economy" has gained a significant presence and has become a central piece of numerous policies in industrialised countries (e.g., the EU Circular Economy Package). Most countries in the world have recognised that the great environmental and societal challenges of our time can only be mastered by a fundamental change in our economic systems towards a truly sustainable approach. Both the 2030 Agenda for Sustainable Development, with its 17 Sustainable Development Goals (SDGs), and the Paris Agreement make clear statements and set clear goals for such a transformation.

A circular economy is a system that creates value by designing out waste and keeping products, materials, and components in use for longer.[1] This entails designing products so that they are used for longer and can easily be reused, repaired, remanufactured, and recycled (Figure 38.1), keeping the materials, energy, and effort put into manufacturing the products in use.

However, design of the products on its own is not enough, and designing a circular system also requires new business models to capture new opportunities, reverse logistics and infrastructure to collect the products and treat them to recover materials in a safe way, and policies and other favourable system conditions to shift incentives (Ellen MacArthur Foundation, 2019).

The role of off-grid solar products and the societal benefits

The societal benefits of off-grid solar solutions are undeniable. Off-grid solar is a healthier, safer, and cheaper alternative to kerosene, batteries, and candles and has the potential to lead to job creation both within the off-grid solar value chain and for people using solar home systems in small businesses. This will also lead to broader societal benefits, including improved education, health, employment opportunities, and the environment.

Off-grid solar systems have the potential to change the lives of hundreds of millions of people in remote areas. While approximately 789 million people lack access to electricity globally (IEA et al., 2020), it is estimated that off-grid solar systems benefit over 120.3 million people, also leading to the phase-out of 18.6 million sources of traditional lighting (e.g., candles, kerosene, battery-powered torches); 28.6 metric tonnes (Mt) of greenhouse gases have been avoided, and $5.2 billion has been saved on lighting and phone charging costs (Lighting Global and GOGLA, 2018).

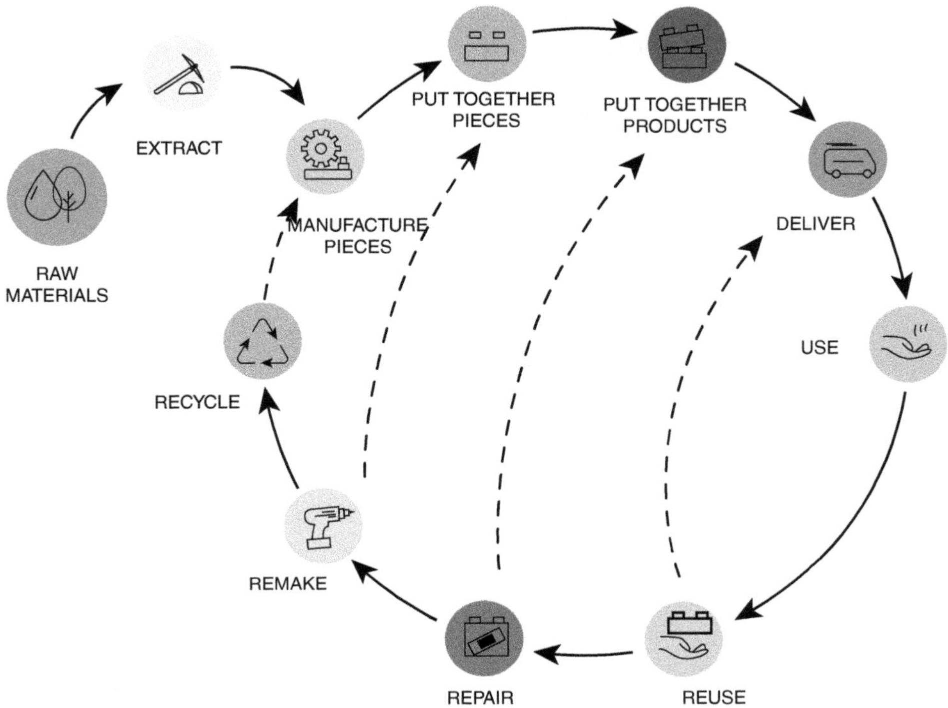

Figure 38.1 Circular economy system diagram

Source: Circulab

Millions of products are being placed in the market every year (GOGLA, 2019), and while these applications are relatively durable, inappropriate disposal or treatment after use can result in the exposure of people and the environment to harmful substances. For example, components such as batteries have been found to be frequently disposed of informally into the surrounding environment (Bensch et al., 2015), and such informal practices also have direct health and safety impacts on the recyclers themselves (WHO, 2017). At the same time, there is huge value in keeping these products and the materials in them in use. Looking only at materials, off-grid solar products contain a wide range of valuable metals, such as copper, steel, and aluminium, and also many critical raw materials[2] (Magalini et al., 2016). The rapidly increasing rate of off-grid solar system production highlights the need to create a system where those products and the resources that compose them are kept in use for as long as possible: a circular economy of off-grid solar applications.

Key drivers for adoption and market penetration

Policy is one of the key drivers in the emerging market, especially if we consider Africa, as off-grid technologies are central in achieving the ambitious access to energy plans of local governments: for example, in Kenya, the national government has set out plans to achieve universal energy access by 2020 and has made significant headway towards this goal, with access to grid connections nearly doubling from 25% to 46% from 2011 to 2015 (USAID, 2016). However, with 64% of Kenyans not having access to grid electricity and a strong private sector with a stable and supportive government, Kenya is one of the most vibrant markets for the solar industry in Africa, with off-grid solar products reaching 52% (GOGLA, 2018) of off-grid and unreliable-grid households. Between 2014 and 2018, the Global Off-Grid Lighting Association (GOGLA) members alone (who account for more than 60% of the market) sold more than 4 million products in Kenya, which is almost equivalent to sales in the other East African countries (Ethiopia, Rwanda, Tanzania, and Uganda) combined.

At the same time, Kenya is also leading the way in introducing innovations and new business models, such as mobile money or pay-as-you-go (PAYG) models, which address the affordability of higher-quality off-grid equipment for domestic use and have seen a huge rise in the uptake of solar powered equipment such as radios, torches, televisions, and recently even white goods.

While in Europe, lease, pay-per-use, and other business models are currently discussed and only tested in very few cases on a larger scale, the solar power industry in developing countries and emerging economies has proved very flexible. It has had to overcome various difficulties related to marketing of products and systems with high upfront costs (e.g., solar home systems) and in direct competition with other, partly subsidised energy options (e.g., kerosene) in rural contexts where the purchasing power, access to credit, and credit history of customers was and still is close to zero. In this context, the industry developed alternative business models compared to the direct sale of equipment typical of the electronics sector.

Two main business models can be identified:

- Models where the appliance is purchased. This includes direct cash purchases or sales "aided" by financing institutions or by providing access to credit mechanisms. This business model is typically applied for smaller systems such as solar lanterns and small solar home systems, which often represent the entry point on the energy ladder.
- Models where the appliance is leased or the ownership is progressively transferred after a period of time (lease-to-own). Those models where customers pay for the product over time are commonly referred as PAYG. All these models usually use a mechanism that allows remotely disabling a system when payment is overdue, and devices are engineered with specific hardware and software solutions to enable this. In addition, payment is usually done by mobile money.

Opportunities enabled by business models in off-grid solar sector

PAYG models have been widely accepted in East Africa, particularly in Kenya (730,000 sales), compared to West Africa (30,000) and South Asia (20,000), where the market is characterised by a strong concentration, with five companies responsible for more than 90% of sales (GOGLA, 2017). PAYG models are increasingly becoming mainstream, especially in those countries where the enabling ecosystem (i.e., mobile money) is favourable.

PAYG models are certainly creating stronger and more stable relationships with the customers, at least until the ownership of the product is finally transferred to the consumer. In this context, PAYG models offer an opportunity to better integrate take-back programs and operations, particularly in the case of consumers who are moving to newer products at the end of the pay-back period. In addition, during maintenance and warranty periods, companies have access to faulty products and components, including batteries, and channel them to their warehouses.

This offers a good opportunity to integrate pick-up services for battery waste and e-waste in more remote areas. And even beyond PAYG, the off-grid solar sector has developed a wide variety of sales and distribution options, including common retail and shops, local agents, NGOs, and

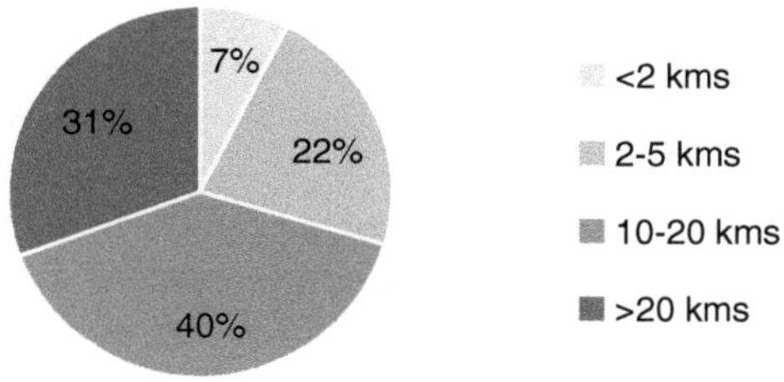

Figure 38.2 Distance of SHS users from nearest service centre

Source: Results from ongoing project by Sofies, funded by UKAid from the UK government via CDC Plus

partnerships with mobile phone operators. It is estimated that up to 50% of sales come from such distribution partnerships and not from traditional sales and distribution models (GOGLA, 2017).

All in all, these new distribution models that create a stronger relationship with the customers can contribute, at least partially, to overcome the typical difficulties of structuring a collection and recycling system for waste batteries and e-waste in the context of developing countries. In many cases, off-grid solar companies represent the most trusted and close options for consumers to properly dispose of end-of-life products. This is illustrated by a survey conducted in Kenya in 2018 (REWMOS, 2018) among users of off-grid products, which revealed that 50% seek advice from the seller on disposal and 5% directly take back to the company they purchased from, and only 1% reported having consulted the National Environment Management Authority for advice.

However, lack of proximity to a service centre where users can drop their products is often prohibitive, as results of a consumer survey[3] run in Kenya in November 2019 revealed (Figure 38.2). The results are from an ongoing project by Sofies, funded by UKAid from the UK government via CDC Plus.

Proactive activities of industry and challenges at end of life

Irrespective of the business model, all companies or energy-access projects placing batteries and solar power equipment on the market are from an ethical perspective responsible for providing sound end-of-life management. While in most developing countries, there is no legal obligation covering this responsibility, it results from the prevalence of extended producer responsibility (EPR), which represents a globally accepted concept for products and waste streams posing significant risks to human health and the environment. In 2014, GOGLA committed to have the EPR principle as a cornerstone for the activities of its members (GOGLA, 2014). Increasingly, board members and investors in the off-grid energy sector are demanding EPR programmes from the companies active in the space. Thus, various solar power companies are considering variations to their business models to address this (ethical) responsibility in their markets in developing countries and emerging economies, with some already incorporating collection schemes. Analysis of return stream handled by the WEEE Centre in 2017 (Magalini et al., 2017), one of the main e-waste recyclers in Kenya, revealed how nearly 50 tonnes of the 750 tonnes of e-waste treated in the previous 3.5 years was made up of off-grid solar products.

Off-grid solar products are widespread in rural areas, where waste collection is often limited or non-existent, and are often stored or kept at home even after the end of their useful life. Despite the common image of many African rural households, many houses contain electronic and electrical equipment even when ignoring the prevalence of off-grid solar products. Figure 38.3 shows the percentage of households with a variety of electrical equipment. As shown, relatively high numbers have TVs, radios, and phones beyond the expected quantities for these households.[4]

Figure 38.3 describes a selection of off-grid energy consumers' attitudes towards their products at their end of life. While many consumers plan to return their products to the seller, the majority plan to keep it at home – a trend observed in recent studies conducted across Europe as well (Magalini et al., 2012, 2019) – and a notable number plan to discard them in unsafe and environmentally unfriendly ways, as shown in Figure 38.4. While attitudes vary, a large number of users opt to keep products at home.

Figure 38.5 shows consumer disposal practices. Disposal practices are often unsafe and environmentally harmful.

A relevant share of off-grid solar products enters the informal repair economy. Repairers have a nationwide presence and are often closer to end users, especially if located in rural areas. Customers with defective solar lamps are likely to try to have them repaired at one of these repair shops rather than at distributors that are located in urban or peri-urban centres. Some off-grid solar companies have also developed internal repair and refurbishment operations, sometimes creating low-cost entry-level products. Most informal repairers do not have formal training, especially on off-grid solar products, and often operate in unsafe environments. Informal repairers will typically discard both products that cannot be repaired and remaining fractions, which have little to no value either in repair or for recycling.

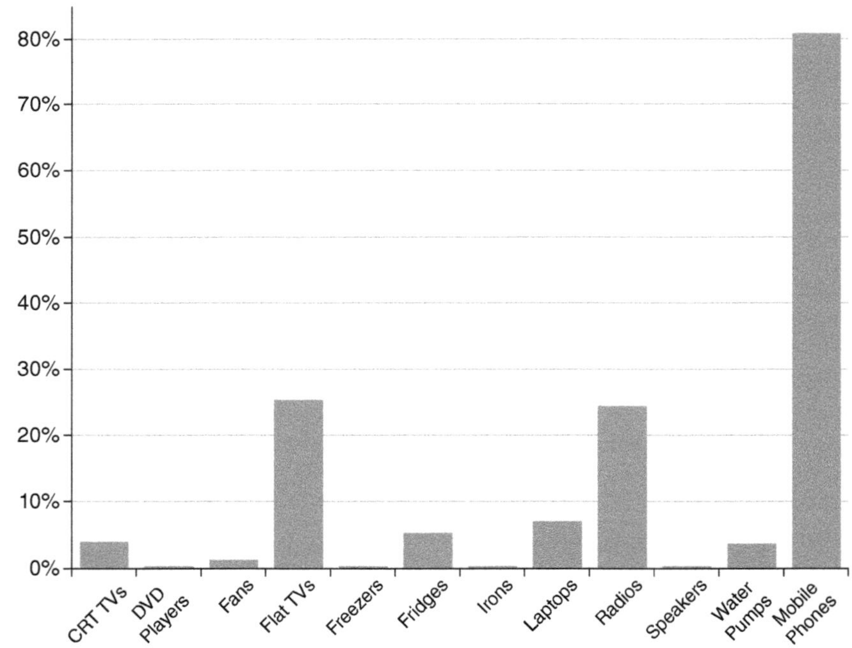

Figure 38.3 Percentage of households containing different types of electronic products. Relatively high numbers of homes have TVs, radios, and mobile phones

Source: Results from ongoing project by Sofies, funded by UKAid from the UK government via CDC Plus

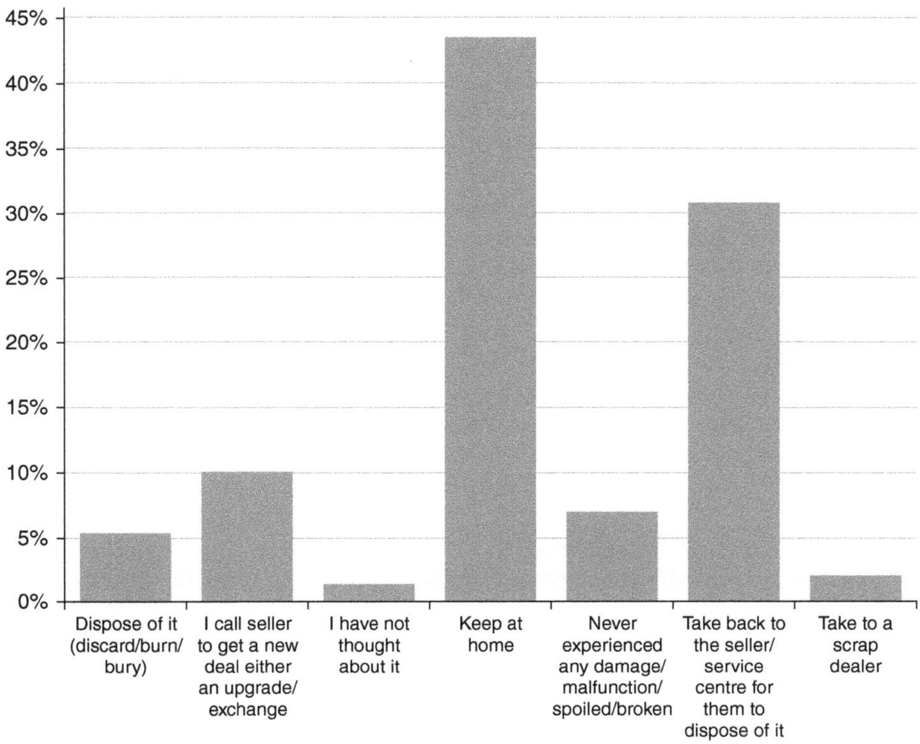

Figure 38.4 Consumer attitudes to off-grid products at the end of life

Source: Results from ongoing project by Sofies, funded by UKAid from the UK government via CDC Plus

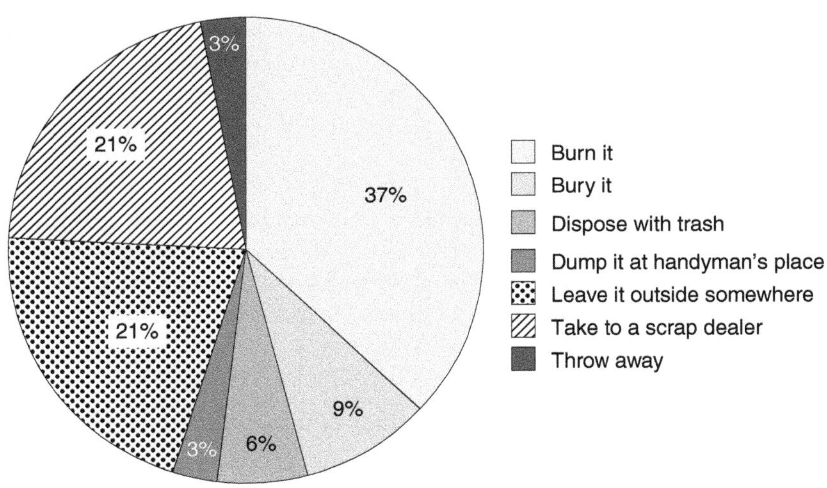

Figure 38.5 Consumer disposal practices

Source: Results from ongoing project by Sofies, funded by UKAid from the UK government via CDC Plus

In Kenya, as in most of the East Africa region, collection of e-waste is done largely by small and medium-sized informal collectors. There is no public collection infrastructure for e-waste, including for off-grid solar products. Only seven centres have been licensed by the National Environment Management Agency (NEMA) for hazardous waste collection. Some formal collectors have set up collection points, a few off-grid solar companies leverage the network of their service centres or tested alternative outlets (such as petrol stations or mobile phones shops), but last-mile collection remains the biggest challenge.

The informal, small-scale collectors separate e-waste from general solid waste and dismantle or channel it to dismantlers who, often with rudimentary approaches, extract components of interest such as copper wires, steel, aluminium, plastic, and printed circuit boards. These practices present risks to local environments and communities and to the recyclers themselves, including air pollution by burning of plastic cabling or soil and groundwater contamination from draining of lead acid, also presenting high health and safety risks to the recyclers.

In this context, the main challenges are represented by the distribution of batteries and equipment to rural areas, which creates logistical challenges and cost implications for sound collection and recycling at the end of life. Apart from this, the following issues have to be considered:

- In all cases, the willingness of consumers (or waste holders) to give away obsolete batteries and e-waste to avoid improper disposal is fundamental and cannot be bypassed. It might therefore be required to create some forms of incentives (monetary or non-monetary) for waste holders to hand in old batteries or defined e-waste types.
- The role of local repair and refurbishment activities, especially for solar home system (SHS) components and connected appliances, needs to be factored in: the availability of spare parts from discarded appliances is creating an economic incentive for the waste holder to sell appliances and batteries to a local repair industry instead of returning them for recycling purposes. While there are already developed repair and refurbishing industries for lead-acid batteries in many developing countries, it is also likely that comparable industries will develop for Li-ion batteries once sufficient end-of-life volumes are available; currently a pilot project is in progress in Kenya as part of the US-AID–financed Global Leap Award.[5]
- Some companies are trying to proactively develop take-back operations on a voluntary basis. But the cost of proper take-back and recycling was not included in the initial pricing structure of the product or PAYG model. This is particularly problematic for systems using battery types and components with a negative net value (e.g., Lithium manganese oxide and lithium iron phosphate batteries).

In some cases, innovative collection systems have been planned or piloted: amongst others in Kenya, TOTAL tested the use of petrol stations as drop-off points for solar waste, thus leveraging existing and easily accessible infrastructure to be used as collection points. Other companies considered using their "road shows," usually organised to facilitate distribution of products, for waste-related awareness raising and collection (GIZ et al., 2018).

Policy versus voluntary approaches

Activities are happening in Kenya, a country where the current legal framework for waste management only includes the Environmental Management and Coordination Act (1999), amended in 2015, and the Waste Management Regulations (2006). Since 2013, a draft bill for e-waste management, based on the EPR principle, has been available. In 2019, the draft bill was revised and submitted for an open consultation process but has not yet been approved by the National Assembly, and thus no legal obligation for proper take-back and recycling exists.

Kenya also adopted, in 2017, jointly with other East African States (Burundi, Rwanda, Tanzania, and Uganda) the Regional E-waste Management strategy, which also aims at establishing and strengthening EPR to finance proper collection and treatment of e-waste in the region. It also calls for harmonisation of policies and legal frameworks in East African Communications Organisation member states, the establishment of regional infrastructure, and improved transboundary movement of e-waste in the region.

The current debate on national e-waste strategy and the draft e-waste bill still leave unresolved the specific challenges of disposing of e-waste and off-grid solar products at end of life, in particular:

- Off-grid solar products and photovoltaic panels are not clearly mentioned, while batteries and connected devices like TVs, fans, and radio are clearly included in the scope.
- The collection of off-grid solar products is much more challenging than for other types of e-waste, particularly for rural distribution.
- The cornerstone of the financing mechanism foreseen by the e-waste bill is "problematic fractions," defined as "those components or parts of e-waste where the collection and treatment costs outweigh the material recovery value." However, the bill does not specify how producers are required to fulfil the requirements, particularly with respect to their market share. Emerging questions include the calculation of the cost for these fractions, as linked to individual collection versus average collection and collection from remote areas versus collection from urban centres (which is less expensive).

Conclusions and way forward

The off-grid solar industry has shown strong interest in building sustainability into its business models and improving access to energy, as demonstrated by the publication of the GOGLA Industry Opinion on "lifecycle and recycling" in 2014. Since then, events with specific focus on end of life have been organised at industry conferences in Dubai (2015)[6] and Hong Kong (2018).[7] In 2018, GOGLA's Sustainability Working Group was re-launched as the E-waste Working Group, reflecting the importance of e-waste as the main challenge to the environmental credentials of the sector, and in February 2019, this working group launched an e-waste toolkit to offer practical guidance and information to member companies and organisations.

Many companies, also due to increasing investor demand, have started implementing policy and voluntary plans to handle the waste generated by the products they place on the market, particularly batteries. In fact, some companies found this investor demand easy to satisfy, as the business model of the company had been built around circular economy principles. By creating a composite product, disposal of defunct systems was minimised, as repair, or rather replacement of failed components, was both easily achieved and more cost effective. These circular principles were not necessarily created due to environmental ideals but rather to better improve business models in an environment where cost reduction is the number-one focus. By building these business models, off-grid solar companies have created a system in which non-functioning products are replaced with working ones at the last mile, enabling consumers to have a high-quality service while also allowing the businesses to continue to control the e-waste flows.

Manufacturers and distributors have signed agreements with e-waste facilities in Kenya and proactively collect and recycle off-grid solar products; some pilots are ongoing, aimed at testing new and innovative collection strategies, for instance, leveraging a network of petrol stations as drop-off points. Solar home systems are getting more connected, presenting opportunities for geo-localisation to assist with collection.

By applying circular economy principles at the business model level, the off-grid sector has the potential to not only provide access to energy for millions of people but also to do so while keeping valuable products, components, and materials in use for longer.

Notes

1 The Ellen MacArthur Foundation defines the circular economy as a system based on three principles: design out waste and pollution, keep products in use for longer, and regenerate natural systems. See, for example: Ellen MacArthur Foundation (2019). *What is the circular economy?* Available at: www.ellenmacarthurfoundation.org/circular-economy/what-is-the-circular-economy.
2 Such as indium and cobalt, among others – 27 materials have been identified as critical raw materials by the European Commission (EC, 2017). Critical raw materials have been broadly described as raw materials for which there are no viable alternatives and whose supply depends on one or few producers (Overland, 2019).
3 Results from ongoing project by Sofies, funded by UKAid from the UK government via CDC Plus.
4 Results from ongoing project by Sofies, funded by UKAid from the UK government via CDC Plus.
5 Available at: https://globalleapawards.org/e-waste
6 4th International Off-Grid Lighting Conference and Exhibition, Dubai, 2015.
7 2018 Global Off-Grid Solar Forum and Expo, Hong Kong, 2018.

References

Bensch G, Peters J, Sievert M. (2015). *The lighting transition in Africa – from kerosene to LED and the emerging dry-cell battery problem*. SSRN. Available at: https://ssrn.com/abstract=2706526 or http://dx.doi.org/10.2139/ssrn.2706526.

Ellen MacArthur Foundation (2019). *Building blocks: circular economy design, business models, reverse cycles and enabling conditions are essential*. Available at: www.ellenmacarthurfoundation.org/circular-economy/concept/building-blocks.

European Commission (2017). *Communication from the Commission to the European Parliament, the Council, the European Economic and Social Committee and the Committee of the Regions*. Available at: https://ec.europa.eu/transparency/regdoc/rep/1/2017/EN/COM-2017-490-F1-EN-MAIN-PART-1.PDF.

GIZ, Oko-Institut e.V., Sofies, and Endev (2018). *End of life management of batteries in the off-grid solar sector*. Available at: https://www.giz.de/de/downloads/giz2018-en-waste-solar-guide.pdf.

GOGLA (2014). GOGLA industry opinion on lifecycle and recycling. *Cologne*. Available at: https://www.gogla.org/sites/default/files/recource_docs/gogla-industry-opinion-on-lifecycle-and-recycling1.pdf (Accessed 24/7/2018).

GOGLA (2017). *Providing energy access through off-grid solar: guidance for governments*. Utrecht. Available at: https://www.gogla.org/sites/default/files/resource_docs/energy_access_through_off-grid_solar_-_guidance_for_govts.pdf (Accessed 19/7/2018).

GOGLA (2018). *Off-grid solar market trends report 2018*. Available at: www.gogla.org/sites/default/files/resource_docs/2018_mtr_full_report_low-res_2018.01.15_final.pdf.

GOGLA (2019). *Global off-grid solar market report: semi-annual sales and impact data*. Available at: www.gogla.org/sites/default/files/resource_docs/global_off-grid_solar_market_report_h1_2019.pdf.

IEA, IRENA, UNSD, World Bank, WHO. 2020. Tracking SDG 7: the energy progress report. World Bank, Washington DC. © World Bank. License: Creative Commons Attribution—NonCommercial 3.0 IGO (CC BY-NC 3.0 IGO).

Lighting Global, Dalberg Advisors and GOGLA (2018). *Off-grid solar market trends report 2018*. Available at: www.gogla.org/sites/default/files/resource_docs/2018_mtr_full_report_low-res_2018.01.15_final.pdf.

Magalini F, Huisman J, Wang F, Mosconi R, Gobbi A, Manzoni M, Pagnoncelli N, Scarcella G, Alemanno A, Monti I. (2012). *Household WEEE generated in Italy: analysis on volumes & consumer disposal behavior for waste electrical and electronic equipment*. Puebla: Ecodom.

Magalini F, Sinha Khetriwal D, Munyambu S. (2017). *Cost-benefit analysis and capacity assessment for the management of electronic waste (e-waste) in the off-grid renewable energy sector in Kenya*. London: Evidence on Demand.

Magalini F, Sinha-Khetriwal D, Rochat D, Huismann J, Munyambu S, Oliech J, Nnorom IC, Mbera O. (2016). *Electronic waste (e-waste) impacts and mitigation options in the off-grid renewable energy sector*. London: Evidence on Demand with assistance from the UK Department for International Development.

Magalini F, Thiebaud E, Kaddouh S. (2019). *Quantifying WEEE in Romania: 2019 vs 2015*. Ecotic. Available at: https://www.ecotic.ro/wp-content/uploads/2015/11/WEEE-Generated-study-2015.pdf.

Overland I. (2019). The geopolitics of renewable energy: debunking four emerging myths. *Energy Research & Social Science*. 49: 36–40.

REWMOS (2018). *Survey results*. Available at: https://00cd5f9a-e84d-4d6c-9af5-38ee143d8973.filesusr.com/ugd/d414a6_394d9e70be404ed1b46b6e73718bc028.pdf.

USAID (2016). *Development of Kenya's power sector 2015–2020*. Available at: www.usaid.gov/sites/default/files/documents/1860/Kenya_Power_Sector_report.pdf.

World Health Organization (2017). *Recycling used lead acid batteries: health considerations*. Geneva: WHO. Licence: CCBY-NC-SA3.0IGO.

39
CIRCULAR SUPPLY CHAIN
Emerging opportunities and challenges

Uthayasankar Sivarajah and Elizabeth Ragonga

Introduction

The current global patterns of production and consumption remain dangerously unsustainable and with serious impact on the environment (Alhola et al., 2019; Nasir et al., 2017; Preston, 2012). The linear model on which the world has been operating over decades based on 'take, make, consume and dispose' is not sustainable, as the level of natural resource depletion and pollution of the environment is enormous. In a recent report, the World Bank (2020) indicated that 2.01 billion tonnes of waste are generated per year and will increase to 3.4 billion tonnes by 2050. This has resulted in an annual cost of £160.1 billion, with an estimated increase of £292.7 billion in 2025. Moreover, there is clear evidence that since the 21st century, the world has been characterised by high levels of turbulence and volatility, ranging from economic factors, geopolitical upheavals, extended global supply chains, and increased disruptions. This state of affairs is now reshaping competitive priorities for businesses and their supply chains. For instance, increasing resource scarcity has led to rising prices and price volatility, impacting the production side.

Moreover, the current level of consumption, if not controlled, will see the world deplete many of its natural resources (Hazen et al., 2017). Production processes spread around the globe have seen suppliers, companies, and customers linked by information, resources, and capital flows. In the processes of producing goods and services to meet the increasing demand, an environmental and social burden has been incurred (Seuring and Müller, 2008). The rise of the supply chain has been attributed to trends in global sourcing, emphasis on time and quality-based competition, and contributions of each actor to greater environmental uncertainty (Mentzer et al., 2001). The authors further note that increased global sourcing by companies for supplies has forced many companies to look for effective ways of coordinating the flow of materials for production purposes, hence the need for establishing a closer relationship with suppliers and customers.

The supply chain involves all activities associated with extraction and transformation of raw materials and information into finished products that can be consumed by the end user. From several studies (Nasir et al., 2017; Farooque et al., 2019; Du et al., 2010; Govindan and Hasanagic, 2018), there is a clear indication that the supply chain process is almost similar to the linear process where raw materials are extracted, transformed into finished products, and sent to consumers, who then discard it when no longer serves their purpose, or incinerated as waste. In the face of sharp volatility increases across the global economy and proliferating signs of resource depletion, the call for a new economic model is getting louder.

The quest for a substantial improvement in resource performance across the economy has led businesses to explore ways to reuse products or their components and restore more of their precious material, energy, and labour inputs. For this reason, one philosophy that is believed to bring change is the circular economy (CE). It has been increasingly recognised as a better alternative to the wasteful linear economy (Ghisellini et al., 2016). The concept is believed to have a potential solution for the challenges faced across the globe due to its ability to create synergies between environmental and economic goals. It is further seen as a system that minimises resource input and waste, emissions, and energy leakage and is hoped to mitigate negative impacts without jeopardising growth and prosperity (EMF, 2015).

CE advocates for the recovery of value from physical goods through narrow cycles of reuse and refurbishment, recycle and remanufacture. For the effectiveness of these cycles, supply chains are believed to be the critical unit of implementation, since they facilitate joint effort "beyond organisational boundaries to involve external coordination with upstream partners to obtain environmentally friendly inputs and with downstream partners to cooperate for environmental management practises such as product return, reuse, and recycling" (Zhu et al., 2013).

This chapter gives consideration to both linear supply chains and circular supply chain management. Initially, a review of the current model of the linear supply chain is presented by looking at its impact on the environment in terms of production and consumption. The chapter then discusses circular supply chains by looking at the opportunities and challenges. With this, the chapter will have answered questions such as: What entails a linear supply chain? What are the opportunities in a circular supply chain? What are the challenges of transitioning to a circular supply chain?

Linear supply chain

For the past 150 years, industrial evolution has been dominated by the linear economy (Franco, 2017), where activities in the supply chains involve manufacturing of products from extracted virgin resources and materials which are then sold, consumed, and eventually discarded as waste (Jawahir and Bradley, 2016). In this case, supply chains are more 'take–make–use–dispose,' leading to significant generation of waste and an increase in landfills (EMF, 2015; Ghisellini et al., 2016; Murray et al., 2017). The world is beset by environmental problems such as climate change, resource scarcity, biodiversity loss, and increased pollution (Rockström et al., 2009; World Bank, 2020). For instance, businesses at the micro level are experiencing challenges such as volatility in resource prices and vulnerability in supply chains (EMF, 2013).

Again, the growing consumer in the middle level, which is expected to increase to 3 billion people by 2030 (WEF, 2014), means more production and consumption. With natural resources being a key factor in the supply chain and an increase in demand for products, nothing less than increased degradation of the environment is expected, and therefore there is an urgent need for a solution (Franco, 2017). The circular economy (CE) has, therefore, emerged as the model with a solution to the challenges currently faced through its ability to create synergies between environmental, economic, and social goals by replacing the take–make–use–dispose model, since it is restorative and regenerative by intention and design (EMF, 2013; Farooque et al., 2019).

To understand how CE is integrated into supply chain management, it is important to have insight into supply chains and supply chain management. La Londe and Masters (1994) define a supply chain as "a set of firms that pass materials forward." According to the authors, these firms (including raw material producers, assemblers, wholesalers, retailers, and transporters) are involved in the production process before sending the final good to the end user. Similarly, Lambert et al. (1998) defined the supply chain as "the alignment of firms that brings products or services to market." Christopher (2016) noted that the focus of supply chain management is on cooperation and trust and recognition that the "whole can be greater than the sum of its part." With this in mind, the author defined it as "the network of organizations that are involved, through upstream and downstream linkages, in the different processes and activities that produce value in the form of products and services delivered to the ultimate consumer."

Considering the number of players involved (Mentzer et al., 2001) defined the supply chain as "a set of three or more entities (organizations or individuals) directly involved in the upstream and downstream flows of products, services, finances, and/or information from a source to a customer." Similarly, Handfield and Nichols (1999) view it as "a network of organizations that are involved, through upstream and downstream linkages, in the processes and activities that produce value in the form of products and services to the consumers." In a nutshell, a supply chain consists of several players, as indicated in Table 39.1, with operations that involve extraction, flow, and transformation of raw materials into finished products through to the final user.

Table 39.1 Circular supply chain opportunities

Opportunities	*Reference(s)*
Cost savings, differentiation, improved customer relations, reduced environmental impacts, and increased brand protection.	Sharma et al. (2010)
Profit opportunities, economic growth, substantial net material cost savings, job creation potential, innovation, reduced volatility and greater security of supply, new demand for business services.	EMF (2013, 2014, 2015)
Environmental benefits in terms of reduced impacts and resource usage, cost savings from reduced natural resource usage, and creation of new markets, hence additional economic benefits leading to job creation, poverty reduction, and wealth creation.	(Behrens, 2016; EMF, 2013, 2014; Wijkman and Skånberg, 2015).
Reduced use of natural resources, reduced waste amounts, decreased greenhouse gas emissions and usage of hazardous substances, and a move to renewable energy supplies.	Govindan and Hasanagic (2018)
Leveraged access to new markets; product differentiation; lower material, energy, and services costs; lower cost of capital and labour; and better risk management and relations with external stakeholders.	(Ambec and Lanoie, 2008).
Economic growth, job opportunities, saving material costs, dampening price volatility, improved security of supply, reduced environmental pressures.	Kalmykova et al. (2018)
Environmentally sound use of resources, new business models, innovative employment opportunities, improved well-being.	Ghisellini et al. (2016)
Reduced usage of natural resources, reduced waste amounts, decreased greenhouse emissions and hazardous substances, use of renewable and sustainable energy.	Bastein et al. (2013)

A 'traditional'/linear supply chain (Weetman, 2016; Manavalan and Jayakrishna, 2019) comprises a simple flow of raw materials from extraction through manufacturing of finished products and distribution to the final user (customer). In a linear setup, the product ends with the consumer, who can either keep or discard it. This means for the production process to continue, and for the demand of products to be met, the process goes back to extraction of new materials. This 'take–make–consume–dispose' pattern is not sustainable, as it depletes natural resources. In the face of sharp volatility increases across the globe, instability in supply chains, and proliferating signs of resource depletion, the call for a new economic model is inevitable. This has led to exploration by businesses to find out ways to reuse products or their components and restore energy and labour inputs.

Circular supply chains

A supply chain is seen as a step towards the adoption of sustainability, since it considers the product from the initial stages (extraction and processing of raw materials) to delivery to the customer. According to Zhu et al. (2010) and Farooque et al. (2019), supply chains play a critical role in the implementation of CE, since they consist of joint effort of both internal and external players downstream and upstream to obtain environmentally friendly inputs and management of practices such as return, reuse, and recycle or systematically applying CE concepts in all supply chain stages and functions. In this case, organisations collaborate within different sectors to maximise full utility of good and materials, hence minimising the negative environmental, social, and economic impacts. However, for it to be circular, it must integrate issues beyond the supply chain and include product design; remanufacturing; recycling; product life extension and reuse (Linton et al., 2007); and configuration of supply chains to close, narrow, slow, intensify, and dematerialise loops (Geissdoerfer et al., 2018).

A circular supply chain is described by Lacy and Rutqvist (2015) as a "business in which recyclable, biodegradable, or renewable resources are used to replace linear inputs hence providing a competitive edge." The authors believe that in a circular supply chain, companies improve the value they offer to customers by operating in future-proof mode by phasing out scarce resources through using fully renewable, recyclable, or biodegradable resources; removing inefficiencies; and cutting waste. Canning (2006), Du et al. (2010), Genovese et al. (2017), and Nasir et al. (2017) describe it as an "integration of CE and SCM that it involves systematic restoration of technical materials and regenerated biological materials towards zero waste through innovation in business models and supply chains" (Farooque et al., 2019). Based on a review carried out by the authors, they define circular supply chains as "the configuration and coordination of organisational functions marketing, sales, R&D, production, logistics, IT and finance to improve effectiveness and efficiency of the system and generate competitive advantage." Similarly, Batista et al. (2018) define it as "the coordinated forward and reverse supply chains via purposeful business ecosystem integration for value creation from products/services, by-products and useful waste flows through prolonged lifecycles that improve the economic, social and environmental sustainability of organizations," while Geissdoerfer et al. (2018) define it as

> the configuration and coordination of the organisational functions marketing, sales, R&D, production, logistics, IT, finance, and customer service within and across business units and organizations to close, slow, intensify, narrow, and dematerialise material and energy loops to minimise resource input into and waste and emission leakage out of the system, improve its operative effectiveness and efficiency and generate competitive advantages.

From these definitions, it is clear that coordination and configuration of activities and players in the supply chain are required to enable restoration (reuse, repair, refurbish, remanufacture, and recycle) and regeneration and hence less use of natural resources and energy for production. Due to the emerging interest in circular supply chains and knowledge of different disciplines and organisational functions, there is a broad range of definitions and understanding of the topic (Burgess et al., 2006). While the term 'circular supply chain' has been used to link CE to sustainable supply chains (Canning, 2006; Du et al., 2010; Genovese et al., 2017; Mishra et al., 2018), Farooque et al. (2019) observe that it is only recently that a working definition of circular supply chain management appeared in the literature.

Again, in supply chain management literature on sustainability, it is worth noting that a number of concepts such as sustainable supply chains (Carter and Rogers, 2008; Seuring and Müller, 2008), green supply chains (Srivastava, 2007; Fortes, 2009), environmental supply chains (Walker et al., 2008), closed-loop supply chains (Neto et al., 2009), green manufacturing and product recovery (Gungor and Gupta, 1999), and circular supply chains (Farooque et al., 2019) have been used interchangeably to integrate the sustainability concept into supply chain management (Gurtu et al., 2015; Ahi and Searcy, 2015). Just like the linear supply chain, the cycle begins with extraction of materials which are then manufactured into finished products sold to the final end user. However, the process doesn't end there. In a circular supply chain, after product use, it is brought back to the producer to recover value through reuse, remanufacture, redesign, and recycle (EMF, 2015; Guide and Wassenhove, 2006). Circular supply chains thus encourage the flow of products back into productive systems; hence, zero waste is generated.

In adoption of the CE by businesses in their supply chains, several benefits can be experienced, from environmental benefits, in terms of reduced impacts and resource usage and cost savings from reduced natural resource usage to creation of new markets, hence additional economic benefits leading to job creation, poverty reduction, and wealth creation (Behrens, 2016; EMF, nd.; Wijkman and Skånberg, 2015). Similarly, as observed by Govindan and Hasanagic (2018), adoption of the CE by businesses in their supply chains could see them reduce waste amounts, decrease greenhouse gas emissions and usage of hazardous substances, and move to renewable and energy supplies.

These benefits can be experienced with the support of the four fundamental CE building blocks. EMF (2013), acknowledges circular economy design, new business models, reverse cycles, and enablers and favourable system conditions as the four essential building blocks for a circular economy. For efficiency and effectiveness in the supply chain, adopting material and product design would call for companies to build competencies in circular design that facilitates product reuse, recycling, and cascading (EMF, 2015).

Consequently, to be economically successful, EMF (2015) advises companies to consider their material selection, standardised components, design-to-last products, design for easy end-of-life sorting, separation and reuse of products and materials, and design-for-manufacturing that takes into consideration use by-products and waste. To implement reverse cycles, EMF (2015) advocates for use of new and additional skills to enable cascading and return of materials back into industrial systems. Through sorting, warehousing, and risk management, supply chains can embark on cost-efficient, quality collection and treatment and segmentation of end-of-life products and hence decreased leakage of materials in the supply chain (EMF, 2014, 2015).

Additionally, new business models adopted can help businesses to create, deliver, and capture value in the circular economic system if they are designed in such a way that they prevent, postpone, or reverse obsolescence, thus minimizing leakage in the supply chain (Den Hollander and Bakker, 2016). As for the enabling conditions, collaborations/networking across the supply chain and adoption of disruptive technology can create a major impact. Since the supply chain consists of several players (suppliers, distributors, retailers, wholesalers, and customers), collaboration can either be vertical, which could include collaboration with customers, internally (across functions), and with suppliers, or horizontal, which could include collaboration with competitors, internally, and with non-competitors, for example, sharing manufacturing capacity (Simatupang and Sridharan, 2002).

In the supply chain that involves flow of materials and information both upstream and downstream, efficiency in use of resources, reduction in energy consumption, and efficiency in the utilization of logistical routes can only be enabled by the use of digital technologies (Salminen et al., 2017). Guided by circular economy principles, technological advances can create ever-greater opportunities for society. These advances, according to EMF (2015), allow more efficient collaboration and knowledge sharing, better tracking of materials, improved forward and reverse logistics setups, and increased use of renewable energy in the supply chain.

There is no doubt of the significant role digital technologies have played in the way businesses operate today. To justify this fact, Bressanelli et al. (2018) carried out a study that identified eight functions that are enabled by digital technologies: improving product design, attracting target customers, monitoring and tracking products, providing maintenance, optimising product usage, upgrading products, enhancing renovation, and end-of-life activities. Christensen et al. (2001), on the other hand, observe that digital technologies create growth in the industries they penetrate or create entirely new industries through introduction of products and services that are cheaper, better, and more convenient.

Opportunities of the circular supply chain

The circular supply chain is being viewed as the solution for the current wasteful model of linear economy. For this reason, it is been gaining prominence among businesses and policymakers, with significant opportunities identified. For instance, according to EMF (2013, 2015), businesses in the EU could benefit from substantial material cost savings amounting to up to USD 630 billion. The linear supply chain has been greatly impacted by price volatilities and supply risks. Circular supply chains, on the other hand, offer businesses an avenue to resilient growth, a systematic answer to reducing dependency on natural resource markets, and a means of reducing exposure to price shocks (Bocken et al., 2016; EMF, 2015). This is supported by a WEF (2014) report that indicated, apart from cost saving, closing production loops, CE would reduce demand for virgin materials and help mitigate demand-driven volatilities on raw material markets and supply chains. The shift to circularity in the supply chain means using less virgin material and more recycled inputs with a higher share of labour costs, hence reducing the businesses' exposure to volatile raw materials prices and threats of disruption by natural disasters (EMF, 2014). Moreover, elimination of waste from the industrial chain by reusing and recycling enables cost savings and less resource dependence (EMF, 2014).

To achieve economic growth, increased revenue generated from emerging circular activities and lower production cost through utilisation of inputs is suggested by EMF (2015). Businesses that are innovative have been associated with higher rates of technological development, improved materials, energy efficiency, and more profit opportunities (EMF, 2015). Innovation provides room for creative solutions and new business models that give them a competitive advantage. Furthermore, technological and organisational innovations underpinning circularity would all increase resource productivity (EMF, 2015). For instance, EMF reports resource productivity growth of 3% by 2030, equivalent to 1.5 trillion benefits from mobility, food, and built environment.

Creation of demand for new business opportunities in the supply chain, such as collection and reverse logistics companies, part and component remanufacturing, product marketers, and creation of employment are among other benefits associated with the circular supply chain (EMF, 2015; Bastein et al., 2013). For instance, a study by Morgan and Mitchell (2015) revealed that approximately 54,000 jobs will be created in the United Kingdom by 2030. A study by EMF, McKinsey, and SUN attributes an increase in jobs to increased spending fuelled by lower prices expected across different sectors and labour-intensive quality recycling activities. They further note that these jobs will not be limited to remanufacturing and large corporations but across industrial sectors through development of reverse logistics, within SMEs, and through innovation and entrepreneurship.

The benefits and opportunities are many, and as studies are being conducted, several have been discovered, as summarised in Table 39.1.

From the table, economic growth, saving material costs, job creation, innovation, new business models, reduced usage of natural resources, improved supply chain security, and reduced waste amounts are some of the prominent benefits and opportunities associated with integrating the circular economy into supply chain management.

Challenges of circular supply chain

While the previous findings indicate clear benefits for supply chains that adopt a circular economy, there are several challenges standing in the way of full implementation, making transition far from complete (van Loon et al., 2018). For example, studies have cited lack of capital as one of the most salient challenges to the integration of a circular economy in supply chains (Veleva and Bodkin, 2018; Geng and Doberstein, 2008; Masi et al., 2017). Shifting from a linear supply chain to a circular supply chain involves activities that are cost related, for instance, raw material extraction, transportation, production planning, distribution, inventory management, and management of reverse logistics networks (Kok et al., 2013).

For any investment, upfront costs and anticipated payback period are factors that have hindered many businesses, especially small and medium-sized enterprises, from implementing CE in their supply chains, since they are more sensitive to additional costs as compared to large firms (Lee et al., 2011; Rademaekers et al., 2011). Furthermore, high costs are related to recycled materials in the supply chain, and therefore they tend to be more expensive than virgin resources (Lieder and Rashid, 2016).

Apart from the direct costs, hidden costs such as time and human resources are needed, since they play a critical role in the implementation of circular supply chains (Pan et al., 2015). This confirms findings by Rizos et al. (2015) that "the lack of government support and encouragement including provision of funding opportunities, training, effective taxation policy, and import duties is recognised as a significant barrier in the uptake of environmental investments."

Uncertainty in the return flows when it comes to the quantity, mix, quality, time, and place of returns for products after use, as identified by Kumar and Putnam (2008), reduces the probability of achieving economic scale in reverse logistics and product renovation. Without proper planning in place, there are challenges experienced in developing a proper take-back mechanism from other companies (Ghisellini et al., 2016). Market-related challenges have also been identified when it comes to ownership issues in product reuse (Singh and Ordoñez, 2016); limited availability of reused product due to lack of refurbished products (Weelden et al., 2016); lack of established standards for refurbished products, leading to low-quality products (Weelden et al., 2016); lack of market demand for circular products (Rizos et al., 2015); and legal issues related to retention of products sold (Tukker, 2015).

Lack of suitable supply chain partners is seen as a challenge, since finding those with appropriate skills and the same level of commitment required for CE implementation is an uphill battle (Walker et al., 2008). According to Barratt (2004), partnership in the supply chain is not all about developing closer relationships, integrating processes between supply chain functions, or developing close information exchange but also needs implementation of integrated strategies both at the tactical and strategic levels in the organisations. For this to work, a number of elements, trust, mutuality, information exchange, and openness and communication, must be considered (Barratt, 2004).

Effective and efficient supply chains require coordination and information sharing (Govindan and Hasanagic, 2018). A number of authors have highlighted the fundamental need for information sharing in the supply chain (Barratt, 2004; Lambert and Cooper, 2000; Lau and Lee, 2000), for instance, improved allocation of resources, improved production planning, lower inventory costs, increased customer service, and reduced lead times (Hill et al., 2012). Therefore, lack of these pieces of information may hinder the integration of CE into supply chains (Rizos et al., 2015). Challenges experienced in information sharing include quality of information, quality and inadequate information systems, fear of opportunistic behaviour, lack of common goals, and fear of losing sensitive data (Porterfield et al., 2010; Barratt, 2004; Harland et al., 2007).

For the implementation of CE, technology is one of the key enablers. Therefore, lack of it can be a stumbling block to the integration of CE into the supply chain. For instance, one of the biggest benefits of technology in the supply chain is allowing businesses to track and optimise production activities in real time, use smart solutions to reduce energy consumption, and reach full capacity utilisation (Kagermann, 2015). However, due to financial constraints, not all businesses, especially SMEs, are in the position of investing in technology; hence, they are hindered from accessing data/information in the supply chain, making decisions about how waste materials can be reused, virtualisation of the distribution chains, and coordination of material flows (Salminen et al., 2017; Kagermann, 2015).

Internal resistance to change or lack of commitment by top management and employees makes it difficult to formulate or implement CE strategies (Wang et al., 2016). According to Su et al. (2013), Liu and Bai (2014), and Shahbazi et al. (2016), organisations with poor leadership in most cases are not interested in CE implementation or may not be able to articulate a clear vision for what the CE means and align it with current sustainability goals (Veleva and Bodkin, 2018). The aforementioned are just a small sample of the challenges faced in the circular supply chain. Studies conducted in different industries have identified these challenges, as summarised in Table 39.2.

From the previous circular supply chain challenges summary, the most-cited barriers include financial/economic challenges due to limited financial capability for environmental investments; high cost associated with collection and segregation during product returns and high costs of environmentally friendly products; informational barriers associated with poor communication, lack of organisational cooperation and coordination, and lack of information; organisational culture in terms of attitudes towards change and lack of clear incorporation of CE strategies into the organisation's vision; technological challenges due to lack or scarcity of advanced technology and equipment, aversion to innovation and technological change, and high costs associated with technology upgrade; user behaviour towards recycled, remanufactured, products; and incorrect perception of these products.

Table 39.2 Circular supply chain challenges

Challenges	*Reference (s)*
Lack of regulation and incentive, data and indicators, cost of product/ waste take-back, lack of awareness and market demand, access to financing, and complex product or packaging design.	Valeva and Bodkin (2018)
Lack of management commitment to make strategic transformation within the organisation.	Zhu and Geng (2013)
Lack of government assistance, that is, funding options, efficient taxation, import duty.	Geng and Doberstein (2008)
Lack of technical capabilities.	Mangla et al. (2018)
Lack of stakeholder environmental awareness.	Luthra et al. (2015), Tukker (2015)
Size and limited bargaining power (most SMEs).	Mangla et al. (2018)
Lack of standard systems for performance, lack of recycling policies, regulatory barriers, lack of supportive laws on waste management.	Su et al. (2013),
Policy barriers, technology barriers, and public participation barriers.	Geng and Doberstein (2008)
Economic and financial viability (e.g., financial and operation risks), market and competition (e.g., brand and reputation issues, cannibalism), product characteristics (e.g., product customisation), standards and regulations (e.g., lack of standards, taxation, and policy misalignment), supply chain management (e.g., return flow uncertainty, coordination and information sharing, organisational cultural issues), technology (e.g., data privacy and security, product technology improvement), and user behaviour (e.g., user willingness to pay, changing user preferences).	Bressanelli et al. (2018)
Lack of information about specific products, perceived risks associated with refurbishing or reusing materials, quality of materials used, processes and product design that support reuse.	Van Weelden et al. (2016)
Financial (significant upfront investment needed), technological (lack of advanced support), societal (negative perception of CE), informational (lack of environmental information disclosure), institutional (lack of regulatory environment).	Masi et al. (2017)
Technological barriers (aversion to innovation and technological changes, lack of technical knowledge), economic barriers (material and product costs), organisational barriers (management resistance to change, poor partnership, and/or uncooperative suppliers), legal barriers (lack of legislation and regulation), informational barriers (poor communication, limited intra-organisational cooperation and interaction), and societal barriers (lack of market preference and customer demand).	Shahbazi et al. (2016)
Political obstacles, traditional development model, high up-front cost, complex international supply chains, challenges of company–company cooperation, lack of consumer enthusiasm, innovation challenges.	Preston (2012)
Government, economic, technological, knowledge and skills, management, circular economy framework, culture and social, and market issues.	Govindan and Hasanagic (2018)

Conclusions

The supply chain is the flow of information and raw materials form the producer through to the final consumer. Currently, businesses are operating in an environment where, to meet consumer demands, raw materials are extracted, energy and labour applied to manufacture products, and the product is sold to the consumer, who then discards it when it no longer serves their purpose. The continuation of this cycle has led to depletion of natural resources, environmental degradation, solid waste generation and landfills, increased greenhouse gas emission, and usage of hazardous substances. The impact of these is currently felt by businesses across the globe, as they now have to deal with price volatilities, supply chain risks, environmental pressures, and scarcity in natural resources.

Furthermore, recent trends in increasing population, urbanisation and growth in the middle class, a new generation of consumers, and tightening of legislation and technology have been a wakeup call for businesses to rethink their business strategies to meet the rise in consumption. For this reason, a new model that is believed to have a solution to the current linear model and that can save the globe from further distraction has been advocated by EMF. This model, the circular economy, is described as "an economy which is restorative and regenerative by design and which aims at keeping products, components, and materials at their highest utility and value at all times" (EMF, 2013, 2014). The circular economy represents a huge opportunity for businesses to create competitive advantage, a 'circular advantage,' disrupting the way production and consumption are done through innovative business models, digital technologies, and the enabling capabilities that support these systems (Cheshire, 2016).

By adopting this model, supply chains will be able to enjoy benefits such as reduced impacts on the environment, cost savings from reduced natural resource usage, creation of new markets and hence additional economic benefits leading to job creation, poverty reduction and wealth creation, innovation, reduced volatility and greater security of supply, and new demand for business services. According to Mishra et al. (2018), businesses must develop competencies in integrating product design, business models, and reverse networks to enable product reuse, cascading, and recycling. Moreover, capturing value through improved material efficiency advances CE transition and hence the reduction in waste generation and extraction, consumption of resources, energy demands, and carbon emissions (Shahbazi et al., 2016).

However, since challenges are inevitable, several of them have been met in the transition process, for instance, financial challenges in acquiring the right products and materials for environmental investment, technological barriers such as lack of expertise to manage advanced technology or lack of proper equipment, lack of supportive laws and regulations, lack of information sharing and cooperation among players in the supply chain, among other challenges. Some of these challenges could easily be resolved without requiring much investment. For instance, lack of information can be fixed by enhancing interactions with key players in the supply chain. This way, it is possible to identify economic and environmental solutions, identify disposal options, increase awareness regarding material efficiency improvements, and identify alternative disposal options (Simpson, 2010).

References

Ahi P, Searcy C. (2015). An analysis of metrics used to measure performance in green and sustainable supply chains. *Journal of Cleaner Production*. 86: 360–377.

Alhola K, Ryding SO, Salmenperä H, Busch NJ. (2019). Exploiting the potential of public procurement: opportunities for circular economy. *Journal of Industrial Ecology*. 23 (1): 96–109.

Ambec S, Lanoie P. (2008). Does it pay to be green? A systematic overview. *Academy of Management Perspective*. 22 (4): 45–62.

Barratt M. (2004). Understanding the meaning of collaboration in the supply chain. *Supply Chain Management: An International Journal*. 9 (1): 30–42.

Bastein AGTM, Roelofs E, Rietveld E, Hoogendoorn A. (2013). *Opportunities for a circular economy in the Netherlands*. Delft: TNO. 1–13.

Batista L, Gong Y, Pereira S, Jia F, Bittar A. (2018). Circular supply chains in emerging economies – a comparative study of packaging recovery ecosystems in China and Brazil. *International Journal of Production Research*. 1–21.

Behret H., Öztayşi B, Kahraman C. (2016). A fuzzy inference system for supply chain risk management. *Practical Applications of Intelligent Systems*. Springer, Heidelberg. Berlin, Germany. 429 – 438

Bocken NM, De Pauw I, Bakker C, van der Grinten B. (2016). Product design and business model strategies for a circular economy. *Journal of Industrial and Production Engineering*. 33 (5): 308–320.

Bressanelli G, Perona M, Saccani N. (2018). Challenges in supply chain redesign for the circular economy: a literature review and a multiple case study. *International Journal of Production Research*. 1–28.

Burgess K, Prakash JS, Rana K. (2006). Supply chain management: a structured literature review and implications for future research. *International Journal of Operations & Production Management*. 26 (7): 703–729.

Canning L. (2006). Rethinking market connections: mobile phone recovery, reuse, and recycling in the UK. *Journal of Business and Industrial Marketing*. 21 (5): 320–329.

Carter CR, Rogers DS. (2008). A framework of sustainable supply chain management: moving toward new theory. *International Journal of Physical Distribution and Logistics Management*. 38 (5): 360–387.

Cheshire D. (2016). *What principles of the circular economy can be carried across to the built environment?* #BuildCircular learning hub @ EcoBuild 2016, 8th–10th March 2016. London, UK.

Christopher M. (2016). *Logistics & supply chain management*. New York, NY: Pearson.

Christensen CM, Anthony SD, Roth EA. (2001). *Innovation in the telecommunications industry: separating hype from reality*. Cambridge, MA: Harvard Business School.

Den Hollander M, Bakker C. (2016). Mind the gap exploiter: circular business models for product lifetime extension. In: *Proceedings of electronic goes green 2016+*. Berlin, Germany: Fraunhofer IZM. 1–8.

Du L, Yu L, Cheng R. (2010). The construction research on rapid-response eco-supply chain of the textile industry based on the circular economy. *2010 International Conference on E-Health Networking Digital Ecosystems and Technologies (EDT), IEEE*. 1: 248–251.

EMF (Ellen MacArthur Foundation) (2013). *Towards the circular economy Vol. 2: opportunities for the consumer goods sector*. Available at: https://www.ellenmacarthurfoundation.org/publications/towards-the-circular-economy-vol-2-opportunities-for-the-consumer-goods-sector

EMF (Ellen MacArthur Foundation) (2014). *Towards the circular economy Vol 3: accelerating the scale-up across global supply chains*. Available at: https://www.ellenmacarthurfoundation.org/publications/towards-the-circular-economy-vol-3-accelerating-the-scale-up-across-global-supply-chains

EMF (Ellen MacArthur Foundation) (2015). *Towards a circular economy: business rationale for an accelerated transition*. Available at: www.ellenmacarthurfoundation.org/assets/downloads/TCE_Ellen-MacArthur-Foundation_9-Dec-2015.pdf (Accessed on 4/10/2019).

Farooque M, Zhang A, Thurer M, Qu T, Huisingh D. (2019). Circular supply chain management: a definition and structured literature review. *Journal of Cleaner Production*. 228: 882–900.

Fortes J. (2009). Green supply chain management: a literature. *Otago Management Graduate Review*. 7 (1): 51–62.

Franco MA. (2017). Circular economy at the micro level: a dynamic view of incumbents' struggles and challenges in the textile industry. *Journal of Cleaner Production*. 241 (20): 118327.

Geissdoerfer M, Morioka SN, de Carvalho MM, Evans S. (2018). Business models and supply chains for the circular economy. *Journal of Cleaner Production*. 190: 712–721.

Geng Y, Doberstein B. (2008). Developing the circular economy in China: challenges and opportunities for achieving 'leapfrog development'. *The International Journal of Sustainable Development and World Ecology*. 15 (3): 231–239.

Genovese A, Acquaye AA, Figueroa A, Koh SL. (2017). Sustainable supply chain management and the transition towards a circular economy: evidence and some applications. *Omega*. 66*:* 344–357.

Ghisellini P, Cialani C, Ulgiati S. (2016). A review on circular economy: the expected transition to a balanced interplay of environmental and economic systems. *Journal of Cleaner Production*. 114: 11–32.

Govindan K, Hasanagic M. (2018). A systematic review on drivers, barriers, and practices towards circular economy: a supply chain perspective. *International Journal of Production Research*. 56 (1–2): 278–311.

Guide VDR, Van Wassenhove LN. (2006). Closed-loop supply chains: an introduction to the feature issue (part 1). *Production and Operations Management*. 15 (3): 345–350.

Gungor A, Gupta SM. (1999). Issues in environmentally conscious manufacturing and product recovery: a survey. *Computers and Industrial Engineering*. 36 (4): 811–853.

Gurtu A, Searcy C, Jaber MY. (2015). An analysis of keywords used in the literature on green supply chain management. *Management Research Review*. 38 (2): 166–194.

Handfield RB, Nichols EL. (1999). *Introduction to supply chain management.* Upper Saddle River, NJ: Prentice Hall.
Harland CM, Caldwell ND, Powell P, Zheng J. (2007). Barriers to supply chain information integration: SMEs adrift of e-lands. *Journal of Operations Management.* 25 (6): 1234–1254.
Hazen BT, Mollenkopf DA, Wang Y. (2017). Remanufacturing for the circular economy: an examination of consumer switching behavior. *Business Strategy and the Environment.* 26 (4): 451–464.
Hill A, Doran D, Stratton R. (2012). How should you stabilise your supply chains? *International Journal of Production Economics.* 135 (2): 870–881.
Kagermann H. (2015). Change through digitization – value creation in the age of industry 4.0. In: *Management of permanent change.* Wiesbaden: Springer Gabler. 23–45.
Kalmykova Y, Sadagopan M, Rosado L. (2018). Circular economy – from review of theories and practices to development of implementation tools. *Resources, Conservation and Recycling.* 135: 190–201.
Kok L, Wurpel G, Ten Wolde A. (2013). *Unleashing the power of the circular economy. Amsterdam.* Amsterdam, The Netherlands: IMSA Amsterdam for Circle Economy.
Kumar S, Putnam V. (2008). Cradle to cradle: Reverse logistics strategies and opportunities across three industry sectors. *International Journal of Production Economics.* 115 (2): 305–315.
Jawahir IA, Bradley RT. (2016). Technological elements of circular economy and the principles of 6R-based closed-loop material flow in sustainable manufacturing. *Procedia CIRP.* 40: 103–108.
La Londe BJ, Masters JM. (1994). Emerging logistics strategies: blueprint for the next century. *International Journal of Physical Distribution and Logistics Management.* 24 (7): 35–47.
Lacy P, Rutqvist J. (2015). *Waste to wealth: the circular economy advantage.* London: Palgrave Macmillan.
Lambert DM, Cooper MC, Pagh JD. (1998). Supply chain management: implementation issues and research opportunities. *The International Journal of Logistics Management.* 9 (2): 1–20.
Lambert DM, Cooper MC. (2000). Issues in supply chain management. *Industrial Marketing Management.* 29 (1): 65–83.
Lau HC, Lee WB. (2000). On a responsive supply chain information system. *International Journal of Physical Distribution and Logistics Management.* 30(7/8): 598 – 610
Lee P, MacGregor A, Willis P. (2011). *The further benefits of business resource efficiency.* London: Oakdene Hollins for the Department for Environment, Food and Rural Affairs (Defra).
Lieder M, Rashid A. (2016). Towards circular economy implementation: a comprehensive review in context of manufacturing industry. *Journal of Cleaner Production.* 115 (1): 36–51.
Linton JD, Klassen R, Jayaraman V. (2007). Sustainable supply chains: an introduction. *Journal of Operations Management.* 25 (6): 1075–1082.
Liu Y, Bai Y. (2014). An exploration of firms' awareness and behavior of developing circular economy: An empirical research in China. *Resources, Conservation and Recycling.* 87: 145–152.
Luthra S, Garg D, Haleem A. (2015). Critical success factors of green supply chain management for achieving sustainability in Indian automobile industry. *Production Planning and Control.* 26 (5): 339–362.
Manavalan E, Jayakrishna K. (2019). An analysis on sustainable supply chain for circular economy. *Procedia Manufacturing.* 33: 477–484.
Mangla SK, Luthra S, Mishra N, Singh A, Rana NP, Dora M, Dwivedi Y. (2018). Barriers to effective circular supply chain management in a developing country context. *Production Planning and Control.* 29 (6): 551–569.
Masi D, Day S, Godsell J. (2017). Supply chain configurations in the circular economy: a systematic literature review. *Sustainability.* 9 (9): 1602.
Mentzer JT, DeWitt W, Keebler JS, Min S, Nix NW, Smith CD, Zacharia ZG. (2001). Defining supply chain management. *Journal of Business Logistics.* 22 (2): 1–25.
Mishra JL, Hopkinson PG, Tidridge G. (2018). Value creation from circular economy-led closed loop supply chains: a case study of fast-moving consumer goods. *Production Planning and Control.* 29 (6): 509–521.
Morgan J, Mitchell P. (2015). *Employment and the circular economy: job creation in a more resource efficient Britain.* London: Green Alliance.
Murray A, Skene K, Haynes K. (2017). The circular economy: an interdisciplinary exploration of the concept and application in a global context. *Journal of Business Ethics.* 140 (3): 369–380.
Nasir MHA, Genovese A, Acquaye AA, Koh SCL, Yamoah F. (2017). Comparing linear and circular supply chains: a case study from the construction industry. *International Journal of Production Economics.* 183: 443–457.
Neto JQF, Walther G, Bloemhof J, Van Nunen JAEE, Spengler T. (2009). A methodology for assessing eco-efficiency in logistics networks. *European Journal of Operational Research.* 193 (3): 670–682.

Pan SY, Du MA, Huang IT, Liu IH, Chang EE, Chiang PC. (2015). Strategies on implementation of waste-to-energy (WTE) supply chain for circular economy system: a review. *Journal of Cleaner Production*. 108: 409–421.

Porterfield TE, Bailey JP, Evers PT. (2010). B2B e-commerce: an empirical investigation of information exchange and firm performance. *International Journal of Physical Distribution & Logistics Management*. 40 (6): 435–455.

Preston F. (2012). *A global redesign? Shaping the circular economy*. London: Chatham House.

Rademaekers K, Asaad SSZ, Berg J. (2011). *Study on the competitiveness of the European Companies and resource efficiency*. Brussels, Belgium: ECORYS, Teknologisk Institut, Cambridge Econometrics, CES info and Idea Consult. The European Commission, DG Enterprise and Industry.

Rizos V, Behrens A, Kafyeke T, Hirschnitz-Garbers M. (2015). *The circular economy: barriers and challenges for SMEs*. Available at: https://www.researchgate.net/publication/283121970_The_Circular_Economy_Barriers_and_Opportunities_for_SMEs

Rockström J, Steffen WL, Noone K, Persson Å, Chapin III FS, Lambin E, Lenton TM, Scheffer M, Folke C, Schellnhuber HJ, Nykvist B. (2009). Planetary boundaries: exploring the safe operating space for humanity. *Ecology and Society*. 14 (2): 32.

Salminen V, Ruohomaa H, Kantola, J. (2017). Digitalization and big data supporting responsible business co-evolution. In: *Advances in human factors, business management, training and education*. Cham: Springer. 1055–1067.

Seuring S, Müller M. (2008). From a literature review to a conceptual framework for sustainable supply chain management. *Journal of Cleaner Production*. 16 (15): 1699–1710.

Shahbazi S, Wiktorsson M, Kurdve M, Jönsson C, Bjelkemyr M. (2016). Material efficiency in manufacturing: Swedish evidence on potential, barriers and strategies. *Journal of Cleaner Production*. 127: 438–450.

Sharma A, Iyer GR, Mehrotra A, Krishnan R. (2010). Sustainability and business-to-business marketing: a framework and implications. *Industrial Marketing Management*. 39 (2): 330–341.

Simatupang TM, Sridharan R. (2002). *The collaborative supply chain: a scheme for information sharing and incentive alignment*. Available at: https://pdfs.semanticscholar.org/7541/dd6b26d1194b61c96e8eab588b126dc60081.pdf

Simpson D. (2010). Use of supply relationships to recycle secondary materials. *International Journal of Production Research*. 48 (1): 227–249.

Singh J, Ordoñez I. (2016). Resource recovery from post-consumer waste: important lessons for the upcoming circular economy. *Journal of Cleaner Production*. 134: 342–353.

Srivastava SK. (2007). Green supply-chain management: a state-of-the-art literature review. *International Journal of Management Reviews*. 9 (1): 53–80.

Su B, Heshmati A, Geng Y, Xiaoman Y. (2013). A review of the circular economy in China: Moving from rhetoric to implementation. *Journal of Cleaner Production*. 42: 215–227.

Tukker A. (2015). Product services for a resource-efficient and circular economy – a review. *Journal of Cleaner Production*. 97: 76–91.

Van Loon P, Van Wassenhove LN. (2018). Assessing the economic and environmental impact of remanufacturing: a decision support tool for OEM suppliers. *International Journal of Production Research*. 56 (4): 1662–1674.

Van Weelden E, Mugge R, Bakker C. (2016). Paving the way towards circular consumption: exploring consumer acceptance of refurbished mobile phones in the Dutch market. *Journal of Cleaner Production*. 113: 743–754.

Veleva V, Bodkin G. (2018). Corporate-entrepreneur collaborations to advance a circular economy. *Journal of Cleaner Production*. 188: 20–37.

Walker H, Di Sisto L, McBain D. (2008). Drivers and barriers to environmental supply chain management practices: lessons from the public and private sectors. *Journal of Purchasing and Supply Management*. 14 (1): 69–85.

Wang Z, Mathiyazhagan K, Xu L, Diabat A. (2016). A decision making trial and evaluation laboratory approach to analyze the barriers to green supply chain management adoption in a food packaging company. *Journal of Cleaner Production*. 117: 19–28.

Weetman C. (2016). *A circular economy handbook for business and supply chains: repair, remake, redesign, rethink*. New York: Kogan Page Publishers.

WEF (World Econcomic Forum) (2014). *Towards the circular economy: accelerating the scale-up across global supply chains*. Available at: https://reports.weforum.org/toward-the-circular-economy-accelerating-the-scale-up-across-global-supply-chains/?doing_wp_cron=1598570672.5027139186859130859375

Wijkman A, Skånberg K. (2015). *The circular economy and benefits for society jobs and climate clear winners in an economy based on renewable energy and resource efficiency*. Available at: https://s3-eu-west-1.amazonaws.com/avfall-norge-no/dokumenter/The-Circular-Economy-and-Benefits-for-Society.pdf

World Bank (2020). *What a waste 2.0*. Available at: http://datatopics.worldbank.org/what-a-waste/trends_in_solid_waste_management.html (Accessed 28/10/2019).

Zhu Q, Geng Y. (2013). Drivers and barriers of extended supply chain practices for energy saving and emission reduction among Chinese manufacturers. *Journal of Cleaner Production*. 40: 6–12.

Zhu Q, Geng Y, Lai KH. (2010). Circular economy practices among Chinese manufacturers varying in environmental-oriented supply chain cooperation and the performance implications. *Journal of Environmental Management*. 91 (6): 1324–1331.

CONCLUSIONS

Terry Tudor and Cleber JC. Dutra

Introduction

Increasing societal concerns around issues such as the impacts of climate change and public health, as well as resource security, are driving greater awareness and actions to manage our resources more sustainably, as well as the need for more environmentally and socially responsible products. The concept of the circular economy (CE) offers a blueprint to address these issues. The concept seeks to minimise the quantity of waste generated, recover value from that produced, and decouple economic growth from natural resource use (Blomsma and Brennan, 2017; Homrich et al., 2018; EASAC, 2016; EEA, 2016; Pauliuk, 2018).

This *Handbook* has taken us on a journey of the CE concept. A journey not only of countries from the Global North and Global South but also of cultures, theories, subjects, perspectives, and CE initiatives. Having outlined the context for and concept of the CE, we examined some of the policy and legislative issues governing circularity and ended with a number of case studies from a range of sectors. Evidently, there are a number of examples globally of best practices being undertaken. For example, use of the Internet of Things (IoT) for the provision of washing machine services in Italy and the Netherlands (Chapter 36); the use of cooperatives in Brazil (Chapter 23); a number of examples of recycling, repair, and reuse in Africa, India, and Europe (e.g., Chapters 30, 33, 34, and 37); initiatives from the informal sector (e.g., Chapters 24, 25, and 30); and the use of technologies in Europe and Africa to improve efficiencies and connectivity (e.g., Chapters 35–39).

However, overall, there are significant issues to be overcome. For example, there are evident challenges in the interpretation and implementation of circularity, as has been highlighted throughout the *Handbook*. Without a clear understanding and interpretation of circularity, it is difficult, for example, to measure progress or to benchmark or effectively utilise the benefits (Moraga et al., 2019). Indeed, while the challenges and dichotomies in the interpretation and implementation of circularity are specifically discussed in Part II, these issues form an underlying and recurring theme throughout the *Handbook*. For example, Bicket, as well as Monteiro and Dutra, in Chapters 14 and 15, respectively, contend that there is a need for a systems approach to deeply embed CE principles. Sheeran in Chapter 18 contends that while there are elements of circularity emerging in international trade agreements, thus far, "substantive adaptation is limited." Similarly, Mahon in Chapter 16 notes that "The use of policy measures to drive greater circularity is in its relative infancy." Mahon concludes that rather than policies, perhaps implementation of circularity might take place through green public procurement and the use of fiscal measures such as a reduction in value-added or corporation tax for qualifying businesses. There are also inconsistencies in policies and legislative measures, even within countries. A key policy issue is with regard to the definition of waste (i.e., when does a resource become waste and when does waste cease to be waste?). While evidence of best practices abounds, it is piecemeal. In addition, initiatives often tend to be short-term, and there is limited transfer of knowledge and capacity. Therefore, two key questions that arise are: How best to encourage collaboration amongst stakeholders, as well as long-term, sustainable change in practices? Chapter 28 notes that for industrial symbiosis, key barriers included limitations in capacity, coordination, and communication, as well as in the predictability of materials and geographies. At the operational level, Webster in Chapter 24 outlines some of the challenges faced by those in the informal sector in their quest to simply maintain their livelihoods, much less to generate any kind of profitability. These challenges include the vagaries of the global recycling market and the simple act of economic mobility. For refugees (Chapter 25), these challenges are magnified even further, and guaranteeing the long-term sustainability and viability of projects is difficult.

Key messages going forward

The chapters have outlined several messages. However, there are six key overarching factors that are crucial if we are to enable a more circular approach globally.

First, while it is complex (e.g., see Chapters 14, 15, and 33), and there are inherent environmental and public health risks (Chapters 2 and 6), there are undoubtedly vast opportunities for the implementation of CE. Indeed, Chapters 10 and 30 discuss some of the opportunities that exist in Africa, the world's second-largest continent after Asia, composed of 54 sovereign countries and a population of approximately 1.26 billion. The continent will overtake Asia as the world's most rapidly urbanising region by 2025. Evidently, therefore, the potential for uptake and change exists, and the impacts are significant. However, effectively implementing the concept of a CE is complex. Nevertheless, various chapters in this *Handbook* have offered practical examples of how this complexity could best be managed. For example, Chapters 19, 23, 24, 27, 29, 32, 35, and 39 propose innovative systems approaches to enable the transition to a CE. Chapter 24 refers to the 'uberisation' of the informal sector to connect stakeholders in the plastics supply chain. Chapter 25 outlines the need for strategic links, including with the informal sector, and for the utilisation of financial mechanisms to spur more long-term investments. Chapter 28 provides a number of enablers for industrial symbiosis, including leadership, governments, the use of intermediaries, and technology, while Chapter 31 examines how best to implement modularity to reduce waste electrical and electronic waste (WEEE). The onset of the COVID-19 pandemic has led to significant changes in consumption and production, as well as in supply chain management. Fears over the disease have, for example, led to changes in shopping practices (e.g., increased online purchases), increased focus on healthy foods, and buying local. However, COVID-19 also offers opportunities for the use of 'greener,' healthier approaches.

Second, countries in the Global North need to provide greater support to those in the Global South. Indeed, Chapters 24 and 25 outline the severe human challenges faced by those in the informal sector and those who are refugees, while Chapter 5 outlines the environmental injustices associated with the export of waste. It is heartening, therefore, to see that European countries are considering regulatory measures (e.g., bans on plastics waste) to reduce the shipment of waste materials. For example, the United Kingdom looks set to introduce a bill that will allow the government to ban the export of plastic waste to non-Organisation for Economic Cooperation and Development (OECD) countries. The work being done around the implementation of the Sustainable Development Goals (SDGs) would appear to be a sensible contextual 'home' for the mechanisms that should be employed to facilitate this collaboration and support. Wilson in Chapter 7 provides a fuller discussion on the links between the SDGs and the CE. With the necessary support to build institutional and technological capacity, and access to sustainable finance, the large informal sectors that already practice 'circular' activities in the Global South (see, for example, Chapters 24, 25, and 30), could engage in higher-value CE supply chains and activities, thereby significantly upscaling revenues, growth, and job creation. Leveraging opportunities for innovation in sectors such as agriculture and construction will require foreign investment. Greater cooperation would open up opportunities for knowledge exchange and capacity building between the Global North and Global South. Indeed, it is evident from the case studies in this *Handbook* that there could be lessons learnt and knowledge and expertise shared on both sides.

Third, there is a need for greater research and an evidence base for the development and effective implementation of CE initiatives. For example, as Kasner and O'Carroll discuss in Chapter 30, research and development is crucial to the exploitation of the many CE opportunities across Africa, for example, in technologies and Agritech. An evidence base would facilitate the more effective development and implementation of policies and legislation. In this way, there could be greater consistency in interpretation and implementation, and ultimately there could also be increased cooperation across borders. In addition, there is a need for more extended producer responsibility (EPR) legislation and initiatives to encourage greater action along the value chain of products and to ensure that producers are held responsible for managing the products that they place on the market when they do eventually become waste. There are already several EPR schemes in place for a range of items, including WEEE and end-of-life vehicles (ELVs). However, often these schemes work best in the Global North countries but less so in the Global South. This limitation in regulation and enforcement results in disastrous impacts on public health and the environment, as discussed in Global South countries (see, e.g., Chapters 5, 6, and 7). Crucially, EPR would facilitate greater shifts towards the design of products to facilitate increased recycling, reuse, and remanufacture, thereby increasing their circularity. As noted in Chapter 26, there is also a need to better understand how best to engage with the public and other key stakeholders about the CE concept.

Fourth, organisations will increasingly be expected to demonstrate compliance with more stringent legislation and regulations. Compliance will require innovation. Indeed, as seen in Chapters 20, 23, and 33–38, companies are not only seeking to meet legislative and policy measures but also view the shift as an opportunity to develop new business models. Thus, the CE is driving innovation and creativity and by extension enabling competitive advantage for those entities that best meet customer needs.

Fifth, to ensure that this innovation continues at pace, there is a need for the facilitation of new skills and ways of thinking if the opportunities presented by a CE are to be effectively exploited. There is a need for skills in a number of other areas, including communications, (sustainable) financing, behaviour change, biochemistry, policy development and implementation, data science, and materials science, just to name a few. For example, there is a need for persons who have skills and knowledge in 4th Industrial Technologies (e.g., blockchain, artificial intelligence, and the IoT). In addition, however, there are many opportunities for entrepreneurship through the development of novel and new products and services (e.g., Chapters 30, 31, 35, 36, and 39). Therefore, there will need to be development of the skills and knowledge to develop and effectively utilise the technologies required. However, there will also need to be changes in policies and behaviours to accompany and facilitate these changes. Even though there has been an evident shift in perspectives towards more environmentally friendly products and services, increasing resource consumption suggests that overall, there is still much work to be done. Getting buy-in from decision-makers, industry, and the public will be vital if meaningful progress is to be made. Thus, developing the capacities to effectively develop and implement appropriate policies, as well as to facilitate behaviour change, will be crucial going forward.

Sixth, the need for sustainable packaging (including plastics) is evidently a key area of focus, as noted in Chapters 4 and 30. This has been driven, for example, by rising incidences of plastics in the seas and rivers, not only causing increasing visible pollution but also resulting in the death of fish and other species. The blocking of drains by plastics waste also results in flooding in some cities. Given the existing and rising quantities of packaging (with production expected to quadruple by the year 2050), it is likely that plastics will be a concern for some time to come. However, there are a number of legislative measures, as well as innovations, being put in place to address the issue. For example, manufacturers are producing packaging materials from biodegradable materials such as corn starch and sugarcane waste. The UK Plastics Pact (which is discussed in Chapters 16 and 30), a programme involving 85+ members, who combined make up two-thirds of the UK consumer packaging market, aims to make 70% of plastic packaging recyclable or compostable and to introduce 30% average recycled content across all plastic packaging by the year 2025. It is important to note that plastics are not the villain. Indeed, they play a vital role as a lightweight, flexible packaging material and as protection for the items inside. Rather it is what we do with the packaging at the end of its life that is more important. As with any other commodity, recyclables are subject to market conditions, and their prices fluctuate with demand. This, therefore, has an impact on the degree of circularity within the system. For example, if there is no market for used plastics or glass, there is limited incentive to recycle them. Thus, measures to create a viable and sustainable end market for secondary materials are vital.

In conclusion, there is no one-size-fits-all solution. In each country and within different regions, there are cultural, logistical (e.g., infrastructure), fiscal, and monetary legislative and policy differences that make it exceedingly challenging for a single option to work in all contexts. That said, it does not mean that these issues cannot be overcome. The key underlying message, therefore, is not for a utopia with a single option but rather the need for greater harmonisation and sharing of resources, expertise, and knowledge to enable individuals, households, communities, countries, and regions to design appropriate strategies. There is evidently much good practice of CE both in the Global North and (contrary to the often-held view) in the Global South. However, the initiatives are often piecemeal. In some cases, these initiatives are 'reinventing the wheel' due to a lack of knowledge. In other cases, the ability to scale and exploit potential opportunities and create value and jobs is limited by the need for greater support. A more harmonised, shared, and holistic approach would facilitate the building of capacity, value chains, and markets and also facilitate the sharing of knowledge and experiences. A logical framework for such an approach would appear to be the SDGs, given the range of environmental, economic, health and well-being, and equality issues that a truly circular economy world should embrace.

References

Blomsma F, Brennan G. (2017). The emergence of circular economy: a new framing around prolonging resource productivity. *Journal of Industrial Ecology*. 21: 603–614.

EASAC (2016). *Indicators for a circular economy*. Halle: European Academies' Science Advisory Council.

EEA (2016). *Circular economy in Europe – developing the knowledge base: report 2*. European Environment Agency. Available at: www.eea.europa.eu/publications/circular-economyin-europe.

Homrich AS, Galvão G, Abadia LG, Carvalho MM. (2018). The circular economy umbrella: trends and gaps on integrating pathways. *Journal of Cleaner Production*. 175: 525–543.

Moraga G, Huysveld S, Mathieux F, Blenginic GA, Alaerts L, Acker KV, de Meester S, Dewulf J. (2019). Circular economy indicators: what do they measure? *Resources, Conservation and Recycling*. 146: 452–461.

Pauliuk S. (2018). Critical appraisal of the circular economy standard BS8001:2017 and a dashboard of quantitative system indicators. *Resources, Conservation and Recycling*. 129: 81–92.

INDEX

Note: Page numbers in *italic* indicate a figure and page numbers in **bold** indicate a table on the corresponding page. Page numbers followed by 'n' indicate a note.

For Product Safety Concerns and Information please contact our EU
representative GPSR@taylorandfrancis.com
Taylor & Francis Verlag GmbH, Kaufingerstraße 24, 80331 München, Germany

www.ingramcontent.com/pod-product-compliance
Ingram Content Group UK Ltd.
Pitfield, Milton Keynes, MK11 3LW, UK
UKHW052056230926
475702UK00032B/2961

* 9 7 8 0 3 6 7 6 3 7 5 9 0 *